Emmanuil G. Sinaiski

Hydromechanics

Related Titles

Leonov, E.G., Isaev, V.I.

Applied Hydroaeromechanics in Oil and Gas Drilling

2009
ISBN: 978-0-470-48756-3

Sinaiski, E.G., Zaichik, L.I.

Statistical Microhydrodynamics

2008
ISBN: 978-3-527-40656-2

Zaichik, L.I., Alipchenkov, V.M., Sinaiski, E.G.

Particles in Turbulent Flow

2008
ISBN: 978-3-527-40739-2

Sinaiski, E.G., Lapiga, E.J.

Separation of Multiphase, Multicomponent Systems

2007
ISBN: 978-3-527-40612-8

Kee, R.J., Coltrin, M.E., Glarborg, P.

Chemically Reacting Flow
Theory and Practice

2003
ISBN: 978-0-471-26179-7

Gregory, G.A., Radus, C. (Eds.)

Multiphase Technology (British Hydromechanics Research (BHR) Group), Publication 40

2000
ISBN: 978-1-86058-252-3

Emmanuil G. Sinaiski

Hydromechanics

Theory and Fundamentals

WILEY-VCH Verlag GmbH & Co. KGaA

The Author

Prof. Dr. Emmanuil Sinaiski
Leipzig, Germany

The Translator

Prof. Moritz Braun
University of South Africa, Dept. of Physics
Unisa, Republic of South Africa

Cover picture

by Volker Weitbrecht, Zürich

Influence of Dead-Water Zones on the Dispersive Mass Transport in Rivers

Library of Congress Card No.: applied for

British Library Cataloguing-in-Publication Data:
A catalogue record for this book is available from the British Library.

Bibliographic information published by the Deutsche Nationalbibliothek
The Deutsche Nationalbibliothek lists this publication in the Deutsche Nationalbibliografie; detailed bibliographic data are available on the Internet at http://dnb.d-nb.de.

Typesetting le-tex publishing services GmbH, Leipzig
Printing and Binding Fabulous Printers Pte Ltd, Singapore
Cover Design Schulz Grafik-Design, Fußgönheim

Printed in Singapore
Printed on acid-free paper

ISBN 978-3-527-41026-2

Dedication

Devoted to the memory of my teachers, academicians of the USSR Academy of Science, Leonid Ivanivitch Sedov and Georgiy Ivanivitch Petrov

Hydromechanics. Theory and Fundamentals. Emmanuil G. Sinaiski

ISBN: 978-3-527-41026-2

Contents

Hydromechanics. Theory and Fundamentals. Emmanuil G. Sinaiski

ISBN: 978-3-527-41026-2

Preface

Continuum mechanics is a part of mechanics devoted to the motion of gaseous, liquid and solid deformable matters. In theoretical mechanics, one studies the motion of mass points, of a discrete system of mass points and of absolute solid bodies. Continuum mechanics addresses the motion of the mass point filling the space continuously with the change of distances between mass points during the motion with the help of, and on the basis of methods developed in theoretical mechanics,

In addition to ordinary material media (gas, fluid, solid body), continuum mechanics is concerned with special (peculiar) media, for example, electromagnetic field, radiation field and so on.

Let us indicate some developed problems of continuum mechanics which gave rise to a self-reliant line of investigation.

The forces acting on a body as viewed from the surrounding fluid or gas are determined by the motion of fluid or gas. Thus, the study of motion of the body in fluid or gas is connected with the motion of a fluid. The last problem is the essence of one of the areas in continuum mechanics, that is, *hydrodynamics*. The problems of motion of gas form the field of *gas dynamics*. A special impetus in the rapid progression of both of these parts of a continuum medium provide problems regarding the motion of airplanes, helicopters, rockets, ships, submarines and so on.

The problems regarding the motion of fluid and gas in pipes and inside different machines are directly relevant to the designing and calculation of the gas- and oil-pipelines, the turbines, the compressors, and other hydraulic engines. The applied problems in this region form the basis of *hydraulics*.

The motion of fluid and gas through soil and other porous media (*filtration*) plays a great role in oil- and gas-field problems. The field devoted to investigations of fluid and gas flows in porous media is called *underground hydromechanics*.

Great theoretical and applied importance has the problem of wave propagation in various media: in solid bodies, in fluid and in gas. This field of continuum mechanics is called *wave mechanics*.

One task of continuum mechanics includes facing problems of flows of multi-component, multi-phase mixtures with regard to interaction between phases, with mass- and heat-exchanges, phase transitions, chemical reactions. These queries form the basis of *physicochemical hydrodynamics*.

Hydromechanics. Theory and Fundamentals. Emmanuil G. Sinaiski

ISBN: 978-3-527-41026-2

The problems of motion of conducting and charged mixtures (fluid, gas, plasma) in electric and magnetic fields enter into the direction of *magneto- hydrodynamics* and *electro-hydrodynamics.*

The study of motion of bodies in rarefied gas, in space, in the atmosphere of stars and planets is also possible with methods of continuum mechanics. The problems associated with such motions are confronted by the *mechanics of rarefied gas and space hydrodynamics.*

Recently, a new direction within mechanic has been formed, that is, biomechanics, in which one investigates the mechanics of biological objects, in particular, the motion of blood in human organs, the contraction of muscles and, the task of constructing mechanical models of internal human organs.

Other areas that employ the methods of continuum mechanics are the weather forecast, the theory of turbulence, the motion of sand in a desert, the motion of snow-slip, the motion of mixtures with chemical reactions, the theory of detonation and many others.

A major part of continuum mechanics is devoted to the investigation of motion and the equilibrium of deformable bodies. The *theory of elasticity* is a basis for calculations regarding construction and machinery design. Ever-growing importance is focused on the realms of continuum mechanics dedicated to studying the elastic properties of the body of complex composition, accounting for inelastic properties in solid bodies. The *theory of plasticity* examines residual deformations in a body. The hereditary property of a substance is an internal property that in some types of matter is retained throughout its motion. In this sense, it is analogous to the hereditary properties in living organisms that are retained throughout their growth and development. With the advent of new materials with complex internal structure, for example, polymers, composites and others, came into demand to develop models of these materials with regard to their internal structure.

It is impossible to list all of the problems and applications of continuum mechanics. However, even the above-mentioned examples are enough to make a conclusion that continuum mechanics embraces a great range of theoretical and applied problems of science and engineering.

In a theoretical course in continuum mechanics, the motions of deformable media are considered. A set of notions which characterizes and uniquely determines the motion of a continuum medium are introduced. As examples serve concepts of velocity field, pressure, temperature and so on. In continuum mechanics, there are derived methods to reduce mechanical problems to the mathematical ones. The solution of the latter allows one to obtain the general properties and the laws of motion of deformable bodies.

Continuum mechanics has had a profound impact on the development of a variety of lines in mathematics. For example, the *theory of airfoil* has a pronounced effect on the progress of some divisions of the *functions of complex variables*, the problem of viscous fluid flow has stimulated investigations in the field of *boundary-value problems of the equations with partial derivatives*, some problems of the elasticity theory – in the line of the *theory of the integral equations* and many others. A great influence on the advance of the numerical methods has served the solution of *boundary-value problems of Navier-Stokes equations* describing the flow of viscous fluids. The solution of the majority of applied problems in continuum mechanics are at present impossible without use of numerical methods and high-performance computers.

The book, bringing to the reader's notice, contains theoretical fundamentals of a part of continuum mechanics, *hydromechanics*. It is based on lecture course, delivered for students of Gubkin State University of Oil and Gas, Russia, Moscow.

Leipzig, October 2010 *Emmanuil G. Sinaiski*

List of Symbols

a	Velocity of sound
a_0	Velocity of sound in the medium at rest
a^*	Stagnation speed of sound
$[a]$	Dimension of quantity a
$\boldsymbol{a}$	Acceleration vector
a_∞	Velocity of sound at infinity
a_r, a_φ, a_z	Components of acceleration in cylindrical coordinates
$a_r, a_\varphi, a_\vartheta$	Components of acceleration in spherical coordinates
A	Work done by the system on the external body
$\boldsymbol{A}$	Vector field, driving force
$A_m^{(e)}$	Work done by external forces
$A_m^{(i)}$	Work done by internal forces
$A^i_{\ j}$	Jacobian
A^{ijkl}	Components of a tensor of fourth order
b	Width of a beam
b_{ij}	Structure function
$b_{LL}(r)$	Longitudinal structure function
$b_{NN}(r)$	Transverse structure function
B	Point
$\boldsymbol{B}$	Matrix, solenoidal vector field ($\nabla \times \boldsymbol{B} = 0$)
$B_{ij}(\boldsymbol{r}, t)$	Components of the correlation tensor of second order
$B_{ij,k}(\boldsymbol{r}, t)$	Correlation tensor of third order
B^{ijkl}	Components of a tensor of fourth order
$B_{LL}(r, t)$	Longitudinal correlation function
$B_{NN}(r, t)$	Transverse structure function
$B_{u_1,u_2,\ldots,u_n}$	Moment of N-th degree
$\mathrm{Ber}(z), \mathrm{Bei}(z)$	Kelvin function
c	Velocity, profile chord
c_f	Friction coefficient
c_p	Specific heat at constant pressure
c_v	Specific heat at constant volume
c_y	Lift coefficient

Hydromechanics. Theory and Fundamentals. Emmanuil G. Sinaiski

ISBN: 978-3-527-41026-2

c_{y0}	Lift coefficient in an incompressible fluid
C	Point, contour, concentration of a passive additive
$\mathbf{C}$	Matrix
C_1, C_2, C_3	Constants of integration, parameters
C^i_j	Elements of an orthogonal matrix
C_k	Contour
C_p	Pressure coefficient
D	Current volume, velocity of the shock wave
D	Molecular diffusion coefficient
D_0	Initial volume
d/dt, D/Dt	Substantial/total derivative
$dA^{(e)}$	Infinitesimal work done by external macroscopic mass and surface forces, Infinitesimal work done by external macroscopic forces
dA^e_s	Infinitesimal work done by external surface forces
$dA^{(i)}$	Infinitesimal work done by internal mass and surface forces
$dA^{(i)}_s$	Infinitesimal work done by internal surface forces
dE	Change of kinetic energy
$d\mathbf{P}$	Force acting on the surface element $d\Sigma$
$dq^{(e)}$	Infinitesimal inflow of external energy
$dq^{(i)}$	Density of work done by internal surface forces
dq^{**}	Energy inflow per mass unit
dq^*_{mas}/dt	External energy inflow per mass unit
dQ'	Non-compensated heat, dissipation of mechanical energy
$dQ^{(e)}$	Infinitesimal heat flow from/to the outside
dQ^*	Infinitesimal heat inflow from the outside
dQ^{**}	External energy inflow
$d\mathbf{s}$	Linear vector element along a curve
$d_e S$	Increment of entropy due to external processes
$d_i S$	Increment of entropy due to internal processes
dT/ds	Derivative in the direction $\mathbf{s}$
dU	Density of internal energy
dU_m	Change of internal energy
$d\sigma_k$	Infinitesimal surface elements
$d\Sigma$	Surface element
$d\tau$	Infinitesimal volume
$\mathbf{D}$	Displacement velocity of the discontinuity surface
D_t	Coefficient of turbulent diffusion
E	Young's modulus of elasticity, kinetic energy of the medium volume V, explosive energy
$E = V^2/2$	Density of kinetic energy
$E(k, t)$	Spectrum of mean energy
E_t	Turbulent energy
e	Density of internal energy

e_i	Main components of the strain velocity tensor
$\boldsymbol{e}_i$	Basis vectors of the observer's system of coordinates, basis vectors of the Cartesian system of coordinates
e_{ij}	Components of the strain velocity tensor
e_{ki}	Components of a symmetric tensor, components of velocity of relative extension of a distance along the X^i axis, velocity of angular reduction for a right angle
e_ρ	Velocity of relative extension of a distance within a deforming body
$\boldsymbol{E}$	Strain tensor
$\dot{\boldsymbol{E}}$	Strain velocity tensor
$\boldsymbol{E}^\circ$	Initial strain tensor
$\hat{\boldsymbol{E}}$	Current strain tensor
E_k	Density of kinetic energy
E_s	Density of the kinetic energy of averaged turbulent movement
E_t	Average density of the kinetic energy of the fluctuating movement
Eu	Euler's number
F	Free energy
$F(\boldsymbol{x})$	Harmonic function
F^*	Mean value of the function F
$\boldsymbol{F}$	Force, main vector of mass forces
$\boldsymbol{F}_i$	Effective force on i-th material point
$\boldsymbol{F}_i^{(e)}$	i-th component of external force
$\boldsymbol{F}_i^{(i)}$	i-th component of internal force
F_α	Generalized thermodynamic force
$F_{ij}(\boldsymbol{k}, t)$	Components of the spectral tensor
$F_{LL}(k, t)$	Longitudinal spectrum
$F_{NN}(k, t)$	Transverse spectrum
$F_{ij,k}(k)$	Spectrum of the correlation tensor of third order of an isotropic velocity field
$\boldsymbol{F}_{\text{surf}}$	Force due to surface tension
$\boldsymbol{f}_{\text{in}}$	Density of inertial force
$\boldsymbol{f}_m$	Force per unit mass
$\boldsymbol{f}_V$	Force per unit volume
f_X, f_Y, f_Z	Cartesian components of the force acting on the unit volume
$\lvert\boldsymbol{f}_\omega\rvert$	Amplitude
Fr	Froude number
g	Acceleration due to gravity
$g, g^\circ, \hat{g}, \hat{g}^\circ$	Determinants of the fundamental matrices
g_{ij}, g^{ks}	Components of the metric tensor
g°_{ij}	Components of the metric tensor in the initial state
g_ω	Amplitude at frequency ω
G	Mass flow of fluid in beam, shear modulus
$\boldsymbol{G}$	Metric tensor, shear force, Archimedes buoyancy

$\boldsymbol{G}'$	Centripetal attractive force
$\boldsymbol{G}_B$	Gravity of a body
G_D	Green's function for the Dirichlet problem
G_N	Green's function for the Neumann problem
$\boldsymbol{h}$	Mass force pairs per unit mass
i	Enthalpy
i^*	Stagnation enthalpy
$\boldsymbol{i}, \boldsymbol{j}, \boldsymbol{k}$	Basis unit vectors of the Cartesian system of coordinates
I	Moment of intertia of a sphere
I_1, I_2, I_3	Invariants of the stress tensor
$I_1^\circ, I_2^\circ, I_3^\circ$	Invariants of the strain tensor $\boldsymbol{E}^\circ$
$\hat{I}_1, \hat{I}_2, \hat{I}_3$	Invariants of the strain tensor $\hat{\boldsymbol{E}}$
$I_1(e)$	First invariant of the strain velocity tensor
$I_1(\varepsilon)$	First invariant of the strain tensor
I^α	Generalized thermodynamic currents
$\boldsymbol{I}$	Unit tensor
$\boldsymbol{I}_i$	Diffusive current of the i-th component of a mixture
$\mathrm{Im}(z)$	Imaginary part of a complex number
$\mathrm{Im}(\omega)$	Imaginary part of angular frequency
J_0	Beam momentum
k	Scalar, Boltzmann constant, wave number
k, n	Constants in the Ostwald–Reiner rheological equation
k	Permeability of a layer
$\boldsymbol{k}$	Density of proper or internal angular momenta
$k^1, k^2, \ldots, k^n$	Physical constants
$\boldsymbol{K}$	Angular momentum of a point, angular momentum of a volume
K^*	Inertial system of coordinates
$\boldsymbol{K}^*$	Angular momentum of all points in the volume V relative to the center of mass O^*
l	Coefficient of relative stretching, inertial radius of the sphere relative to center of rotation, mixing length in the Prandtl model, microscale of fluctuations
l, m, n	Direction cosines of the normal vector
l_0	Internal microscale, Kolmogorov microscale
l_1	Characteristic length in the Taylor model
l_i	Coefficients of relative stretching in the direction of the axes ξ^i
L	Current line, characteristic linear scale, linear body measure
L_0	Initial line
L_1	Longitudinal integral scale
L_2	Transverse integral scale
$L^{\alpha\beta}$	Quadratic symmetric matrix
m	Mass, porosity
m_B	Mass of a body
m_i	Mass of the i-th component
m_n	Coefficients of the Laurent series

Δm	Mass of an infinitesimal volume ΔV
M	Mach number
M	Point
$\boldsymbol{M}$	Torque of a force, resulting torque, main torque
M'	Point
M_∞	Mach number at infinity
$m\boldsymbol{V}$	Momentum
$m_i\boldsymbol{V}_i$	Momentum of the i-th material point
n	Number of components, polytropic exponent
$\boldsymbol{n}$	Unit normal vector
$(\boldsymbol{n}, \boldsymbol{t}_1, \boldsymbol{t}_2)$	Orthogonal basis vectors of the local system of coordinates
n_i	Components of the unit normal vector, covariant components of the external unit normal vector
$\boldsymbol{n}_k$	Normal vector to elementary surface $d\sigma_k$
O^*	Center of mass of a body
$\boldsymbol{p}$	Density of surface forces
p	Pressure
$\langle p \rangle$	Mean value of the pressure
p_1, p_2, p_3	Main components of the stress tensor
$\boldsymbol{p}$	Density of surface forces
$\boldsymbol{p}^1, \boldsymbol{p}^2, \boldsymbol{p}^3$	Stresses at the point M on surface elements parallel to coordinate planes
$\boldsymbol{p}^i$	Stress vector on the surface element with unit normal vector $\boldsymbol{e}_i$
p_d	Cavitation pressure
p_{dyn}	Dynamic pressure
p'	Small pressure perturbation
p_{cap}	Capillary pressure
p_{st}	Hydrostatic pressure
p^*	Dynamic pressure
p_i	Momentum
p^{ki}	Components of the expansion of the vector $\boldsymbol{p}^i$ in basis vectors of the Cartesian system of coordinates, contravariant components of the stress tensor
$p_{\min}$	Minimum pressure
$\boldsymbol{p}_n$	Density of surface forces at the point M, stress on the surface element at the point M
p_{nn}	Normal component of the force of internal stresses
p_{nt}	Tangential component of the force of internal stresses
$\boldsymbol{p}_x, \boldsymbol{p}_y, \boldsymbol{p}_z$	Components of the stress vector
$\boldsymbol{P}$	Matrix of p^{ki}, stress tensor, resulting force (main vector)
$\nabla \cdot \boldsymbol{P}$	Divergence of a vector
$\boldsymbol{P} : \boldsymbol{E}$	Total scalar product of the stress tensor and the strain velocity tensor

P	Measure of the probability for a state under consideration, pressure function
∇p	Pressure gradient
Pe_D	Diffusion Peclet number
Pe_T	Thermal Peclet number
Pr	Prandtl number
$\boldsymbol{q}$	Vector of heat flow through an elementary surface
q_ε	Density of internal heat sources
Q	Volume flow per second, strength of a stream tube, flow through a stream tube, volume flow, flow through a contour
$\boldsymbol{Q}$	Momentum of a system of material points
$\boldsymbol{Q} = m\boldsymbol{V}^*$	Momentum of the center of mass
q	Surface mass density
$\boldsymbol{q}_n$	Surface force pair per unit surface
r	Radial coordinate, polar coordinate
$\boldsymbol{r}$	Radius vector
$\boldsymbol{R}$	Main vector
$\boldsymbol{r}^*$	Radius vector of the center of mass
R	Radius of a continuum sphere, gas constant
$\boldsymbol{r}_{ic}$	Radius vector of the i-th point relative to the center of mass
r, φ, Z	Cylindrical coordinates
r, φ, ϑ	Spherical coordinates
$d\boldsymbol{r}$	Infinitesimal element of the radius vector, differential
$\Delta\boldsymbol{r}$	Increment of the radius vector
R_1, R_2	Main radii of curvature
R_y	Buoyancy
Re	Reynolds number
R_{cr}	Critical Reynolds number
$\mathrm{Re}(z)$	Real part of a complex number
Re_l	Reynolds number of the microscale l
s	Density of entropy
$\boldsymbol{s}$	Vector
$\boldsymbol{s}^\circ$	Unit vector
S	Entropy, current surface
$\boldsymbol{S}$	Vector of an entropy current
S_0	Initial surface
$S_{\min}$	Minimal cross section of a tube
Sc	Schmidt number
Sh	Strouhal number
t	Time
$\boldsymbol{t}$	Tangential vector
$\boldsymbol{t}_1, \boldsymbol{t}_2$	Unit tangential vectors
T	Temperature
$\boldsymbol{T}$	Spherical tensor
T^*	Stagnation temperature

∇T	Gradient of a scalar function, vector gradient
u, v, w	Velocity components in the space point (X, Y, Z)
$\langle u \rangle$	Mean value of the velocity
$\langle \boldsymbol{u} \rangle$	Mean value of the velocity vector
$\boldsymbol{u}(\boldsymbol{X}, t)$	Velocity vector
u_i	Components of the velocity vector in Cartesian coordinates
$\boldsymbol{u}'$	Small perturbation of the velocity vector
u_l	Velocity on the micro scale l
u'_i	Small perturbations of the velocity components
U	Density of internal energy
U_∞	Flow velocity at infinity
U, V, W	Velocity components in Cartesian coordinates
U_∞, V_∞	Velocity components at infinity
U^i	Physical components of the velocity vector
U_r, U_φ, U_z	Velocity components in cylindrical coordinates
$\boldsymbol{V}$	Velocity vector, velocity vector of the whole mixture
V	Volume, volume in the current state
V^*	Moving volume
V'	Volume
$\boldsymbol{V}^*$	Velocity of the center of mass, velocity relative to the moving system of coordinates K^*
V_0	Volume in the initial state
$\boldsymbol{V}_0$	Translatory velocity
V^i	Components of the velocity vector
$\boldsymbol{V}_i$	Velocity of the i-th component of the mixture, $(i = 1, 2, \ldots, n)$
$\boldsymbol{V}^*$	Velocity of pure deformation, velocity of the center of mass
$\boldsymbol{V}_c$	Velocity of the center of mass of a system of material points, velocity of the point M relative to O^*
V_{cr}	Critical velocity
V_{void}	Pore volume
W	Total energy flow
$\boldsymbol{V}_{def}$	Velocity of the pure deformation
$\boldsymbol{V}_{ic}$	Velocity of the i-th point relative to the center of mass system of coordinates
V_{max}	Maximum flow velocity
V_n	Normal component of the velocity
$\boldsymbol{V}_{rot}$	Rotating velocity
V_S	Component of the vector $\boldsymbol{V}$ in the direction $\boldsymbol{s}$
w	Complex potential
$W(Z)$	Characteristic function
$\boldsymbol{W}$	Displacement vector, resistance force of the body
W_{av}	Average velocity
W^k	Contravariant component of the displacement vector
$\boldsymbol{X}$	Space point
X, Y, Z	Coordinates of a point in Cartesian coordinates

X^i	Coordinates of a point
$\{X^i\}$	Cartesian system of coordinates
$Y = h(X)$	Contour of a body
z	Complex variable
α	Angle
β	Angle of inclination of the shock wave
Γ	Circulation of the vector along the contour, velocity circulation along a closed contour, circulation of the vector along the contour
Γ_C	Strength of a vortex tube
Γ^j_{ki}	Christoffel symbol of the second kind
γ	Adiabatic exponent, density of vortices per unit length
Δ	Determinant
δ	Angular aperture of a wedge, thickness of a boundary layer
δ_*	Thickness of a laminar layer
$\partial/\partial t$	Local derivative
$\delta^i_j, \delta^i_k, \delta_{ik}$	Kronecker symbol
δ_{ij}	Components of the unit tensor
ε	Infinitesimal quantity, dissipative function, small parameter, angle of attack of a profile, shear strain
$\langle\varepsilon\rangle$	Mean specific energy dissipation
ε_0	Initial deformation
$\boldsymbol{\varepsilon}_1, \boldsymbol{\varepsilon}_2$	Basis of the accompanying system of coordinates
$\boldsymbol{\varepsilon}_i$	Basis vectors of the accompanying (deformed, frozen, curvilinear) coordinate systems
$\boldsymbol{\varepsilon}^\circ_i$	Basis vector
ε_i	Main components (main values) of the strain tensor
$\boldsymbol{\varepsilon}_i$	Vectors of the covariant basis
ε_{ij}	Components of the strain tensor
$\dot{\varepsilon}$	Shear velocity
ε_{kji}	Components of the permutation symbol
ε°_i	Main values in an initial state
$\hat{\varepsilon}_i$	Main values in a current state
ε_s	Specific energy dissipation of the average turbulent movement due to viscous forces
$\langle\varepsilon_t\rangle$	Mean specific energy dissipation of the fluctuating movement
ς	Coefficient of volume viscosity
η	Similar variable, dimensionless coordinate
η^i	Coordinates in an orthogonal system of coordinates
ϑ	Coefficients of volume expansion, angle, inclination angle to the X-axis of the velocity behind the shock wave, inclination angle of the tangent to the profile
κ	Heat conductivity
κ_t	Coefficient of turbulent heat conductivity
Λ	Function of the state parameters of a system, total energy of the system, Loizianski integral, Loizianski invariant

λ	Eigenvalue, scalar, Lamé elasticity constant, dimensionless parameter, Taylor microscale, microscale of energy dissipation
λ_1	Parameter of the Navier–Stokes law, longitudinal differential scale
λ_2	Transverse differential scale
λ_{ik}	Virtual mass coefficients
μ	Coefficient of dynamic viscosity, Lamé elasticity constant, virtual mass of a sphere
$\mu(M)$	Density of a double layer
μ'	Coefficient of structural viscosity
μ_1	Parameter of the Navier–Stokes law
$\mu^1, \mu^2, \dots, \mu^n$	State parameters, variable parameters
ν	Kinematic viscosity coefficient, complex velocity
$\bar{\nu}$	Conjugate complex variable
ν_*	Velocity at the boundary between a lower layer and a turbulent core
ν_t	Turbulent viscosity coefficient
ξ, η, ς	New coordinates
ξ^1, ξ^2, ξ^3	Coordinates of a point in a curvilinear system of coordinates
ξ^i	Lagrange coordinates
$\{\xi^i\}$	Curvilinear (accompanying) system of coordinates
Π	Dimensionless quantity
ρ	True (local) mass density, density of a whole mixture
ρ_{av}	Average mass density
ρ_i	Mass density of the i-th component ($i = 1, 2, \dots, n$) of the mixture
$\wp_i$	Mass change of the i-th component of the mixture per unit time and unit volume
ρ^i	Infinitesimal distance
ρ_B	Mass density of the body
ρ^*	Stagnation pressure
$\boldsymbol{\rho}$	Radius vector $\boldsymbol{OO}_1$
$\boldsymbol{\rho'}_1$	Radius vector $\boldsymbol{O'}_1\boldsymbol{O'}_1$
ρ_∞	Mass density at infinity
$\Delta\boldsymbol{\rho}$	Change of a radius vector
Σ	Surface, cross-sectional surface of the stream tube
Σ'	Surface
σ	Tangential surface, Poisson number, coefficient of surface tension
σ_{ij}	Components of the viscous stress tensor
τ	Characteristic period of fluctuations, tangential stress on the surface of the body, tangential stress, shear stress, time
τ_0	Friction force on the wall per surface unit, limiting tension
τ_{ij}	Reynolds stresses
$\tau_{ij}^{(1)}$	Components of the turbulent additional stresses
τ_f	Friction force on the wall per surface unit
τ'	Reynolds stress (shear stress)

τ_w	Stress on the wall
$\bar{\tau}_w$	Average stress due to friction
τ_i	Parameters of the Navier–Stokes Law, Main components of the viscous stress tensor
τ^{ij}	Components of the viscous stress tensor
τ_h	Characteristic hydrodynamic relaxation time
v	Specific volume
v_i	Components of the displacement velocity
v^*	Small volume element
$\boldsymbol{\Phi}$	Quadratic form, potential, physical characteristic, dissipation function, quadratic form of the generalized thermodynamic forces, potential of the external mass forces, Airy function, Newton's potential
Φ_{in}	Potential of the inertial force
ϕ	Density of the physical characteristic
ϕ	Flow potential, polar coordinate, scalar potential
φ_B	Potential value at point B
φ_A	Potential value at point A
$\varphi_i^{(D)}$	Potential of the internal Dirichlet problem
$\varphi_i^{(N)}$	Potential of the internal Neumann problem
$\varphi_e^{(D)}$	Potential of the external Dirichlet problem
$\varphi_e^{(N)}$	Potential of the external Neumann problem
φ_n	Potential of a multipole flow
φ_{tr}	Potential for a translatory flow
$\partial\varphi/\partial n$	Normal derivative
χ	Cavitation number
χ	Coefficient of heat conductivity
χ_i	Physiochemical parameters
χ_{ij}	Difference between ψ_{ij} and the right angle
Ψ	Quadratic form
ψ	Gibbs thermodynamic potential, angle
ψ	Stream function
ψ_{ij}	Angle between basis vectors $\boldsymbol{\varepsilon}_i$ and $\boldsymbol{\varepsilon}_j$
$\nabla^4\psi = 0$	Biharmonic equation
$\boldsymbol{\Omega}$	Asymmetric tensor, curl of vector $\boldsymbol{A}$
Ω_1, Ω_2	Functions
$\boldsymbol{\omega}$	Vector of the instantaneous angular velocity of the body, axial vector, vortex vector, angular velocity of the body
ω	Angular velocity, complex frequency
ω_i	Components of the vector $\boldsymbol{\omega}$
ω_{ij}	Components of the antisymmetric tensor
ω_n	Normal component of the vector $\boldsymbol{\omega}$
∇_i	Covariant derivative

$\nabla \cdot \boldsymbol{V}$	Divergence of the velocity vector $\boldsymbol{V}$, velocity of a relative volume change, first invariant of the tensor of strain velocities
$\nabla \times \boldsymbol{A}$	Curl of vector $\boldsymbol{A}$
Δ	Laplace operator
$\Delta = \frac{\partial^2}{\partial X^2} + \frac{\partial^2}{\partial Y^2} + \frac{\partial^2}{\partial Z^2}$	Laplace operator in Cartesian coordinates

1
Introduction

1.1
Goals and Methods of Continuum Mechanics

Continuum mechanics is that part of mechanics that deals with the movement of deformed gaseous, liquid and solid bodies. In theoretical mechanics, one studies the movement of a mass point, of a discrete system of a such points and of rigid bodies. On the other hand, in continuum mechanics, using the methods and results obtained in theoretical mechanics, one deals with the movement of such material bodies that fill the space continuously, and where the distance between points changes during the movement.

In addition to the usual media, that is, gas, fluid and solid, we also consider unusual media such as the electromagnetic field, the gravitational field and the radiation field.

In the following, we will discuss the most important problems that have given rise to the independent branches of continuum mechanics.

Since the forces that are exerted on a solid body by the gas or fluid that surrounds it depend on the movement of the gas or fluid, the study of the movement of solid bodies in a gas or a fluid is therefore closely linked with the investigation of the movement of a gas or a fluid. Hydromechanics as one of several branches of continuum mechanics deals with the movement of a fluid. Finally, the movement of a gas is discussed in gas dynamics. The problems of the movement of planes, helicopters, rockets, ships, submarines and so on gave the impetus for the massive boost in this research area.

The problems of movement of gas and fluid in tubes and different machines are of great relevance for the planning and calculation of natural gas pipelines, oil pipelines, pumps, compressors, turbines and other hydraulic machines. Applied problems in this field are the foundation for hydraulics.

The movement of gas and fluids through the ground and other porous media (filtration) is a very hot topic in the production of oil and natural gas. Thus, this field that deals with movement of gas and fluids through porous bodies is called subterranean hydromechanics.

Hydromechanics. Theory and Fundamentals. Emmanuil G. Sinaiski

ISBN: 978-3-527-41026-2

Large theoretical and applied importance has the problem of wave propagation in the media mentioned above, that is, gases, fluids and solid bodies. This field is referred to as wave mechanics.

Finally, another part of continuum mechanics deals with the movement of multicomponent and multiphase mixtures caused by different interactions between phases, exchange of heat and matter, phase transitions and chemical reactions. These problems are the foundation of physicochemical hydrodynamics.

The problems connected to the movement of conducting and charged media in magnetic and electric fields are dealt with by the field of magneto and electrohydrodynamics.

The methods of continuum mechanics can also be applied to the movement of bodies in a gas of low density, in outer space, and in stellar and planet atmospheres. The fields of continuum mechanics dealing with these problems are known as the mechanics of diluted gases and cosmic hydrodynamics.

Recently, a new field of biomechanics has come into the forefront which investigates the mechanics of biological objects, the flow of blood in living organisms, the contraction of muscles and which attempts to construct mechanicals models of the internal organs.

Among other important problems, the following needs to be mentioned: weather forecasting, theory of turbulence, movement of sand dunes in the desert, avalanches, burning and detonation as well as the theory of explosions.

Large parts of continuum mechanics are devoted to the investigation of the dynamics and the equilibrium of deformed solid bodies. The theory of elasticity provides the computational foundation for the planning of machines and buildings. Those parts of continuum mechanics that deal with the elastic properties of bodies having a complicated composition and with making provision for inelastic effects in solid bodies become more and more important. The theory of plasticity investigates the behavior of a solid body beyond the elastic limit. Of high importance is also the investigation of various types of material fatigue and taking the the memory effect for the dynamics and equilibrium of a solid body into account. With the invention of new compounds and polymers with composite internal structures, the need to develop new models for these compounds based on their internal structures has become evident.

It is impossible to enumerate all problems and applications of continuum mechanics. However, the above-mentioned examples are sufficient to make the conclusion that continuum mechanics is involved when dealing with a large set of theoretical and applied problems in science and engineering.

Continuum mechanics has also had a large influence on the development of a number of areas of in mathematics. For example, the theory of the wing of a plane influenced the theory of functions of a complex variable, the movement of a viscous fluid gave impetus to research about boundary conditions of partial differential equations and some problems of elasticity theory influenced the research of the theory of integral equations.

The solution of many important questions for applied problems has always necessitated the use of numerical algorithms. Therefore, continuum mechanics has boosted the development of numerical methods.

The following series of lectures about hydromechanics as part of continuum mechanics covers the research methods applicable to the dynamics of deformed bodies. We start by introducing a number of concepts, that is, the fields of velocity, pressure and temperature. We further show how mechanical problems can be reduced to mathematical ones whose solutions determine the properties and dynamics of the deformed bodies.

1.2 The Main Hypotheses of Continuum Mechanics

When investigating the dynamics of bodies, one has to exploit their real properties that depend on their internal structure. Every material body consists of different molecules and atoms which are in constant irregular movement. Certain interactions exist between the particles. For a gas, those are mostly determined by collisions. For a fluid, the particles are closer together than in a gas and thus the molecular forces, that is, London–Van der Waals attractive forces and repulsive electrostatic forces are important in this case. The strength and elasticity of solids are due to forces that are electrostatic in nature. If all forces are known, then it is, in principle, possible to create a theory of the dynamics of the material body. However, the complicated internal structure of the body and the huge number of molecules in the volume of interest are substantial hurdles for creating models of the medium. Therefore, it is practically impossible to consider the movement of one particle while taking into account the interactions with all others. Fortunately, however, when dealing with technical applications, there is no need to know the movement of all particles. One only needs to know mean values. In this connection, there are two methods to investigate the dynamics of the medium. The one is the statistical method that has been developed in physics and where the concept of a probability distribution is applied to the system under investigation, and one considers mean values of system data taken over all possible realizations of the system. The second method consists of arriving at a phenomenological macroscopic theory based on experimental observations and laws. In continuum mechanics, the second method is mostly used. However, in some areas, that is, when investigating the dynamics of heterogeneous multiphase media (emulsion, suspensions etc.), both methods are combined.

The methods of continuum mechanics are founded on two hypotheses. The first hypothesis is the continuum hypothesis which assumes that since the number of particles in any volume of practical interest of each material body is extremely large, the body can be considered as a medium that fills the volume in a continuous manner. Such a medium is referred to as a continuous medium. This idealization of the real medium makes it possible to employ the mathematical devices of continuous functions, and differential and integral equations for modeling the processes of interest in the medium.

The second hypothesis is the space-time hypothesis. In continuum mechanics, one normally considers the Euclidean space concept to be applicable so that a homogeneous global system of coordinates can be introduced. The version of mechanics that has been developed on the basis of this hypothesis is called Newtonian mechanics. In general, however, the time does depend on the coordinate system if relativistic effects are taken into account. By neglecting relativity, the time passes in the same manner for all observers. Such a time is referred to as absolute time. Normally, continuum mechanics investigates the dynamics of the continuum in Euclidean space, that is, continuum mechanics is based on Newtonian mechanics. However, in some cases, it is necessary to use the methods on non-Euclidean geometry and the theory of relativity, for example, when investigating the movement and deformation of boundary between different phases of a medium and also when considering the movements of objects in outer space at very high velocity.

2
Kinematics of the Deformed Continuum

2.1
Dynamics of the Continuum in the Lagrangian Perspective

A system of coordinates is introduced in which the point coordinates are denoted by X^i, with $i = 1, 2, 3$ for the three dimensional space. This system of coordinates is not necessarily orthogonal. In the following, the coordinates of a point will be denoted by X, Y, Z or ξ^1, ξ^2, ξ^3. If the coordinates of a point are changing according to the law

$$X^i = X^i(t) , \qquad (i = 1, 2, 3) , \tag{2.1}$$

one concludes that the point is moving relative to the system of coordinates and that (2.1) is the law governing that movement. To describe the movement of the continuum as a whole, we need to know the movements of all points. It is therefore necessary to include additional information in order to identify the point being considered. If the coordinates ξ^i of the material point are known at $t = 0$, the law describing the movement of the point must also depend on ξ^i, that is,

$$X^i = X^i(\xi^1, \xi^2, \xi^3, t) , \qquad (i = 1, 2, 3) . \tag{2.2}$$

For constant ξ^j and changing t, (2.2) describes the movement of a particular point. For constant t and changing ξ^j, (2.2) describes the distribution of continuum points in space. Finally, if both ξ^j and t are changing, (2.2) describes the movement of the medium as a whole. The coordinates ξ^i that label the points are known as Lagrangian coordinates. The main problem of mechanics consists of the determination of (2.2). We now assume that the functions on the right-hand side are continuous, and partial derivatives exist with respect to all arguments. We will later relax these strong assumptions when necessary to deal with boundary layers. In addition, we assume that the functions defined in (2.2) are bijective for all values of t. This is equivalent to the Jacobian being nonzero everywhere, that is,

$$\Delta = \left| A^i_{\ j} \right| = \left| \frac{\partial X^i}{\partial \xi^j} \right| \neq 0 ,$$

Hydromechanics. Theory and Fundamentals. Emmanuil G. Sinaiski

ISBN: 978-3-527-41026-2

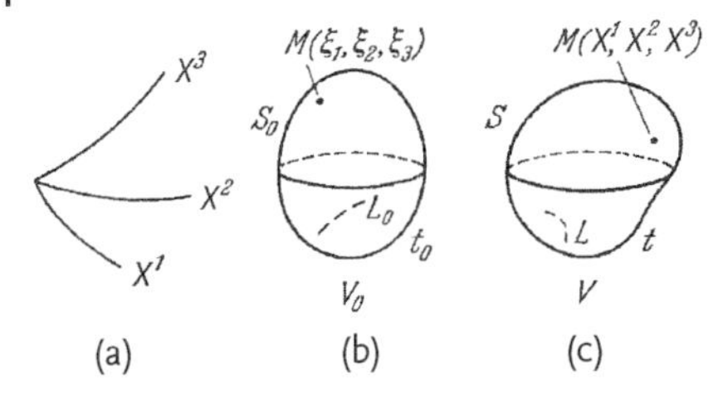

Figure 2.1 Movement of an element of the continuum. For $t = t_0$, we have $X^1 = \xi^1$, $X^2 = \xi^2$ and $X^3 = \xi^3$.

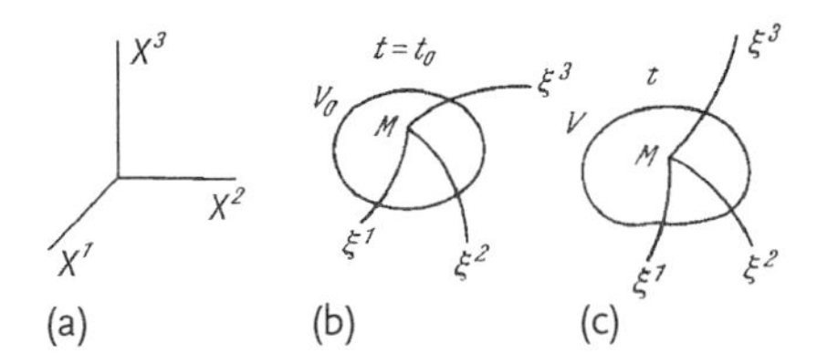

Figure 2.2 (X^1, X^2, X^2) is the system of coordinates of the observer and $(\xi^1$ (a), ξ^2 (b), ξ^3 (c)) is the associated system of coordinates.

and the inverse relationship between the ξ^i and X^i can be obtained from (2.2), that is,

$$\xi^i = \xi^i(X^1, X^2, X^3, t) , \qquad (i = 1, 2, 3) , \tag{2.3}$$

where the functions on the right-hand side of (2.3) are continuous.

If the coordinates of a point M are ξ^i at $t = 0$, then D_0 is the volume that the body occupies at $t = 0$. In this case, (2.3) is a bijective and continuous mapping of the volumes D and D_0. Topological arguments show that the volume D becomes the volume D_0, the surface S becomes the surface S_0 and the closed loop L becomes the closed loop L_0 (see Figure 2.1).

In addition to the system of coordinates $\{X^i\}$ that is associated with the observer, we introduce another one, namely, $\{\xi^i\}$ which is connected to the deformed material body. Provided the coordinate lines within the continuum are connected to an initial point of the continuum, they become the coordinate lines in the associated system of coordinates that moves with the medium. Even if the coordinates lines were straight in the beginning, they will become curved at a later point in time. While the continuum points are moving in the system of coordinates $\{X^i\}$ of the observer, they are at rest in the associated system of coordinates $\{\xi^i\}$ (see Figure 2.2).

Using both ξ and t as variables, define the Lagrangian approach for the investigation of the dynamics of a continuous medium. Let us define the notions of velocity and acceleration for the continuum as follows: a material point is at time t at space point $\boldsymbol{M}$ and at $t + \Delta t$ at space point $\boldsymbol{M}'$, and we define $\boldsymbol{MM}' = \Delta \boldsymbol{r}$ with the radial vector $\boldsymbol{r}$ from the origin of the system of coordinates $\{X^i\}$ to the position of the material point. Since the vector $\boldsymbol{r}$ is defined at the individual point

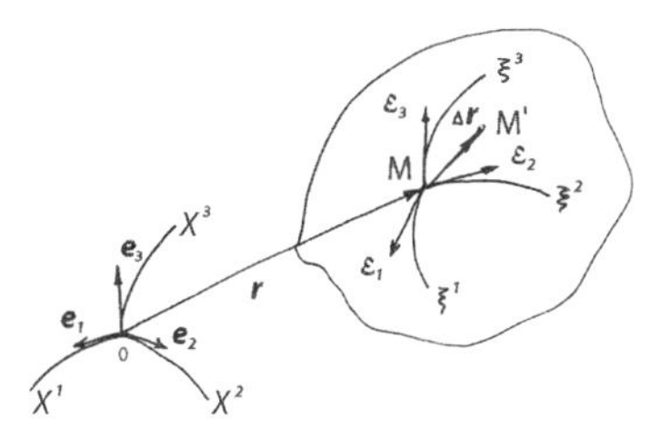

Figure 2.3 Bases of the observer and co-moving systems of coordinates.

of the continuum, it depends on both X^i and t. The velocity of the individual point of continuum in the observer system of coordinates is defined as

$$V = \left(\frac{\partial r}{\partial t}\right)_{\xi^i} . \tag{2.4}$$

Since the medium does not move relative to the co-moving system of coordinates, the velocity with respect to this system of coordinates will vanish.

We denote the basis vectors of the observer system of coordinates $\{X^i\}$ and of the co-moving system of coordinates $\{\xi_j\}$ as e_i and ε_i. According to Figure 2.3, coordinates are generally curvilinear as the basis vectors are given by the tangents along the respective coordinate lines. The base vectors are given by the following obvious relationships:

$$e_i = \frac{\partial r}{\partial X^i}, \qquad \varepsilon_i = \frac{\partial r}{\partial \xi^i} . \tag{2.5}$$

For an infinitesimal shift from M to M', the corresponding change in the radial vector becomes

$$\Delta r = \Delta X^i e_i = \Delta \xi^j \varepsilon_j . \tag{2.6}$$

In the above equation and the following summation over identical upper and lower indices is assumed, that is the Einstein sum convention.

From (2.4) and (2.6), it follows that

$$V = V^i e_i , \quad V^i = \left(\frac{\partial X^i}{\partial t}\right)_{\xi^i} . \tag{2.7}$$

For a individual point in the continuum, the acceleration is defined as

$$a = \left(\frac{\partial V}{\partial t}\right)_{\xi^i} = \left[\frac{\partial (V^i e_i)}{\partial t}\right]_{\xi^i} = \frac{\partial V^i}{\partial t} e_i + V^i \frac{\partial e_i}{\partial t} . \tag{2.8}$$

In a Cartesian system of coordinates, we have $\partial e_i / \partial t = 0$ and therefore

$$a = \frac{\partial V^i}{\partial t} e_i = a^i e_i . \tag{2.9}$$

In a curvilinear system of coordinates, not only V^i, but also e_i changes. Therefore, (2.8) has to be used to define the acceleration.

2.2
Dynamics of the Continuum in the Eulerian Perspective

In the following, we are not interested in the history of the movement of an individual particle in the continuum. We focus instead on what happens at a fixed point in space. This approach defines the Eulerian perspective. Therefore, the coordinates X^i and time t are referred to as Eulerian variables.

The movement is known when velocity, acceleration, temperature and other parameters are given as functions of X^i and t. For constant t and variable X^i, these functions describe the spatial distribution of the parameters at the fixed point in time. For constant X^i and variable t, these functions describe the development of the parameters in this point.

From a mathematical perspective, the difference between the Lagrangian and Eulerian approaches is that in the first case, the points of the continuum are labeled by ξ^i. However, in the second case, the independent variables are X^i and t. If the relationships $X^i = X^i(\xi^j, t)$ are known, the movement of the continuum is completely described. By inverting these relationships, one obtains $\xi^i = \xi^i(X^j, t)$. Thus, it is possible to effect a transition from Lagrangian to Eulerian Variables.

We now assume that the velocity, acceleration and temperature are given as functions $\boldsymbol{V} = \boldsymbol{V}(\xi^j, t)$, $\boldsymbol{a} = \boldsymbol{a}(\xi^j, t)$ and $T = T(\xi^j, t)$ of the Lagrangian variables. Then, it is possible by using the relationships $\xi^i = \xi^i(X^j, t)$ to obtain the dependencies of these quantities from the Eulerian variables, that is, $\boldsymbol{V} = \boldsymbol{V}(\xi^j, t) = \boldsymbol{V}(\xi^j(X^i, t), t) = \boldsymbol{V}(X^i, t)$. To achieve a transition from the Lagrangian to the Eulerian variables, one has to use the expression for the velocities. We assume, for simplicity, a Cartesian system of coordinates in which the components of the velocity u, v, w in a space point X, Y, Z are given by

$$\frac{dX}{dt} = u(X, Y, Z, t) , \quad \frac{dY}{dt} = v(X, Y, Z, t) , \quad \frac{dY}{dt} = w(X, Y, Z, t) . \tag{2.10}$$

After the solution of this system of ordinary differential equations, one obtains

$$X = X(C_1, C_2, C_3, t) , \quad Y = Y(C_1, C_2, C_3, t) , \quad Z = Z(C_1, C_2, C_3, t) . \tag{2.11}$$

The parameters C_1, C_2 and C_3 are uniquely defined if X, Y and Z are defined at an initial time t_0. This transition from Lagrangian to Eulerian variables for a given velocity field is thus reduced to the integration of ordinary differential equations.

2.3
Scalar and Vector Fields and Their Characteristics

When discussing the dynamics of the continuum, one introduces scalar fields, that is, pressure, temperature and so on, and vector fields, that is, velocity, acceleration, and so on. These fields can be examined in different systems of coordinates: in the observer system of coordinates X^i and the media-fixed system of coordinates ξ^i. By using the example of the temperature field, we will discuss some common characteristics in the following.

If the temperature field is considered in the Lagrangian perspective, we have $T = T(\xi^i, t)$ and the change of T per unit time interval for the individual particle in the continuum will be given by the quantity $(\partial T/\partial t)_{\xi^i}$. By considering the temperature field in the Eulerian perspective, we then have $T = T(X^i, t)$. By using the relationship between X^i and ξ^i according to (2.10) and by employing the chain rule one, obtains

$$\left(\frac{\partial T}{\partial t}\right)_{\xi^i} = \left(\frac{\partial T}{\partial t}\right)_{X^i} + \frac{\partial T}{\partial X^i}\frac{\partial X^i}{\partial t} = \left(\frac{\partial T}{\partial t}\right)_{X^i} + V^i\frac{\partial T}{\partial X^i} . \tag{2.12}$$

The derivative on the left-hand side of (2.12) characterizes the temperature change in the given point of the continuum and is referred to as the total derivative. It is normally denoted by dT/dt or DT/Dt. The first term on the right-hand side is called the local derivative and is denoted by $\partial T/\partial t$, and the second term is called the convective derivative. Equation 2.12 can be rewritten as

$$\frac{dT}{dt} = \frac{\partial T}{\partial t} + V^i\frac{\partial T}{\partial X^i} = \frac{\partial T}{\partial t} + \boldsymbol{V}\cdot\nabla T . \tag{2.13}$$

Since $\boldsymbol{V}\cdot\nabla T$ is a scalar and therefore an invariant, (2.13) is valid not only in Cartesian, but also in curvilinear coordinate systems.

We now consider the temperature as a function of the Eulerian variables. The surfaces defined by $F(X, Y, Z) =$ const are called isosurfaces. Let us consider the isosurface for $T = T_1$ and a point M on this surface. We define a normal vector $\boldsymbol{n}$ and a vector $\boldsymbol{s}$ such that the angle between them is α. Then, dT/ds is defined as the derivative of T in the direction $\boldsymbol{s}$. The derivatives in the directions of $\boldsymbol{n}$ and $\boldsymbol{s}$ are related by

$$\frac{dT}{ds} = \frac{dT}{dn}\cos\alpha . \tag{2.14}$$

If $\boldsymbol{s}$ lies in the tangential plane, then $dT/ds = 0$. From (2.14), it follows that the maximum of dT/ds is attained in the direction of the normal to the surface. With the unit vector $\boldsymbol{n}$, the gradient of the temperature is

$$\nabla T = \frac{dT}{dn}\boldsymbol{n} . \tag{2.15}$$

The gradient vector points in the direction of the increase of the temperature and its absolute value is large according to the density of the isosurfaces. The projections of ∇T on the axes are

$$(\nabla T)_{X^i} = \frac{\partial T}{\partial X^i} . \tag{2.16}$$

A process is stationary if all parameters in the Eulerian perspective are explicitly independent of time, otherwise it is non-stationary. However, it must be noted that the same process can be stationary as well as non-stationary. This depends on the choice of the system of coordinates in which the movement is observed. For example, the waves created in the water behind a ship which is moving with constant velocity are stationary from the perspective of an observer on the ship while they are non-stationary from the perspective of an observer on the beach.

For every vector field, it is possible to define the notion of a vector curve as a curve for which the tangent at every point is parallel to the vector field at that point. As an example, we will consider the velocity field. We assume that the vector field $\boldsymbol{V} = V^i \boldsymbol{e}_i$ is defined in every point X^i at a given time. The vector curves of the velocity field at this point in time are then referred to as stream lines.

Let us derive the differential equation for the stream lines. We consider an infinitesimal element of the stream line $d\boldsymbol{r} = dX^i \boldsymbol{e}_i$ and the corresponding velocity vector $\boldsymbol{V}$. Because of the collinearity it follows that $d\boldsymbol{r} = \boldsymbol{V} d\lambda$ and that $d\lambda$ is a scalar. By projecting this expression on the coordinate axes, we obtain

$$\frac{dX^i}{d\lambda} = V^i(X^1, X^2, X^3, t) , \qquad (i = 1, 2, 3) . \tag{2.17}$$

The family of stream lines $X^i = X^i(C^1, C^2, C^3, \lambda, t)$ is found after solving (2.17). We now compare (2.17) with the equation of particle trajectories

$$\frac{dX^i}{dt} = V^i(X^1, X^2, X^3, t) , \qquad (i = 1, 2, 3) . \tag{2.18}$$

For stationary movements, we have $V^i = V^i(X^j)$ and the differential equations of stream lines and particle trajectories coincide, though this is generally not the case for non-stationary movements. However, there are some non-stationary movements for which the two differential equations coincide, that is, the linear movement of a solid body with variable velocity.

By rewriting the equations for stream lines (2.17), we obtain

$$\frac{dX^2}{dX^1} = \frac{V^2}{V^1} , \quad \frac{dX^3}{dX^1} = \frac{V^3}{V^1} , \tag{2.19}$$

for which we consider the Cauchy initial value problem. It is known from the theory of differential equations that the Cauchy problem has a unique solution if the right-hand sides of (2.19) and their derivatives are continuous. Under these conditions, it is possible to draw a single stream line through every point X_0^i. If any of the velocities vanishes or diverges, the conditions of the uniqueness theorem for the Cauchy initial problem are violated and such points are referred to as singular points. These singular points are classified into knots, focal points, saddle points or center points (see Figure 2.4). The singular points in the flow of a fluid are also called critical points. For example, when air flows around a hydrofoil profile (see Figure 2.5), the point A, in which the airflow hits the profile and the stream line L bifurcates, is a critical point.

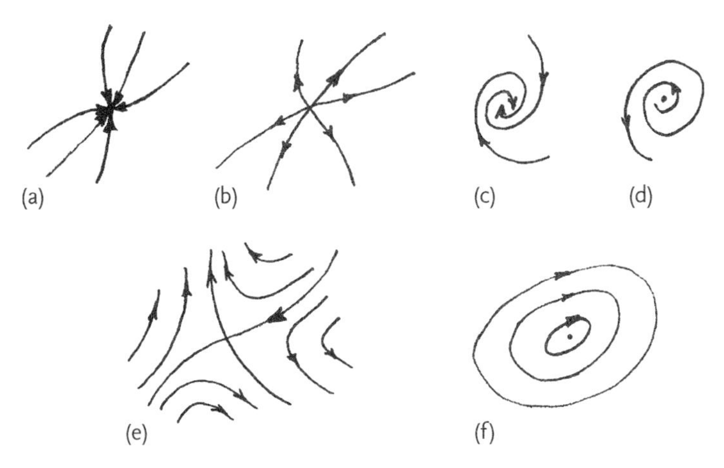

Figure 2.4 Singular points: (a) stable knot, (b) unstable knot, (c) stable focal point, (d) unstable focal point, (e) saddle point, and (f) center point.

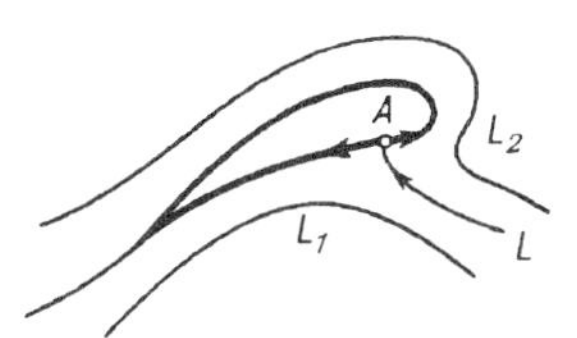

Figure 2.5 Example of a singular point for the flow of a fluid around the hydrofoil profile.

Let us consider an arbitrary curve C that does not coincide with a stream line, and connect its stream line to every point on this curve. This defines what we call a stream surface. In every point of this surface, the velocity vector lies in the tangential plane that includes this point. In differential geometry, such vectors are called surface vectors (see Section B.12).

Let the stream surface be defined via $f(X^i) = \text{const}$. Since the vector ∇f is collinear with the normal direction on the stream surface, we have $\nabla f \cdot V = 0$ or in terms of the components,

$$V^i \frac{\partial f}{\partial X^i} = 0 . \tag{2.20}$$

Equation 2.20 can be considered as the defining equation for the function $f(X^i)$. The corresponding mathematical task can be formulated as follows: the family of surfaces that include the contour C has to be determined from the given V^i. This problem has a unique solution, provided that C does not coincide with a stream line. Equation 2.20 can be solved using the method of characteristics. If C is a closed curve, we refer to the corresponding stream surface as a stream tube.

We now introduce the notion of a potential vector field. If the velocity vector can be written as $V = \nabla \phi$, it is said that the vector field V is a potential vector field. The scalar ϕ is then referred to as the stream potential. The surfaces defined by $\phi = \text{const}$ are called equipotential surfaces and V is orthogonal to them. The condition for the velocity field to be orthogonal is

$$V \cdot dX = \frac{\partial \varphi}{\partial X^i} dX^i = d\varphi \,, \qquad (dX = (dX^1, dX^2, dX^3)) \,.$$

The necessary conditions for the above expression to be a total differential are

$$\frac{\partial V^1}{\partial X^2} = \frac{\partial V^2}{\partial X^1} \,, \quad \frac{\partial V^1}{\partial X^3} = \frac{\partial V^3}{\partial X^1} \,, \quad \frac{\partial V^2}{\partial X^3} = \frac{\partial V^3}{\partial X^2} \,. \tag{2.21}$$

These equations are necessary and sufficient conditions for a fluid flow to be a potential flow.

Let us now consider some examples of potential flows.

1. Translational Flow
   We now consider a fluid flow with constant velocity u_0 along the X^1-axis. This means that the components of the velocity are given by $V^1 = u_0$, $V^2 = V^3 = 0$. From the conditions $\partial\varphi/\partial X^1 = u_0$, $\partial\varphi/\partial X^2 = \partial\varphi/\partial X^3 = 0$, we obtain $\varphi = u_0 X^1 + C$. The translational flow is a potential flow and the stream potential is defined up to an additional constant. For the general cases of the translational flow, we have $V^1 = u_0$, $V^2 = v_0$, $V^3 = w_0$ and $\varphi = u_0 X^1 + v_0 X^2 + w_0 X^1 + C$. This shows that a potential flow is much easier to work with because only one unknown is needed, while three functions V^i are generally necessary.
2. Source and Sink
   We now consider the fluid flow associated with the following potential

$$\varphi = -\frac{Q}{4\pi r} \,, \qquad r = \sqrt{\sum_i X^{i2}} \,, \tag{2.22}$$

with the Cartesian coordinates X^i and $Q = \text{const}$ or $Q = Q(t)$. The equipotential surfaces are spheres $r = \text{const}$ with their center at the origin. The velocity $V = \nabla\varphi$ is orthogonal to these spheres and directed along the radius. The stream lines are rays that are converging in the origin for $Q < 0$ and diverging from the origin for $Q > 0$. The velocity is

$$V = (\nabla\varphi)_r = \left|\frac{\partial\varphi}{\partial r}\right| = \frac{|Q|}{4\pi r^2} \,. \tag{2.23}$$

From (2.23), it follows that $|V| \to 0$ when $r \to \infty$ and that $|V| \to \infty$ when $r \to 0$. The origin as well as the point at infinity are critical points. For $Q > 0$, there is a fluid flow emanating from the origin and such a fluid flow is called the source. For $Q < 0$, the fluid flows into the origin and this flow is called a sink. For $Q > 0$, the origin is the source and the point at infinity is the sink. For $Q > 0$, the origin is the sink while the point at infinity is the source.
For the volume flow per second through the surface of a sphere with radius r, we obtain

$$\int_S V d\sigma = \frac{Q}{4\pi r^2} 4\pi r^2 = Q \,.$$

2.4 Theory of Strains

We consider a totally rigid body that is moving with respect to the observers system of coordinates. Two configurations of this body for $t = t_0$ and an arbitrary later time $t > t_0$ are shown in Figure 2.6. We now attach to a point M of the body the co-moving associated system of coordinates $\{\xi^i\}$. The basis set of this body-fixed system of coordinates depends on M as well as on t. For a totally rigid body, the transition from the the basis $\boldsymbol{\varepsilon}_i^{\circ}$ to the basis $\hat{\boldsymbol{\varepsilon}}_i$ can be effected via rotation and translation such that the lengths of the basis vectors as well as the angles between them are not changed. Therefore, the scalar products $\boldsymbol{\varepsilon}_i^{\circ} \cdot \boldsymbol{\varepsilon}_j^{\circ}$ and $\hat{\boldsymbol{\varepsilon}}_i \cdot \hat{\boldsymbol{\varepsilon}}_j$ which define the components of the metric tensor (see Section A.2) do not change during the movement of a totally rigid body, that is, we have $g_{ij}^{\circ} = \hat{g}_{ij}$.

We now consider the movement of a deforming body. The main difference to the movement of a totally rigid body lies in the change of the distance between the points M and M', and the distortion of the coordinate lines of the co-moving system of coordinates. Also, the lengths and directions of the basis vectors change.

We consider two configurations of the body at times t' and t (see Figure 2.7). Since the body is undergoing a continuous distortion during the movement, an infinitisemal element of the continuum $\Delta \boldsymbol{r}'$ is transformed into $\Delta \boldsymbol{r}$ with

$$\Delta \boldsymbol{r}' = d\xi^i \boldsymbol{\varepsilon}_i' , \quad \Delta \boldsymbol{r} = d\xi^i \hat{\boldsymbol{\varepsilon}}_i .$$

Here, we take into account that the coordinates of M and M' in the body fixed system of coordinates are the same.

We are interested in the change of distances from M to M'. Therefore, we introduce the metric tensor for the two system of coordinates, that is $g_{ij}' = \boldsymbol{\varepsilon}_i' \cdot \boldsymbol{\varepsilon}_j'$ and $\hat{g}_{ij} = \hat{\boldsymbol{\varepsilon}}_i \cdot \hat{\boldsymbol{\varepsilon}}_j$. Thus, we have

$$|d\boldsymbol{r}'|^2 = g_{ij}' d\xi^i d\xi^j , \quad |d\boldsymbol{r}|^2 = \hat{g}_{ij} d\xi^i d\xi^j . \tag{2.24}$$

We define the coefficient of relative strain via:

$$l = \frac{|d\boldsymbol{r}| - |d\boldsymbol{r}'|}{|d\boldsymbol{r}'|} = \frac{|d\boldsymbol{r}|}{|d\boldsymbol{r}'|} - 1 . \tag{2.25}$$

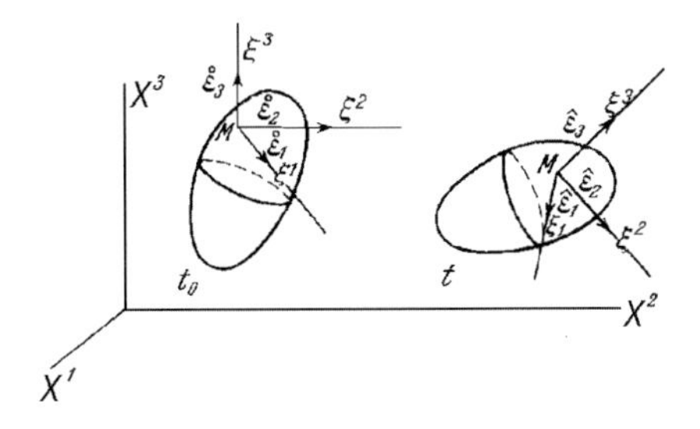

Figure 2.6 Movement of a totally rigid body with respect to the observers system of coordinates.

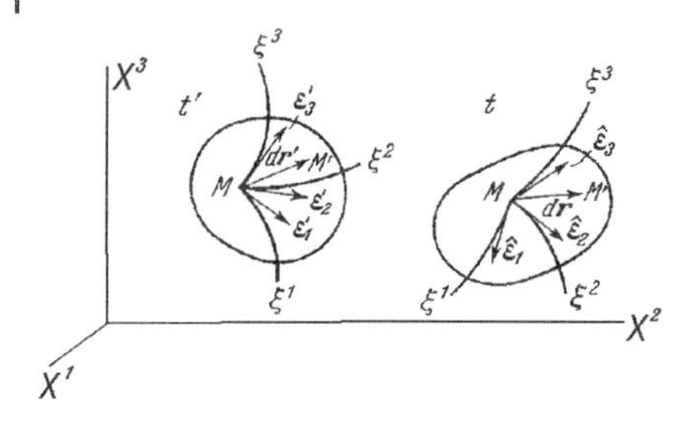

Figure 2.7 Movement of a deforming body.

It is obvious that $l = 0$ for a totally rigid body, while for a deforming body, $l \neq 0$. If l is infinitesimal in every point, these deformations are referred to as infinitesimal strains, and for finite deformations they are called finite strains.

Let us consider the following expression:

$$\varepsilon_{ij} = \frac{1}{2}(\hat{g}_{ij} - g'_{ij}) . \tag{2.26}$$

The quantities ε_{ij} are components of a second order tensor, the so-called strain tensor. Since the metric tensor is symmetric, the strain tensor is also symmetric. From (2.24), it follows that

$$|d\boldsymbol{r}|^2 - |d\boldsymbol{r}'|^2 = 2\varepsilon_{ij} d\xi^i d\xi^j .$$

We now elucidate the geometric meaning of the components of the strain tensor. If we denote the components of the metric tensor and the basis vectors for the initial state of the continuum by $\overset{\circ}{g}_{ij}$ and $\overset{\circ}{\boldsymbol{\varepsilon}}_i$, then the metric tensor corresponding to this initial state is

$$\overset{\circ}{g}_{ij} = \overset{\circ}{\boldsymbol{\varepsilon}}_i \cdot \overset{\circ}{\boldsymbol{\varepsilon}}_j = \left|\overset{\circ}{\boldsymbol{\varepsilon}}_i\right| \left|\overset{\circ}{\boldsymbol{\varepsilon}}_j\right| \cos \overset{\circ}{\psi}_{ij} , \tag{2.27}$$

where $\overset{\circ}{\psi}_{ij}$ are the angles between the vectors $\overset{\circ}{\boldsymbol{\varepsilon}}_i$ and $\overset{\circ}{\boldsymbol{\varepsilon}}_j$.

At time t, the metric tensor is given by

$$\hat{g}_{ij} = \hat{\boldsymbol{\varepsilon}}_i \hat{\boldsymbol{\varepsilon}}_j = |\hat{\boldsymbol{\varepsilon}}_i| \left|\hat{\boldsymbol{\varepsilon}}_j\right| \cos \hat{\psi}_{ij} . \tag{2.28}$$

Let us calculate the following ratio:

$$\frac{|\hat{\boldsymbol{\varepsilon}}_i|}{\left|\overset{\circ}{\boldsymbol{\varepsilon}}_i\right|} = \frac{\left|\frac{\partial \boldsymbol{r}}{\partial \xi^i}\right|}{\left|\frac{\partial \boldsymbol{r}_0}{\partial \xi^i}\right|} = \frac{|d\boldsymbol{r}_i|}{|d\boldsymbol{r}_{0i}|} = l_i + 1 . \tag{2.29}$$

Here, l_i are the coefficients of relative strain in the direction of the ξ^i-axis. From (2.26)–(2.29), one obtains

$$2\varepsilon_{ij} = \left[(1 + l_i)\left(1 + l_j\right) \cos \hat{\psi}_{ij} - \cos \overset{\circ}{\psi}_{ij}\right] \left|\overset{\circ}{\boldsymbol{\varepsilon}}_i\right| \left|\overset{\circ}{\boldsymbol{\varepsilon}}_j\right| .$$

For components of the strain tensor with identical indices, we have

$$2\boldsymbol{\varepsilon}_{ii} = \left[(1 + l_i)^2 - 1\right] g^{\circ}_{ii} ,$$

from which we obtain

$$l_i = \sqrt{1 + \frac{2\boldsymbol{\varepsilon}_{ii}}{g^{\circ}_{ii}}} - 1$$

for the coefficients of relative strain in the direction of the axes ξ^i.

In the case of small deformations, we have

$$l_i \approx \frac{\boldsymbol{\varepsilon}_{ii}}{g_{ii}} .$$

If the co-moving system of coordinates is initially Cartesian, then $g^{\circ}_{ii} = 1$ and

$$l_i = \boldsymbol{\varepsilon}_{ii} . \tag{2.30}$$

This means that the covariant components of the strain tensor are equal to the coefficients of relative strain along the corresponding axes.

We now discuss the meaning of the off-diagonal components of the strain tensor. Using an orthogonal basis in the initial configuration, that is, $\psi^{\circ}_{ij} = \pi/2$ and substituting $\hat{\psi}_{ij} = \pi/2 - \chi_{ij}$ in (2.26)–(2.28), one obtains

$$2\boldsymbol{\varepsilon}_{ij} = \left|\hat{\boldsymbol{\varepsilon}}_i\right| \left|\hat{\boldsymbol{\varepsilon}}_j\right| \sin\chi_{ij} = \sqrt{\hat{g}_{ii}\hat{g}_{jj}} \sin\chi_{ij}$$

and

$$\sin\chi_{ij} = \frac{2\varepsilon_{ij}}{\sqrt{\hat{g}_{ii}\hat{g}_{jj}}} . \tag{2.31}$$

An angle that was initially a right angle will cease to be such after the deformation and the off-diagonal components of the strain tensor are related to the deviation from the right angle.

If the deformations are infinitesimal and if the initial system of coordinates is Cartesian, we have $g^{\circ}_{ij} = 0$, $\hat{g}_{ii} \approx 1$, $\sin\chi_{ij} \approx 2\boldsymbol{\varepsilon}_{ij}$ and

$$\chi_{ij} \approx 2\boldsymbol{\varepsilon}_{ij} . \tag{2.32}$$

Since the strain tensor is symmetric, it is connected to the quadratic form $\boldsymbol{\varepsilon}_{ij} d\xi^i d\xi^j$.

In every point, an orthogonal system of coordinates η^i exists such that the symmetric form can be transformed into the canonical form

$$\boldsymbol{\varepsilon}_{ij} d\xi^i d\xi^j = \boldsymbol{\varepsilon}_i \left(d\eta^i\right)^2 .$$

The transformation of the system of coordinates from ξ^i to η^i depends on the components of the strain tensor $\boldsymbol{\varepsilon}_{ij}$. If an orthogonal system of coordinates ξ^i is initially used for which the corresponding quadratic form is canonical, it is transformed into another orthogonal system of coordinates for which the quadratic form is also canonical. These axes of the system of coordinates are referred to as the main axes of the strain tensor.

This means that it is possible to bind an orthogonal basis to the main axes of the strain tensors such that this orthogonal basis behaves as a totally rigid body. In other words, one can use at every point of the body the usual Cartesian axes that are directed along the main axes of the strain tensor. In such a system of coordinates, the strain tensor is transformed into a diagonal matrix with components $\boldsymbol{\varepsilon}_i$ that are called the main components or the main values of the strain tensor.

The strain tensors in the initial state $\boldsymbol{E}^\circ$ and the current state $\hat{\boldsymbol{E}}$ have different main values, namely, $\varepsilon_i^\circ \neq \hat{\varepsilon}_i$. In spite of that, they are still connected. To prove this, we consider the directions $d\boldsymbol{r}_i$ along the main axes. For the current and and the initial point in time, we have (Sedov L.I. (1973), Mechanics of Continuous Media, Vol. 1 Nauka, Moscow)

$$|d\boldsymbol{r}_i|^2 = \hat{g}_{ii}\left(d\eta^i\right)^2\,, \qquad |d\boldsymbol{r}|^2 = \sum_i |d\boldsymbol{r}_i|^2\,,$$

$$|d\boldsymbol{r}_{0i}|^2 = g^\circ_{ii}\left(d\eta^i\right)^2\,, \qquad |d\boldsymbol{r}_0|^2 = \sum_i |d\boldsymbol{r}_{0i}|^2\,.$$

There is no summation over i implied in the first expression on each line!

We now consider the following difference:

$$|d\boldsymbol{r}|^2 - |d\boldsymbol{r}_0|^2 = \left(\hat{g}_{ii} - g^\circ_{ii}\right)\left(d\eta^i\right)^2 = 2\sum_i \varepsilon'_{ii}\frac{|d\boldsymbol{r}_i|^2}{\hat{g}_{ii}} = 2\sum_i \varepsilon'_{ii}\frac{|d\boldsymbol{r}_{0i}|^2}{g^\circ_{ii}}\,. \tag{2.33}$$

The primes next to the strain tensors indicate that they have been transformed into the system of coordinates of the main axes. The matrices $\hat{g}_{ij}$ and g°_{ij} are diagonal and therefore their inverse matrices are also diagonal. In fact, both metric tensors satisfy $g_{ij} = \delta_{ij}$. Thus, we have

$$\varepsilon'_{ii}/g^\circ_{ii} = \varepsilon'_{ii}g^\circ_{ii} = \varepsilon^\circ_i\,, \quad \varepsilon'_{ii}/\hat{g}_{ii} = \varepsilon'_{ii}\hat{g}_{ii} = \hat{\varepsilon}_i\,. \tag{2.34}$$

Here, ε°_i and $\hat{\varepsilon}_i$ are the main components of the strain tensor for the initial and current configurations.

From (2.33) and (2.34), we obtain that the following applies along any of the main axes

$$|d\boldsymbol{r}_i|^2 - |d\boldsymbol{r}_{0i}|^2 = 2\hat{\varepsilon}_i\,|d\boldsymbol{r}_i|^2 = 2\varepsilon^\circ_i\,|d\boldsymbol{r}_{0i}|^2 \tag{2.35}$$

or

$$2\varepsilon^\circ_i = \frac{|d\boldsymbol{r}_i|^2}{|d\boldsymbol{r}_{0i}|^2} - 1\,, \quad 2\hat{\varepsilon}_i = 1 - \frac{|d\boldsymbol{r}_{0i}|^2}{|d\boldsymbol{r}_i|^2}\,.$$

It is obvious that $\varepsilon_i^\circ \neq \hat{\varepsilon}_i$ since

$$2\hat{\varepsilon}_i = \frac{2\varepsilon_i^\circ}{1 + 2\varepsilon_i^\circ} . \tag{2.36}$$

Using the previously obtained expressions, the relative distortion coefficients in the directions of the main axes can be obtained as

$$l_i = \sqrt{\frac{1}{1 + 2\varepsilon_i^\circ}} - 1 = \sqrt{\frac{1}{1 - 2\hat{\varepsilon}_i}} - 1 . \tag{2.37}$$

The equations (2.37) are also valid for finite deformations. For infinitesimal deformations, we have

$$l_i \approx -\varepsilon_i^\circ \approx \hat{\varepsilon}_i . \tag{2.38}$$

Thus, the coefficients of relative distortion approach the main elements of the infinitesimal strain tensors both for the current as well as for the initial configuration. For infinitesimal deformations, that difference between the tensors $\boldsymbol{E}^\circ$ and $\hat{\boldsymbol{E}}$ vanishes.

To determine the main components of the strain tensor, one has to find the eigenvalues λ_i of a symmetric matrix. To obtain those, one has to find the roots of the characteristic equation

$$\left| \varepsilon^i_{\ j} - \lambda \delta^i_{\ j} \right| = 0 \tag{2.39}$$

with $\varepsilon^i_{\ j} = \varepsilon_{jk} g^{ki}$.

Since the matrix $\varepsilon^i_{\ j}$ is diagonal with respect to the main axes, this leads to the following expression

$$(\lambda - \varepsilon_1)(\lambda - \varepsilon_2)(\lambda - \varepsilon_3) = 0$$

or

$$\lambda^3 - I_1 \lambda^2 + I_2 \lambda - I_3 = 0 . \tag{2.40}$$

The roots of this equation are the main components of the strain tensor such that

$$I_1 = \varepsilon_1 + \varepsilon_2 + \varepsilon_3 = \varepsilon^i_{\ i} ,$$

$$I_2 = \varepsilon_1 \varepsilon_2 + \varepsilon_1 \varepsilon_3 + \varepsilon_2 \varepsilon_3 = \frac{1}{2} \left[\left(\varepsilon^i_{\ i} \right)^2 - \varepsilon^i_{\ j} \varepsilon^j_{\ i} \right] ,$$

$$I_3 = \varepsilon_1 \varepsilon_2 \varepsilon_3 = \left| \varepsilon^i_{\ j} \right| . \tag{2.41}$$

It is known that a matrix C under linear transformation with a matrix B is transformed according to the rule $C^* = BCB^{-1}$ and that the determinant stays the same, that is, $|C^*| = |B| |C| \left| B^{-1} \right| = |C|$. Therefore, (2.40) is valid for any system of coordinates and I_1, I_2, I_3 are invariants. The invariants for the strain tensors $\boldsymbol{E}^\circ$ and $\hat{\boldsymbol{E}}$ are denoted by I_1°, I_2°, I_3° and $\hat{I}_1$, $\hat{I}_2$, $\hat{I}_3$. Since the main components ε_i° and $\hat{\varepsilon}_i$ of the strain tensors are connected via (2.36), the invariants are connected as well.

It follows from (2.36) and (2.41) that for infinitesimal deformation $\varepsilon_i^\circ = \hat{\varepsilon}_i$ and $\hat{I}_i = I_i^\circ$. However, for finite deformations:

$$\hat{I}_i = \hat{\varepsilon}_1 + \hat{\varepsilon}_2 + \hat{\varepsilon}_3 = \frac{I_1^\circ + 4I_2^\circ + 12I_3^\circ}{1 + 2I_1^\circ + 4I_2^\circ + 8I_3^\circ} ,$$
$$\hat{I}_2 = \hat{\varepsilon}_1\hat{\varepsilon}_2 + \hat{\varepsilon}_1\hat{\varepsilon}_3 + \hat{\varepsilon}_2\hat{\varepsilon}_3 = \frac{I_2^\circ + 6I_3^\circ}{1 + 2I_1^\circ + 4I_2^\circ + 8I_3^\circ} ,$$
$$\hat{I}_3 = \hat{\varepsilon}_1\hat{\varepsilon}_2\hat{\varepsilon}_3 = \frac{I_3^\circ}{1 + 2I_1^\circ + 4I_2^\circ + 8I_3^\circ} . \tag{2.42}$$

We now discuss the connection between the elementary volumes dV and dV_0 in the current and initial state. In the main axes of the strain tensor, the volumes of the elementary parallelepipeds are

$$dV = |d\boldsymbol{r}_1|\,|d\boldsymbol{r}_2|\,|d\boldsymbol{r}_3| , \quad dV_0 = |d\boldsymbol{r}_{01}|\,|d\boldsymbol{r}_{02}|\,|d\boldsymbol{r}_{03}| .$$

The quantity

$$\vartheta = \frac{dV - dV_0}{dV_0} = \frac{|d\boldsymbol{r}_1|}{|d\boldsymbol{r}_{01}|}\frac{|d\boldsymbol{r}_2|}{|d\boldsymbol{r}_{02}|}\frac{|d\boldsymbol{r}_3|}{|d\boldsymbol{r}_{03}|} - 1$$
$$= \sqrt{\left(1 + 2\varepsilon_1^\circ\right)\left(1 + 2\varepsilon_2^\circ\right)\left(1 + 2\varepsilon_3^\circ\right)} - 1 \tag{2.43}$$

is called the coefficient of cubic (spatial) expansion. By using (2.35) and (2.41), one obtains

$$\vartheta = \sqrt{\left(1 + 2\varepsilon_1^\circ\right)\left(1 + 2\varepsilon_2^\circ\right)\left(1 + 2\varepsilon_3^\circ\right)} - 1 = \sqrt{1 + 2I_1^\circ + 4I_2^\circ + 8I_3^\circ} - 1 . \tag{2.44}$$

Since ϑ consists of the invariants, it is itself an invariant. Therefore, (2.44) is valid in every system of coordinates and the coefficient of cubic expansion does not depend on the initial volume dV_0. For the case of infinitesimal deformations, we have

$$\vartheta \approx \sqrt{1 + 2\left(\varepsilon_1^\circ + \varepsilon_2^\circ + \varepsilon_3^\circ\right)} - 1 \approx \varepsilon_1^\circ + \varepsilon_2^\circ + \varepsilon_3^\circ = I_1^\circ$$

and

$$\vartheta \approx I_1 = \varepsilon^{\circ}{}^i_i = \hat{\varepsilon}^i_i . \tag{2.45}$$

Thus, for infinitesimal deformations, the first invariant of the strain tensor has a simple physical meaning: it is equal to the coefficient of cubic expansion.

We now derive expressions for ε_{ij} when the movement of the continuum is specified in the form of (2.2) or (2.3) and the metric of the observers system of coordinates is given. For this purpose, we introduce the displacement vector $\boldsymbol{W}$ from

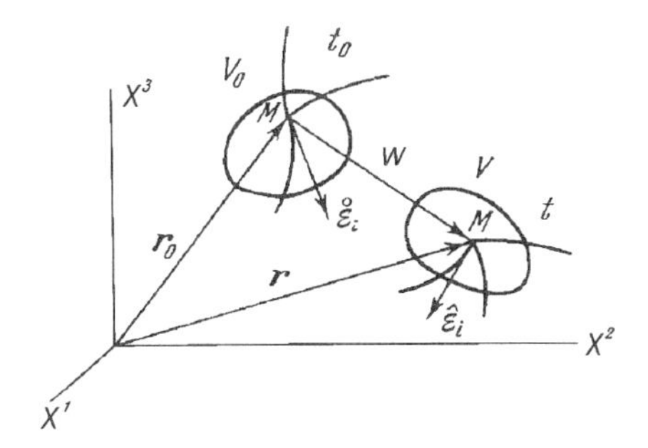

Figure 2.8 Displacement vector.

the initial to the current configuration (see Figure 2.8). Here, $\boldsymbol{r}$ and $\boldsymbol{r}_0$ are the radius vectors of the same point M as seen from the observers system of coordinates in the initial and current configurations. Thus, we have $\partial \boldsymbol{W}/\partial \xi^i = \partial \boldsymbol{r}/\partial \xi^i - \partial \boldsymbol{r}_0/\partial \xi^i = \hat{\boldsymbol{\varepsilon}}_i - \boldsymbol{\varepsilon}_i^\circ$ and $\hat{\boldsymbol{\varepsilon}}_i = \boldsymbol{\varepsilon}_i^\circ + \partial \boldsymbol{W}/\partial \xi^i$.

Thus, the components of the metric tensor are

$$\hat{g}_{ij} = \hat{\boldsymbol{\varepsilon}}_i \cdot \hat{\boldsymbol{\varepsilon}}_j = \boldsymbol{\varepsilon}_i^\circ \cdot \boldsymbol{\varepsilon}_j^\circ + \boldsymbol{\varepsilon}_i^\circ \cdot \frac{\partial \boldsymbol{W}}{\partial \xi^j} + \boldsymbol{\varepsilon}_j^\circ \cdot \frac{\partial \boldsymbol{W}}{\partial \xi^i} + \frac{\partial \boldsymbol{W}}{\partial \xi^i} \cdot \frac{\partial \boldsymbol{W}}{\partial \xi^j},$$

$$g_{ij}^\circ = \boldsymbol{\varepsilon}_i^\circ \cdot \boldsymbol{\varepsilon}_j^\circ = \hat{\boldsymbol{\varepsilon}}_i \cdot \hat{\boldsymbol{\varepsilon}}_j - \hat{\boldsymbol{\varepsilon}}_i \cdot \frac{\partial \boldsymbol{W}}{\partial \xi^j} - \hat{\boldsymbol{\varepsilon}}_j \cdot \frac{\partial \boldsymbol{W}}{\partial \xi^i} + \frac{\partial \boldsymbol{W}}{\partial \xi^i} \cdot \frac{\partial \boldsymbol{W}}{\partial \xi^j}.$$

Since $g_{ij}^\circ = \boldsymbol{\varepsilon}_i^\circ \cdot \boldsymbol{\varepsilon}_j^\circ$, we have

$$\varepsilon_{ij} = \frac{1}{2}\left(\hat{g}_{ij} - g_{ij}^\circ\right) = \frac{1}{2}\left(\boldsymbol{\varepsilon}_i^\circ \cdot \frac{\partial \boldsymbol{W}}{\partial \xi^j} + \boldsymbol{\varepsilon}_j^\circ \cdot \frac{\partial \boldsymbol{W}}{\partial \xi^i} + \frac{\partial \boldsymbol{W}}{\partial \xi^i} \cdot \frac{\partial \boldsymbol{W}}{\partial \xi^j}\right)$$
$$= \frac{1}{2}\left(\hat{\boldsymbol{\varepsilon}}_i \cdot \frac{\partial \boldsymbol{W}}{\partial \xi^j} + \hat{\boldsymbol{\varepsilon}}_j \cdot \frac{\partial \boldsymbol{W}}{\partial \xi^i} - \frac{\partial \boldsymbol{W}}{\partial \xi^i} \cdot \frac{\partial \boldsymbol{W}}{\partial \xi^j}\right). \tag{2.46}$$

We will now discuss the partial derivatives of $\boldsymbol{W} = W^k \boldsymbol{\varepsilon}_k$ that appear in (2.46). In the Cartesian system of coordinates X^i, the basis vectors do not change from point to point and we therefore have

$$\frac{\partial \boldsymbol{W}}{\partial X^i} = \frac{\partial \left(W^k \boldsymbol{e}_k\right)}{\partial X^i} = \frac{\partial W^k}{\partial X^i} \boldsymbol{e}_k.$$

However, in a curvilinear system of coordinates, the basis vectors change from point to point. As an example, we consider the constant vector field $\boldsymbol{A}$ (see Figure 2.9).

When moving from one point to the other, the vector $\boldsymbol{A}$ does not change. Therefore, its derivative must vanish. For curvilinear coordinates, we take the polar coordinates r and ϕ, and for the local basis of the co-moving system of coordinates, we use the vectors $\boldsymbol{\varepsilon}_1$ and $\boldsymbol{\varepsilon}_2$ that are parallel to the radius and the tangent on the circle. Since these are different in the points B and C, the coordinates of $\boldsymbol{A}$ change and the derivatives of its components do not vanish. Therefore, we have for a curvilinear system of coordinates:

$$\frac{\partial \boldsymbol{W}}{\partial \xi^i} = \frac{\partial W^k}{\partial \xi^i} \boldsymbol{\varepsilon}_k + W^k \frac{\partial \boldsymbol{\varepsilon}_k}{\partial \xi^i}.$$

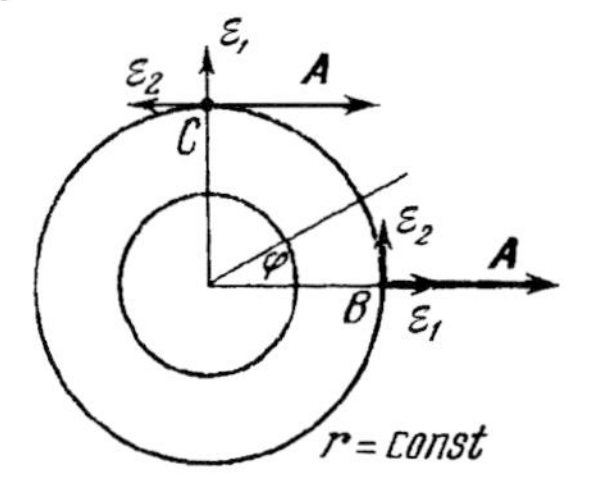

Figure 2.9 System of polar coordinates in the plane.

According to (A27), we have

$$\frac{\partial \boldsymbol{\varepsilon}_k}{\partial \xi^i} = \Gamma^j_{ki} \boldsymbol{\varepsilon}_j \,, \tag{2.47}$$

with the Christoffel symbol of the second kind Γ^j_{ki}, and

$$\frac{\partial \boldsymbol{W}}{\partial \xi^i} = \left(\frac{\partial W^k}{\partial \xi^i} + W^j \Gamma^k_{ji} \right) \boldsymbol{\varepsilon}_k = \nabla_i W^k \boldsymbol{\varepsilon}_k \,. \tag{2.48}$$

Here, the covariant derivative of the contravariant components of the vector $\boldsymbol{W}$ is denoted by $\nabla_i W^k$. The properties of the Christoffel symbol of the second kind and of the covariant derivative are discussed in Appendix A. In continuum mechanics, we often use special curvilinear systems of coordinates, that is, cylindrical and spherical, in addition to the Cartesian system of coordinates. Therefore, the geometric and tensorial characteristics of these systems of coordinates are discussed in Appendix A.

Since the displacement vector can be described both in the current basis $\hat{\boldsymbol{\varepsilon}}_i$ as well as in the initial basis $\boldsymbol{\varepsilon}^{\circ}_i$,

$$\boldsymbol{W} = W^{\circ k} \boldsymbol{\varepsilon}^{\circ}_k = \hat{W}^k \hat{\boldsymbol{\varepsilon}}_k \,,$$

it is possible to define two types of covariant derivatives:

$$\frac{\partial \boldsymbol{W}}{\partial \xi^i} = \nabla^{\circ}_i \left(W^{\circ k} \boldsymbol{\varepsilon}^{\circ}_k \right), \qquad \frac{\partial \boldsymbol{W}}{\partial \xi^i} = \hat{\nabla}_i \left(\hat{W}^{\circ k} \hat{\boldsymbol{\varepsilon}}^{\circ}_k \right). \tag{2.49}$$

In the first expression, the Christoffel symbols are calculated from g°_{ij} and in the second from $\hat{g}_{ij}$.

From (2.46) and (2.49), one obtains

$$\varepsilon_{ij} = \frac{1}{2} \left[\left(\nabla^{\circ}_i W^{\circ k} \right) g^{\circ}_{kj} + \left(\nabla^{\circ}_j W^{\circ k} \right) g^{\circ}_{ki} + \left(\nabla^{\circ}_i W^{\circ k} \nabla^{\circ}_j W^{\circ l} \right) g^{\circ}_{kl} \right].$$

In the last equation, it is possible to move g°_{kj} under the derivative (see (A34)). Using $W^{\circ k} g^{\circ}_{ki} = W^{\circ}_j$, the result is

$$\varepsilon_{ij} = \frac{1}{2} \left[\nabla^{\circ}_i W^{\circ}_j + \nabla^{\circ}_j W^{\circ}_j + \nabla^{\circ}_i W^{\circ}_k \nabla^{\circ}_j W^{\circ k} \right]. \tag{2.50}$$

A similar expression can be obtained for the components of the strain tensor in terms of the current basis by replacing ∘ by ^ in the above equation.

For infinitesimal deformations, the components of the displacement vector are also infinitesimal, and thus the third term in (2.50) can be neglected:

$$\varepsilon_{ij} = \frac{1}{2}\left[\nabla_i^\circ W_j^\circ + \nabla_j^\circ W_j^\circ\right] = \frac{1}{2}\left[\hat{\nabla}_i \hat{W}_j + \hat{\nabla}_j \hat{W}_j\right]. \tag{2.51}$$

In a Cartesian system of coordinates, the components of the strain tensor are

$$\varepsilon_{ij} = \frac{1}{2}\left[\frac{\partial W_j^\circ}{\partial X^i} + \frac{\partial W_i^\circ}{\partial X^j}\right]. \tag{2.52}$$

The components of the strain tensor are influenced by the components of the displacement vector $\boldsymbol{W}$ that transforms the initial continuum point with the metric g_{kj}° into the final continuum point with the metric $\hat{g}_{ij}$. We assume that the initial space is Euclidean. Then, the current space must also be Euclidean in order for the $\boldsymbol{W}$ to exist. The necessary condition for this space to be Euclidean is that the following equations must be realized:

$$\Gamma_{\gamma\beta}^{\omega}\frac{\partial\xi^\gamma}{\partial\eta^i}\frac{\partial\xi^\beta}{\partial\eta^j} + \frac{\partial^2\xi^\omega}{\partial\eta^i\partial\eta^j} = 0\,, \qquad \omega, i, j = 1, 2, 3\,. \tag{2.53}$$

On the other hand, the solution of the system of partial differential equations (2.53) serves as the transformation $\xi^i(\eta^j)$ of the curvilinear system of coordinates ξ^j to the Cartesian system of coordinates η^j. Therefore, the conditions for the space to be Euclidean are equivalent to the integrability conditions of the system (2.53), which we give here without derivation, namely,

$$R_{\beta s\alpha.}^{\cdots\omega} = \frac{\partial\Gamma_{\alpha\beta}^{\omega}}{\partial\xi^s} - \frac{\partial\Gamma_{\alpha s}^{\omega}}{\partial\xi^\beta} + \Gamma_{\lambda s}^{\omega}\Gamma_{\alpha\beta}^{\lambda} - \Gamma_{\lambda\beta}^{\omega}\Gamma_{\alpha s}^{\lambda} = 0 \tag{2.54}$$

and

$$\frac{\partial^2\varepsilon_{\nu i}}{\partial\xi^j\partial\xi^\mu} + \frac{\partial^2\varepsilon_{\mu j}}{\partial\xi^i\partial\xi^\nu} - \frac{\partial^2\varepsilon_{\mu i}}{\partial\xi^j\partial\xi^\nu} - \frac{\partial^2\varepsilon_{\nu j}}{\partial\xi^i\partial\xi^\mu} = 0\,. \tag{2.55}$$

Direct substitution of (2.51) into (2.55) shows that the expression (2.51) is a general integral of (2.55). If the displacement vector from the initial to the current state of the continuum exists, the compatibility conditions (2.55) must be valid and the expressions for the components of the strain tensor in terms of the displacement vector can be considered as the general solutions of those equations.

Let us now consider the transformations that happen during the movement of a small continuum particle. We start with the movement of a totally rigid body and take two arbitrary configurations, I and II (see Figure 2.10). In configuration I, we have an arbitrary point M in the body and a system of coordinates which is fixed to the body and whose origin is at M. In configuration II, the point M becomes M' and

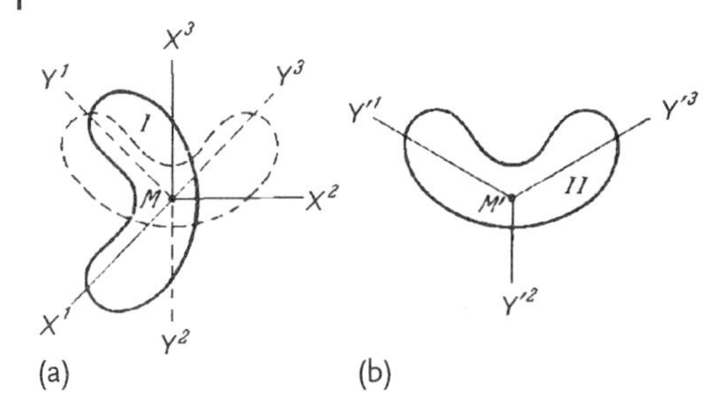

Figure 2.10 Transformation of the system of coordinates during the movement of a continuum particle.

the system of coordinates X^i becomes Y'^i. We denote the system of coordinates by Y^i, which is obtained by displacing Y'^i from M' back to M.

The transformation from the system of coordinates X^i to the system of coordinates Y^i consists of the rotation $Y^i = C^i_j X^j$ with an orthogonal matrix C^i_j, that is, $C^{\mathrm{T}} C = C C^{\mathrm{T}} = \boldsymbol{E}$. The arbitrary movement of the rigid body consists of the above rotation combined with a forward displacement.

We now consider the displacement of a deforming body. We also assume that the transformations in this case are bijective, continuous and differentiable. We further consider an infinitesimal neighborhood of the continuum point M. At time t_0, we denote the basis vectors of the Lagrangian system of coordinates in M by $\boldsymbol{\varepsilon}''^{\circ}_i$ and at time t as $\hat{\boldsymbol{\varepsilon}}''_i$. The positions of all points in the neighborhood of the point M at the times t_0 and t are denoted by the vectors $d\boldsymbol{r}_0 = d\xi^i \boldsymbol{\varepsilon}''^{\circ}_i$ and $d\boldsymbol{r} = d\xi^i \hat{\boldsymbol{\varepsilon}}''_i$. It must be noted that the coordinates of both vectors are identical since the Lagrangian system of coordinates is frozen to the body. If we superimpose the points M and M' and decompose the vector $d\boldsymbol{r}$ using the basis $\boldsymbol{\varepsilon}''^{\circ}_i$, the components of this decomposition will be different from $d\xi^i$. We call them $d\eta^i$, such that $d\boldsymbol{r} = d\eta^i \hat{\boldsymbol{\varepsilon}}''_i$. The connection between $d\eta^i$ and $d\xi^i$ defines the transformation of the continuum particle. From the equation

$$d\boldsymbol{r} = d\eta^i \hat{\boldsymbol{\varepsilon}}''_i = d\xi^i \hat{\boldsymbol{\varepsilon}}''_i$$

and the connection between the vectors $\boldsymbol{\varepsilon}''^{\circ}_i$, $\hat{\boldsymbol{\varepsilon}}''_i$ and $\boldsymbol{W}$, it follows that

$$\hat{\boldsymbol{\varepsilon}}''_i = \boldsymbol{\varepsilon}''^{\circ}_i + \frac{\partial \boldsymbol{W}}{\partial \xi^i} = \boldsymbol{\varepsilon}''^{\circ}_i + \nabla^{\circ}_i \left(\delta^k_i W^{\circ k} \boldsymbol{\varepsilon}''^{\circ}_k \right) = \left(\delta^k_i + \nabla^{\circ}_i W^{\circ k} \right) \boldsymbol{\varepsilon}^{\circ}_k = \boldsymbol{\varepsilon}''^{j}_k$$

and

$$d\boldsymbol{r} = d\xi^i \hat{\boldsymbol{\varepsilon}}''_i = d\xi^i C^k_i \hat{\boldsymbol{\varepsilon}}''_k = d\eta^k \hat{\boldsymbol{\varepsilon}}''_k \,.$$

From this equation, one obtains

$$d\eta^k = C^k_i d\xi^i \,. \tag{2.56}$$

Thus, (2.56) provides the transformation from $d\xi^i$ zu $d\eta^i$ with the matrix

$$C^k_i = \delta^k_i + \nabla^{\circ}_i W^{\circ k} \tag{2.57}$$

whose elements depend on the coordinates of the point M. Thus, for the infinitesimal neighborhood of the point M, the values of C_i^k in M are used. Therefore, they are continuous and the transformation (2.56) is an affine transformation that maps a point to a point, a line to line, a plane to a plane, parallel lines to parallel lines and parallel planes to parallel planes. This means that all lines of equal length and direction are equally contracted or expanded and for each line, the ratio between lengths before and after the transformation will not depend on the initial length. This ratio only depends on the direction. The coefficient of relative distortion only depends on the direction and an algebraic curve or surface will be transformed into another such curve or surface. For example, an algebraic surface of second order becomes a surface of the same type and a sphere becomes a sphere or an ellipsoid, while the conjugate diameters of the sphere become the conjugate diameters of the ellipsoid. Since all conjugate diameters of a sphere are orthogonal to each other and since each ellipsoid has a unique triplet of conjugates diameters, there will be at least one such triplet, that is, the main directions exist.

Although the volume does change during an affine transformation, the relative change of volume does not depend on the initial form or size. Since the relative volume change is calculated from the components of the strain tensor according to (2.43), an arbitrary volume can be replaced by an elementary parallelepiped. Therefore, it follows that a sphere in the continuum is transformed into an ellipsoid. If the directions of the main axes of the ellipsoid stay the same, we refer to the process as a pure deformation that is equivalent to a contraction or expansion. If the directions of the main axes change, we have an affine transformation which consists of a pure deformation and a rotation in space

It must be noted that in the case of a deformation, all lines in a continuum particle that do not coincide with the main axes change their direction in space. For a totally rigid body, however, we can only have a pure rotation. Thus, a sphere will be changed to a sphere of the same radius.

The matrix C_i^k for an affine transformation consists of nine elements that are expressed in terms of the derivatives of the displacement vector $\boldsymbol{W}$ with respect to the coordinates ξ^i. A pure deformation is characterized by six parameters, three main components of the strain tensors and three parameters, that determine the directions of the main axes in space. For a pure rotation, the matrix C_i^k is orthogonal and only has three independent parameters. Thus, it follows that the arbitrary displacement of a continuum particle consists of a translation, a rotation and a pure deformation.

Geometrical characteristics are generally important for solid bodies. For fluids and gases, they do not have the same importance. They are only affected indirectly via the volume change. Thus, the strain tensor plays the leading role in the deformation theory of the solid, that is, in the theory of elasticity. In the theory of fluid and gas movements, that is, in hydroaerodynamics, the deformations as such are actually unimportant. What is important, however, is how fast they happen. The speed of the deformation change is characterized by the tensor of the strain velocities.

2.5 The Tensor of Strain Velocities

Let us consider two continuum states at times t and $t + \Delta t$. We denote the metrics at these times by $\hat{g}_{ij}$ and $\hat{g}'_{ij}$, and the corresponding strain tensors by ε_{ij} and ε'_{ij}, such that

$$\varepsilon_{ij} = \frac{1}{2}\left(\hat{g}_{ij} - \overset{\circ}{g}_{ij}\right), \quad \varepsilon'_{ij} = \frac{1}{2}\left(\hat{g}'_{ij} - \overset{\circ}{g}_{ij}\right).$$

The change of the components of the strain tensor after the transition from the state at time t to that at time $t + \Delta t$ is given by

$$\Delta\varepsilon_{ij} = \frac{1}{2}\left(\hat{g}'_{ij} - g_{ij}\right) = \frac{1}{2}\left(\nabla_i W_j + \nabla_j W_i\right).$$

The corresponding displacement vector is given by

$$\mathbf{W} = W_i \hat{\boldsymbol{\varepsilon}}^i = v_i \Delta t \hat{\boldsymbol{\varepsilon}}^i ,$$

where v_i are the components of the displacement velocity. The components of the tensor of the strain velocity are defined as follows:

$$e_{ij} = \dot{\varepsilon}_{ij} = \lim_{\Delta t \to 0} \frac{\Delta\varepsilon_{ij}}{\Delta t} = \frac{1}{2}\left(\nabla_i v_j + \nabla_j v_i\right). \tag{2.58}$$

Since the tensor e_{ij} is obviously symmetric, we have $(e_{ij} = e_{ji})$. Therefore, if the metric tensor $\overset{\circ}{g}_{ij}$ in the initial state is time independent, we have, because of (2.26),

$$\hat{e}_{ij} = \frac{1}{2}\frac{dg_{ij}}{dt} = \frac{d\varepsilon_{ij}}{dt} . \tag{2.59}$$

It is important to stress that the first equality in (2.59) is only valid when $\overset{\circ}{g}_{ij} =$ const, while the second one is always valid. In addition, the strain tensor is introduced by comparing two continuum states, while the strain velocity is a characteristic of the given state.

The components of the tensor of strain velocity as well as those of the strain tensor have to fulfill the compatibility conditions. Let us consider the tensor $\hat{e}_{ij} = \dot{\varepsilon}_{ij}\Delta t$. For $\Delta t \to 0$, one can consider this to be a tensor of an infinitesimal deformation whose components satisfy the compatibility conditions (2.54). If one takes the limit $\Delta t \to 0$, the following compatibility conditions for the components of the strain velocity tensor are obtained:

$$\frac{\partial^2 e_{\nu i}}{\partial \xi^j \partial \xi^\mu} + \frac{\partial^2 e_{\mu j}}{\partial \xi^i \partial \xi^\nu} - \frac{\partial^2 e_{\mu i}}{\partial \xi^j \partial \xi^\nu} - \frac{\partial^2 e_{\nu j}}{\partial \xi^i \partial \xi^\mu} = 0 . \tag{2.60}$$

Thus, (2.58) for three arbitrary functions v_i give a general integral of the system (2.60) of six linear partial differential equations of second order.

2.6
The Distribution of Velocities in an Infinitesimal Continuum Particle

We consider an infinitesimal continuum particle (see Figure 2.11) and examine the distribution of velocities in this particle. We will define an infinitesimal continuum particle as a set of continuum points with coordinates $\xi^i + d\xi^i = \xi^i + \rho^i$ that are at an infinitesimal distance ρ^i from the point O. We assume that the velocity field is continuous and has derivatives up to at least first order.

We shall denote the velocities in the points O and O_1 as $\boldsymbol{V}_0$ and $\boldsymbol{V}_1$. During the infinitesimal time interval Δt, the vector $\boldsymbol{O}\boldsymbol{O}_1$ is transformed into the vector $\boldsymbol{O}'\boldsymbol{O}'_1 = \boldsymbol{\rho}'$ where

$$\boldsymbol{\rho}' = \boldsymbol{\rho} + (\boldsymbol{V}_1 - \boldsymbol{V}_0)\,\Delta t\,. \tag{2.61}$$

From the continuity of the vector $\boldsymbol{V}$, it follows that it can be expanded into a series in the neighborhood of the point O. Thus, to first order in $\rho = |\boldsymbol{\rho}|$, one obtains

$$\boldsymbol{V}_1 = \boldsymbol{V}_0 + \left(\frac{\partial \boldsymbol{V}}{\partial \xi^i}\right)_0 \rho^i + \boldsymbol{\rho}\, O(\rho)\,. \tag{2.62}$$

By substituting (2.59) into (2.61), one obtains

$$\boldsymbol{\rho}' = \boldsymbol{\rho} + \left(\frac{\partial \boldsymbol{V}}{\partial \xi^i}\right)_0 \rho^i \Delta t + \boldsymbol{\rho}\, O(\rho)\Delta t\,. \tag{2.63}$$

According to (2.49), (2.62) can be represented as follows:

$$\begin{aligned}
\boldsymbol{V}_1 &= \boldsymbol{V}_0 + (\nabla_i)\left(\frac{\partial \boldsymbol{V}}{\partial \xi^i}\right)_0 \rho^i + \boldsymbol{\rho}\, O(\rho)\\
&= \boldsymbol{V}_0 + \frac{1}{2}\left(\nabla_i V_k + \nabla_k V_i\right)\rho^i \boldsymbol{\varepsilon}^k + \frac{1}{2}\left(\nabla_i V_k - \nabla_k V_i\right)\rho^i \boldsymbol{\varepsilon}^k + \boldsymbol{\rho}\, O(\rho)\\
&= \boldsymbol{V}_0 + e_{ki}\rho^i \boldsymbol{\varepsilon}^k + \omega_{ki}\rho^i \boldsymbol{\varepsilon}^k + \boldsymbol{\rho}\, O(\rho)\,.
\end{aligned} \tag{2.64}$$

In the last equation, the terms that contain the symmetric tensor e_{ki} and the antisymmetric tensor ω_{ki} have been separated where

$$e_{ki} = \frac{1}{2}\left(\nabla_i V_k + \nabla_k V_i\right) = \frac{1}{2}\left(\frac{\partial V_i}{\partial \xi^k} - V_j \Gamma^j_{ik} + \frac{\partial V_k}{\partial \xi^i} - V_j \Gamma^j_{ki}\right),$$

$$\omega_{ki} = \frac{1}{2}\left(\nabla_i V_k - \nabla_k V_i\right) = \frac{1}{2}\left(\frac{\partial V_k}{\partial \xi^i} - V_j \Gamma^j_{ki} - \frac{\partial V_i}{\partial \xi^k} + V_j \Gamma^j_{ik}\right).$$

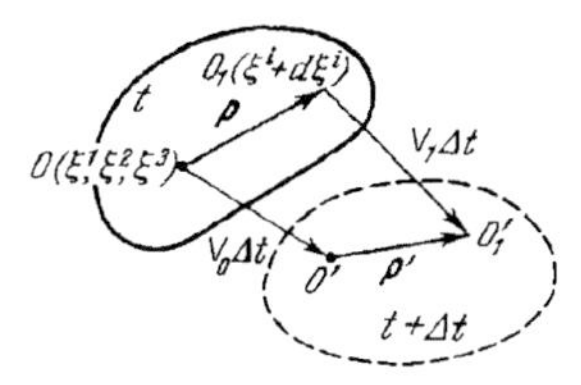

Figure 2.11 Displacement of a small continuum particle within a time interval of Δt.

Since $\Gamma^j_{ki} = \Gamma^j_{ik}$, we have

$$e_{ki} = \frac{1}{2}\left(\frac{\partial V_i}{\partial \xi^k} + \frac{\partial V_k}{\partial \xi^i}\right), \quad \omega_{ki} = \frac{1}{2}\left(\frac{\partial V_k}{\partial \xi^i} - \frac{\partial V_i}{\partial \xi^k}\right). \tag{2.65}$$

We now determine the mechanical meaning of the terms in (2.64). For simplicity, we use the Cartesian system of coordinates. Since $\boldsymbol{\rho} = X^i \boldsymbol{e}_i = X\boldsymbol{i} + Y\boldsymbol{j} + Z\boldsymbol{k}$, we have

$$V_k = V_{k0} + e_{ki} X^i + \omega_{ki} X^i . \tag{2.66}$$

The components e_{ki} and ω_{ki} are independent of X^i. We now consider the following quadratic form

$$\Phi = \frac{1}{2} e_{pq} X^p X^q , \quad \frac{\partial \Phi}{\partial X^k} = e_{ki} X^i .$$

Then, (2.66) can be written as

$$V_k = V_{k0} + \frac{\partial \Phi}{\partial X^k} + \omega_{ki} X^i .$$

We define the antisymmetric matrix

$$\|\omega_{ki}\| = \begin{pmatrix} 0 & \omega_{12} & \omega_{13} \\ \omega_{21} & 0 & \omega_{23} \\ \omega_{31} & \omega_{32} & 0 \end{pmatrix} = \begin{pmatrix} 0 & \omega_{12} & \omega_{13} \\ -\omega_{12} & 0 & \omega_{23} \\ -\omega_{13} & \omega_{32} & 0 \end{pmatrix}$$
$$= \begin{pmatrix} 0 & -\omega_3 & \omega_2 \\ \omega_3 & 0 & -\omega_1 \\ -\omega_2 & \omega_1 & 0 \end{pmatrix}$$

here as $\omega_1 = \omega_{32}, \omega_2 = \omega_{13}, \omega_3 = \omega_{21}$. Using (2.65), one obtains

$$\omega_1 = \frac{1}{2}\left(\frac{\partial V_3}{\partial Y} - \frac{\partial V_2}{\partial Z}\right), \quad \omega_2 = \frac{1}{2}\left(\frac{\partial V_1}{\partial Z} - \frac{\partial V_3}{\partial X}\right),$$
$$\omega_e = \frac{1}{2}\left(\frac{\partial V_2}{\partial X} - \frac{\partial V_1}{\partial Y}\right).$$

The quantities ω_i can be considered as components of

$$\boldsymbol{\omega} = \frac{1}{2}\begin{Vmatrix} \boldsymbol{i} & \boldsymbol{j} & \boldsymbol{k} \\ \partial/\partial X & \partial/\partial Y & \partial/\partial Z \\ V_1 & V_2 & V_3 \end{Vmatrix} = \frac{1}{2}\nabla \times \boldsymbol{V} .$$

The last term in (2.66) can be written as

$$\omega_{ki} X^i = (\boldsymbol{\omega} \times \boldsymbol{\rho})_k .$$

Now, the velocity of every point within the infinitesimal particle can be written as

$$\boldsymbol{V}_1 = \boldsymbol{V}_0 + \nabla \Phi + \boldsymbol{\omega} \times \boldsymbol{\rho} + \boldsymbol{\rho} O(\rho) . \tag{2.67}$$

Thus, the velocity of the points of the infinitesimal particle consists of three terms: the first one, V_0, is independent of the coordinates of the points X^i and represents the translational velocity of the whole particle and is equal to the velocity of the center of mass; the second term is related to the potential Φ and the third one characterizes the rotation with the instantaneous angular velocity ω. For totally rigid bodies, the Euler formula is valid:

$$V_1 = V_0 + \boldsymbol{\omega} \times \boldsymbol{\rho} \,. \tag{2.68}$$

Here, V_0 is the velocity of a certain point in the body, V_1 is the velocity of an arbitrary point in the body, $\boldsymbol{\omega}$ is the vector of the instantaneous angular velocity and $\boldsymbol{\rho}$ is the radius vector $\boldsymbol{O O_1}$. For a deforming particle, in contrast to the totally rigid body, we have the additional term $\nabla\Phi + \rho O(\rho)$ in the expression for the velocity in which the last term can be neglected since it is very small. We now consider the role of the term $\nabla\Phi$. During the movement of the continuum, the vector $\boldsymbol{\rho}$ becomes the vector $\boldsymbol{\rho}'$. The change of the radius vector $\Delta\boldsymbol{\rho} = \boldsymbol{\rho}' - \boldsymbol{\rho}$ is caused by different points of the body having different velocities.

From (2.61), we obtain that

$$\frac{d\boldsymbol{\rho}}{dt} = V_1 - V_0 \,. \tag{2.69}$$

We now calculate the quantity that is referred to as the speed of relative distance change in the continuum, that is,

$$e_\rho = \frac{1}{|\boldsymbol{\rho}|}\frac{d\,|\boldsymbol{\rho}|}{dt} = \frac{1}{\rho}\frac{d\,|\boldsymbol{\rho}|}{dt} = \frac{1}{2}\frac{1}{\rho^2}\frac{d\rho^2}{dt} = \frac{1}{2}\frac{1}{\rho^2}\frac{d(\boldsymbol{\rho}\cdot\boldsymbol{\rho})}{dt} = \frac{1}{\rho^2}\left(\boldsymbol{\rho}\cdot\frac{d\boldsymbol{\rho}}{dt}\right).$$

Since $\boldsymbol{\rho}\cdot\boldsymbol{\omega}\times\boldsymbol{\rho} = 0$, we have

$$\begin{aligned} e_\rho &= \frac{1}{\rho^2}\left[\boldsymbol{\rho}\cdot(V_1 - V_1)\right] = \frac{1}{\rho^2}\left[\boldsymbol{\rho}\cdot(\nabla\Phi + \boldsymbol{\omega}\times\boldsymbol{\rho}_1)\right] = \frac{1}{\rho^2}\left[\boldsymbol{\rho}\cdot(\nabla\Phi)\right] \\ &= \frac{1}{\rho^2}\left(X\frac{\partial\Phi}{\partial X} + Y\frac{\partial\Phi}{\partial Y} + Z\frac{\partial\Phi}{\partial Z}\right) = \frac{2\Phi}{\rho^2} = \frac{e_{ij}X^iX^j}{\rho^2} = e_{ij}\alpha^i\alpha^j \,, \end{aligned} \tag{2.70}$$

where $\alpha^i = X^i/\rho = \cos(\boldsymbol{\rho}, X^i)$. Thus, one can calculate from the known components of the tensor e_{ij} of strain velocities and for a given vector $\boldsymbol{\rho}$, the velocity of relative distance changes in the direction of $\boldsymbol{\rho}$. From (2.61), we obtain the kinematic interpretation of the components e_{ij}. Taking $\boldsymbol{\rho}$ parallel to the axis X^i, we have $\cos(\boldsymbol{\rho}, X^i) = 1$ and $\cos(\boldsymbol{\rho}, X^j) = 0$ for $j \neq i$. From (2.70), we obtain

$$e_x = e_{11} \,, \quad e_y = e_{22} \,, \quad e_z = e_{33} \,. \tag{2.71}$$

This means, that the components e_{ii} are the velocities of relative distance change along the axes X^i. Since for infinitesimal deformations $e_{ij} = \Delta\varepsilon_{ij}/\Delta t$ and since the ε_{ij} determine the change of angles away from a right angle, the quantities

$e_{ij} = \Delta\varepsilon_{ij}/\Delta t$ for $j \neq i$ characterize the velocities of angular change between the axes with indices i and j. From the above, it follows that the term $\nabla\Phi$ in (2.67) corresponds to the deformation of the continuum particle. We call it the velocity of pure deformation and refer to it by $\boldsymbol{V}^* = \nabla\Phi$.

Since e_{ij} is a symmetric tensor of second order, one can introduce main axes for it so that the matrix $E = \|e_{ij}\|$ will assume a diagonal form in the Cartesian coordinate system using the these main axes, that is,

$$E = \begin{pmatrix} e_1 & 0 & 0 \\ 0 & e_2 & 0 \\ 0 & 0 & e_3 \end{pmatrix} .$$

The quantities e_i are referred to as main components of the tensor of the strain velocities. To find the main axes and main components, one has to bring the corresponding quadratic form $\Phi(X^i)$ into its canonical form. For $e_i > 0$, we have an expansion along the axis, X^i, while it will be a contraction along this axis for $e_i < 0$.

We now consider the third term in (2.67). According to (2.64), it is possible to introduce the asymmetric tensor

$$\boldsymbol{\omega} = \omega_{ik}\boldsymbol{e}^i\boldsymbol{e}^k .$$

This tensor corresponds to an axial vector $\boldsymbol{\omega}(\omega_1, \omega_2, \omega_3)$ whose components can be expressed in terms of the ω_{ik}. We now give the kinematic explanation for the vector $\boldsymbol{\omega}$. From (2.61) and (2.67), it follows that

$$\boldsymbol{\rho}' = \boldsymbol{\rho} + \nabla\Phi\, dt + (\boldsymbol{\omega} \times \boldsymbol{\rho})\, dt + \rho O(\rho) dt .$$

This expression can be considered as an infinitesimal transformation in an infinitesimal time interval dt. The total transformation thus consists of the transformation that is determined via the tensor of strain velocities

$$\boldsymbol{\rho}' = \boldsymbol{\rho} + \nabla\Phi\, dt \tag{2.72}$$

and the transformation that is determined via the vector $\boldsymbol{\omega}$, that is,

$$\boldsymbol{\rho}' = \boldsymbol{\rho}^* + (\boldsymbol{\omega} \times \boldsymbol{\rho}^*)\, dt = \boldsymbol{\rho}^* + d\boldsymbol{\rho}^* . \tag{2.73}$$

The transformation (2.72) is connected to a quadratic form Φ which can, via a rotation, be brought into the canonical form

$$\Phi = \frac{1}{2}\left(e_1 X^2 + e_2 Y^2 + e_3 Z^2\right) .$$

Therefore, the transformation (2.72) corresponds to pure contractions or expansions along the three main axes. This means that the expansions for vectors that are parallel to the main axes X^i per length unit in the time interval dt are $e_i dt$. For vectors in an arbitrary direction, $\boldsymbol{\rho}$ are equal to $e_\rho dt$. We now consider that part of the transformation that corresponds to $\boldsymbol{\omega}$.

By calculating the scalar product, we obtain

$$\boldsymbol{\rho}^* \cdot d\boldsymbol{\rho}^* = \boldsymbol{\rho}^* \cdot \left(\boldsymbol{\omega} \times \boldsymbol{\rho}^*\right) = 0 \,.$$

This means that the change of the vector $\boldsymbol{\rho}^*$ is orthogonal to the vector $d\boldsymbol{\rho}^*$. Thus, the distance vector is rotated according to the transformation (2.73) without change in length, that is, a small particle behaves as a totally rigid body and the vector $\boldsymbol{\omega}$ can be interpreted as the angular velocity of the infinitesimal particle that is fixed during the time interval dt. In other words, the vector $\boldsymbol{\omega}$ is the instantaneous angular velocity of the trihedron angle that is constructed from the main axes of the tensor of strain velocities. The vector $\boldsymbol{\omega}$ is called the vorticity vector.

The above kinematical explanation of (2.67) provides the following formulation of the Cauchy–Helmholtz theorem regarding the decomposition of the velocity of a point within an infinitesimal continuum particle. The velocity $\boldsymbol{V}_1$ of a point O_1 inside of the infinitesimal continuum particle with its center in point O is given by

$$\boldsymbol{V}_1 = \boldsymbol{V}_0 + \nabla\Phi + (\boldsymbol{\omega} \times \boldsymbol{\rho}) = \boldsymbol{V}_0 + \boldsymbol{V}_{\text{rot}} + \boldsymbol{V}_{\text{def}} \,, \tag{2.74}$$

which is a sum of the velocity for translational movement $\boldsymbol{V}_0$, the velocity for rotating movement $\boldsymbol{V}_{\text{rot}} = \boldsymbol{\omega} \times \boldsymbol{\rho}$ for a totally rigid body and of the velocity of pure deformation $\boldsymbol{V}_{\text{def}} = \nabla\Phi$.

We now introduce the notion of the divergence of the velocity vector and elucidate its mechanical meaning. At $t = 0$, we consider an infinitesimal sphere defined by $X^2 + Y^2 + Z^2 = R^2$ that consists of continuum points. After the time interval Δt, this sphere is deformed into an ellipsoid. Since the main axes of the sphere becomes the main axes of the ellipsoid, the equation defining the ellipsoid assumes the following form:

$$\frac{(X^*)^2}{(1 + e_1\Delta t)^2} + \frac{(Y^*)^2}{(1 + e_2\Delta t)^2} + \frac{(Z^*)^2}{(1 + e_3\Delta t)^2} = R^2 \,.$$

We now consider the change of volume during the time interval Δt. While the initial volume of the sphere is $V_0 = 4\pi R^3/3$, the volume at time $t = \Delta t$ is

$$V_1 = \frac{4}{3}\pi R^3 (1 + e_1\Delta t)(1 + e_2\Delta t)(1 + e_3\Delta t) \,,$$

and the speed of relative volume change is

$$\lim_{\Delta t \to 0, v \to v_0} \frac{V - V_0}{V_0 \Delta t} = e_1 + e_2 + e_3 \,. \tag{2.75}$$

The sum on the right-hand side of this expression is equal to the first invariant of the tensor of strain velocities, which for an arbitrary system of coordinates takes the form

$$I_1(e) = e_1 + e_2 + e_3 = e_\alpha^\alpha = g^{\alpha\beta} e_{\alpha\beta} \,, \tag{2.76}$$

where according to definition (2.58), $e_\alpha^\alpha = \nabla_\alpha v^\alpha$.

This quantity is called the divergence of the velocity vector $\boldsymbol{V}$ and is denoted by $\nabla \cdot \boldsymbol{V}$. In the Cartesian system of coordinates, the divergence of the velocity vector is given by

$$I_1(e) = \nabla \cdot V = \frac{\partial U}{\partial X} + \frac{\partial V}{\partial Y} + \frac{\partial W}{\partial Z} ,$$

while, for a curvilinear system of coordinates, it becomes

$$I_1(e) = \nabla \cdot V = \frac{\partial V^k}{\partial \xi^k} + V^i \Gamma^k_{ki} = \frac{\partial V^k}{\partial \xi^k} + \frac{V^i}{\sqrt{g}} \frac{\partial \sqrt{g}}{\partial X^i} = \frac{1}{\sqrt{g}} \frac{\partial \left(V^i \sqrt{g}\right)}{\partial X^i} . \quad (2.77)$$

From (2.75), it follows that the mechanical meaning of the divergence of the velocity lies in the relative change of the elementary continuum volume per time interval.

2.7
Properties of Vector Fields. Theorems of Stokes and Gauss

The discussion leading to (2.67) can be repeated for an arbitrary vector $\boldsymbol{A}$ that is continuous and whose first derivatives exist. Thus, for every vector field $\boldsymbol{A}$ with those properties, we have

$$A_1 = A_0 + \nabla \Psi + \boldsymbol{\omega} \times \boldsymbol{\rho} + \rho O(\rho) . \quad (2.78)$$

Here,

$$\Psi = \frac{1}{2} a_{ij} X^i X^j , a_{ij} = \frac{1}{2} \left(\nabla_i A_j + \nabla_j A_i\right) ,$$

and the vector $\boldsymbol{\omega}$ is in Cartesian coordinates

$$\boldsymbol{\omega} = \frac{1}{2} \Omega = \frac{1}{2} \begin{Vmatrix} \boldsymbol{i} & \boldsymbol{j} & \boldsymbol{k} \\ \partial/\partial X & \partial/\partial Y & \partial/\partial Z \\ A_1 & A_2 & A_3 \end{Vmatrix} . \quad (2.79)$$

The vector $\boldsymbol{\Omega}$ is called the curl of the vector $\boldsymbol{A}$ and is denoted by $\nabla \times \boldsymbol{A}$. By comparison of (2.67) and (2.78), one obtains

$$\boldsymbol{\omega} = \frac{1}{2} \boldsymbol{\Omega} = \frac{1}{2} \nabla \times V , \quad (2.80)$$

that is, the vector of eddy velocity $\boldsymbol{\omega}$ is equal to half the the curl of the velocity.

We now introduce the concept of the circulation of a vector along a contour. This contour can be closed or open. In the latter case, we denote the end points of the contour by A and B. When the integration takes place along the contour from A to B, the contour is denoted by AB and in the opposite case by BA. We denote the directed contour element by ds. The circulation of a vector $\boldsymbol{A}$ along the contour AB is defined as

$$\Gamma = \int_{AB} A \cdot ds = - \int_{BA} A \cdot ds . \quad (2.81)$$

The sign of the circulation depends on the direction of the contour integration. One defines the sign as positive for clockwise circulation and negative for counter clockwise circulation. For the continuum velocity vector $\boldsymbol{V}$ with components U, V, W in a Cartesian system of coordinates, we have

$$\Gamma = \int_{AB} \boldsymbol{V} \cdot d\boldsymbol{s} = \int_{AB} U dX + V dY + W dZ \,. \tag{2.82}$$

The expression (2.82) is called the velocity circulation along the contour AB. For $\boldsymbol{V} = \nabla\varphi$, the velocity field is a potential field and we have

$$\Gamma = \int_{AB} \boldsymbol{V} \cdot d\boldsymbol{s} = \int_{AB} \frac{\partial\varphi}{\partial s} ds = \varphi_B - \varphi_A \,. \tag{2.83}$$

From the last expression, we deduce that the velocity circulation in case of a potential flow does not depend on the form of the contour, and that it vanishes, $\Gamma = 0$, for a closed contour when φ is a unique function.

We now consider the general case of the non-potential flow. We use a closed contour C and a smooth surface S, such that C is the boundary of S and that the vector $\boldsymbol{V}$ is continuously differentiable on S (see Figure 2.12). By decomposing the surface into many small surface elements and taking into account that the circulation integrals compensate on all common contour segments, one obtains

$$\Gamma = \int_{C} \boldsymbol{V} \cdot d\boldsymbol{s} = \sum_k \int_{C_k} \boldsymbol{V} \cdot d\boldsymbol{s} \,. \tag{2.84}$$

Assuming that the contours C_k are sufficiently small, the velocity on each contour with its center in O_k is determined via the Cauchy–Helmholtz theorem (2.74):

$$\boldsymbol{V}_{C_k} = \boldsymbol{V}_{O_k} + \nabla\Phi + \boldsymbol{\omega} \times \boldsymbol{\rho} + \rho O(\rho) \,. \tag{2.85}$$

By evaluating the circulation via (2.84), we note that the integrals of $\boldsymbol{V}_{O_k}$ and $\nabla\Phi$ over the closed contour vanish. In addition, the integral over the third term can be rewritten using some vector algebra and thus

$$\int_{C_k} (\boldsymbol{\omega} \times \boldsymbol{\rho}) \cdot d\boldsymbol{s} = \int_{C_k} (\boldsymbol{\omega} \times \boldsymbol{\rho}) \cdot d\boldsymbol{\rho} = \int_{C_k} \boldsymbol{\omega} \cdot (\boldsymbol{\rho} \times d\boldsymbol{\rho}) \approx 2 d\sigma_k \boldsymbol{\omega} \cdot \boldsymbol{n}$$
$$= 2\omega_n d\sigma_k \,.$$

Here, we have also used that for an infinitesimal contour O_k, the elementary surface $d\sigma$ can be considered as a plane, the vector $\boldsymbol{\omega}$ can be considered as continuous, the integral $\int_{C_k} (\boldsymbol{\rho} \times d\boldsymbol{\rho})$ is equal to $2 d\sigma_k$ in absolute value and its direction is parallel to the normal vector $\boldsymbol{n}_k$ on the surface. Since $\boldsymbol{V} \cdot d\boldsymbol{s} = V_S ds$, in the limit $k \to \infty$ and $d\sigma \to 0$, one obtains the result

$$\int_{C_k} V_S ds = 2 \int_{\Sigma} \omega_n d\sigma \,, \tag{2.86}$$

which is referred to as the Stokes theorem.

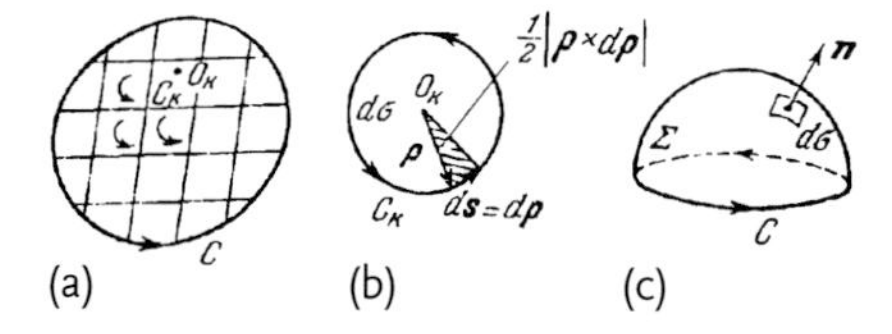

Figure 2.12 Derivation of the Stokes theorem.

The circulation of a velocity vector along the closed contour C is equal to twice the flow of the vorticity vector $\boldsymbol{\omega}$ through the surface S that is spanned by the contour. The direction of the normal vector is chosen such that the contour direction is seen to be counterclockwise from the tip of the normal. Equation 2.86 is valid for every continuous and differentiable vector $\boldsymbol{A}$. We now give a different formulation of the Stokes theorem:

$$\int_C A_S ds = \int_C A_i dX^i = \int_\Sigma (\nabla \times \boldsymbol{A})_n \, d\sigma$$
$$= \int \left\{ \left(\frac{\partial A_3}{\partial Y} - \frac{\partial A_2}{\partial Z} \right) \cos(n, x) + \left(\frac{\partial A_1}{\partial Z} - \frac{\partial A_3}{\partial X} \right) \cos(n, y) \right.$$
$$\left. + \left(\frac{\partial A_2}{\partial X} - \frac{\partial A_1}{\partial Y} \right) \cos(n, z) \right\} d\sigma \,. \tag{2.87}$$

A movement of the continuum is called eddy free when $\boldsymbol{\omega} = 0$. Otherwise, it is an eddy movement. A potential flow, that is, $\boldsymbol{V} = \nabla\varphi$, is eddy free since $\boldsymbol{\omega} = 1/2\nabla \times \boldsymbol{V} = 1/2\,(\nabla \times \nabla\varphi) = 0$. This conclusion is also valid in the opposite direction: an eddy free flow is also a potential flow, and these two concepts are equivalent.

The vector field $\boldsymbol{B}$ is called a solenoidal vector field when $\nabla \cdot \boldsymbol{B} = 0$. It is easy to check that if $\boldsymbol{B} = \nabla \times \boldsymbol{A}$, we have $\nabla \cdot \boldsymbol{B} = 0$. Since the vorticity vector $\boldsymbol{\omega}$ satisfies $\nabla \cdot \boldsymbol{B} = 0$, this vector field is always solenoidal. The converse statement is also true: if $\nabla \cdot \boldsymbol{B} = 0$, then $\boldsymbol{B}$ can be represented in the form $\boldsymbol{B} = \nabla \times \boldsymbol{A}$, and $\boldsymbol{A}$ is unique up to the gradient of an arbitrary function, that is, $\boldsymbol{A}' = \boldsymbol{A} + \nabla\Psi$ leads to the same $\boldsymbol{B}$. We also note that for an incompressible fluid, due to the property (2.75), we have $\nabla \cdot \boldsymbol{V} = 0$ and therefore the velocity field of an incompressible fluid is solenoidal.

Using the vorticity vector $\boldsymbol{\omega}$ as an example, we consider a few properties of solenoidal fields. We now introduce the notions of a vortex line, a vortex surface and vortex tube for a vector field. A line for which the tangent has in all points the same direction as $\boldsymbol{\omega}$ is called a vortex line. Using this definition, the differential equation for a vortex line is given by

$$\frac{dX}{\omega_1} = \frac{dY}{\omega_2} = \frac{dZ}{\omega_3} \,.$$

A surface that only consists of vortex lines is referred to as a vortex surface. The vortex surface $f(X, Y, Z) = \text{const}$ is defined by the following differential equation:

$$\omega_1 \frac{df}{\partial X} + \omega_2 \frac{df}{\partial Y} + \omega_3 \frac{df}{\partial z} = 0 \,.$$

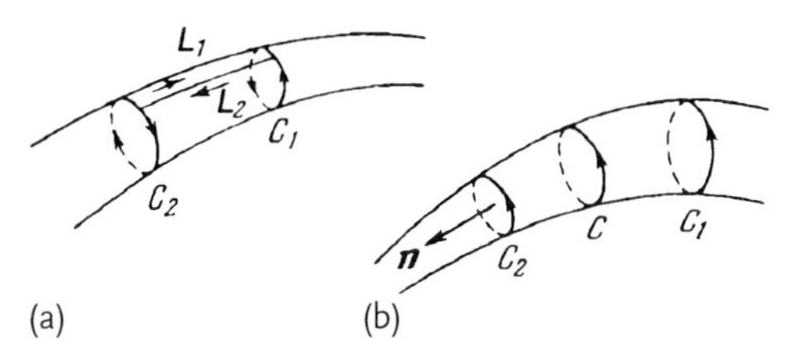

Figure 2.13 Properties of vortex tubes.

A vortex tube is constructed by considering a closed contour which is not a vortex line, and connecting a vortex line to each point of this contour (see Figure 2.13). Any surface segment of the vortex tube is a vortex surface and we therefore have $\omega_n = 0$ on this surface.

We now consider the properties of a vortex tube. We consider the part of the vortex tube that is situated between the contours C_1 and C_2, connect these contours via a cut $L_1 L_2$ and apply the Stokes theorem to this contour.

Since we have $\omega_n = 0$ on the surface, we obtain, according to (2.86),

$$\int\limits_{C_1+C_2+L_1+L_2} V \cdot ds = 0 .$$

Although the cuts L_1 and L_2 are identical, they are transversed in opposite directions. Thus, the corresponding integrals drop out and the above equation becomes

$$\int\limits_{C_1} V \cdot ds = \int\limits_{C_2} V \cdot ds .$$

The contours C_1 and C_2 are chosen at will and they orbit the vortex once. Thus, the circulation around any arbitrary contour that goes once around the vortex tube is constant.

$$\Gamma_C = \int\limits_{C} V \cdot ds = \text{const}$$

and is referred to as the strength of the vortex tube.

From the above, we can obtain the two kinematic theorems of Helmholtz:

1. The strength of a vortex tube is constant along the tube and is a characteristic number for a given tube.
2. Vortex tubes cannot start or end inside of the medium.

The first theorem follows from (2.87), the second one from the continuity of the vortex field $\boldsymbol{\omega}$ and the impossibility of vortex lines crossing each other. Thus, the vortex tubes are either closed, begin or terminate at the boundaries of the moving medium, or for an infinite medium they have to continue to infinity.

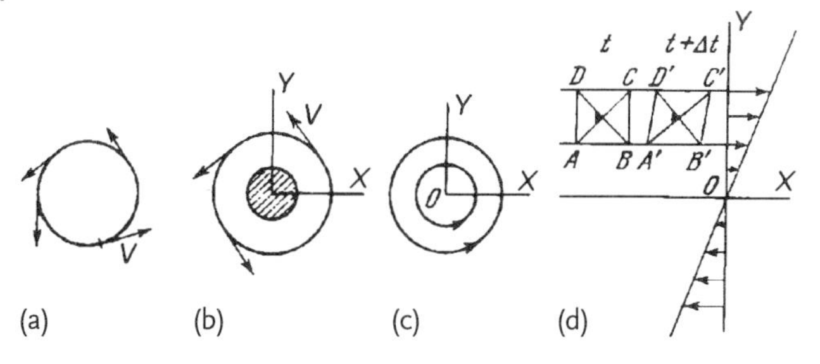

Figure 2.14 Examples of possible vortex flows.

We now consider some examples of vortex flows and start with a flow for which the stream lines are circles (see Figure 2.14a). For this flow, we have

$$\Gamma_C = \int\limits_C V_S \, ds = 2 \int\limits_\Sigma \omega_n d\sigma \neq 0$$

since the sign of V_S does not change. On first sight, it seems that the flow will remain a vortex flow if the stream lines are closed. However, such a deduction is only valid under the conditions for the Stokes theorem, namely, when the contour C can be connected to a surface Σ for which the velocity field $\boldsymbol{V}$ and its partial derivatives are continuous. Otherwise, it is impossible to forecast whether the flow is a vortex flow. This is shown in Figure 2.14b where the contour surrounds a solid cylindrical body with an axis parallel to the coordinate axis Z, or a flow with peculiarities for $\boldsymbol{V}$ and $\boldsymbol{\omega}$ in 2.14c.

We now consider a potential flow with

$$\varphi = k\vartheta = k \arctan \frac{Y}{X} \, .$$

This flow is a potential flow everywhere except for $X = Y = 0$. The velocity field is given by the following expression:

$$U = \frac{\partial \phi}{\partial X} = -\frac{kY}{X^2 + Y^2} \, , \qquad V = \frac{\partial \phi}{\partial Y} = \frac{kX}{X^2 + Y^2} \, .$$

The stream lines of this flow are orthogonal to the equipotential lines $\varphi = \text{const}$ or $Y = \alpha X$. Therefore, the stream lines are circles (see Figure 2.14c). For $k > 0$, the velocity points in the direction of the increase in ϕ, that is, counter-clockwise,

$$\Gamma_C = \int\limits_C V_S \, ds = V \int\limits_0^{2\pi} R d\vartheta = 2\pi R V \frac{k}{R} = 2\pi k \neq 0 \, .$$

However, on the other hand,

$$\boldsymbol{\omega} = \frac{1}{2} \nabla \times \boldsymbol{V} = \frac{1}{2} \begin{Vmatrix} \boldsymbol{i} & \boldsymbol{j} & \boldsymbol{k} \\ \partial/\partial X & \partial/\partial Y & 0 \\ U & V & 0 \end{Vmatrix} = \frac{1}{2} \left(\frac{\partial V}{\partial X} - \frac{\partial U}{\partial Y} \right) \boldsymbol{k} = 0$$

everywhere except on the coordinate axis Z ($X = Y = 0$). Therefore, such a flow can be considered as being induced by an isolated vortex flow along the coordinate axis Z with finite strength $\Gamma = 2\pi k$. Vortex flows are not necessarily caused by stream lines. An example for this is given by the shear flow (see Figure 2.14) which is a vortex flow.

We now derive the Gauss theorem. We consider within the moving medium at time t an individual continuum volume V (see Figure 2.15) which is bounded by its surface Σ. For every point of this surface, there exists a external normal vector $\boldsymbol{n}$. At time $t + \Delta t$, the volume V becomes the volume V' and the surface Σ becomes Σ'. The change in volume during the time interval Δt is given by

$$V - V' = \int_{\Sigma} V_n \Delta t d\sigma .$$

The velocity of volume change is

$$\frac{dV}{dt} = \lim_{\Delta t \to 0} \frac{V - V_0}{\Delta t} = \int_{\Sigma} V_n d\sigma .$$

Assuming that V is continuous and differentiable with respect to time, we decompose the volume into many small volume elements v^*. For every volume element, one can write down a similar expression for the velocity change. By taking the mechanical meaning of the divergence of the velocity vector as a velocity change for the infinitesimal continuum element into account according to (2.75), one obtains

$$\int_{\Sigma^*} V_n d\sigma = \int_{\Sigma^*} [U \cos(n, x) + V \cos(n, y) + W \cos(n, z)]\, d\sigma = v^* \nabla \cdot \boldsymbol{V} ,$$

with an infinitesimal quantity v^*. By summing over all elementary volumes and taking into account that the integrals over surfaces which touch each other fall away since the normal vectors have opposite directions, we obtain in the limit $v^* \to 0$

$$\int_{\Sigma} V_n d\sigma = \int_{\Sigma} [U \cos(n, x) + V \cos(n, y) + W \cos(n, z)]\, d\sigma$$
$$= \int_{V} \nabla \cdot \boldsymbol{V} d\tau . \qquad (2.88)$$

Equation 2.88 is called the Gauss theorem. It can also be written in the following vector form

$$\int_{\Sigma} \boldsymbol{V} \cdot \boldsymbol{n} d\sigma = \int_{V} \nabla \cdot \boldsymbol{V} d\tau \qquad (2.89)$$

or in the coordinate form

$$\int_{\Sigma} V^k n_k d\sigma = \int_{V} \nabla_k V^k d\tau . \qquad (2.90)$$

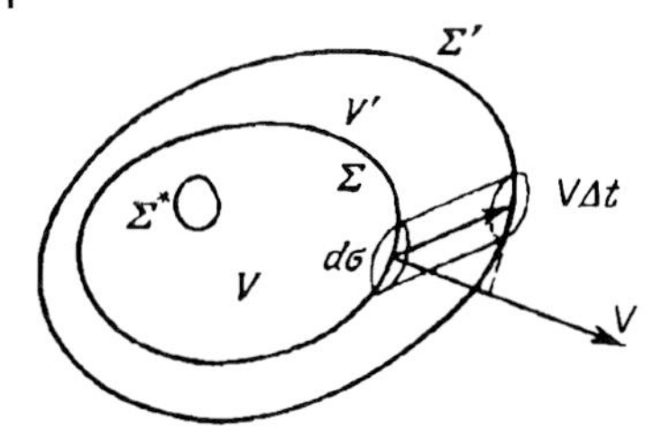

Figure 2.15 The Gauss theorem.

It must be noted that (2.90) is valid in an arbitrary curvilinear system of coordinates for each vector field V that is continuous and has continuous partial derivatives on $V + \Sigma$, while (2.88) is written in a Cartesian system of coordinates.

We now derive another urgently required formula for the derivative with respect to the time of an integral over a moving volume, the Reynolds formula. For this, we consider the following expression (see Figure 2.15)

$$\begin{aligned}
&\frac{d}{dt}\int\limits_{V(t)} f(X,Y,Z,t)d\tau \\
&= \lim_{\Delta t\to 0}\frac{1}{\Delta t}\left(\int\limits_{V'} f(X,Y,Z,t+\Delta t)d\tau - \int\limits_{V} f(X,Y,Z,t)d\tau\right) \\
&= \lim_{\Delta t\to 0}\frac{1}{\Delta t}\left[\left(\int\limits_{V} f(X,Y,Z,t+\Delta t)d\tau - \int\limits_{V} f(X,Y,Z,t)d\tau\right)\right. \\
&\quad \left. + \int\limits_{V'-V} f(X,Y,Z,t)d\tau\right] \\
&= \int\limits_{V(t)}\frac{\partial f}{\partial t}d\tau + \frac{1}{\Delta t}\int\limits_{V'-V} f(X,Y,Z,t+\Delta t)d\tau\,.
\end{aligned} \tag{2.91}$$

Since the volume $V' - V$ consists of elementary cylinders, we have $d\tau = V_n d\sigma\Delta t$ and $\Sigma' \to \Sigma$ for $\Delta t \to 0$. By using the Gauss theorem for the last integration, one obtains:

$$\frac{1}{\Delta t}\int\limits_{V'-V} f(X,Y,Z,t+\Delta t)d\tau = \int\limits_{\Sigma} f V_n d\sigma = \int\limits_{V} \nabla_i\left(f V^i\right)d\tau\,.$$

We finally obtain

$$\frac{d}{dt}\int\limits_{V(t)} f d\tau = \int\limits_{V(t)}\frac{\partial f}{\partial t}d\tau + \int\limits_{\Sigma} f V_n d\sigma = \int\limits_{V(t)}\left[\frac{\partial f}{\partial t} + \nabla_i\left(f V^i\right)\right]d\tau\,. \tag{2.92}$$

Equation 2.92 is the Reynolds formula. It can be written in a different form if the total derivative $d f/dt$ via (2.13) is used, that is,

$$\frac{\partial f}{\partial t} + \nabla_i \left(f V^i \right) = \frac{\partial f}{\partial t} + V^i \nabla_i f + f \nabla_i V^i = \frac{d f}{dt} + f \nabla_i V^i ,$$

yielding the following final form:

$$\frac{d}{dt} \int\limits_{V(t)} f \, d\tau = \int\limits_{V(t)} \left(\frac{d f}{dt} + f \nabla_i V^i \right) . \tag{2.93}$$

It must be noted that the Reynolds formula is not only valid for a scalar function, but also for a tensor function.

3
Dynamic Equations of Continuum Mechanics

3.1
Equation of Continuity

The law of conservation of mass within each individual volume, that is, for a volume containing the same particles, is a fundamental law of Newtonian mechanics. This law can be stated as follows:

$$\frac{dm}{dt} = 0 . \tag{3.1}$$

We now introduce the notion of mass density. The average mass density for a volume that is filled with a mass ΔM is defined as the ratio

$$\rho_{\text{av}} = \frac{\Delta m}{\Delta V} ,$$

and the true (local) mass density is defined as the limit of the ratio

$$\rho = \lim_{\Delta V \to 0} \frac{\Delta m}{\Delta V} . \tag{3.2}$$

The mass of an infinitesimal volume ΔV is equal to $\Delta m \approx \rho \Delta V$, while the mass of a finite volume V is given by

$$m = \int\limits_V \rho d\tau , \tag{3.3}$$

leading to the following form for the law of mass conservation (3.1)

$$\frac{d}{dt} \int\limits_V \rho d\tau = 0 . \tag{3.4}$$

As time passes, the volume $V = V(t)$ changes and (3.4) can therefore be transformed using (2.90) and (2.92), yielding

$$\frac{d}{dt} \int\limits_{V(t)} \rho d\tau = \int\limits_V \left(\frac{\partial \rho}{\partial t} + \nabla \cdot (\rho \mathbf{V}) \right) d\tau = \int\limits_V \left(\frac{d\rho}{dt} + \rho \nabla \cdot \mathbf{V} \right) d\tau = 0 . \tag{3.5}$$

Hydromechanics. Theory and Fundamentals. Emmanuil G. Sinaiski

ISBN: 978-3-527-41026-2

Since this equation is valid for every arbitrary volume, one obtains

$$\frac{\partial \rho}{\partial t} + \nabla \cdot (\rho \boldsymbol{V}) = 0 \,, \tag{3.6}$$

which is referred to as the equation of continuity in Eulerian variables.

Using (2.13), this equation can also be written as

$$\frac{d\rho}{dt} + \rho \nabla \cdot \boldsymbol{V} = 0 \,. \tag{3.7}$$

In addition to mass, there are other physical quantities of scalar, vector or tensor nature that are conserved in every volume of the continuum during its movement. We denote such a quantity by Φ and introduce the associated density as $\phi = \lim_{\Delta V \to 0} \frac{\Delta \Phi}{\Delta V}$. Then, Φ and ϕ will satisfy the same equations of conservation, that is,

$$\frac{d\Phi}{dt} = 0 \,, \quad \Phi = \int_V \phi \, d\tau \,, \quad \frac{d\phi}{dt} + \phi \nabla \cdot \boldsymbol{V} = 0 \,.$$

We now derive the equation of continuity for a mixture of multiple components. We consider a mixture that consists of n components. Every such mixture of multiple components can be represented by the complete set of n continua that occupy the same mixture volume. For each of these continua, one can define its respective mass density ρ_i and velocity $\boldsymbol{V}_i$ so that at each point of the mixture, we have n mass densities and velocities.

By using this approach for the equation of continuity, every component fills the same volume. Such an approximation is commonly used in the mechanics of multiphase media. Sometimes, this approximation is referred to as mechanics of mutually penetrating media. Provided there are no chemical reactions between the components or other processes that contribute to the transition from one to the other component and thus to the exchange of mass, the equation of continuity must be satisfied for every component separately, that is,

$$\frac{dm_i}{dt} = 0 \,, \quad \frac{\partial \rho_i}{\partial t} + \nabla \cdot (\rho_i \boldsymbol{V}_i) = 0 \,. \tag{3.8}$$

We now consider the case where chemical reactions and similar processes take place in the mixture that result in changing the component masses m_i. We denote by $\wp_i$ the change of mass for the i-th component of the mixture per unit time and unit volume. These quantities are defined in chemical kinetics. The law of conservation of mass has, for each component, the following form

$$\frac{dm_i}{dt} = \int_V \wp_i \, d\tau \,.$$

For $\wp_i > 0$, the mass of the i-th component increases, while it decreases for $\wp_i < 0$. In this case, we now obtain, instead of (3.8), the following equations for the ρ_i

$$\frac{\partial \rho_i}{\partial t} + \nabla \cdot (\rho_i \boldsymbol{V}_i) = \wp_i \,. \tag{3.9}$$

Considering the mixture as a whole, the fundamental law of chemical reactions is given by the conservation of the mass of the whole mixture and therefore the following relationship must hold:

$$\sum_{i=1}^{n} \wp_i = 0 \,. \tag{3.10}$$

In addition to the quantities ρ_i and $\boldsymbol{V}_i$ for each component of the mixture, one can also define a density as well as a velocity for the whole mixture, that is,

$$\rho = \frac{m}{V} = \frac{1}{V}\sum_{i=1}^{n} m_i = \sum_{i=1}^{n} \rho_i \,, \quad \boldsymbol{V} = \frac{1}{m}\sum_{i=1}^{n} m_i \boldsymbol{V}_i = \frac{1}{\rho}\sum_{i=1}^{n} \rho_i \boldsymbol{V}_i \,, \tag{3.11}$$

that serve as average quantities. Summation of (3.9) over i and using (3.10) results in the equation of continuity (3.6) with ρ and $\boldsymbol{V}$ defined above as averages over the mixture.

If all components are moving with the same velocity that coincides with the velocity of the total mixture, that is $\boldsymbol{V}_i = \boldsymbol{V}$, one refers to this process as taking place without diffusion. If the velocities of the components are different, that is, $\boldsymbol{V}_i \neq \boldsymbol{V}$, a diffusion process takes place and some components are moving relative to the others. For this diffusion process, the vectors of diffusion currents are introduced as

$$\boldsymbol{I}_i = \rho_i \left(\boldsymbol{V}_i - \boldsymbol{V}\right) \,. \tag{3.12}$$

The current $\boldsymbol{I}_i$ characterizes the movement of the i-th component relative to the whole mixture. Using (3.12), (3.9) can be written in the following form:

$$\frac{\partial \rho_i}{\partial t} + \nabla \cdot (\rho_i \boldsymbol{V}_i) = \wp_i - \nabla \cdot \boldsymbol{I}_i \,. \tag{3.13}$$

To determine the the diffusion currents $\boldsymbol{I}_i$, it is necessary to use the physical laws or phenomenological relationships that make it possible to express the diffusion currents $\boldsymbol{I}_i$ in terms of the gradients of the concentration of the corresponding components. From the definition of the current (3.12), it follows that

$$\sum_{i=1}^{n} \boldsymbol{I}_i = 0 \,. \tag{3.14}$$

When solving problems involving the movements of multicomponent mixtures, one can, instead of using n equations (3.13), use $n - 1$ equations (3.13) and one equation of continuity (3.6).

If the medium is incompressible, that is, if every elementary volume stays constant during the movement, the equation of continuity is simplified. Since then $d\rho/dt = 0$, we obtain from (3.6) that

$$\nabla \cdot \boldsymbol{V} = 0 \,. \tag{3.15}$$

A medium is called homogeneous when the mass density is the same for all particles of the continuum, that is, when ρ does not depend on the spatial coordinates. If the mass density does depend on the coordinates, the medium is called inhomogeneous. Equation 3.15 is valid both for homogeneous as well as inhomogeneous fluids.

It follows from (3.15) that the velocity field of an incompressible fluid is solenoidal. Therefore, the stream tubes have the same properties as the vortex tubes discussed in connection to the kinematical theorems of Helmholtz. Therefore, the strength Q of the stream tube is given by

$$Q = \int_{\Sigma} V_n d\sigma \,. \tag{3.16}$$

Q is called the flow of the stream tube and is constant along a given stream tube, and Σ is the cross sectional surface. Like the vortex tubes, the stream tubes of an incompressible fluid can not start or end inside of the volume.

The equation of continuity (3.6) has been derived in Euler coordinates. We now derive the equation of continuity in Lagrange coordinates. For this, we consider two infinitesimal individual volumes that have the shape of a general non-rectangular parallelepiped. Both volumes consist of the same particles. The first one is considered at time t_0 and is defined by the infinitesimal vectors $\boldsymbol{\varepsilon}_i^\circ d\xi^i$ in the co-moving system of coordinates ξ^i. The second one is considered at $t = t_0 + \Delta$ and is defined by the infinitesimal vectors $\hat{\boldsymbol{\varepsilon}}_i d\xi^i$. Since the system of coordinates is co-moving and the vectors are taken at the same point M, the coordinates of the two volumes are the same. We denote the mass densities of the medium at the times t_0 and t by ρ_0 and ρ. The volumes of the two parallelepipeds V_0 and V are given by the mixed products of the vectors that define the parallelepipeds, that is,

$$V_0 = \left(\boldsymbol{\varepsilon}_1^\circ \times \boldsymbol{\varepsilon}_2^\circ\right) \cdot \boldsymbol{\varepsilon}_3^\circ d\xi^1 d\xi^2 d\xi^3 \,, \quad V = (\hat{\boldsymbol{\varepsilon}}_1 \times \hat{\boldsymbol{\varepsilon}}_2) \cdot \hat{\boldsymbol{\varepsilon}}_3 d\xi^1 d\xi^2 d\xi^3 \,.$$

Since $\left(\boldsymbol{\varepsilon}_1^\circ \times \boldsymbol{\varepsilon}_2^\circ\right) \cdot \boldsymbol{\varepsilon}_3^\circ = \sqrt{g^\circ}$ and $(\hat{\boldsymbol{\varepsilon}}_1 \times \hat{\boldsymbol{\varepsilon}}_2) \cdot \hat{\boldsymbol{\varepsilon}}_3 = \sqrt{\hat{g}}$, where $g^\circ = \left\| g_{ij}^\circ \right\| = \left\| \boldsymbol{\varepsilon}_i^\circ \cdot \boldsymbol{\varepsilon}_j^\circ \right\|$ and $\hat{g} = \left\| \hat{g}_{ij} \right\| = \left\| \hat{\boldsymbol{\varepsilon}}_i \hat{\boldsymbol{\varepsilon}}_j \right\|$ are the determinants of the metric tensors for t_0 and t (s. Section A.2), we have

$$V_0 = \sqrt{g_0} d\xi^1 d\xi^2 d\xi^3 \,, \quad V = \sqrt{\hat{g}} d\xi^1 d\xi^2 d\xi^3 \,.$$

Due to the law of mass conservation, we have $\rho_0 V_0 = \rho V$ and thus

$$\rho = \rho_0 \frac{V_0}{V} = \frac{\sqrt{g^\circ}}{\sqrt{\hat{g}}} \rho_0 \,. \tag{3.17}$$

Equation 3.17 is the equation of continuity in Lagrange variables. It must be noted that the density φ for every quantity Φ that is conserved in all individual volumes of the medium satisfies (3.17).

3.2 Equations of Motion

The movement of the continuous medium takes place under the influence of forces. We now consider the forces that act on a given individual volume. In theoretical mechanics, one deals with point forces, that is, forces that act in a point. For a continuous medium, one mainly deals with distributed forces, that is, with forces that act in every partial volume V or on every surface element Σ where, if V or Σ goto to zero, the resulting vector of the acting forces also goes to zero. The forces that are distributed throughout the volume are referred to as volume forces or mass forces. We denote by $\boldsymbol{F}$ the main vector of the force that acts on the mass unit ΔM or volume unit ΔV in the given point and define the following quantities as the force per mass unit $\boldsymbol{f}_m$ and the force per volume unit $\boldsymbol{f}_V$ by

$$\boldsymbol{f}_m = \lim_{\Delta m \to 0} \frac{\boldsymbol{F}}{\Delta m}\,, \quad \boldsymbol{f}_V = \lim_{\Delta V \to 0} \frac{\boldsymbol{F}}{\Delta V}\,. \tag{3.18}$$

We obviously have $\boldsymbol{f}_m = \rho \boldsymbol{f}_V$. Examples of mass forces are gravity, electromagnetic forces and so on. In the mechanics of a totally rigid body, the effect of every configuration of forces is equivalent to its main vector and main moment. In the mechanics of a deforming body, the details of the force distribution are very important. In the mechanics of the continuum, the mass forces do not play a major role, but rather the surface forces which are distributed on the surface of the continuum. We now consider a surface element $d\Sigma$, denote the force that acts on this element by $\boldsymbol{P}$ and introduce as for the mass forces the density of the surface force as

$$\boldsymbol{p} = \lim_{\Delta m \to 0} \frac{\Delta \boldsymbol{P}}{\Delta m}\,. \tag{3.19}$$

All forces are either internal or external forces. The internal forces are such forces that are induced via objects that belong to the system that is considered. External forces are due to objects that are not part of the continuous medium. The notions of internal and external forces are of a relative nature. Depending on the circumstances, one and the same force can either be considered as internal or external. For example, when considering the movement of a mixture of a fluid and solid particles, the forces exerted by the fluid on the particles are internal forces relative to the mixture. However, if the movement of only one of this particles is considered, the force exerted by the fluid is external relative to the particle.

We now introduce the concept of internal stress. We consider an arbitrary volume V within the continuous medium and cut it along the surface S into two parts, V_1 and V_2 (see Figure 3.1). When considering the movement of V_1, the effect of V_2 on V_1 consists both of volume forces inside of V_1 as well as surface forces on S. These forces are external relative to V_1. Considering the entire volume $V_1 + V_2$, these forces are internal with respect to the V.

We now consider a point M on the surface S and consider a surface element $d\sigma$ with external normal vector $\boldsymbol{n}$ relative to the volume V_1.

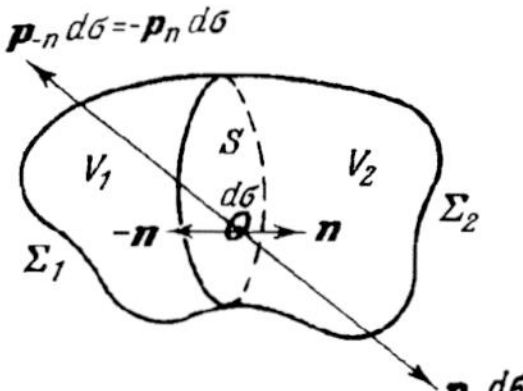

Figure 3.1 Internal stress forces.

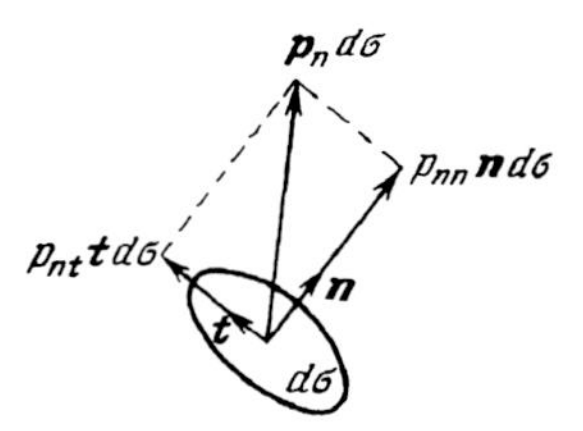

Figure 3.2 Internal stress forces.

We denote the surface force per unit area by $\boldsymbol{p}_n$ such that the surface force acting on the surface element is given by

$$d\boldsymbol{P} = \boldsymbol{p}_n d\sigma \,. \tag{3.20}$$

These surface forces introduced in every internal point of the continuum are referred to as internal stress forces. These forces can be decomposed into components parallel to the normal vector and along a tangent of the surface (see Figure 3.2), that is,

$$\boldsymbol{p}_n d\sigma = p_{nn} d\sigma \boldsymbol{n} + p_{nt} d\sigma \boldsymbol{t} \,. \tag{3.21}$$

Here, p_{nn} and p_{nt} are the normal and tangential components of the internal stress forces. If the medium is a fluid, p_{nt} is called the viscous force. It must be noted that the surface force $\boldsymbol{p}_n d\sigma$ generally contributes to a force external to the whole volume and it acts on the surface S that is the boundary to the continuum.

The movement of a material point is governed by the fundamental relationship

$$\boldsymbol{F} = m\boldsymbol{a} \,.$$

If the mass of the material point is constant, this law can be written as

$$\boldsymbol{F} = m\frac{d\boldsymbol{V}}{dt} = \frac{d\,(m\boldsymbol{V})}{dt} \,. \tag{3.22}$$

Since $m\boldsymbol{V}$ is the quantity of movement, (3.22) is also called the equation of motion. For a system of material points, we obtain, instead of (3.22),

$$\boldsymbol{F}_i = \frac{d\,(m_i \boldsymbol{V}_i)}{dt} \,, \quad (i = 1, 2, \ldots, n) \,.$$

The forces $\boldsymbol{F}_i$ consist of both internal forces $\boldsymbol{F}_i^{(i)}$ as well as the external forces $\boldsymbol{F}_i^{(e)}$ for the system of material points considered here. After summation over i and taking into account that the sum of internal forces must vanish, one obtains

$$\frac{d}{dt}\sum_i m_i \boldsymbol{V}_i = \sum_i \boldsymbol{F}_i^{(e)} .$$

We now define

$$\boldsymbol{Q} = \sum_i m_i \boldsymbol{V}_i = m\boldsymbol{V}_c , \quad m = \sum_i m_i , \quad \boldsymbol{V}_c = \frac{\sum_i m_i \boldsymbol{V}_i}{m} ,$$

where $\boldsymbol{Q}$ is the quantity describing the movement of the system as a whole and $\boldsymbol{V}_c$ is the velocity of the center of mass. Thus, we finally obtain the law of conservation of the center of mass momentum in the absence of external forces

$$\frac{d\boldsymbol{Q}}{dt} = m\frac{d V_c}{dt} = \sum_i F_i^{(e)} . \qquad (3.23)$$

To derive the equation of motion for a continuum volume, we divide the volume into volumes $d\tau$, each of which can be considered as a material point. Then, we use (3.22), substituting summation by integration and introduce the density of mass forces $\boldsymbol{f}_m$. Finally, we obtain the following integral form of the equation of motion for a finite continuum volume V that is bounded by a surface Σ:

$$\frac{d}{dt}\int_V \rho \boldsymbol{V} d\tau = \int_V \rho \boldsymbol{f}_V d\tau + \int_\Sigma \boldsymbol{p}_n d\Sigma . \qquad (3.24)$$

It is noteworthy that (3.24) is not derived, but postulated in the same way as Newton's second law for a material point. Since the masses of elementary volumes are constant, we have

$$\int_V \rho \frac{d\boldsymbol{V}}{dt} d\tau = \int_m \frac{d\boldsymbol{V}}{dt} dm = \frac{d}{dt}\int_m \boldsymbol{V} dm = \frac{d}{dt}\int_V \rho \boldsymbol{V} d\tau = \frac{d\boldsymbol{Q}}{dt} .$$

Thus, (3.24) can be written in the following form:

$$\int_V \rho \frac{d\boldsymbol{V}}{dt} d\tau = \int_V \rho \boldsymbol{f}_V d\tau + \int_\Sigma \boldsymbol{p}_n d\Sigma . \qquad (3.25)$$

We now consider the properties of the internal stress $\boldsymbol{p}_n$. By applying (3.25) to both subvolumes V_1 and V_2 as well as to the total volume V, one obtains the following equations:

$$\int_{V_1} \rho \frac{d\boldsymbol{V}}{dt} d\tau = \int_{V_1} \rho' \boldsymbol{f}'_V d\tau + \int_{\Sigma_1} \boldsymbol{p}_n d\Sigma + \int_S \boldsymbol{p}_n d\Sigma ,$$

$$\int_{V_2} \rho \frac{d\boldsymbol{V}}{dt} d\tau = \int_{V_2} \rho \boldsymbol{f}''_V d\tau + \int_{\Sigma_2} \boldsymbol{p}_n d\Sigma + \int_S \boldsymbol{p}_{-n} d\Sigma ,$$

$$\int_V \rho \frac{d\boldsymbol{V}}{dt} d\tau = \int_V \rho \boldsymbol{f}_V d\tau + \int_\Sigma \boldsymbol{p}_n d\Sigma ,$$

where we denote the density of mass forces in the volumes V_1 and V_2 by $\boldsymbol{f}'_V$ and $\boldsymbol{f}''_V$. By adding the first two equations and subtracting from the result the third and taking into account the following obvious equation

$$\int_{V_1} \rho \boldsymbol{f}'_V d\tau + \int_{V} \rho \boldsymbol{f}''_V d\tau = \int_{V} \rho \boldsymbol{f}_V d\tau ,$$

we obtain

$$\int_{S} (\boldsymbol{p}_n + \boldsymbol{p}_{-n}) d\Sigma = 0 .$$

Since the volumes V, V_1, V_2 and the intersecting surface S are arbitrary, we have

$$\boldsymbol{p}_n = -\boldsymbol{p}_{-n} . \tag{3.26}$$

In order for (3.26) to be valid, the movement of the volume V must be continuous. However, if there are surfaces in the volume V, across which the velocity is discontinuous and the continuum particles cross these surfaces, (3.26) is not valid. In Chapter 6, we will formulate the conditions applicable to surfaces of discontinuity.

We now consider an arbitrary point M of the continuum and construct an infinitesimal tetrahedron having three sides that are parallel to the Cartesian coordinate axes (see Figure 3.3). We further define $dX = MA, dY = MC, dZ = MB$ and denote the volume of the tetrahedron by V. The normal vector for the surface ABC is

$$\boldsymbol{n} = n^i \boldsymbol{e}_i = \cos(n, X)\boldsymbol{i} + \cos(n, Y)\boldsymbol{j} + \cos(n, Z)\boldsymbol{k} .$$

By denoting the stresses on the side surfaces with normal vectors $\boldsymbol{i}, \boldsymbol{j}, \boldsymbol{k}$ by $\boldsymbol{p}^1, \boldsymbol{p}^2, \boldsymbol{p}^3$, the area of ABC as S and the height of the tetrahedron by h, we write the equation of motion for the tetrahedron as

$$\begin{aligned} \frac{1}{3}\mathrm{Sh}\left(\rho\frac{d\boldsymbol{V}}{dt}\right)_M &= \frac{1}{3}\mathrm{Sh}\left(\rho \boldsymbol{f}_V\right)_M + \boldsymbol{p}_n S - \boldsymbol{p}^1 \cos(n, X) S \\ &\quad - \boldsymbol{p}^2 \cos(n, Y) S - \boldsymbol{p}^3 \cos(n, Z) S + O\left(h^{2+\lambda}\right) . \end{aligned}$$

In the limit of $h \to 0$ and $S \to 0$, that is, when the tetrahedron approaches a point, we obtain from the above equation

$$\boldsymbol{p}_n = \boldsymbol{p}^1 \cos(n, X) + \boldsymbol{p}^2 \cos(n, Y) + \boldsymbol{p}^3 \cos(n, Z) . \tag{3.27}$$

After substitution of (3.27) in (3.25) and application of Gauss's theorem, one obtains

$$\int_V \rho \frac{d\boldsymbol{V}}{dt} d\tau = \int_V \rho \boldsymbol{f}_V d\tau + \int_V \nabla \cdot \boldsymbol{P} d\tau , \tag{3.28}$$

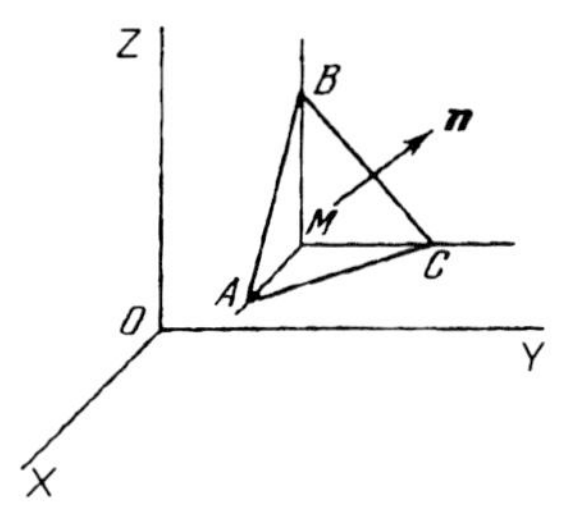

Figure 3.3 Properties of internal stress.

with

$$\nabla \cdot \boldsymbol{P} = \frac{\partial \boldsymbol{p}^1}{\partial X} + \frac{\partial \boldsymbol{p}^2}{\partial Y} + \frac{\partial \boldsymbol{p}^3}{\partial Z} = \nabla_i \boldsymbol{p}^i .$$

Since (3.28) is valid for any volume V, one can change to the differential form, that is,

$$\rho \frac{d\boldsymbol{V}}{dt} = \rho \boldsymbol{f}_V + \nabla \cdot \boldsymbol{P} . \tag{3.29}$$

This equation is only valid if $\boldsymbol{V}$ and the $\boldsymbol{p}^i$ are continuous and differentiable.

We now write the vectors $\boldsymbol{p}^i$ in terms of the basis vectors of the Cartesian system of coordinates, that is,

$$\boldsymbol{p}^i = p^{ki} \boldsymbol{e}_i \tag{3.30}$$

and introduce a matrix $\boldsymbol{P} = \left\| p^{ki} \right\|$ which consists of nine numbers. According to (3.27), the stress on an arbitrary surface element with unit normal vector $\boldsymbol{n}$ can be written as

$$\boldsymbol{p}_n = p_n^j \boldsymbol{e}_j , \quad p_n^j = p^{ji} n_i . \tag{3.31}$$

Here, $n_i = \cos(n, X^i)$ are the components of the unit normal vector.

According to (3.31), the matrix $\boldsymbol{P}$ can be considered as a transformation matrix from the components of the unit normal vector n_i to the components p_n^i of the vector $\boldsymbol{p}_n$. Therefore, the equation of motion (3.29) takes on the following form in Cartesian coordinates:

$$\rho \frac{dV^i}{dt} = \rho f_V^i + \frac{\partial p^{ij}}{\partial X^j} . \tag{3.32}$$

The relationship (3.27) between the stress vector $\boldsymbol{p}_n$ on an arbitrarily oriented surface element with unit normal vector $\boldsymbol{n}$ and the stress vectors on surface elements with unit normal vectors $\boldsymbol{e}_i$ is only valid in the Cartesian system of coordinates. For an arbitrary curvilinear system of coordinates, one obtains the following relationship:

$$\boldsymbol{p}_n = \boldsymbol{p}^i n_i p^{ki} \boldsymbol{\varepsilon}_k n_i = \boldsymbol{p}^i (\boldsymbol{\varepsilon}_i \cdot \boldsymbol{n}) = p^{ki} \boldsymbol{\varepsilon}_k (\boldsymbol{\varepsilon}_i \cdot \boldsymbol{n}) , \tag{3.33}$$

where the p^{ki} are the contravariant components of the stress tensor, that is,

$$P = p^{ki}\boldsymbol{\varepsilon}_k \cdot \boldsymbol{\varepsilon}_i \; . \tag{3.34}$$

The equation of motion (3.29) takes on the following form in an arbitrary curvilinear system of coordinates:

$$\rho \frac{d\boldsymbol{V}}{dt} = \rho \boldsymbol{f} + \nabla_i \boldsymbol{P}^i \,, \tag{3.35}$$

with the substantive derivative $\partial V^k/\partial t$ and the covariant gradient ∇_i

$$\frac{dV^k}{dt} = \frac{\partial V^k}{\partial t} + V^i \nabla_i V^k = \frac{\partial V^k}{\partial t} + V^i \left(\frac{\partial V^k}{\partial \xi^i} + V^s \Gamma^k_{si} \right) ,$$

$$\nabla_i \boldsymbol{P}^i = \frac{\partial \boldsymbol{p}^i}{\partial \xi^i} + \boldsymbol{p}^s \Gamma^i_{si} = \nabla_i p^{ki} \boldsymbol{\varepsilon}_k = \left(\frac{\partial p^{ki}}{\partial \xi^i} + p^{li} \Gamma^k_{li} + p^{kl} \Gamma^i_{li} \right) \boldsymbol{\varepsilon}_k \, .$$

If the medium is at rest and there are no mass forces, we obtain from (3.35)

$$\nabla_i \boldsymbol{P}^i = 0 \quad \text{or} \quad \nabla_i p^{ki} = 0 \, . \tag{3.36}$$

This equation forms the foundation of the theory of elasticity, which often deals with examining the equilibrium of different objects that are only subject to external surface forces.

As noted in Appendix A.3, using a curvilinear system of coordinates has the result that the tensor components have different physical dimensions. In this case, one has to consider the physical components of the stress tensor that are defined in terms of the usual components of the stress tensor as

$$\tilde{p}^{ki} = p^{ki} \sqrt{g_{ii} g_{kk}} \,,$$

where g_{ij} are the components of the corresponding metric tensor. By using the Cartesian coordinates, such a problem does not occur since $g_{ii} = 1$ and $\tilde{p}^{ki} = p^{ki}$.

Every stress tensor corresponds to one stress tensor surface. We now construct such a surface for a given tensor. We consider an arbitrary continuum point O and look at the different surface elements $d\Sigma$ including this point. These are characterized by different unit normal vectors. On each of these surfaces, a surface force $\boldsymbol{p}_n$ is acting, (see Figure 3.1), whose projection on the unit normal is given by

$$p_{nn} = \boldsymbol{p}_n \cdot \boldsymbol{n} = \left(\boldsymbol{p}^i \cdot \boldsymbol{n} \right) n_i = p^{ki} n_k n_i \, .$$

For simplicity, we use the Cartesian system of coordinates for which the order of indices for vector and tensor components does not matter. We consider the radius vector $\boldsymbol{r} = X_i \boldsymbol{e}^i$ which emanates from O and is parallel to $\boldsymbol{n}$. Then, we have $n_i = \cos(n X_i) = X_i/r$. We now choose the length of the vector $\boldsymbol{r}$ such that

$$p_{nn} r^2 = p^{ki} X_i X_k = 2\Phi(X, Y, Z) = \text{const} \, . \tag{3.37}$$

Here, $\Phi(X, Y, Z)$ is the quadratic form that corresponds to the symmetric tensor p^{ki}. Equation 3.37 results in a surface of second order, that is, the surface of the stress tensor. Choosing the length r of the radius vector as in (3.37) allows us to construct this surface as follows. For the vector of internal stress, we have

$$\boldsymbol{p}_n = \boldsymbol{p}^i n_i = \boldsymbol{p}^i \frac{X_i}{r} \, .$$

By projecting this vector equation on the coordinate axes, one obtains:

$$r p_n^k = p^{ki} X_i \, .$$

It follows from (3.37), that $\partial \Phi / \partial X_k = p^{ki} X_i$. Therefore we have $r p_n^k = \partial \Phi / \partial X_k$ or

$$r \boldsymbol{p}_n = \nabla \Phi \, . \tag{3.38}$$

With the help of this expression, it is possible to find, for a given tensor surface, the direction of the stress vector $\boldsymbol{p}_n$ acting on the elementary surface $d\Sigma$ with the normal vector $\boldsymbol{r}$. We first define $\boldsymbol{r}$ as the vector emanating from O and having a right angle to the given elementary surface $d\Sigma$. Then, we define the surface Σ as the tangential surface in the point where $\boldsymbol{r}$ crosses the tensor surface. According to (3.38), the vector $\boldsymbol{p}_n$ is orthogonal to the tangential surface σ. For a surface of second order, there exist at least three directions $\boldsymbol{r}$ that are at a right angle to the corresponding tangential surface σ. These directions are called main directions for which the the vector $\boldsymbol{p}_n$ is orthogonal to the tangential surface σ. If there are three such directions, they are referred to as the main axes of the stress tensor. From these, an orthogonal system of coordinates can be constructed. To determine the main axes, the following method is used. Since the vector $\boldsymbol{p}_n$ is collinear with the $\boldsymbol{n}$ normal vector for the main directions, we have

$$\boldsymbol{p}_n = p^{ki} n_i \boldsymbol{e}_k = \lambda \boldsymbol{n} = \lambda n_i \boldsymbol{e}^k \, . \tag{3.39}$$

In coordinate form, this equation can be written as

$$p^{ki} n_i \boldsymbol{e}_k = p_k^i n_i \boldsymbol{e}^k = \lambda n_i \delta_k^i \boldsymbol{e}^k$$

or

$$\left(p_k^i - \lambda \delta_k^i\right) n_i = 0 \, . \tag{3.40}$$

The last equation is a system of three algebraic equations for the three unknowns n_i which are equal to the directional cosines of the normal vector $\boldsymbol{n}$. This system has a non-trivial solution under the condition $\left\| p_k^i - \lambda \delta_k^i \right\| = 0$. Expanding this condition yields the cubic equation

$$-\lambda^3 + I_1 \lambda^2 - I_2 \lambda + I_3 = 0 \, , \tag{3.41}$$

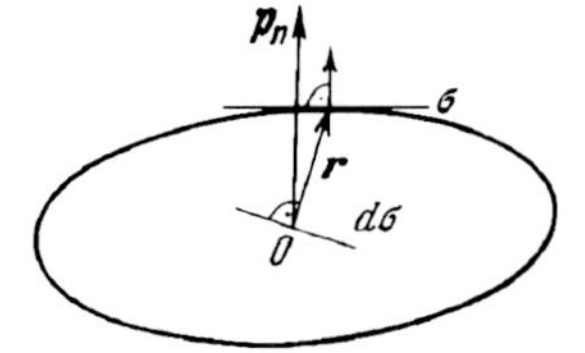

Figure 3.4 Tensor surface of the stress tensor.

where

$$I_1 = p^i_i = p^1_1 + p^2_2 + p^3_3 ,$$

$$I_2 = \begin{vmatrix} p^2_2 & p^3_2 \\ p^2_3 & p^3_3 \end{vmatrix} + \begin{vmatrix} p^3_3 & p^1_3 \\ p^3_1 & p^1_1 \end{vmatrix} + \begin{vmatrix} p^1_1 & p^2_1 \\ p^1_2 & p^2_2 \end{vmatrix} ,$$

$$I_3 = \left| p^i_k \right| .$$

For a symmetric tensor p^{ik}, (3.41) has three real roots, $\lambda_1, \lambda_2, \lambda_3$, that define according to (3.39) the stresses on the surfaces that are orthogonal to the main axes, that is,

$$\lambda_1 = p_{n1} = p_1 , \quad \lambda_2 = p_{n2} = p_2 , \quad \lambda_3 = p_{n3} = p_3$$

and that are called the main components of the stress tensor. Then, one can find from (3.40) and using the additional conditions $\boldsymbol{n}^i \cdot \boldsymbol{n}_k = \delta^i_k$ the components of three orthogonal unit normal vectors $\boldsymbol{n}_k$ that correspond to the main axes. In a system of coordinates with axes along the main axes, the equation for the tensor surface is reduced to its canonical form

$$2\Phi = p_1 X^2 + p_2 Y^2 + p_3 Z^2 = \text{const}$$

and the components of the stress tensor are given by

$$p^{ii} = \lambda_i = p_i , \quad p^{ki} = 0 , \quad (k \neq i) .$$

This means that on the surfaces that are perpendicular to the main axes, the normal components p_{nn} of the stresses $\boldsymbol{p}_n$ are unequal to zero, while the tangential components $p_{nt}$of the stresses vanish.

For $p_1 = p_2 = p_3$, the tensor surface is a sphere. For $p_1 \neq 0$ and $p_2 = p_3 = 0$, there is a pure stretching force for $p_1 > 0$ and a pure compressional force for $p_1 < 0$.

From the above, it follows that in a given point of the continuum, each stress state is equivalent to the combination of three stretching or compressing forces along the main axes of the stress tensor. Since the coefficients I_i in (3.21) are invariants of the strain tensor, they can be calculated from the main components of the stress tensor, that is,

$$I_1 = p_1 + p_2 + p_3 , \quad I_2 = p_1 p_2 + p_1 p_3 + p_2 p_3 , \quad I_3 = p_1 p_2 p_3 .$$

3.3
Equation of Motion for the Angular Momentum

In theoretical mechanics, one obtains the equation of motion for the angular momentum of a material point from Newton's second law, (3.22), via vector multiplication of both sides with the radius vector $\boldsymbol{r}$, which is taken relative to the origin O of the system of coordinates, that is,

$$\frac{d\boldsymbol{K}}{dt} = \boldsymbol{M} . \tag{3.42}$$

Here, $\boldsymbol{K} = \boldsymbol{r} \times m\boldsymbol{V}$ is the angular momentum of the material point and $\boldsymbol{M} = \boldsymbol{r} \times \boldsymbol{F}$ is the torque. For a system of n material points, (3.42) is also valid, with

$$\boldsymbol{K} = \sum_i (\boldsymbol{r}_i \times m_i \boldsymbol{V}_i) , \quad \boldsymbol{M} = \sum_i \left(\boldsymbol{r}_i \times \boldsymbol{F}_i^{(e)}\right) .$$

It must be noted that $\boldsymbol{K}$ can also be written in the following form:

$$\boldsymbol{K} = \boldsymbol{r}^* \times m\boldsymbol{V}^* + \sum_i (\boldsymbol{r}_{ic} \times m_i \boldsymbol{V}_{ic}) ,$$

where $m = \sum_i m_i$, $\boldsymbol{V}^*$ is the velocity of the center of mass, $\boldsymbol{r}_{ic}$ – the radius vector of the i-th point relative to the center of mass and $\boldsymbol{V}_{ic}$ the velocity of the i-th point relative to the center of mass.

We now consider a volume V in the continuous medium and introduce the angular momentum of this volume as

$$\boldsymbol{K} = \int_V (\boldsymbol{r} \times \boldsymbol{V}) \rho d\tau , \tag{3.43}$$

where $\boldsymbol{r}$ is the radius vector relative to a point O which is at rest, that is, not moving with the medium (see Figure 3.5).

The velocity of an arbitrary point M can be written as $\boldsymbol{V} = \boldsymbol{V}^* + \boldsymbol{V}_c$ where $\boldsymbol{V}^*$ is the velocity of the center of mass of the body and $\boldsymbol{V}_c$ is the velocity of the point M relative to O^*. Then, the angular momentum can be rewritten as

$$\boldsymbol{K} = \boldsymbol{r}^* \times \boldsymbol{Q} + \int_V (\boldsymbol{r}_c \times \boldsymbol{V}_c) \rho d\tau = \boldsymbol{r}^* \times \boldsymbol{Q} + \boldsymbol{K}^* .$$

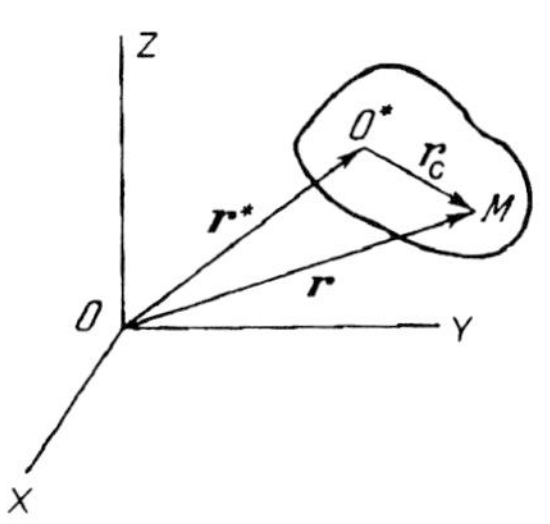

Figure 3.5 Derivation of the equation of motion for the angular momentum of a finite volume.

Here, $\boldsymbol{Q} = m\boldsymbol{V}^*$ is the linear momentum of the center of mass and $\boldsymbol{K}^* = \int_V (\boldsymbol{r}_c \times \boldsymbol{V}_c)\,\rho d\tau$ is the angular momentum of all points in the volume V relative to the center of mass O^*. The last term can also be written as

$$\boldsymbol{K}^* = \int\limits_V \rho \boldsymbol{k} d\tau\,,$$

where $\boldsymbol{k}$ is the density of internal angular momentum.

For the most part, the internal angular momentum $\boldsymbol{K}^*$ can be neglected relative to $\boldsymbol{r}^* \times \boldsymbol{Q}$. Sometimes, however, it is necessary to take the internal angular momentum into account. As an example, we consider a infinitesimal homogeneous continuum sphere with radius R and volume $d\tau$ that rotates around an axis going through its center O^* with angular velocity ω. Then, $\boldsymbol{K}^* = I\omega = ml^2\omega$ where I is the sphere's moment of inertia, m is its mass and l is the inertial radius with respect to O^*. Since $ml^2 \sim R^5$ and $\boldsymbol{r}^* \times \boldsymbol{Q} \sim R^3$, the quantity $\boldsymbol{K}^* \sim R^5\omega$ can be neglected relative to $\boldsymbol{r}^* \times \boldsymbol{Q}$ for finite angular velocity. Therefore, we have

$$\boldsymbol{K} = \int\limits_V (\boldsymbol{r} \times \boldsymbol{V})\,\rho d\tau\,.$$

However, if the continuous volume contains small volumes that are rotating with such a large angular velocity ω that $l^2\omega$ does not vanish, both quantities $\boldsymbol{K}^*$ and $\boldsymbol{r}^*\times$ will be of order R^3. In this case, the quantity $\boldsymbol{K}^*$ may not be neglected and the distributed internal angular momenta must be taken into account. By denoting the internal angular momentum density by $\boldsymbol{k}$, the angular momentum of the continuous medium becomes

$$\boldsymbol{K} = \int\limits_V (\boldsymbol{r} \times \boldsymbol{V})\,\rho d\tau + \int\limits_V \boldsymbol{k}\rho d\tau\,. \tag{3.44}$$

Since we have introduced internal angular momenta into the discussion, we also need to take into account the distributed mass and surface force pairs that act from the outside on the continuum volume. We denote by $\boldsymbol{h}$ and $\boldsymbol{q}_n$ the mass and surface force pairs per mass and source unit respectively. Then, the generalized angular momentum equation for a finite volume V that is bounded by Σ takes the following form:

$$\begin{aligned}\frac{d}{dt}&\left[\int\limits_V (\boldsymbol{r} \times \boldsymbol{V})\,\rho d\tau + \int\limits_V \boldsymbol{k}\rho d\tau\right]\\ &= \int\limits_V (\boldsymbol{r} \times \boldsymbol{f}_V)\,\rho d\tau + \int\limits_\Sigma (\boldsymbol{r} \times \boldsymbol{p}_n)\,d\Sigma\\ &\quad+ \int\limits_V \boldsymbol{h}\rho d\tau + \int\limits_\Sigma \boldsymbol{q}_n d\Sigma\,.\end{aligned} \tag{3.45}$$

In the treatment of continuous media by classical mechanics, the internal torques $(\boldsymbol{e}_i \times \boldsymbol{e}_k)\,p^{ki}$ as well as the moment densities of the external forces $\boldsymbol{h}$ and $\boldsymbol{q}_n$ are not taken into account, and the equation of motion for the angular momentum takes on the following form:

$$\frac{d}{dt}\int_V (\boldsymbol{r}\times \mathbf{V})\,\rho d\tau = \int_V (\boldsymbol{r}\times \boldsymbol{f}_V)\,\rho d\tau + \int_\Sigma (\boldsymbol{r}\times \boldsymbol{p}_n)\,d\Sigma\ . \tag{3.46}$$

We now transform (3.45) or (3.46) from the integral formulation into a differential formulation. According to Gauss's theorem (2.90) we can transform the surface integral into a volume integral, that is,

$$\int_\Sigma (\boldsymbol{r}\times \boldsymbol{p}_n)\,d\Sigma = \int_\Sigma (\boldsymbol{r}\times \boldsymbol{p}^i)\,n_i d\Sigma = \int_V \nabla_i (\boldsymbol{r}\times \boldsymbol{p}^i)\,d\tau\ ,$$

$$\int_\Sigma \boldsymbol{q}_n d\Sigma = \int_\Sigma \boldsymbol{q}^i_n n_i d\Sigma = \int_V \nabla_i \boldsymbol{q}^i_n d\tau\ .$$

Since $\nabla_i \boldsymbol{r} = \partial \boldsymbol{r}/\partial X^i = \boldsymbol{\varepsilon}_i$, one obtains, via (3.29),

$$\begin{aligned}\int_V \nabla_i (\boldsymbol{r}\times \boldsymbol{p}^i)\,d\tau &= \int_V (\boldsymbol{r}\times \nabla_i \boldsymbol{p}^i)\,d\tau + \int_V (\nabla_i \boldsymbol{r}\times \boldsymbol{p}^i)\,d\tau \\ &= \int_V (\boldsymbol{r}\times \nabla_i \boldsymbol{p}^i)\,d\tau + \int_V (\boldsymbol{e}_i\times \boldsymbol{e}_k)\,p^{ki} d\tau\ .\end{aligned}$$

Since the mass of the volume element dm is constant, (3.45) can be rewritten as

$$\begin{aligned}&\int_V \boldsymbol{r}\times\left(\frac{d\mathbf{V}}{dt} - \boldsymbol{f}_V - \frac{1}{\rho}\nabla_i \boldsymbol{p}^i\right)\rho d\tau + \int_V \frac{d\boldsymbol{k}}{dt}\rho d\tau \\ &\quad = \int_V \boldsymbol{h}\rho d\tau + \int_V \nabla_i \boldsymbol{q}^i_n d\tau + \int_V (\boldsymbol{e}_i\times \boldsymbol{e}_k)\,p^{ki} d\tau\ .\end{aligned}$$

Since the first term on the left-hand side vanishes according to (3.28) and the previous equation is valid for any volume, one obtains

$$\frac{d\boldsymbol{k}}{dt} = \boldsymbol{h}\rho + \nabla_i q^i + (\boldsymbol{e}_i\times \boldsymbol{e}_k)\,p^{ki}\ . \tag{3.47}$$

Equation 3.47 follows from the conservation of angular momentum under the conditions that the movement is continuous and that the distributed mass and surface force pairs are taken into account.

If the internal moments $\boldsymbol{k}$ and the disturbed force pairs $\boldsymbol{h}$ and $\boldsymbol{q}$ are not present, (3.47) is equivalent to the following:

$$(\boldsymbol{e}_i\times \boldsymbol{e}_k)\,p^{ki} = 0\ .$$

We rearrange this equation by splitting the left-hand side into two terms, thereby obtaining

$$\underset{(k<i)}{(\boldsymbol{e}_i\times \boldsymbol{e}_k)\,p^{ki}} + \underset{(k>i)}{(\boldsymbol{e}_i\times \boldsymbol{e}_k)\,p^{ki}} = 0\ .$$

Since we sum over indices k and i in both terms, we can swap the names of indices i and k, resulting in

$$\underset{(k<i)}{(e_i \times e_k)\, p^{ki}} + \underset{(k>i)}{(e_k \times e_i)\, p^{ik}} = 0 \,.$$

Since $(e_i \times e_k) = -(e_k \times e_i)$, the last equation results in

$$(e_i \times e_k)\left(p^{ki} - p^{ik}\right) = 0 \,,$$

which leads to

$$p^{ki} = p^{ik} \,. \tag{3.48}$$

The equation of continuity (3.6), the equation of motion (3.28), the angular momentum equation (3.47) and the conditions (3.48) provide seven equations for the 13 unknown quantities ρ, V^i and p^{ik}. To close this system, we need additional equations.

We note that (3.48) does not contribute new unknowns, but reduces the number of components of the stress tensor p^{ik} from nine to six. In the next chapter, it is shown that the further reduction of unknowns can be achieved via assuming a model for the continuous medium.

4
Closed Systems of Mechanical Equations for the Simplest Continuum Models

Creating mathematical models for a given continuum medium allows us to arrive at a closed system of equations together with the corresponding initial and boundary conditions. Equations 3.6, 3.28, 3.47 and 3.48 are valid for all continuous movements of a continuum medium. However, these equations alone are insufficient to determine the movement of a concrete continuum medium since there are fewer equations than unknowns, that is, the system of equations is underdetermined or not closed. In order to close the system, it needs be to supplemented by additional relationships between the parameters of the continuum medium. In other words, the necessity of creating a model for the continuum medium arises. The area of continuum mechanics that deals with creating such models of the continuum is called rheology. To create a model for a continuum medium, one has to use the results of experimental studies of material properties as well as the general principles of mechanics, physics and chemistry.

In this chapter, we restrict our investigation to such media and processes for which the thermodynamical circumstances are not needed to construct mathematical models of the movement, that is, we only consider the movement of such media whose mechanical system of equations can be closed without using the equations of thermodynamics.

4.1
Ideal Fluid and Gas

An ideal fluid or ideal gas is a continuum medium for which the stress vector $\boldsymbol{p}_n$ for each surface element is collinear with the unit normal vector on the surface element. In this case, the main components of the stress vector are given by $p_1 = p_2 = p_3 = -p$ and the tensor surface is a sphere. The quantity p is called pressure. Since an ideal fluid or ideal gas normally exists in a compressed state, we have $p_i < 0$ and $p > 0$. For an ideal fluid or ideal gas, each set of mutually orthogonal directions can be used as main axes and therefore the components of the stress tensor in a Cartesian system of coordinates are given by

$$p_k^i = -p\delta_k^i \,. \tag{4.1}$$

Hydromechanics. Theory and Fundamentals. Emmanuil G. Sinaiski

ISBN: 978-3-527-41026-2

Since the components of the unit tensor $\delta_k^i = \boldsymbol{e}^k \boldsymbol{e}_i$ do not change under a coordinate transformation, for mixed tensor components, (4.1) is also valid in every orthogonal curvilinear system of coordinates. The corresponding covariant components p_{ki} and contravariant components p^{ki} can be obtained using the metric tensor $\boldsymbol{G}$ as

$$p_{ki} = g_{ks} p_i^s = -p g_{ks} \delta_i^s = -p g_{ki} ,$$
$$p^{ki} = g^{ks} p_s^i = -p g^{ks} \delta_s^i = -p g^{ki} .$$

This means that the assumption of an ideal continuum medium makes it possible to reduce the number of stress tensor components to one and to write the stress tensor $\boldsymbol{P}$ in the following form:

$$\boldsymbol{P} = -p \boldsymbol{G} . \tag{4.2}$$

The tensor surface for an ideal fluid or an ideal gas is a sphere. Such a tensor is called a spherical tensor. Thus, every tensor $\boldsymbol{T}$ that has the form $\boldsymbol{T} = k\boldsymbol{G}$ with a scalar k and the metric tensor $\boldsymbol{G}$ is a spherical tensor. The equation of motion (3.35) now becomes

$$\rho \frac{d V^k}{dt} = \rho f_V^k - g^{ki} \nabla_i p . \tag{4.3}$$

To derive (4.3), it was taken into account that the components of the metric tensor behave under covariant differentiation like a constant under differentiation. On the left-hand side of this equation, we have the substantial derivative which, using (2.13), can be written as

$$\frac{d V^k}{dt} = \frac{\partial V^k}{\partial t} + V^i \nabla_i V^k .$$

Thus, (4.3) becomes

$$\rho \left(\frac{\partial V^k}{\partial t} + V^i \nabla_i V^k \right) = \rho f_V^k - g^{ki} \nabla_i p . \tag{4.4}$$

By writing equations (4.3) or (4.4) in vector form, we obtain

$$\rho \frac{d \boldsymbol{V}}{dt} = \rho \left(\frac{\partial \boldsymbol{V}}{\partial t} + \boldsymbol{V} \nabla \cdot \boldsymbol{V} \right) = \rho \boldsymbol{f}_V - \nabla p . \tag{4.5}$$

In the Cartesian system of coordinates, (4.5) has the following form:

$$\frac{dU}{dt} = \frac{\partial U}{\partial t} + U \frac{\partial U}{\partial X} + V \frac{\partial U}{\partial Y} + W \frac{\partial U}{\partial Z} = f_X - \frac{1}{\rho} \frac{\partial p}{\partial X} ,$$
$$\frac{dV}{dt} = \frac{\partial V}{\partial t} + U \frac{\partial V}{\partial X} + V \frac{\partial V}{\partial Y} + W \frac{\partial V}{\partial Z} = f_Y - \frac{1}{\rho} \frac{\partial p}{\partial Y} ,$$
$$\frac{dW}{dt} = \frac{\partial W}{\partial t} + U \frac{\partial W}{\partial X} + V \frac{\partial W}{\partial Y} + W \frac{\partial W}{\partial Z} = f_Z - \frac{1}{\rho} \frac{\partial p}{\partial Z} . \tag{4.6}$$

These equations are called the Euler equations.

We now write these equations in a slightly different form. We first note that

$$\begin{aligned}\frac{dU}{dt} &= \frac{\partial U}{\partial t} + \frac{1}{2}\frac{\partial}{\partial X}\left(U^2 + V^2 + W^2\right) \\ &\quad - \left(\frac{\partial V}{\partial X} - \frac{\partial U}{\partial Y}\right) V + \left(\frac{\partial U}{\partial Z} - \frac{\partial W}{\partial X}\right) W \\ &= \frac{\partial U}{\partial t} + \frac{1}{2}\frac{\partial \boldsymbol{V}^2}{\partial X} + 2\left(\omega_Y W - \omega_Z V\right) \\ &= \frac{\partial U}{\partial t} + \frac{1}{2}\frac{\partial \boldsymbol{V}^2}{\partial X} + 2\left(\boldsymbol{\omega} \times \boldsymbol{V}\right)_X . \end{aligned} \tag{4.7}$$

In a similar fashion, we have

$$\begin{aligned}\frac{dV}{dt} &= \frac{\partial V}{\partial t} + \frac{1}{2}\frac{\partial \boldsymbol{V}^2}{\partial X} + 2\left(\boldsymbol{\omega} \times \boldsymbol{V}\right)_Y , \\ \frac{dW}{dt} &= \frac{\partial W}{\partial t} + \frac{1}{2}\frac{\partial \boldsymbol{V}^2}{\partial Z} + 2\left(\boldsymbol{\omega} \times \boldsymbol{V}\right)_Z . \end{aligned} \tag{4.8}$$

Using this equation, (4.5) becomes

$$\frac{\partial \boldsymbol{V}}{\partial t} + \frac{1}{2}\nabla \boldsymbol{V}^2 + 2\left(\boldsymbol{\omega} \times \boldsymbol{V}\right) = \boldsymbol{f}_V - \frac{1}{\rho}\nabla p . \tag{4.9}$$

The above equation is called the equation of motion in Gromeka–Lambs form.

The system of equations that consists of the equation of continuity

$$\frac{\partial \rho}{\partial t} + \nabla \cdot (\rho \boldsymbol{V}) = 0 \tag{4.10}$$

and the equation of motion

$$\rho \frac{d\boldsymbol{V}}{dt} = \rho\left(\frac{\partial \boldsymbol{V}}{\partial t} + \boldsymbol{V}\nabla \cdot \boldsymbol{V}\right) = \rho \boldsymbol{f}_V - \nabla p \tag{4.11}$$

is a system of four equations for five unknowns: the velocity $\boldsymbol{V}$, the mass density ρ and the pressure p. It is not closed. However, if the fluid is not only incompressible, but also homogeneous, we have $\rho =$ const for every fluid particle and the equation of continuity becomes

$$\nabla \cdot \boldsymbol{V} = \nabla_i V^i = 0 . \tag{4.12}$$

Thus, the system of equations (4.5) and (4.12) is closed with respect to the unknowns $\boldsymbol{V}$ and p. For a incompressible but inhomogeneous fluid, the density is constant for every individual moving particle of the fluid, but the mass density of different particles in the fluid is different. In this case, on has to add to the system of equations, (4.5) and (4.12), the conservation of mass for every individual fluid particle, that is,

$$\frac{d\rho}{dt} = \frac{\partial \rho}{\partial t} + \boldsymbol{V} \cdot \nabla \rho = 0 . \tag{4.13}$$

Thus, for an incompressible inhomogeneous fluid, we have the closed system of (4.11)–(4.13) with unknowns $\boldsymbol{V}$, p and ρ. Therefore, the system of equations is closed for the ideal incompressible homogeneous as well as for the inhomogeneous ideal fluid.

A different situation arises for the ideal compressible fluid which is then referred to as a gas, and the system of equations is not closed. It is then necessary to add another equation that characterizes the continuum medium, that is, the gas. One often adds the condition

$$p = f(\rho) \,. \tag{4.14}$$

A process for which the pressure is only a function of the density ρ is called barotropic. The movement of a gas with isothermal change of state and with the Clapeyron equation of state $p = \rho R T$, with the temperature T and the gas constant R is an example of such a process. Thus, the closed system of (4.10)–(4.12) describes the movement of an ideal compressible barotropic gas with respect to the unknowns $\boldsymbol{V}$, p and ρ.

For a non-isothermal process ($T \neq \text{const}$), the equation of state has the general form

$$p = f(\rho, T) \,. \tag{4.15}$$

Since a new unknown T appears, the system of equation is now not closed. In order to close it, one has to add the equation of conservation of energy. Chapter 5 is dedicated to the discussion of this problem.

4.2
Linear Elastic Body and Linear Viscous Fluid

We now consider other models of continuum media: the model of a linear elastic body and the model of a linear viscous fluid. Although these models represent two very different behaviors of real media, the study of their behavior is done jointly since the methods needed to introduce them are formally identical.

An elastic body is referred to as a medium for which the components of the stress tensor p^{ki} for every point of the continuum medium are functions of the components of the strain tensor ε_{ij}, of the metric tensor g_{ij} of the temperature T and possibly of other physicochemical parameter χ_i, for example, phase concentrations, that is

$$p^{ij} = f^{ij}(\varepsilon_{kl}, g^{kl}, T, \chi_1, \chi_2, \ldots, \chi_n) \,. \tag{4.16}$$

A viscous fluid is called a medium for which the components of the stress tensor can, for each particle in the continuum medium, be written in the following form:

$$p^{ij} = -p g^{ij} + \tau^{ij} \,, \tag{4.17}$$

where

$$p = p(\rho, T, \chi_1, \chi_2, \dots, \chi_n), \tau^{ij} = \varphi^{ij}(e_{kl}, g^{kl}, T, \chi_1, \chi_2, \dots, \chi_n),$$

and where e_{kl} are the the components of the strain velocity tensor. For simplicity, we assume that φ^{ij} does not depend on T and χ_i.

Experimental investigations show that for small temperatures and stresses, the components of the stress tensor p^{ij} and the components of the strain tensor ε^{ij} are related by Hooke's law and the components of the tensor of viscous stresses τ^{ij} and the components of the strain velocity tensor e_{kl} are related by Navier–Stokes law. Formally, these laws follow by expanding the components of the stress tensor p^{ij} into a Taylor series in ε^{ij} or e_{kl} depending on media type. For small deformations and strain velocities, one can restrict these series to their linear terms such that

$$p^{ij} = A^{ijkl}\varepsilon_{kl} \tag{4.18}$$

and

$$\tau^{ij} = B^{ijkl}e_{kl}. \tag{4.19}$$

Equation 4.18 is called Hooke's law and (4.19) is called Navier–Stokes law. It must be noted that although Navier–Stokes law is only supposed to be valid for small strain velocities, it is actually valid for many fluids and gases for the case of larger strain velocities. The part of mechanics that is devoted to the behavior of a continuum medium subject to Hooke's law is referred to as theory of elasticity, while the part of mechanics that deals with the behavior of a continuum medium subject to Navier–Stokes law is the theory of the viscous fluid.

Since (4.18) and (4.19) are invariant with respect to the choice of the system of coordinates, it follows that the quantities A^{ijkl} and B^{ijkl} are components of tensors of fourth order. Each tensor has $3^4 = 81$ components. However, in some cases, the number of independent components is much smaller. The symmetry of the strain tensor and strain velocity tensor allows the reduction of the number of independent components to 36.

A further reduction of the number of independent components is connected with the notion of the isotropic medium. A medium is called isotropic if all its properties are the same in all directions. If the properties of the medium in different directions are different, the medium is called anisotropic. The properties of isotropy and anisotropy follow from the symmetry of the medium. One refers to a medium as isotropic if a group of coordinates transformations exists such that the components of the tensors that define the properties of the medium are unchanged for all transformations that belong to the group. For example, a medium is isotropic if the components of the tensors defining the properties of the medium are unchanged for any orthogonal transformation, that is, for any transformation that leaves the components of the metric tensor intact. The full orthogonal group contains the rotations for which the determinant of the transformation matrix is one and the combinations of rotations with reflections for which the determinant of the transformation matrix is -1. If the properties of the medium are invariant with respect to the rotation group but not with respect to reflections, such a medium is called gyrotropic. Examples of isotropic media in which there is no preferred system of coordinates are water and other amorphous media, and media with small randomly orientated particles. Crystalline solids and fibrous compounds are examples of anisotropic media. We now consider the details of whether an elastic medium subject to Hooke's law will be isotropic or gyrotropic. We consider two Cartesian systems of coordinates X^i and X'^i at the same point of the continuum medium at a given point in time. The second system of coordinates will be rotated relative to the first one. When moving from the first system of coordinates to the second one, the components of the tensor $\boldsymbol{A}$ are transformed according to the rule

$$A'^{i'j'k'l'} = \frac{\partial X'^{i'}}{\partial X^i}\frac{\partial X'^{j'}}{\partial X^j}\frac{\partial X'^{k'}}{\partial X^k}\frac{\partial X'^{l'}}{\partial X^l}A^{ijkl} . \tag{4.20}$$

In the system of coordinates X^i, Hooke's law is expressed via the coefficients A^{ijkl}. It is expressed by the coefficients A'^{ijkl} in the transformed coordinate system X'^i, while the form of the Hooke's law stays the same. If $A'^{ijkl} = A^{ijkl}$, that is, the coefficients of Hooke's law are the same in both system of coordinates, and the medium is both isotropic and gyrotropic. For a fourth order tensor, the notions of isotropy and gyrotropy are identical. For $A'^{ijkl} \neq A^{ijkl}$, the medium is anisotropic. When choosing the coordinate axes along the main axes of the strain tensor ε_{ij}, the only non-vanishing coefficients of the Hooke's law are A^{ijaa}.

We now rotate the system of coordinates by 180° around the i-axis. This will leave the i-coordinate unchanged, while both of the other two coordinates pick up a factor of −1. By using (4.20), we obtain $A'^{ijaa} = -A^{ijaa}$ for $j \neq i$. Since the tensor A is preserved under any rotation, this also means $A^{ijaa} = -A^{ijaa}$ for $j \neq i$ and thus $A^{ijaa} = 0$ at $i \neq j$. Thus, the number of nonzero components has been reduced from 81 to nine, only two of which are independent, namely,

$$A^{1111} = A^{2222} = A^{3333} = 2\mu + \lambda ,$$
$$A^{1122} = A^{1133} = A^{2233} = \lambda , \quad A^{iiaa} = A^{aaii} .$$

The constants λ and μ are called the Lamé parameters.

In an isotropic medium subject to Hooke's law, the main axes of the strain tensor and stress tensor will be identical. Using these as the axes of the Cartesian system of coordinates, Hooke's law is written as follows:

$$p^{ij} = \lambda I_1(\varepsilon)\delta_{ij} + 2\mu\varepsilon_{ij} . \tag{4.21}$$

In the theory of elasticity, one normally introduces, instead of the Lamé parameters, the following two medium parameters, that is, Young's elasticity model

$$E = \mu\frac{3\lambda + 2\mu}{\lambda + \mu}$$

and the Poisson number

$$\sigma = \frac{\lambda}{2(\lambda + \mu)} .$$

A similar argument can be made for an isotropic medium under the influence of the Navier–Stokes law. In the Cartesian system of coordinates, whose axes are given by the main axes of the strain tensor and stress tensor, the Navier–Stokes law has the form

$$\tau_i = \lambda_1 (e_1 + e_2 + e_3) + 2\mu_1 e_i , \quad (i = 1, 2, 3) . \tag{4.22}$$

In the theory of the movement of viscous fluids, it is customary to use, instead of λ_1 and μ_1, the coefficients of dynamic viscosity $\mu = \mu_1$ or kinematical viscosity ν/ρ and volume viscosity $\varsigma = \lambda_1 + \frac{2}{3}\mu$. Sometimes, one assumes $\varsigma = 0$, which then gives $\lambda = \lambda_1 = -2/3\mu$. For an isotropic medium, Hooke's law has the following form in a curvilinear system of coordinates:

$$p^{ij} = \lambda I_1(\varepsilon) g^{ij} + 2\mu g^{ik} g^{jl} \varepsilon_{kl} . \tag{4.23}$$

The Navier–Stokes law can be written in a similar fashion as

$$\tau^{ij} = \lambda_1 I_1(e) g^{ij} + 2\mu_1 g^{ik} g^{jl} e_{kl} . \tag{4.24}$$

Here, $I_1(\varepsilon)$ and $I_1(e)$ are the first invariants of the strain and strain velocity tensor.

Using (4.17) and (2.77), one obtains the following expression for the stress tensor of a viscous fluid in a curvilinear system of coordinates:

$$p^{ij} = -p g^{ij} + \lambda_1 g^{ij} \nabla \cdot \boldsymbol{V} + 2\mu_1 g^{ik} g^{jl} e_{kl} . \tag{4.25}$$

In a Cartesian system of coordinates, the two laws can be written as follows:

Hooke's law:

$$p_{ii} = \lambda I_1(\varepsilon) + 2\mu \varepsilon_{ii} \quad \text{and} \quad p_{ij} = 2\mu \varepsilon_{ij} \quad \text{for} \quad i \neq j \tag{4.26}$$

Navier–Stokes law:

$$\begin{aligned} p_{ii} &= -p + \lambda_1 \nabla \cdot \boldsymbol{V} + 2\mu \frac{\partial V_i}{\partial X^i} \quad \text{and} \\ p_{ij} &= 2\mu_1 e_{ij} = \mu_1 \left(\frac{\partial V_i}{\partial X^j} + \frac{\partial V_j}{\partial X^i} \right) \quad \text{for} \quad i \neq j . \end{aligned} \tag{4.27}$$

We now derive the equation of motion for a viscous fluid subject to Navier–Stokes law. Using (3.35), we obtain

$$\begin{aligned} \nabla_j p_{ij} &= \nabla_j \left(-p g^{ij} + \lambda g^{ij} \nabla \cdot \boldsymbol{V} + 2\mu e^{ij} \right) \\ &= -g^{ij} \nabla_j p + \lambda g^{ij} \nabla_j (\nabla \cdot \boldsymbol{V}) + 2\mu \nabla_j \left(e^{ij} \right) \\ &= -g^{ij} \nabla_j p + \lambda g^{ij} \nabla_j \nabla_\alpha V^\alpha + \mu \nabla_j g^{i\alpha} g^{j\beta} \left(\nabla_\alpha V_\beta + \nabla_\beta V_\alpha \right) \\ &= -g^{ij} \nabla_j p + \lambda g^{ij} \nabla_j \nabla_\alpha V^\alpha + \mu g^{i\alpha} \nabla_j \nabla_\alpha V^j + \mu g^{j\beta} \nabla_j \nabla_\beta V^i \\ &= -g^{ij} \nabla_j p + (\lambda + \mu) g^{ij} \nabla_j \nabla_\alpha V^\alpha + \mu \nabla^\beta \nabla_\beta V^i \\ &= -\nabla^i p + (\lambda + \mu) \nabla^i (\nabla \cdot \boldsymbol{V}) + \mu \Delta V^i . \end{aligned} \tag{4.28}$$

Here, we take into account that $\nabla^i = gij\nabla_j$, that the Laplace operator has the form $\varDelta = \nabla^\beta \nabla_\beta$, that for covariant differentiation, g^{ij} can be permuted with the gradient operator ∇ and that the covariant differentiation can be permuted in Euclidean space, that is, $\nabla_j \nabla_\alpha V^\alpha = \nabla_\alpha \nabla_j V^\alpha$.

In Cartesian coordinates, the Laplace operator is

$$\Delta V^i = \frac{\partial^2 V^i}{\partial X^2} + \frac{\partial^2 V^i}{\partial Y^2} + \frac{\partial^2 V^i}{\partial Z^2} .$$

Equation 4.28 can be written in vector form as

$$\nabla_j \boldsymbol{p}^i = -\nabla p + (\lambda + \mu) \nabla (\nabla \cdot \boldsymbol{V}) + \mu \Delta \boldsymbol{V} . \tag{4.29}$$

Thus, the equation of motion takes on the following coordinate form

$$\frac{d V^i}{dt} = f_V^i - \frac{1}{\rho} g^{ij} \frac{\partial p}{\partial X^i} + \frac{(\lambda + \mu)}{\rho} g^{ij} \frac{\partial (\nabla \cdot \boldsymbol{V})}{\partial X^j} + \nu \Delta V^i \tag{4.30}$$

or vector form

$$\frac{d\boldsymbol{V}}{dt} = \boldsymbol{f}_V - \frac{1}{\rho} \nabla p + \frac{(\lambda + \mu)}{\rho} \nabla (\nabla \cdot \boldsymbol{V}) + \nu \Delta \boldsymbol{V} . \tag{4.31}$$

These equations are called the Navier–Stokes equations. In the case of an incompressible fluid, the Navier–Stokes equations are simplified to

$$\frac{d\boldsymbol{V}}{dt} = \boldsymbol{f}_V - \frac{1}{\rho} \nabla p + \nu \Delta \boldsymbol{V} . \tag{4.32}$$

Together with the equation of continuity,

$$\nabla \cdot \boldsymbol{V} = 0 , \tag{4.33}$$

these equations constitute a full system of equations describing the movement of a homogeneous, viscous and incompressible fluid for $\nu = \text{const}$. If the fluid is incompressible though inhomogeneous, one must additionally use the equation

$$\frac{d\rho}{dt} = \frac{\partial \rho}{\partial t} + \boldsymbol{V} \nabla \rho = 0 . \tag{4.34}$$

We now write the full system of equations for the movement of a homogeneous, viscous and incompressible fluid in Cartesian coordinates:

$$\frac{\partial U}{\partial X} + \frac{\partial V}{\partial Y} + \frac{\partial W}{\partial Z} = 0 ,$$

$$\frac{\partial U}{\partial t} + U \frac{\partial U}{\partial X} + V \frac{\partial U}{\partial Y} + \frac{\partial U}{\partial Z} = f_{Vx} - \frac{1}{\rho} \frac{\partial p}{\partial X} + \nu \left(\frac{\partial^2 U}{\partial X^2} + \frac{\partial^2 U}{\partial Y^2} + \frac{\partial^2 U}{\partial Z^2} \right) ,$$

$$\frac{\partial V}{\partial t} + U\frac{\partial V}{\partial X} + V\frac{\partial V}{\partial Y} + \frac{\partial V}{\partial Z} = f_{VY} - \frac{1}{\rho}\frac{\partial p}{\partial Y} + \nu\left(\frac{\partial^2 V}{\partial X^2} + \frac{\partial^2 V}{\partial Y^2} + \frac{\partial^2 V}{\partial Z^2}\right),$$

$$\frac{\partial W}{\partial t} + U\frac{\partial W}{\partial X} + V\frac{\partial W}{\partial Y} + \frac{\partial W}{\partial Z} = f_{VZ} - \frac{1}{\rho}\frac{\partial p}{\partial Z} + \nu\left(\frac{\partial^2 W}{\partial X^2} + \frac{\partial^2 w}{\partial Y^2} + \frac{\partial^2 W}{\partial Z^2}\right),$$

$$\frac{\partial \rho}{\partial t} + U\frac{\partial \rho}{\partial X} + V\frac{\partial \rho}{\partial Y} + W\frac{\partial \rho}{\partial Z} = 0\,. \tag{4.35}$$

The simple linear models for the continuum medium listed above are used in large parts of hydrodynamics and theory of elasticity. However, for many real media, these linear laws are not applicable. Thus, when subjecting certain solids such as asphalt, putty or metal to large stresses, it is observed that some deformations remain even after the external force has been removed. Therefore, it is necessary to construct new models, taking into account the plasticity, the creep and other properties of the materials. Many fluids and mixtures show properties that deviate from Navier–Stokes law. For such fluids, models with a nonlinear dependence of the stress tensor on the strain velocity tensor are created. Fluids that do not obey the Navier–Stokes law are called non-Newtonian. The research area which deals with different deformation behaviors, taking into account the relationships between the strain tensor, stress velocity tensor and stress tensor, is called rheology.

Thus, in order to describe the movements of real media, one must create complicated models. In order to do so, the notions and characteristics of the movements and states of continuum particles introduced above are insufficient. It is necessary to introduce new notions and characteristics such as temperature, entropy, energy, internal energy and so on. In order to introduce these characteristics, one has to add new equations to those introduced above. This necessitates using additional equations from physics and thermodynamics.

4.3 Equations in Curvilinear Coordinates

The equations of motions and the equation of continuity (4.35) that were obtained in the previous section are written using the Cartesian system of coordinates (X, Y, Z).

For many applications, one needs to know the form of these equations in an orthogonal curvilinear system of coordinates, especially in the cylindrical and the spherical systems of coordinates.

4.3.1
Equation of Continuity

We now write the expression for the divergence of the velocity vector in a curvilinear system of coordinates as

$$\nabla \cdot V = \nabla_i V^i = \frac{\partial V^i}{\partial X^i} + V^j \Gamma^i_{ij} = \frac{\partial V^i}{\partial X^i} + \frac{V^j}{\sqrt{g}} \frac{\partial \sqrt{g}}{\partial X^i} = \frac{1}{\sqrt{g}} \frac{\partial \left(V^i \sqrt{g}\right)}{\partial X^i} . \quad (4.36)$$

By substituting (4.36) into (3.6), one obtains the equation of continuity in curvilinear coordinates,

$$\sqrt{g}\frac{\partial \rho}{\partial t} + \frac{\partial \left(\rho V^i \sqrt{g}\right)}{\partial X^i} = 0 . \quad (4.37)$$

It must be noted that V^i are the components of the vector V with respect to the covariant basis ε_i, whose basis vectors are not necessarily normalized. Therefore, the ε_i are not suitable to be used here. Instead, we introduce the normalized vectors $\varepsilon_i/\sqrt{g_{ii}}$. Then, we have

$$V = V^i \varepsilon_i = U^i \frac{\varepsilon_i}{\sqrt{g_{ii}}} .$$

The components U^I are the physical components of the velocity vector,

$$U^i = V^i \sqrt{g_{ii}} . \quad (4.38)$$

In Cartesian coordinates, we have $g_{ii} = 1$ and $U^i = V^i$. Therefore, the physical and usual components are the same.

In the physical components of the velocity U^i, the equation of continuity is given by:

$$\sqrt{g}\frac{\partial \rho}{\partial t} + \frac{\partial \left(\rho U^1 \sqrt{g_{22} g_{33}}\right)}{\partial X^1} + \frac{\partial \left(\rho U^2 \sqrt{g_{11} g_{33}}\right)}{\partial X^2} + \frac{\partial \left(\rho U^3 \sqrt{g_{11} g_{22}}\right)}{\partial X^3} = 0 . \quad (4.39)$$

To write down the equation of continuity in a concrete curvilinear system of coordinates, one has to use the corresponding expressions for the components of the metric tensor (see Appendix A.3). In cylindrical coordinates $X^1 = r$, $X^2 = \varphi$, $X^3 = Z$, we have

$$g_{11} = g_{33} = 1 , \quad g_{22} = r^2 , \quad U^1 = U_r , \quad U^2 = U_\varphi , \quad U^3 = U_z ,$$

$$r\frac{\partial \rho}{\partial t} + \frac{\partial (\rho U_r r)}{\partial r} + \frac{\partial (\rho U_\varphi)}{\partial \varphi} + r\frac{\partial (\rho U_z)}{\partial Z} = 0 . \quad (4.40)$$

In spherical coordinates $X^1 = r$, $X^2 = \varphi$ (latitude) and $X^3 = \vartheta$ (longitude), we have

$$g_{11} = 1 \,, \quad g_{22} = r^2 \,, \quad g_{33} = r^2 \sin^2 \varphi \,,$$
$$U^1 = U_r \,, \quad U^2 = U_\varphi \,, \quad U^3 = U_\vartheta$$

and the equation of continuity becomes

$$r^2 \sin\varphi \frac{\partial \rho}{\partial t} + \sin\varphi \frac{\partial \left(\rho U_r r^2\right)}{\partial r} + r \frac{\partial \left(\rho U_\varphi \sin\varphi\right)}{\partial \varphi} + r \frac{\partial \left(\rho U_\vartheta\right)}{\partial \vartheta} = 0 \,. \tag{4.41}$$

4.3.2
Equation of Motion

To obtain the equation of motion in curvilinear coordinates, one first has to obtain the corresponding expressions for the acceleration, the gradient of a scalar function and the Laplace operator. The components of the acceleration are given by

$$a^i = \frac{\partial V^i}{\partial t} + V^j \nabla_j V^i = \frac{\partial V^i}{\partial t} + V^j \frac{\partial V^i}{\partial X^j} + V^j V^k \Gamma^i_{jk} = 0 \,.$$

The last term on the right-hand side can be rewritten in the following form, where no summation over i takes place:

$$\begin{aligned} V^j V^k \Gamma^i_{jk} &= \underset{(k \neq i)}{2 V^i V^k \Gamma^i_{ki}} + \left(V^k\right)^2 \Gamma^i_{kk} \\ &= \underset{(k \neq i)}{2 V^i V^k \Gamma^i_{ki}} + \left(V^i\right)^2 \Gamma^i_{ii} + \underset{(i \neq k)}{\left(V^k\right)^2 \Gamma^i_{kk}} \,. \end{aligned}$$

For an orthogonal curvilinear system of coordinates, we have $g_{ij} = 0$ for $i \neq j$. By using (A29), one obtains, without summation over i,

$$a^i = \frac{\partial V^i}{\partial t} + V^j \frac{\partial V^i}{\partial X^j} + \underset{(k \neq i)}{\frac{V^i V^k}{g_{ii}} \frac{\partial g_{ii}}{\partial X^k}} + \frac{1}{2} \frac{\left(V^i\right)^2}{g_{ii}} \frac{\partial g_{ii}}{\partial X^i} - \frac{1}{2} \frac{\left(V^k\right)^2}{g_{ii}} \frac{\partial g_{ii}}{\partial X^k} \,.$$

By using the physical components U^i of the velocity instead of V^i according to (4.38) in the above equation, the components of the acceleration in cylindrical coordinates take on the following form:

$$\begin{aligned} a_r &= \frac{\partial U_r}{\partial t} + U_r \frac{\partial U_r}{\partial r} + \frac{U_\varphi}{r} \frac{\partial U_r}{\partial \varphi} + U_z \frac{\partial U_r}{\partial Z} - \frac{U_\varphi^2}{r} \,, \\ a_\varphi &= \frac{\partial U_\varphi}{\partial t} + U_r \frac{\partial U_\varphi}{\partial r} + \frac{U_\varphi}{r} \frac{\partial U_\varphi}{\partial \varphi} + U_z \frac{\partial U_\varphi}{\partial Z} + \frac{U_\varphi U_r}{r} \,, \\ a_z &= \frac{\partial U_z}{\partial t} + U_r \frac{\partial U_z}{\partial r} + \frac{U_\varphi}{r} \frac{\partial U_z}{\partial \varphi} + U_z \frac{\partial U_z}{\partial Z} \,. \end{aligned} \tag{4.42}$$

The components of the acceleration in spherical coordinates are

$$
\begin{aligned}
a_r &= \frac{\partial U_r}{\partial t} + U_r \frac{\partial U_r}{\partial r} + \frac{U_\varphi}{r}\frac{\partial U_r}{\partial \varphi} + \frac{U_\vartheta}{r\sin\vartheta}\frac{\partial U_r}{\partial \vartheta} - \frac{U_\varphi^2 + U_\vartheta^2}{r} , \\
a_\varphi &= \frac{\partial U_\varphi}{\partial t} + U_r \frac{\partial U_\varphi}{\partial r} + \frac{U_\varphi}{r}\frac{\partial U_\varphi}{\partial \varphi} + \frac{U_\vartheta}{r\sin\varphi}\frac{\partial U_\varphi}{\partial \vartheta} + \frac{U_r U_\varphi - U_\vartheta^2 \cot\varphi}{r} , \\
a_\vartheta &= \frac{\partial U_\vartheta}{\partial t} + U_r \frac{\partial U_\vartheta}{\partial r} + \frac{U_\varphi}{r}\frac{\partial U_\vartheta}{\partial \varphi} + \frac{U_\vartheta}{r\sin\varphi}\frac{\partial U_\vartheta}{\partial \vartheta} + \frac{U_r U_\vartheta + U_\varphi U_\vartheta \cot\varphi}{r} .
\end{aligned}
\tag{4.43}
$$

4.3.3
Gradient of a Scalar Function

The physical components A^i of the gradient of a scalar function p in a curvilinear system of coordinates are obtained from the obvious equation

$$
\nabla p = \frac{\partial p}{\partial X^i}\boldsymbol{\varepsilon}_i = (\nabla_i p)\,\boldsymbol{\varepsilon}_i = A_i \frac{\boldsymbol{\varepsilon}_i}{\sqrt{g_{ii}}} .
$$

Therefore, the physical components of the gradient of the scalar function p, that is, the pressure is

$$
A_i = \frac{\partial p}{\partial X^i}\sqrt{g_{ii}} .
$$

In cylindrical coordinates, they are given by

$$
\begin{aligned}
A_1 &= \frac{\partial p}{\partial r}\sqrt{g_{11}} = \frac{\partial p}{\partial r} , \\
A_2 &= \frac{\partial p}{\partial \varphi}\sqrt{g_{22}} = \frac{1}{r}\frac{\partial p}{\partial \varphi} , \\
A_3 &= \frac{\partial p}{\partial Z}\sqrt{g_{33}} = \frac{\partial p}{\partial Z} .
\end{aligned}
\tag{4.44}
$$

In spherical coordinates, they are

$$
A_1 = \frac{\partial p}{\partial r} , \quad A_2 = \frac{1}{r}\frac{\partial p}{\partial \varphi} , \quad A_3 = r\sin\varphi \frac{\partial p}{\partial \vartheta} .
\tag{4.45}
$$

4.3.4
Laplace Operator

We now put $\boldsymbol{V} = \nabla \Phi$. Using (2.77) for the divergence of a vector $\boldsymbol{V}$, we obtain

$$
\begin{aligned}
\nabla \cdot \nabla \Phi = \Delta \Phi = \frac{1}{\sqrt{g}} \bigg[& \frac{\partial}{\partial X^1}\left(\sqrt{\frac{g_{22}g_{33}}{g_{11}}}\frac{\partial \Phi}{\partial X^1}\right) \\
& + \frac{\partial}{\partial X^2}\left(\sqrt{\frac{g_{11}g_{33}}{g_{22}}}\frac{\partial \Phi}{\partial X^2}\right) + \frac{\partial}{\partial X^3}\left(\sqrt{\frac{g_{11}g_{22}}{g_{33}}}\frac{\partial \Phi}{\partial X^3}\right)\bigg] .
\end{aligned}
\tag{4.46}
$$

In cylindrical coordinates, we have

$$\Delta \Phi = \frac{1}{r}\frac{\partial}{\partial r}\left(r\frac{\partial \Phi}{\partial r}\right) + \frac{1}{r^2}\frac{\partial^2 \Phi}{\partial \varphi^2} + \frac{\partial^2 \Phi}{\partial Z^2} \,. \tag{4.47}$$

In spherical coordinates, the Laplace operator is given in a similar fashion by:

$$\begin{aligned}\Delta \Phi &= \frac{1}{r^2}\frac{\partial}{\partial r}\left(r^2\frac{\partial \Phi}{\partial r}\right) \\ &+ \frac{1}{r^2 \sin\varphi}\frac{\partial}{\partial \varphi}\left(\sin\varphi\frac{\partial \Phi}{\partial \varphi}\right) + \frac{1}{r^2 \sin^2\varphi}\frac{\partial^2 \Phi}{\partial \vartheta^2} \,. \end{aligned}\tag{4.48}$$

4.3.5
Complete System of Equations of Motion for a Viscous, Incompressible Medium in the Absence of Heating

Using the Navier–Stokes equations (4.32), the equation of continuity (4.33) and the expressions obtained above for the differential operators in curvilinear coordinates, it is easy to record the complete system of equations of motion for a viscous and incompressible fluid. Thus, in cylindrical coordinates, one obtains

$$\begin{aligned}
&\frac{\partial U_r}{\partial t} + U_r\frac{\partial U_r}{\partial r} + \frac{U_\varphi}{r}\frac{\partial U_r}{\partial \varphi} + U_z\frac{\partial U_r}{\partial Z} - \frac{U_\varphi^2}{r} \\
&= f_{Vx} - \frac{1}{\rho}\frac{\partial p}{\partial r} + \nu\left(\frac{\partial^2 U_r}{\partial r^2} + \frac{1}{r^2}\frac{\partial^2 U_r}{\partial^2 \varphi} + \frac{\partial^2 U_r}{\partial Z^2}\right. \\
&\left. + \frac{1}{r}\frac{\partial U_r}{\partial r} - \frac{2}{r^2}\frac{\partial U_\varphi}{\partial \varphi} - \frac{U_r}{r^2}\right), \\
&\frac{\partial U_\varphi}{\partial t} + U_r\frac{\partial U_\varphi}{\partial r} + \frac{U_\varphi}{r}\frac{\partial U_\varphi}{\partial \varphi} + U_z\frac{\partial U_\varphi}{\partial Z} + \frac{U_\varphi U_r}{r} \\
&= f_{Vy} - \frac{1}{\rho r}\frac{\partial p}{\partial r} + \nu\left(\frac{\partial^2 U_\varphi}{\partial r^2} + \frac{1}{r^2}\frac{\partial^2 U_\varphi}{\partial^2 \varphi} + \frac{\partial^2 U_\varphi}{\partial Z^2} + \frac{1}{r}\frac{\partial U_\varphi}{\partial r}\right. \\
&\left. + \frac{2}{r^2}\frac{\partial U_r}{\partial \varphi} - \frac{U_\varphi}{r^2}\right), \\
&\frac{\partial U_z}{\partial t} + U_r\frac{\partial U_z}{\partial r} + \frac{U_\varphi}{r}\frac{\partial U_z}{\partial \varphi} + U_z\frac{\partial U_z}{\partial Z} \\
&= f_{Vz} - \frac{1}{\rho}\frac{\partial p}{\partial Z} + \nu\left(\frac{\partial^2 U_z}{\partial r^2} + \frac{1}{r^2}\frac{\partial^2 U_z}{\partial^2 \varphi} + \frac{\partial^2 U_z}{\partial Z^2} + \frac{1}{r}\frac{\partial U_z}{\partial r}\right),
\end{aligned}\tag{4.49}$$

and in spherical coordinates

$$\frac{\partial U_r}{\partial t} + U_r \frac{\partial U_r}{\partial r} + \frac{U_\varphi}{r}\frac{\partial U_r}{\partial \varphi} + \frac{U_\vartheta}{r\sin\varphi}\frac{\partial U_r}{\partial \vartheta} - \frac{U_\varphi^2 + U_\vartheta^2}{r}$$
$$= f_{Vr} - \frac{1}{\rho}\frac{\partial p}{\partial r} + \nu\left(\frac{\partial^2 U_r}{\partial r^2} + \frac{1}{r^2}\frac{\partial^2 U_r}{\partial \varphi^2} + \frac{1}{r^2\sin^2\varphi}\frac{\partial^2 U_r}{\partial \vartheta^2} + \frac{2}{r}\frac{\partial U_r}{\partial r} - \right.$$
$$\left. + \frac{\cot\varphi}{r^2}\frac{\partial U_r}{\partial \varphi} - \frac{2}{r^2}\frac{\partial U_\varphi}{\partial \varphi} - \frac{2}{r^2\sin^2\varphi}\frac{\partial U_\vartheta}{\partial \vartheta} - \frac{2U_r}{r^2} - \frac{2\cot\varphi}{r^2}U_\varphi\right),$$

$$\frac{\partial U_\varphi}{\partial t} + U_r \frac{\partial U_\varphi}{\partial r} + \frac{U_\varphi}{r}\frac{\partial U_\varphi}{\partial \varphi} + \frac{U_\vartheta}{r\sin\varphi}\frac{\partial U_\varphi}{\partial \vartheta} + \frac{U_r U_\varphi - U_\vartheta^2 \cot\varphi}{r}$$
$$= f_{V\varphi} - \frac{1}{\rho r}\frac{\partial p}{\partial \varphi} + \nu\left(\frac{\partial^2 U_\varphi}{\partial r^2} + \frac{1}{r^2}\frac{\partial^2 U_\varphi}{\partial \varphi^2} + \frac{1}{r^2\sin^2\varphi}\frac{\partial^2 U_\varphi}{\partial \vartheta^2} + \frac{2}{r}\frac{\partial U_\varphi}{\partial r}\right.$$
$$\left. + \frac{\cot\varphi}{r^2}\frac{\partial U_\varphi}{\partial \varphi} - \frac{2\cos\varphi}{r^2\sin^2\varphi}\frac{\partial U_\vartheta}{\partial \vartheta} + \frac{2}{r^2}\frac{\partial U_r}{\partial \varphi} - \frac{U_\varphi}{r^2\sin^2\varphi}\right),$$

$$\frac{\partial U_\vartheta}{\partial t} + U_r \frac{\partial U_\vartheta}{\partial r} + \frac{U_\varphi}{r}\frac{\partial U_\vartheta}{\partial \varphi} + \frac{U_\vartheta}{r\sin\varphi}\frac{\partial U_\vartheta}{\partial \vartheta} + \frac{U_r U_\vartheta + U_\varphi U_\vartheta \cot\varphi}{r}$$
$$= f_{V\vartheta} - \frac{1}{\rho r\sin\varphi}\frac{\partial p}{\partial \vartheta} + \nu\left(\frac{\partial^2 U_\vartheta}{\partial r^2} + \frac{1}{r^2}\frac{\partial^2 U_\vartheta}{\partial \varphi^2} + \frac{1}{r^2\sin^2\varphi}\frac{\partial^2 U_\vartheta}{\partial \vartheta^2} + \frac{2}{r}\frac{\partial U_\vartheta}{\partial r}\right.$$
$$\left. + \frac{\cot\varphi}{r^2}\frac{\partial U_\vartheta}{\partial \varphi} + \frac{2}{r^2\sin\varphi}\frac{\partial U_r}{\partial \vartheta} + \frac{2\cos\varphi}{r^2\sin^2\varphi}\frac{\partial U_\varphi}{\partial \vartheta} - \frac{U_\vartheta}{r^2\sin^2\varphi}\right). \tag{4.50}$$

Equations 4.49 are only valid for constant viscosity, that is, if $\nu = \text{const}$. Since ν often depends on the temperature, this condition means that $T = \text{const}$, and thus the process is isothermal. If $T = \text{const}$, but $\nu \neq \text{const}$, for example, $\nu = \nu(X, Y, Z)$, the ν can be moved under the differentiation operators so that the second to the last term in (4.35) takes on the following form:

$$\frac{\partial}{\partial X}\left(\nu\frac{\partial W}{\partial X}\right) + \frac{\partial}{\partial Y}\left(\nu\frac{\partial W}{\partial Y}\right) + \frac{\partial}{\partial Z}\left(\nu\frac{\partial W}{\partial Z}\right).$$

5
Foundations and Main Equations of Thermodynamics

5.1
Theorem of the Living Forces

We consider an arbitrary volume V that is bounded by the surface Σ and moves together with the particles of the continuum medium. We assume that the components of the stress tensor $\boldsymbol{P} = p^{ij}\boldsymbol{\varepsilon}_i\boldsymbol{\varepsilon}_j$ and of the velocity vector $\boldsymbol{V} = V^i\boldsymbol{\varepsilon}_i$ are continuously differentiable functions. We take the displacement vector $d\boldsymbol{r} = \boldsymbol{V}dt$, evaluate the scalar product of the equation of motion (3.29) with this vector and integrate the result over the volume V. Thus, one obtains

$$\int_V \rho\frac{d\boldsymbol{V}}{dt}\cdot\boldsymbol{V}dtd\tau = \int_V \rho\boldsymbol{f}_V\cdot d\boldsymbol{r}d\tau + \int_V \left(\nabla_j p^{ij}\right)V_i dtd\tau\,. \tag{5.1}$$

Since the mass element $dm = \rho d\tau$ is constant, the integral on the left-hand side is transformed in the following form:

$$\int_V \rho\frac{d\boldsymbol{V}}{dt}\cdot\boldsymbol{V}dtd\tau = \int_V \rho d\left(\frac{\boldsymbol{V}^2}{2}\right)dm = d\int_V \frac{\boldsymbol{V}^2}{2}dm = dE\,.$$

By definition, the quantity

$$E = \int_V \frac{\boldsymbol{V}^2}{2}dm \tag{5.2}$$

is called the kinetic energy of the volume V in the medium.

Since the mass forces can be divided into the internal forces $\boldsymbol{f}_m^{(i)}$ and the external forces $\boldsymbol{f}_m^{(e)}$, relative to the volume V, we have

$$\int_V \rho\boldsymbol{f}\cdot d\boldsymbol{r}d\tau = \int_V \rho\boldsymbol{f}_m^{(i)}\cdot d\boldsymbol{r}d\tau + \int_V \rho\boldsymbol{f}_m^{(e)}\cdot d\boldsymbol{r}d\tau = dA_m^{(i)} + dA_m^{(e)}\,. \tag{5.3}$$

Here, the infinitesimal amounts of work done by the internal and external forces due to infinitesimal displacements $d\boldsymbol{r}$ in the volume V are denoted by $dA_m^{(i)}$

Hydromechanics. Theory and Fundamentals. Emmanuil G. Sinaiski

ISBN: 978-3-527-41026-2

and $dA_m^{(e)}$. It must be noted that, in spite of their terminology, they are not total differentials. In addition, although the sum of internal forces vanishes, the work done by internal forces in total does not vanishes. We now consider the last integral in (5.1), that is,

$$\int_V \left(\nabla_j p^{ij}\right) V_i dt d\tau = \int_V \nabla_j \left(p^{ij} V_i\right) dt d\tau - \int_V p^{ij} \left(\nabla_j V_i\right) dt d\tau .$$

We transform the first integral on the right-hand side with the help of the Gauss theorem and the expression $\nabla_j V_j$ in the second integral to

$$\nabla_j V_i = \frac{1}{2}\left(\nabla_j V_i + \nabla_i V_j\right) + \frac{1}{2}\left(\nabla_j V_i - \nabla_i V_j\right) = e_{ij} + \omega_{ij} .$$

Thus, we obtain

$$\int_V \left(\nabla_j p^{ij}\right) V_i dt d\tau = \int_\Sigma p^{ij} V_i n_j dt d\sigma - \int_V p^{ij} e_{ij} dt d\tau - \int_V p^{ij} \omega_{ij} dt d\tau , \tag{5.4}$$

where n_j are the covariant components of the external unit normal vector on the boundary Σ. The first integral on the right-hand side of (5.4) is equal to

$$\int_\Sigma p^{ij} V_i n_j dt d\sigma = \int_\Sigma \boldsymbol{p}_n \cdot d\boldsymbol{r} d\sigma = dA_s^e , \tag{5.5}$$

and is referred to as the work done by external surface forces that act on Σ during infinitesimal displacements $d\boldsymbol{r}$ of the points on this surface. The sum of the two last integrals is called the work due to internal surface stress forces, that is,

$$-\int_V p^{ij} e_{ij} dt d\tau - \int_V p^{ij} \omega_{ij} dt d\tau = dA_s^{(i)} . \tag{5.6}$$

Since ω_{ij} are the components of an antisymmetric tensor ($\omega_{ij} = -\omega_{ji}$), the second integral in (5.6) vanishes if the stress tensor p^{ij} is symmetric. This is the case, provided that no distributed volume and surface force pairs exist within V and on Σ. Therefore, we have

$$-\int_V p^{ij} e_{ij} dt d\tau = dA_s^{(i)} . \tag{5.7}$$

Now, we can write (5.1) in the following form:

$$dE = dA_m^{(e)} + dA_m^{(i)} + dA_s^{(e)} + dA_s^{(i)} . \tag{5.8}$$

Equation 5.8 is called the theorem of living forces or theorem of the conservation of energy of a finite volume inside a deformed continuum. It is noted that in (5.8), only dE constitutes a total differential, while the other terms are generally infinitesimal quantities, that is, the infinitesimal amounts of work done by the corresponding forces due to infinitesimal continuous displacements $d\boldsymbol{r}$ that are known in any point of the volume V and the surface Σ of the continuum medium.

In the case of a symmetric stress tensor p^{ij}, the work done by the internal surface forces is determined by (5.7). The antisymmetric tensor ω_{ij} corresponds to an axial vortex velocity vector. Since the last integral in (5.4) vanishes for a symmetric stress tensor, the presence of vortices in a moving continuum medium does not have a direct influence on the infinitesimal work done by surface forces or on a change of the kinetic energy.

For an infinitesimal continuum volume, one can move from the integral form of (5.1) to the differential form. One first uses the mean value theorem in each integral, divides both sides of the equation by the mass of the volume and then takes the limit of $V \to 0$. Thus, one obtains

$$d\left(\frac{V^2}{2}\right) = \boldsymbol{f} \cdot d\boldsymbol{r} + \frac{1}{\rho}\nabla_j \left(p^{ij} V_i\right) dt - \frac{1}{\rho} p^{ij} \nabla_j V_i dt \,. \tag{5.9}$$

The quantity $V^2/2$ is called kinetic energy density and the quantities on the right-hand side are the density of mass forces, the density of external surface forces and the density of internal surface forces, respectively. The infinitesimal work done by internal mass forces does not appear in (5.9) since it approaches zero for $V \to 0$.

The theorem of the living forces is a direct consequence of the conservation of kinetic energy and thus constitutes a balance equation for the kinetic energy. It must be noted that (5.9) only relates to the conservation of mechanical energy and does not apply to the total energy. This equation will only apply to the total energy if mechanical energy is not transformed into heat or other forms of energy. Therefore, the law of conservation of energy has to deal with two kinds of energy: the mechanical energy and the non-mechanical energy.

We finally obtain the equation for the infinitesimal work done by the surface forces on an ideal gas. Since the stress tensor is $p^{ij} = -p\delta^{ij}$, we have

$$\frac{1}{dm} dA_s^{(i)} = -\frac{1}{\rho} p^{ij} e_{ij} dt = \frac{p}{\rho}\delta^{ij} e_{ij} dt = \frac{p}{\rho} e^{i}_{.i} dt = \frac{p}{\rho}\nabla \cdot \boldsymbol{V} dt \,. \tag{5.10}$$

From the equation of continuity (3.7), it follows that $\rho\nabla \cdot \boldsymbol{V} = -d\rho/dt$. Therefore,

$$\frac{1}{dm} dA_s^{(i)} = -\frac{p}{\rho^2}\frac{d\rho}{d\rho} = p d\left(\frac{1}{\rho}\right) = p dv \,, \tag{5.11}$$

with the specific volume $v = 1/\rho$. Using (5.10) and (5.11), one obtains the theorem of the living forces for an infinitesimal volume of an ideal gas, that is,

$$\begin{aligned} d\left(\frac{V^2}{\rho}\right) &= \frac{1}{dm} dA_m^{(e)} + \frac{1}{dm} dA_s^{(i)} + p d\left(\frac{1}{\rho}\right) \\ &= \boldsymbol{f}^{(e)} \cdot \boldsymbol{V} dt - \frac{1}{\rho}\nabla_k \left(p V^k\right) + p d\left(\frac{1}{\rho}\right) . \end{aligned} \tag{5.12}$$

5.2
Law of Conservation of Energy and First Law of Thermodynamics

The notions of the motion state of a physical system and of the state parameters lie at the foundation of both thermodynamics and continuum mechanics. Thus, the motion state of a physical system, that is, of a continuum volume, is specified with the help of parameters $\mu^1, \mu^2, \ldots, \mu^n$ in terms of which all the observed characteristics of the system can be completely defined. The makeup and the number of these parameters $\mu^1, \mu^2, \ldots, \mu^n$ are different for different continuum media.

To specify the state of the system, one needs to know the positions and velocities of all particles that make up the medium at all times. However, this requirement is unpractical since we are dealing with a huge number of state parameters. On the other side, it is well known that one can describe some systems in terms of just a few parameters. For example, the state of air at rest can be described in terms of just two parameters, that is, the pressure p and the mass density ρ or the temperature T.

The transition from a huge number of parameters that defines the state of the medium as a discrete system to a small number that defines the macroscopic state of the medium constitutes an important problem of physics, especially of continuum mechanics. Solving this problem requires additional hypotheses founded on the laws of probability as well as on experimental results. For example, in the case of a gas, it is possible to introduce the velocity $\boldsymbol{V}$, the temperature T and other parameters that are averaged over the statistical ensemble of particles, and over the volume and so on.

When researching the continuum medium, one considers the infinitesimal particles of the continuum as thermodynamic systems. The state parameters of these systems can be of a geometrical, physical or chemical nature. As examples, we refer to some of the most commonly used parameters, that is, the spatial coordinates, velocity, mass density, characteristics of deformation, temperature, the concentration of different components, the phase characteristics of the medium, the coefficient of viscosity, the coefficient of heat conduction, the coefficient of surface tension and many more.

In the following, we denote the variable parameters by μ^i and the physical constants by k^i. It is said that the complete set of quantities $\mu^1, \mu^2, \ldots, \mu^n$ and $k^1, k^2, \ldots, k^m$ constitute a basis for a continuum particle if they can be specified arbitrarily, and the other characteristics of the state of motion can be described as functions of this basis independent of the concrete problem. The density and temperature of a gas particle can, for example, be specified within known limits, but other thermodynamic parameters, such as pressure and entropy, are defined by them.

The description of the movement of a real system begins with the creation of a continuum model. For this, one has to choose a system of parameters that determine the physical state of the continuum particles. From a mathematical point of view, the state parameters μ^i and k^i are the arguments of the functions that appear in the closed system of equations describing the movement of the continuum. Thus, it is possible to introduce a state space whose coordinates are the state parameters. This space is called the phase space.

The set of all states of the medium that corresponds to a sequence of values of the state parameters over time is referred to as a process. If the system returns during a process to its initial state in the phase space, such a process is called a cycle. If the state parameters $\mu^1, \mu^2, \dots, \mu^n$ for a given particle constitute a continuous curve in the phase space, this process is referred to as a continuous process. In the opposite case, it is called a discontinuous process. For the case of a continuous process, a cycle corresponds to a closed curve.

During the course of a process, the system can interact with external bodies and processes. Therefore, in order to create a model of the continuum medium, it is necessary to establish the laws and mechanisms that govern the interactions of the medium particle with external particles and fields. In physics and mechanics, it is important to know the exchange of energy between the medium particle and those particles that surround it. When considering the behavior of continuum medium at the macro level, one has to introduce different forms of energy, although the ideas about those will change during the investigation.

We now consider an infinitesimal process, during which the state parameters $\mu^1, \mu^2, \dots, \mu^n$ change by infinitesimal quantities $d\mu^1, d\mu^2, \dots, d\mu^n$. The total flux of external energy to the infinitesimal particle during this process can be written as

$$d A^{(e)} + d Q^{(e)} + d Q^{**} = d A^{(e)} + d Q^{*} . \tag{5.13}$$

Here, we denote the work done by the external macroscopic mass and surface forces by $d A^{(e)}$, the inflow of heat by $d Q^{(e)}$, and the energy gained due to other interaction mechanisms by $d Q^{**}$, for example, due to the interaction with an external electro-magnetic field that contributes to the energy required for the magnetization and electric polarization of the medium.

Therefore, the last two terms on the left-hand side of (5.13) are combined into one and denoted by $d Q^*$. We note that the energy inflows $d Q^{(e)}$, $d Q^{**}$ and $d Q^*$ as well as the infinitesimal work of the external macroscopic forces $d A^{(e)}$ are generally not differentials of any functions, but infinitesimal quantities.

We now consider a process that takes place in phase space along the phase curve L_1 from point A with the values μ_0^i of the state parameters to point B with the values μ^i of the state parameters (see Figure 5.1). The total inflow of energy into the system during the process is

$$d A^{(e)} + d Q^{*} = \int\limits_{AB(L_1)} A_i d\mu^i + \int\limits_{AB(L_2)} Q_i d\mu^i . \tag{5.14}$$

Here, A_i and Q_i are the coefficients of the expansions $d A^{(e)} = A_i d\mu^i$ and $d Q^* = Q_i d\mu^i$. The first law of thermodynamics or the law of the conservation of energy can be formulated as the impossibility of constructing a perpetuum mobile, that is, to construct a machine that works cyclically and that can serve as a source of useful energy without using any sources of energy that are external to the machine.

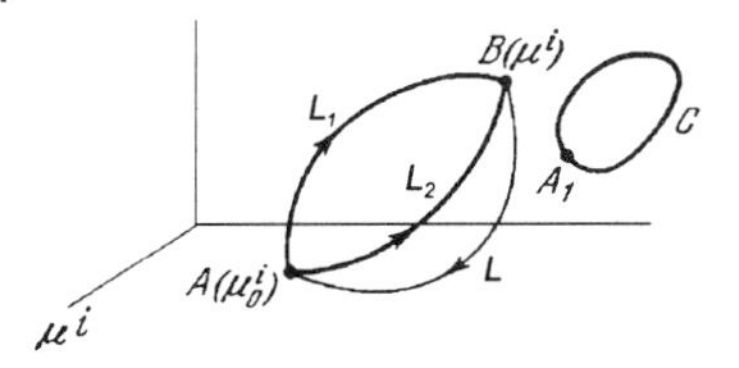

Figure 5.1 The law of conservation of energy.

Therefore, when the system goes through a cycle, the total inflow of energy from the outside is zero, that is,

$$\oint_C (A_i + Q_i)\, d\mu^i = 0\,. \tag{5.15}$$

From (5.14) and (5.15), it follows that

$$\begin{aligned} dA^{(e)} + dQ^* &= \oint_{L_1} (A_i + Q_i)\, d\mu^i \\ &= \oint_{L_2} (A_i + Q_i)\, d\mu^i = -\oint_{L} (A_i + Q_i)\, d\mu^i\,. \end{aligned}$$

Thus, the total inflow of energy from the outside is independent of the path of integration. It only depends on the initial state and final state of the system. If the initial state is fixed, the total inflow of energy will only depend on the final state of the system for all realized processes so that

$$dA^{(e)} + dQ^* = \Lambda\left(\mu^1, \mu^2, \ldots, \mu^n\right) - \Lambda\left(\mu_0^1, \mu_0^2, \ldots, \mu_0^n\right)\,. \tag{5.16}$$

Here, Λ is a function of the state parameters, that is, the total energy. From the fundamental law of thermodynamics (5.15), it follows that a function of state parameters Λ exists, whose total differential is a sum of infinitesimal amounts of work done by the macroscopic mass and surface forces, and infinitesimal energy inflow from the outside, and we have

$$d\Lambda = dA^{(e)} + dQ^* = (A_i + Q_i)\, d\mu^i\,.$$

On the other hand, the total energy of the infinitesimal volume element is equal to the sum of kinetic and internal energies.

$$\Lambda = (E + U)\,\rho d\tau\,,$$

where $E = V^2/2$ is the density of kinetic energy and U is the density of the internal energy.

The density of internal energy as well as the total energy Λ of the system can be specified up to an additive constant (see (5.16)) that can, for example, be chosen as the initial energy. It is noted that the concepts of internal energy and other thermodynamic quantities are generally required for studying the movement of a continuum. However, there are certain models of a continuum medium where the concept of internal energy is not needed to close the system of equations. For example, the concept of internal energy is not needed to close the system of equations for the examination of the movement of an ideal incompressible fluid and for the theory of elasticity without heat transfer.

Using the concept of internal energy introduced above, one can write the law of energy conservation as follows:

$$dE + dU_m = dA^{(e)} + dQ^{(e)} + dQ^{**} , \tag{5.17}$$

where dE is the change of kinetic energy, dU_m the change of internal energy, $dA^{(e)}$ the infinitesimal work done by external macroscopic forces, $dQ^{(e)}$ the infinitesimal heat inflow from the outside and dQ^{**} the infinitesimal inflow of other forms of energy from the outside.

Thus, we have on the one hand, the law of energy conservation (5.17) and on the other hand, the theorem of living forces (5.8). By subtracting the latter equation from (5.17), we obtain

$$dU_m = -dA^{(i)} + dQ^{(e)} + dQ^{**} = -dA^{(i)} + dQ^{*} . \tag{5.18}$$

Here, $dA^{(i)}$ is the infinitesimal work done by the internal mass and surface forces. For an infinitesimal continuum volume, it is equal to the infinitesimal work done by the surface forces alone since, as noted when deriving (5.9), the infinitesimal work done by the internal mass forces goes to zero for $V \to 0$. Therefore, we have

$$dA^{(i)} = dA_s^{(i)} . \tag{5.19}$$

Equation 5.18 is called the law of heat inflow.

If the process takes place continuously and slowly, which is called a quasistationary manner, one can neglect the change in kinetic energy, that is, $dE = 0$. Thus, (5.8) leads to $dA^{(e)} = -dA^{(i)}$ and (5.18) takes on the following form:

$$dU_m = dA^{(e)} + dQ^{*} . \tag{5.20}$$

We now obtain the equation for heat inflow for an infinitesimal particle by introducing the density of change in internal energy dU and the infinitesimal inflows of external energy per mass unit, that is,

$$dU = \lim_{\Delta m \to 0} \frac{dU_{\Delta m}}{\Delta m} , \quad dq^{(e)} = \lim_{\Delta m \to 0} \frac{dQ^{(e)}_{\Delta m}}{\Delta m} , \quad dq^{**} = \lim_{\Delta m \to 0} \frac{dQ^{**}_{\Delta m}}{\Delta m} .$$

Using (5.10), one obtains the density of work done by internal surface forces as

$$dq^{(i)} = -\frac{1}{\rho} p^{ij} e_{ij} dt = -\frac{1}{\rho} p^{ij} d\varepsilon_{ij} . \tag{5.21}$$

Here, we take into account that according to (2.59), $e_{ij} = d\varepsilon_{ij}/dt$ if the metric tensor of the initial state does not depend on the time. By dividing (5.17) by Δm and taking the limit to $\Delta m = 0$, we obtain

$$dU = \frac{1}{\rho} p^{ij} d\varepsilon_{ij} + dq^{(e)} + dq^{**} = \frac{1}{\rho} p^{ij} d\varepsilon_{ij} + dq^* \,. \tag{5.22}$$

5.3
Thermodynamic Equilibrium, Reversible and Irreversible Processes

A state is referred to as a state in thermodynamic equilibrium if the characteristics of this state can maintain their values permanently, provided the external conditions stay the same. The state of thermodynamic equilibrium corresponds to a point in phase space. The substantial derivatives in a state of thermodynamic equilibrium vanish.

Thermodynamic processes can take place infinitely slowly as well as also infinitely quickly. If a process is taking place infinitely slowly such that every intermediate state is an equilibrium state, this process is called an equilibrium process. In the equations that correspond to equilibrium processes, the velocity of parameter changes is of no interest, only the direction of change is important. Those processes that take place with finite velocities and which influence the velocities due to physical interactions are referred to as nonequilibrium processes.

When speaking of a process taking place in a system, one refers to a material object whose parameters are changing. Therefore, the definitions of an equilibrium process and of a stationary process of movement of the medium do not generally coincide. A process can be stationary, that is, its state parameters do not depend on the time in a given point of space ($\partial\mu^i/\partial t = 0$). At the same time, this process can be a nonequilibrium process, that is, the velocities of parameter change ($d\mu^i/dt \neq 0$) are finite.

A process that takes places from state A to state B is called reversible if, for all intermediate states, all equations stay valid when changing the signs of the infinitesimal energy changes. Therefore, if a reversible process takes place over a certain sequence of states, the system can also undergo the same process in the reverse direction, where the external energy inflows $dA^{(e)}$, $dQ^{(e)}$ and dQ^{**} in the forward and backward direction are only different in sign. Any process that does not have the above property is called irreversible. For an irreversible process, it is characteristic that the state parameters also include quantities that specify the direction of change for a few fundamental parameters.

All real macroscopic processes happen with finite velocities and thus their directions are important. Thus, in reality, all processes are irreversible. From a practical point of view, however, real processes can be approximated in many cases by reversible ones. For example, all generally known microscopic laws that describe the movement and interaction of elementary particles, among them Newtons law of gravitation as well as the law describing the electro-magnetic interactions, are reversible. During the transition from the microscopic to the macroscopic description of a process, one takes averages over all particles. This finally results in laws of a statistical nature. Therefore, the irreversibility of macroscopic processes is the price one pays for being able to describe a system consisting of a enormous number of particles in terms of a simple system that can be described in terms of a small number of macroscopic parameters.

One of the most important state parameters of a physical body is its temperature. It is known from the theory of molecular kinetics that the temperature can be considered as a quantity that is proportional to the mean energy of the chaotic heat movement of the molecules per degree of freedom of the molecules.

For equilibrium processes, the temperature of small volumes can be uniquely defined. For nonequilibrium processes, however, the concept of one temperature for the whole medium does not make sense. For example, in the case of a nonequilibrium plasma, one introduces two different temperatures for ions and electrons if these are separately in equilibrium states.

5.4
Two Parameter Media and Ideal Gas

A medium, for which all thermodynamic functions only depend on two thermodynamic state parameters, is called a two parameter medium. We now consider two such parameter media, namely, the pressure p and the mass density ρ. The internal energy of such a medium is a function of p and ρ: $U = U(p, \rho)$.

For an ideal compressible fluid or gas, the the work done by the internal surface forces per mass unit is according to (5.22) given by

$$\frac{1}{dm} dA_s^{(i)} = p\, d\left(\frac{1}{\rho}\right) . \tag{5.23}$$

For $dq^{**} = 0$, the equation (5.22) describing the heat inflow has the form:

$$dU + p\, d\left(\frac{1}{\rho}\right) = dq^{(e)} . \tag{5.24}$$

The gas is called ideal when p, ρ and T are related by Clapeyron's equation of state, that is,

$$p = \rho R T , \tag{5.25}$$

with the gas constant R, which is different for different gases.

All equations of the type (5.25) that relate the pressure, mass density, temperature and often other physical parameters of the medium are called equations of state. For an ideal gas, we have

$$U = c_v T + \text{const} ,$$

where c_v is the specific heat capacity for constant volume.

From (5.24), it follows that for constant specific volume $v = 1/\rho$, we have

$$\left(dq^{(e)}\right)_{v=\text{const}} = dU = c_v dT \, .$$

In the case of a process at constant pressure, we have

$$\left(dq^{(e)}\right)_{p=\text{const}} = c_v dT + d\left(\frac{p}{\rho}\right) = c_v dT + R dT = (c_v + R)\, dT \, . \tag{5.26}$$

We now define

$$\left(\frac{dq^{(e)}}{dT}\right)_{p=\text{const}} = c_p \, .$$

The quantity c_p is called the specific heat capacity at constant pressure. Then, we obtain from (5.26) the Maier formula

$$c_p = c_v + R \, .$$

Equation 5.22, regarding the heat inflow, generally contains the inflow of heat from the outside $dq^{(e)}$ that can be caused by the following physical phenomena:

- Conduction of heat, the process by which medium particles that are in contact align their mean heat energy. The heat transfer during heat conduction is caused by the uneven macroscopic temperature distribution in the continuum volume.
- Heat radiation and absorption of radiation, phenomena that are caused by the state changes of particles such as molecules, atoms, electrons and so on.
- Heat transfer because of the production of Joule's heat, dissipation of energy, condensation, evaporation and the absorption of heat during chemical reactions.

Since the solution of concrete problems using the heat inflow equations is rather difficult, one often uses additional assumptions in applications. In particular, the following ideal processes are frequently used:

- Adiabatic processes, these are processes that take place with out inflow of heat from outside or outflow of heat to the outside, that is, $dq^{(e)} = 0$. The concept of an adiabatic process is connected to a body that is insulated against heat transfer or to a very fast process for which it can be shown that heat transfer can be neglected. For adiabatic processes, the heat inflow relationship (5.24) has the form

$$dU + p d\left(\frac{1}{\rho}\right) = 0 \, . \tag{5.27}$$

For an ideal gas, we have $p = \rho R T$ and the equation of heat inflow becomes

$$\frac{c_v}{R} d\left(\frac{p}{\rho}\right) + p d\left(\frac{1}{\rho}\right) = 0 \, .$$

This gives

$$\frac{dp}{\rho} + \gamma p d\left(\frac{1}{\rho}\right) = 0\,, \quad \gamma = \frac{c_p}{c_v}\,,$$

with the adiabatic exponent γ. Integration yields

$$\frac{p}{\rho^\gamma} = \text{const}\,. \tag{5.28}$$

The plot of this function in the $p, 1/\rho$-plane is called Poisson's adiabatic.

- Processes with isothermic change of state. These are processes for which the temperature in all parts of the volume can be kept constant. Examples of such processes are those taking place in the presence of very intensive heat exchange or heat conduction, or processes for which the state of the volume changes very slowly.
  From the equation of state of the ideal gas, it follows that the isothermic in the $p, 1/\rho$-plane is a hyperbola p/ρ = const. The inflow of heat from the outside that makes the process isothermic can be obtained from (5.24). For an ideal gas, it is given by

$$\left(dq^{(e)}\right)_{T=\text{const}} = p d\left(\frac{1}{\rho}\right) = RT\rho d\left(\frac{1}{\rho}\right). \tag{5.29}$$

  From the above equation, we deduce for an ideal gas: $dq^{(e)} > 0$ for isothermal expansion and $dq^{(e)} < 0$ for isothermal compression. In Figure 5.2, the relative position of the isothermic and adiabatic for the ideal gas are shown.

- Barotropic processes are such processes for which the equation of state has the form $p = p(\rho)$, and where this functional relationship is the same for all particles of the continuum. In this case, one can take the equation $p = p(\rho)$ instead of using the equation of heat inflow. A popular form of this equation is

$$p = C\rho^n\,. \tag{5.30}$$

  Such a process is called a polytropic change of state and the exponent n is referred to as the polytropic exponent. The necessary heat inflow from the outside for a polytropic process can be obtained from (5.24) as

$$dq^{(e)} = dU + C\rho^n d\left(\frac{1}{\rho}\right) = c_v \frac{n - \frac{c_p}{c_v}}{n-1} dT\,. \tag{5.31}$$

For a given relationship $p = p(\rho)$, the work done by internal forces between the points A and B can be calculated for each process L_1. The work done by internal forces is, except for the sign, equal to the work done on the system or the work that the system performs on the outside. Therefore, the integral

$$\frac{1}{m} A = \int\limits_{AB(L_1)} p d\left(\frac{1}{\rho}\right) \tag{5.32}$$

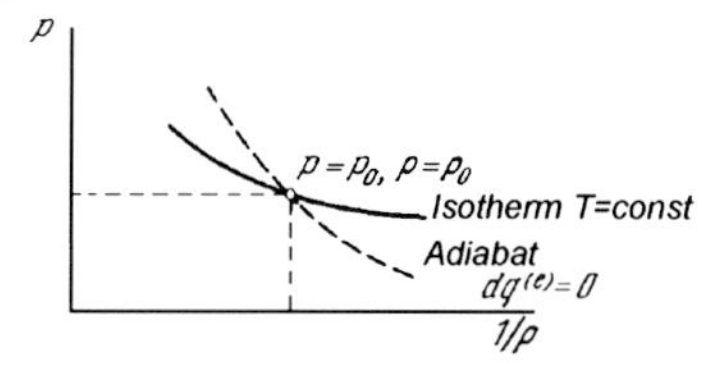

Figure 5.2 Relative position of isothermic and Poisson's adiabatic for ideal gas.

along the path L_1 in the $(p, 1/\rho)$-plane yields the work done per mass unit. For $A > 0$, it is the work done by the thermodynamic system on the outside during the equilibrium process. For $A < 0$, it is the total work that has to be done by the external forces on the thermodynamic system in order to realize the process L_1. From the above, it follows that when the internal energy of the medium $U = U(p, 1/\rho)$ is given, it is possible to calculate the total energy flow $Q^{(e)} > 0$ for every process L_1 with a given function $p = p(\rho)$, and that for $Q^{(e)} > 0$, it has to be added to the system, or for $Q^{(e)} < 0$, it has to flow out of the system to realize the process. Using (5.24) and (5.32), one obtains:

$$
\begin{aligned}
Q^{(e)} &= \int\limits_{L_1} d\,Q^{(e)} = \int\limits_{L_1} \left[d\,U + p\,d\left(\frac{1}{\rho}\right)\right] d\,m \\
&= \int\limits_{A}^{B} d\,U_m + A = U_{mB} - U_{mA} + A\,.
\end{aligned} \tag{5.33}
$$

5.5
The Second Law of Thermodynamics and the Concept of Entropy

The second law of thermodynamics states that there is no procedure that can transfer heat from a body with lower temperature to another body with higher temperature without any change to the other body. For every reversible cycle that a multiparameter system undergoes in phase space and a two parameter system undergoes in the phase plane $(p, 1/\rho)$, along a closed curve L_1, there is the following mathematical formulation of the second law of thermodynamics:

$$
\oint\limits_{L} \frac{d\,Q^{(e)}}{T} = 0\,. \tag{5.34}
$$

This equation follows from a thorough analysis of the Carnot cycle, and can be found in every textbook on thermodynamics. From (5.34), we deduce that for reversible processes, the integral $\int_A^B (d\,Q^{(e)})/T$ does not depend on the path of integration. For a fixed initial state A, the integral only depends on the final state B. We now consider a two-parameter medium and a reversible transition from the initial state A to the final state B. The following function of the state parameters

$$S(B) = S\left(p, \frac{1}{\rho}\right) = \int_A^B \frac{dQ^{(e)}}{T} + S(A) \tag{5.35}$$

is called entropy. It is known up to an additive constant. From (5.35), we know that $dS = dQ^{(e)}/T$. Together with the entropy S, one can also define the entropy density s. Then, we obtain, from the equation of heat inflow, that

$$ds = \frac{dq^{(e)}}{T} = \frac{dU}{T} + \frac{1}{T}p\left(\frac{1}{\rho}\right). \tag{5.36}$$

If the internal energy of the medium is known as a function of the state parameters p and $1/\rho$, one can also obtain the entropy as a function of the state parameters from (5.36). For example, we can consider an ideal gas with $p = \rho RT$, $U = c_v T$ and $c_v = \text{const}$. Then, we have

$$ds = \frac{c_v dT}{T} + \frac{R d(1/\rho)}{1/\rho} = d\ln\left[T^{c_v}\left(\frac{1}{\rho}\right)^R\right]$$

and one obtains after integration

$$\begin{aligned} ds &= c_v \ln\left[\frac{T}{\rho^{\gamma-1}}\right] + \text{const} = c_p \ln\left(\frac{T}{\rho^{(\gamma-1)/\gamma}}\right) + \text{const}_1 \\ &= c_v \ln\left(\frac{T}{\rho^\gamma}\right) + \text{const}_2 = c_v \ln\left(\frac{p}{\rho^\gamma}\right) - c_v \ln\left(\frac{p_0}{\rho_0^\gamma}\right) + s_0 . \end{aligned} \tag{5.37}$$

Equation 5.36 provides restrictions for the choice of the thermodynamic functions $U(p, \rho)$ and $T(p, \rho)$. Since ds is a total differential, the following equation must hold,

$$ds = \frac{\partial s}{\partial p} dp + \frac{\partial s}{\partial \rho} d\rho .$$

On the other hand, from (5.36), we obtain

$$ds = \frac{1}{T}\left(\frac{\partial U}{\partial p} dp + \frac{\partial U}{\partial \rho} d\rho\right) - \frac{p}{T\rho^2} d\rho = \frac{1}{T}\frac{\partial U}{\partial p} dp + \left(\frac{1}{T}\frac{\partial U}{\partial \rho} - \frac{p}{T\rho^2}\right) d\rho .$$

By equating the last two equations, one obtains

$$\frac{\partial s}{\partial p} = \frac{1}{T}\frac{\partial U}{\partial p}, \quad \frac{\partial s}{\partial \rho} = \frac{1}{T}\frac{\partial U}{\partial \rho} - \frac{p}{T\rho^2},$$

and from the obvious relationship $\partial^2 s/(\partial p \partial \rho) = \partial^2 s/(\partial \rho \partial p)$, it follows that

$$\frac{\partial}{\partial \rho}\left(\frac{1}{T}\frac{\partial U}{\partial p}\right) = \frac{\partial}{\partial p}\left(\frac{1}{T}\frac{\partial U}{\partial \rho} - \frac{p}{T\rho^2}\right)$$

or

$$\frac{\partial T}{\partial \rho}\frac{\partial U}{\partial p} = \frac{\partial T}{\partial p}\left(\frac{\partial U}{\partial \rho} - \frac{p}{\rho^2}\right) + \frac{T}{\rho^2} \,. \tag{5.38}$$

For a given $U(p,\rho)$, the function $T(p,\rho)$ must be the solution of (5.38). Therefore, the functions U and T cannot be chosen in an arbitrary manner.

For the case of an irreversible cycle, the second law of thermodynamics is formulated as the inequality

$$\oint_L \frac{dQ^{(e)}}{T} \leq 0 \,. \tag{5.39}$$

We assume $C = L + L_1$. If the process is irreversible along a path L between points A and B, the entropy can be calculated along a corresponding arbitrary reversible process between the points A and B, provided, of course, that such a process exists. Since for a reversible process the integral $\int_A^B (dQ^{(e)})/T$ does not depend on the integration path, it follows from (5.39) that we have along the irreversible path:

$$S(B) - S(A) \geq \int_L \frac{dQ^{(e)}}{T} \,. \tag{5.40}$$

For an irreversible process that connects two infinitely close states A and B, one has, according to (5.40),

$$TdS \geq dQ^{(e)} \,.$$

Therefore, it is possible to write TdS as follows:

$$TdS = dQ^{(e)} + dQ' \,, \tag{5.41}$$

where dQ' is the non-compensated heat. Is is obvious that we have $dQ' = 0$ for a reversible processes. In the following (see Section 5.8), it will be shown that the reverse contention is not generally valid, that is, the statement $dQ' = 0$ is a necessary, though insufficient condition for the process to be reversible.

The entropy can be defined statistically by the probability of the corresponding state. In statistical physics, we have the Boltzmann formula

$$S = k \ln P \,, \tag{5.42}$$

where k is the Boltzmann constant and P is a measure of the probability of the state under consideration, that is, defined as a relative number of the possible microscopic states consistent with the given macroscopic state. From (5.42), we obtain the additivity of entropy if the probability of a state is given by the product of probabilities of the states of individual parts of the system. That is only possible if the probabilities of the individual parts are independent. Equation 5.42 is valid for both equilibrium as well as for nonequilibrium states. Since those states that are realized experimentally have the highest probability, it follows that the entropy grows while the closed system moves to the equilibrium state.

In the case of system that is insulated against heat flow but is subject to external forces, the processes that this system undergoes are called adiabatic processes. For an adiabatic and reversible process, we have $dQ^{(e)} = 0$ and $TdS = 0$ or $S = \text{const.}$ Therefore, these processes are referred to as isentropic state changes. The reverse contention is also true: a process is reversible and isentropic only if it is adiabatic. However, for an adiabatic and irreversible process, we have $dS = dQ'/T$ and since $dQ' > 0$, the entropy, if it does change, can only grow. For non-adiabatic and irreversible processes, the entropy can both grow as well as decrease since the expression $TdS = dQ^{(e)} + dQ'$ can be both positive and negative.

From the above, it follows that the second law of thermodynamics is, in contrast to the first law of thermodynamics, not of a quantitative, but of a qualitative nature since it determines the direction of processes. For example, adiabatic and irreversible processes can only happen in that direction which leads to an increase in entropy. Therefore, states of an isolated system have maximum entropy when they are equilibrium states. As far as non-adiabatic processes are concerned, one cannot give a definitive answer as to whether the entropy grows or decreases. One can only state that the process takes place under the condition $dQ' > 0$.

5.6 Thermodynamic Potentials of Two-Parameter Media

We consider two-parameter media and the processes that take place in them. In Section 5.4, we discussed the case of taking the pressure p and the mass density ρ as the main parameters for a two-parameter medium. It was shown that if the internal energy is given as a function of p and ρ, the other thermodynamic state function T cannot be chosen arbitrarily. It must be a solution of the partial differential equation (5.38). This differential equation has many solutions, that is, for a given function $U(p, \rho)$, the function $T(p, \rho)$ and thus the thermodynamic properties are not uniquely defined. To remove this ambiguity, one has to choose one particular solution of (5.38) that is a concrete equation of state $T(p, \rho)$. Then, the entropy is defined by (5.36) up to an additive constant.

However, one can also take other thermodynamic main state parameters other than the pressure p and the mass density ρ. It turns out, that one can select pairs of state parameters, that is, ρ and s, p and s, and ρ and T and so on. Then, the questions arise as to whether or not one can define the internal energy as a function of these parameters in such a way that the other thermodynamic functions are completely and uniquely defined. This is indeed possible in a couple of cases. We now consider these cases, one after the other.

We assume that the internal energy is given as a function $U = U(\rho, s)$, that is, that the mass density ρ and entropy density s are chosen as main state parameters. Then, we have, because of (5.36),

$$dU = \left(\frac{\partial U}{\partial s}\right)_\rho ds + \left(\frac{\partial U}{\partial \rho}\right)_s d\rho = Tds - pd\left(\frac{1}{\rho}\right) = Tds + \frac{p}{\rho^2}d\rho \,. \quad (5.43)$$

Since ds and $d\rho$ are arbitrary, we have

$$T = \left(\frac{\partial U}{\partial s}\right)_\rho , \quad p = \rho^2 \left(\frac{\partial U}{\partial \rho}\right)_s . \tag{5.44}$$

Therefore, when $U = U(\rho, s)$ is given, one can uniquely determine T and p by (5.44). In this case, the internal energy serves as thermodynamic potential.

We now assume that the entropy density is given as a function $s = s(\rho, U)$, that is, the mass density and the internal energy are chosen as main state parameters. Then, we have, because of (5.43),

$$\frac{1}{T} = \left(\frac{\partial s}{\partial U}\right)_\rho , \quad \frac{p}{T} = \left[\frac{\partial s}{\partial (1/\rho)}\right]_v . \tag{5.45}$$

Thus, if the mass density ρ and internal energy U are chosen as main state parameters, T and p are uniquely defined by (5.45). In this case, the entropy density is a thermodynamic potential.

We now consider the case in which the main state parameters are ρ and T. We rewrite (5.43) as

$$d(U - Ts) = -s dT + \frac{p}{\rho^2} d\rho$$

and define

$$F = U - Ts . \tag{5.46}$$

The state function F is called free energy. Since

$$dF = \left(\frac{\partial F}{\partial T}\right)_p dT + \left(\frac{\partial F}{\partial \rho}\right)_T d\rho = -s dT + \frac{p}{\rho^2} d\rho ,$$

we have

$$s = -\left(\frac{\partial F}{\partial T}\right)_\rho , \quad p = \rho^2 \left(\frac{\partial F}{\partial \rho}\right)_T . \tag{5.47}$$

Therefore, when ρ and T are chosen as main state parameters, the free energy F serves as thermodynamic potential and the thermodynamic functions s and p are uniquely determined by (5.47).

When choosing p and T as main state parameters, it is convenient to rewrite (5.43) as follows:

$$d\left(U + \frac{p}{\rho}\right) = -T ds + \frac{dp}{\rho} .$$

We now introduce the state function i as

$$i(p, s) = U + \frac{p}{\rho} . \tag{5.48}$$

This function is called enthalpy or heat content. Then, we have

$$di = \left(\frac{\partial i}{\partial s}\right)_p ds + \left(\frac{\partial i}{\partial p}\right)_s dp = Tds - pd\left(\frac{1}{\rho}\right) = Tds + \frac{1}{\rho}dp \,.$$

This leads to

$$T = \left(\frac{\partial i}{\partial s}\right)_p \,, \quad \frac{1}{\rho} = \left(\frac{\partial i}{\partial p}\right)_s \,. \tag{5.49}$$

Thus, when s and p are chosen as main state parameters, the enthalpy i serves as thermodynamic potential and the thermodynamic functions T and p are uniquely determined by (5.49).

Finally, when choosing p and T as main state parameters, we rewrite (5.43) as

$$d\left(U - Ts + \frac{p}{\rho}\right) = -sdT + \frac{dp}{\rho}$$

and introduce the function ψ as

$$\psi\,(p, T) = U - sT + \frac{p}{\rho} \,, \tag{5.50}$$

which is referred to as Gibbs thermodynamic potential. Therefore, we have

$$s = \left(\frac{\partial \psi}{\partial T}\right)_p \,, \quad \frac{1}{\rho} = \left(\frac{\partial \psi}{\partial p}\right)_T \,. \tag{5.51}$$

Thus, when p and T are chosen as main state parameters, the thermodynamic functions s and ρ are uniquely determined by the Gibbs thermodynamic potential ψ.

It is important to note that the internal energy U and entropy density s are uniquely defined up to a constant, while the free energy F and Gibbs thermodynamic potential ψ are only defined up to a linear function of the temperature T.

For the special choices of state parameters considered above, the unique determination of state parameters is achieved by determining the corresponding thermodynamic potential. However, if choosing ρ and p, or T and s as main state parameters, no such thermodynamic potential exists. Therefore, the thermodynamic and mechanical properties of the ideal two parameter medium are then determined via a given function $U(p, s)$, $i(p, s)$, $F(p, s)$ or $\psi(\rho, T)$.

If these functions are given as functions of other state parameters, one has to use additional relationships, for example, an equation of state that is the solution of a certain partial differential equation, that is, (5.38). To determine the thermodynamic potentials of real gases and fluids, one uses the data of statistical physics and the results of the corresponding experiments. In the latter case, by measuring the heat capacities c_p and c_v. In terms of the thermodynamic potentials, the heat capacities of a compressible medium are expressed as follows:

$$\begin{aligned} c_p &= \left(\frac{\partial q}{\partial T}\right)_p = \left(\frac{\partial i}{\partial T}\right)_p = \left(\frac{\partial U}{\partial T}\right)_p - \frac{p}{\rho^2}\left(\frac{\partial U}{\partial T}\right)_p \\ &= \left(\frac{\partial U}{\partial \rho}\right)_T \left(\frac{\partial \rho}{\partial T}\right)_p + \left(\frac{\partial U}{\partial T}\right)_\rho - \frac{p}{\rho^2}\left(\frac{\partial \rho}{\partial T}\right)_p , \\ c_v &= \left(\frac{\partial q}{\partial T}\right)_v = \left(\frac{\partial U}{\partial T}\right)_\rho = \left(\frac{\partial i}{\partial \rho}\right)_T \left(\frac{\partial p}{\partial T}\right)_\rho + \left(\frac{\partial i}{\partial T}\right)_p - \frac{1}{\rho}\left(\frac{\partial p}{\partial T}\right)_\rho . \end{aligned} \tag{5.52}$$

5.7 Examples of Ideal and Viscous Media, and Their Thermodynamic Properties, Heat Conduction

The results obtained in the previous sections make it possible to formulate the system of equations describing the movement of a continuum system.

The system consists of:

- The equation of continuity

$$\frac{\partial \rho}{\partial t} + \nabla \cdot V = 0 , \tag{5.53}$$

- the momentum equation or equation of motion

$$\rho a^i = \nabla_j p^{ij} + \rho f^i , \tag{5.54}$$

- the equation of moments in the absence of distributed pair forces

$$p^{ij} = p^{ji} , \tag{5.55}$$

- the equation of heat inflow

$$dU = \frac{p^{ij}}{\rho} d\varepsilon_{ij} + dq^{(e)} + dq^{**} , \tag{5.56}$$

- and the second law of thermodynamics

$$T ds = dq^{(e)} = dq', \quad dq' > 0 . \tag{5.57}$$

To solve the applied problems of the movement of a continuum, one has to supplement this system with relationships that are due to the properties of the concrete continuum medium. We now consider a few important models of the continuum medium with respect to their thermodynamic properties.

5.7.1
The Model of the Ideal, Incompressible Fluid

For an ideal fluid, the equation $p^{ij} = -pg^{ij}$ (see (4.1)) is valid. For an ideal, homogeneous and incompressible fluid, the system of equations consisting of the equation of continuity

$$\nabla \cdot V = 0 \tag{5.58}$$

and the Euler equation

$$\frac{dV}{dt} = \rho f - \nabla p \tag{5.59}$$

form a closed system of equations with respect to the velocity V and the pressure p.

If the fluid is ideal, incompressible and inhomogeneous, a fifth equation has to be added to equations (5.48) and (5.49), namely,

$$\frac{d\rho}{dt} = 0 \,, \tag{5.60}$$

that serves to determine the mass density ρ from the Eulerian perspective. The work done by internal forces vanishes for the ideal incompressible fluid, that is,

$$dA^{(i)} = -\frac{p^{ij}}{\rho} e_{ij} dt = \frac{p}{\rho} g^{ij} e_{ij} dt = \frac{p}{\rho} g^{ij} e_{i} dt = \frac{p}{\rho} \nabla \cdot V = 0 \,.$$

Then, the heat inflow equation (5.22) or (5.56) takes on the following form for $dq^{**} = 0$:

$$dU = dq^{(e)} \,. \tag{5.61}$$

This equation can be considered as an equation for the internal energy U if $q^{(e)}$ is given, or as the heat transport equation within the fluid flow. The reversibility of mechanical processes is one of the assumptions for the model of an ideal, incompressible fluid. Therefore, we have $dq' = 0$ and because of (5.57), it follows that

$$dq^{(e)} = Tds \,. \tag{5.62}$$

The heat inflow equation (5.61) is then rewritten as

$$dU = Tds \,.$$

If the entropy density s is constant, than $U = \text{const}$ or $U = U(s)$, and from $dU/sd = T$, it follows that $T = T(s)$ or $U = U(T)$ and $s = s(T)$. For the heat capacity of the ideal incompressible, we obtain the following equation:

$$c = \frac{dq^{(e)}}{dT} = \frac{dU}{dT} = c(T) \,.$$

This leads to:

$$U = \int c(T) dT , \quad s = \int \frac{c(T)}{T} dT .$$

If $c = \text{const}$, we have

$$U = cT + \text{const} , \quad s = c \ln T + \text{const}$$

and the equation of heat inflow (5.62) results in the following equation:

$$\frac{dq^{(e)}}{dT} = T\frac{ds}{dT} = c(T)\frac{dT}{dt} = c(T)\left(\frac{\partial T}{\partial t} + V^i \frac{\partial T}{\partial X^i}\right) . \tag{5.63}$$

Thus, the model of the ideal incompressible fluid only depends on the two parameters, mass density ρ and heat capacity $c(T)$. To solve a concrete problem, it is necessary to specify, in addition to the external mass forces f, if they exist, the heat inflow from the outside and the initial and boundary conditions for the system.

The solution of the mechanical problem ((5.58)–(5.60)) will be obtained independently from the heat problem and the velocity distribution is obtained. Then, the heat problem is solved that contains the already found $\boldsymbol{V}$. Thus, the mechanical and heat problems have been separated.

5.7.2 The Model of the Ideal, Compressible Gas

We define the ideal gas as a two-parameter medium for which the following applies:

1. The stress tensor $\boldsymbol{P}$ has the form $p^{ij} = -pg^{ij}$, that is, $\boldsymbol{P}$ is a spherical tensor.
2. The internal energy depends on two parameters, namely, p and s.
3. All mechanical processes that are continuous are also reversible, that is, $dq' = 0$.

These three assumptions make it possible to write down the following closed system of equations for $\boldsymbol{V}, p, \rho, T$ and s:

- the equation of continuity

$$\frac{\partial \rho}{\partial t} + \nabla \cdot \boldsymbol{V} = 0 , \tag{5.64}$$

- the Euler equation

$$\frac{d\boldsymbol{V}}{dt} = \rho \boldsymbol{f} - \nabla p , \tag{5.65}$$

- the equation of heat inflow

$$dU = -pd\left(\frac{1}{\rho}\right) + dq^{(e)} , \tag{5.66}$$

- and the equation of state

$$T = \left(\frac{\partial U}{\partial s}\right)_\rho , \quad p = \rho^2 \left(\frac{\partial U}{\partial \rho}\right)_s . \tag{5.67}$$

If the mass forces f and the external heat inflow $dq^{(e)}$ are given, there are seven equations for the seven unknowns V^i, ρ, p, T and s. As an example of an ideal, compressible gas, we consider the ideal gas that is normally defined by the two functions

$$U = c_v T + \text{const} , \quad p = \rho R T$$

where the heat capacity c_v and the gas constant R are constant for a given gas.

The model of the ideal compressible gas can also be analyzed using any of the thermodynamic potentials $U(\rho, s)$, $F(\rho, T)$, $i(p, s)$, $\psi(p, T)$, for example, the internal energy

$$U = c_v T_0 \left(\frac{\rho}{\rho_0}\right)^{\gamma-1} \exp\left(\frac{s - s_0}{c_v}\right) + \text{const} .$$

If the process is reversible and adiabatic in every element of the medium, we have $dq' = 0$. Therefore, $s = s(\xi^1, \xi^2, \xi^3)$ where ξ^i are the Lagrange coordinates of the particle under consideration. This means that the entropy of each continuum particle is preserved. To solve concrete problems, one has to supply the entropy or one has to determine it from additional conditions that follow from the specifics of the problem.

If the entropy density is preserved during an adiabatic process for all particles, we have $s =$ const and from the equation of state it follows that p and T only depend on ρ, that is, the process is barotropic. In this case, the system of equations is closed for a given function $U(\rho, s)$.

If ρ and T are independent parameters, it is convenient to use the free energy $F(\rho, T) = U - Ts$. The corresponding equations of state are then given by (5.47). It is particularly convenient to use these equations for the investigation of a isothermic state change. In that case, if the function $T(\xi^1, \xi^2, \xi^3)$ is given or we have $T =$ const, we can obtain from (5.47) that the pressure is a certain function of ρ for every particle and only of ρ for $\nabla T = 0$.

Thus, if $F(\rho, T)$ is given, the system of equations will also be closed. In this case, the entropy is determined using the first equation of (5.47) and the external heat inflow that is needed to maintain an isothermic process can be determined from the heat inflow equation (5.62) which has, in this case, the following form:

$$dq^{(e)} = T ds = -T d\left(\frac{\partial F}{\partial T}\right)_\rho .$$

5.7.3 The Model of Viscous Fluid

Viscous fluid is defined by the relationship (4.17) between the components of the stress tensor p^{ij} and the components of the strain velocities e^{kl}, that is,

$$p^{ij} = -pg^{ij} + \tau^{ij}(e^{kl})$$

with the components τ^{ij} of the viscosity stress and $2e^{kl} = \nabla_k V_l + \nabla_l V_k$. If the dependence $\tau^{ij}(e^{kl})$ is linear and the fluid is isotropic, we have, due to (4.23) and (4.24),

$$\tau^{ij} = \lambda g^{ij}\nabla \cdot \mathbf{V} + 2\mu g^{ik} g^{jl} e^{kl}$$

and

$$p^{ij} = -pg^{ij} + \lambda g^{ij}\nabla \cdot \mathbf{V} + 2\mu g^{ik} g^{jl} e^{kl} .$$

It is customary to introduce, instead of λ, the coefficient of volume viscosity

$$\varsigma = \lambda + \frac{2}{3}\mu .$$

The viscosity coefficients μ and ς are different for different media. They can be constant or they can be a function of temperature of the scalar invariants of the tensor e^{ij}, or of other thermodynamic parameters. In the following, we will consider them as physical constants. For the model of the viscous fluid, one has to know the internal energy as a function of two parameters, that is, ρ and s. In addition, since the processes in a viscous fluid are normally irreversible, we have $dq' \neq 0$ and it must therefore be known as well.

The equation of motion of a viscous fluid with constant viscosity (Navier–Stokes equation) has the following form:

$$\frac{d\mathbf{V}}{dt} = \mathbf{f} - \frac{1}{\rho}\nabla p + \left(\frac{\varsigma}{\rho} + \frac{\nu}{3}\right)\nabla(\nabla \cdot \mathbf{V}) + \nu\Delta\mathbf{V} , \tag{5.68}$$

where $\nu = \mu/\rho$ is the kinematic coefficient of viscosity.

The equation of heat inflow reads according to the second law of thermodynamics

$$dU = -\frac{1}{dm} dA^{(i)} + T ds - dq' . \tag{5.69}$$

We now calculate the infinitesimal work done by the internal stresses per unit mass of the viscous fluid and obtain

$$\frac{1}{dm} dA^{(i)} = -\frac{p^{ij}}{\rho} e_{ij} dt = \frac{p}{\rho} g^{ij} e_{ij} dt - \frac{\tau^{ij}}{\rho} e_{ij} dt = \frac{p}{\rho}\nabla \cdot \mathbf{V} - \frac{\tau^{ij}}{\rho} e_{ij} dt .$$

Using the equation of continuity (5.64), we transform the right-hand side of the equation, that is,

$$\frac{1}{dm} dA^{(i)} = p d\left(\frac{1}{\rho}\right) - \frac{\tau^{ij}}{\rho} e_{ij} dt$$

and substitute the result into (5.69) to obtain

$$dU = -p d\left(\frac{1}{\rho}\right) + \frac{\tau^{ij}}{\rho} e_{ij} dt + T ds - dq' . \tag{5.70}$$

For a viscous, compressible fluid as well as for an ideal compressible fluid (gas), we postulate that

$$dU = -p d\left(\frac{1}{\rho}\right) + T ds . \tag{5.71}$$

This assumption that is referred to as Gibbs formula immediately allows one to determine the non-compensated heat. By comparing (5.70) and (5.71), one obtains

$$dq' = \frac{\tau^{ij}}{\rho} e_{ij} dt . \tag{5.72}$$

We now show that the existence of non-compensated heat dq' in the form (5.54) leads to the dissipation of mechanical energy during the movement of the viscous fluid. We consider the theorem of living forces for the viscous fluid. In analogy to the ideal fluid (see (5.11)), we have

$$\begin{aligned} d\left(\frac{V^2}{\rho}\right) &= \frac{1}{dm} dA^{(e)} + p d\left(\frac{1}{\rho}\right) - \frac{\tau^{ij}}{\rho} e_{ij} dt \\ &= \frac{1}{dm} dA^{(e)} + p d\left(\frac{1}{\rho}\right) - dq' . \end{aligned}$$

The last term on the right-hand side is negative since $dq' > 0$ and vanishes only for $e_{ij} = 0$. Therefore, the kinetic energy can only decrease due to the work done by viscous stresses.

Consider the process of heat transfer in a continuum volume. As it was noted before, the heat can be transported in different ways through the medium: heat conduction, radiation, electrical current, chemical reactions and so on. We now restrict ourselves to the heat conduction, that is, to the heat transport process due to the unevenness of the temperature distribution. We consider a continuum volume V. By heat conduction, the heat can only be transmitted through the surface of the volume. We denote the heat inflow by $\boldsymbol{q}$ via the surface element $d\Sigma$ that is characterized by the unit normal vector $\boldsymbol{n}(n_1, n_2, n_3)$ (see Figure 5.3). The total heat flow into the volume V is then given by

$$Q^{(e)} = -\int_{\Sigma} \boldsymbol{q} \cdot \boldsymbol{n} d\sigma dt = -\int_{V} \nabla \cdot \boldsymbol{q} d\tau dt .$$

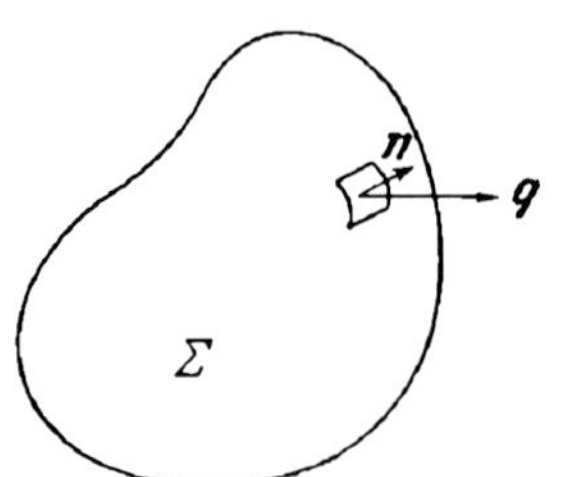

Figure 5.3 The vector of heat flow.

Thus, the heat flow into the infinitely small volume $d\tau$ is equal to

$$dQ^{(e)} = -\nabla \cdot \boldsymbol{q} d\tau dt$$

and per unit mass it is

$$dq^{(e)} = -\frac{1}{\rho}\nabla \cdot \boldsymbol{q} d\tau dt \,. \tag{5.73}$$

To close the system of equations, one needs to know the function $\boldsymbol{q}(T)$. The law that provides for this dependence is called Fourier's heat conduction law:

$$\boldsymbol{q} = -\kappa \nabla T \,. \tag{5.74}$$

Since $\boldsymbol{q}$ and ∇T point in opposite directions, we have $\kappa > 0$. The coefficient κ is called heat conduction coefficient. For a given fluid, it can be constant or depend on the temperature. We consider the particular case $\kappa = \text{const}$. Using (5.73) and (5.74), we then obtain

$$\frac{dq^{(e)}}{dt} = \frac{\kappa}{\rho}\nabla \cdot (\nabla T) = \frac{\kappa}{\rho}\Delta T \,. \tag{5.75}$$

Thus, the heat inflow equation takes on the following form:

$$\frac{dU}{dt} = -p\frac{d\left(\frac{1}{\rho}\right)}{dt} + \frac{\tau^{ij}}{\rho}e_{ij} + \frac{\kappa}{\rho}\Delta T \,. \tag{5.76}$$

By using (5.71), one obtains

$$T\frac{ds}{dt} = \frac{\tau^{ij}}{\rho}e_{ij} + \frac{\kappa}{\rho}\Delta T \,. \tag{5.77}$$

By substituting the expression for τ^{ij} from 4.24 into (5.77), one obtains

$$T\frac{ds}{dt} = \frac{\varsigma}{\rho}(\nabla \cdot \mathbf{V})^2 + \frac{2\mu}{\rho}\left[e^{ij}e_{ij} - \frac{1}{3}(\nabla \cdot \mathbf{V})^2\right]e_{ij} + \frac{\kappa}{\rho}\Delta T \,. \tag{5.78}$$

Equations 5.76, 5.77 or 5.78 now serve to determine the temperature distribution in the fluid, provided that $U(T,\rho)$ or $s(T,\rho)$ are known. For example, if

$$U = \int c_v(T) dT + \text{const} \,,$$

we have in Cartesian coordinates

$$\frac{dU}{dt} = c_v \frac{dT}{dt} = c_v \left(\frac{\partial T}{\partial t} + U \frac{\partial T}{\partial X} + V \frac{\partial T}{\partial Y} + W \frac{\partial T}{\partial Z} \right) . \tag{5.79}$$

Thus, when the external forces f are given, the equation of continuity (5.64), the Navier–Stokes equation (5.68), the equation of heat inflow (5.78) and the equation of state that depends on the choice of main parameters constitute a closed system of equations for the movement of a compressible, viscous and heat conducting fluid or a compressible, viscous, heat conducting and ideal gas.

5.8
First and Second Law of Thermodynamics for a Finite Continuum Volume

Above, the laws of thermodynamics have been formulated for an infinitesimal continuum volume. Assuming that the internal energy and entropy are additive over the particles of the finite continuum volume, one can write the first and second laws of thermodynamics in the following integral forms:

$$\frac{d}{dt} \int_V \rho \left(\frac{\mathbf{V}^2}{\rho} + U \right) d\tau = \int_V \boldsymbol{f} \cdot \mathbf{V} d\tau + \int_\Sigma \boldsymbol{p}_n \cdot \mathbf{V} d\sigma - \int_\Sigma q_n^* d\sigma + \int_V \frac{dq_{mas}^*}{dt} \rho d\tau , \tag{5.80}$$

$$\frac{dS}{dt} = \frac{d}{dt} \int_V \rho s d\tau = \int_V \frac{1}{T} \left(\frac{dq^{(e)}}{dt} + \frac{dq'}{dt} \right) d\tau . \tag{5.81}$$

Here, V is the finite moving continuum volume whose boundary is the surface Σ. The energy inflow consists of the term containing q^*, giving the energy inflow through Σ that is not due to the work done by mechanical forces, and the term containing dq_{mas}^*/dt, which gives the energy inflow due to a different kind of mass energy inflow (Joule's heat or the work of external mass pairs). We now consider the heat conduction process in the non-moving body. Assuming $dq' = 0$, one obtains, from (5.81) and (5.74),

$$\frac{dS}{dt} = -\int_V \frac{1}{T} \nabla \cdot \boldsymbol{q} d\tau = -\int_V \left(\nabla \cdot \frac{\boldsymbol{q}}{T} - \nabla \frac{1}{T} \right) d\tau = -\int_\Sigma \frac{q_n}{T} d\sigma + \int_V \boldsymbol{q} \cdot \nabla \frac{1}{T} d\tau . \tag{5.82}$$

This means that the entropy can both increase or decrease under the condition $q_n \neq 0$ because of the resulting heat flow through the surface Σ in the first term. In the case of an isolated body, we have $q_n = 0$ and thus

$$\frac{dS}{dt} = \int_V \boldsymbol{q} \cdot \nabla \frac{1}{T} d\tau = -\int_V \boldsymbol{q} \cdot \frac{\nabla T}{T^2} d\tau . \tag{5.83}$$

Since $\boldsymbol{q}$ and ∇ are related by (5.75), from (5.83), we obtain

$$\frac{dS}{dt} = \int\limits_V \kappa \frac{|\nabla T|^2}{T^2} d\tau \,. \tag{5.84}$$

Therefore, although the body is isolated against heat flow, we have $dS/dt > 0$ and the entropy grows because of the heat conduction. Thus, the entropy of the body increases in spite of the absence of an external heat inflow for $Tds = dq^{(e)}$ and $dq' = 0$, that is, the condition $dq' = 0$ is necessary, but not sufficient for having a reversible process. We now obtain the sufficient condition for the reversibility of a process. We assume that $dq' = 0$ and

$$dS = d_i S + d_e S \,. \tag{5.85}$$

The first term $d_i S$ is an infinitesimal quantity and represents the increment of the entropy because of internal irreversible processes where $d_i S > 0$ for the irreversible processes and $d_i S = 0$ for the reversible ones. The second term $d_e S$ is an infinitesimal quantity that represents the change in entropy because of the inflow of entropy due to the heat exchange with external bodies. The sign of $d_e S$ can be negative or positive for reversible processes. Therefore, we have two equations for $dq' = 0$, namely, (5.82) and (5.83). Thus, one obtains, using the notation of (5.85),

$$\frac{d_e S}{dt} = -\int\limits_\Sigma \frac{q_n}{T} d\sigma = -\int\limits_V \nabla \cdot \frac{\boldsymbol{q}}{T} d\tau \,,$$

$$\frac{d_i S}{dt} = \int\limits_V \boldsymbol{q} \cdot \nabla \frac{1}{T} d\tau = -\int\limits_V \boldsymbol{q} \cdot \frac{\nabla T}{T^2} d\tau \,.$$

The terms in the integrals represent the velocities of change for the entropy densities, $d_e s$ and $d_i s$, which are

$$d_e s = -\frac{1}{\rho} \nabla \cdot \frac{\boldsymbol{q}}{T} dt \,, \quad d_i s = -\boldsymbol{q} \cdot \frac{\nabla T}{T^2} dt \,. \tag{5.86}$$

This means that $d_i s = 0$, if $\nabla T = 0$. Therefore, the absence of a temperature gradient is the sufficient condition for the reversibility of the heat conduction process.

5.9 Generalized Thermodynamic Forces and Currents, Onsager's Reciprocity Relations

The equations for $d_e s/dt$ and $d_i s/dt$ that were obtained in the last section are only suitable for the heat conduction process. However, it is possible to obtain similar relationships for many other classes of irreversible processes. We write these in the common form

$$\frac{d_e s}{dt} = -\frac{1}{\rho} \nabla \cdot \boldsymbol{S} \,, \quad \rho \frac{d_i s}{dt} = \Phi \sum_k F_\alpha I^\alpha \,. \tag{5.87}$$

Here, $\boldsymbol{S}$ is the entropy current vector, Φ is the dissipation function that influences the value of the irreversible entropy increase due to internal processes, F_α are the generalized thermodynamic forces, and I^α are the generalized thermodynamic currents. By comparing (5.86) and (5.87), for the heat conductions processes, one obtains

$$\boldsymbol{S} = \frac{\boldsymbol{q}}{T}\,, \quad F_\alpha = -\frac{1}{\rho T^2}\frac{\partial T}{\partial X^\alpha}\,, \quad I^\alpha = q^\alpha\,.$$

For the movement of viscous heat conducting fluid, we have

$$\frac{d_i s}{dt} = \frac{1}{\rho}\frac{\tau^{ij} e_{ij}}{T} - \frac{\boldsymbol{q}}{\rho}\cdot\frac{\nabla T}{T^2} = \frac{dq'}{T} - \frac{\boldsymbol{q}}{\rho}\cdot\frac{\nabla T}{T^2}\,, \tag{5.88}$$

$$\Phi = \kappa\frac{(\nabla T)^2}{T^2} + \left\{\varsigma\,(\nabla\cdot\mathrm{V})^2 + 2\mu\left[e^{ij} e_{ij} - \frac{1}{3}(\nabla\cdot\mathrm{V})^2\right]\right\}\frac{1}{T}\,. \tag{5.89}$$

The positivity of the function Φ is ensured by the conditions $\kappa > 0$, $\varsigma > 0$, and $\mu > 0$. The dissipation function Φ can be considered as a function of F_α or I^α since the generalized thermodynamic forces F_α and the generalized thermodynamic currents I^α are related. In the case of heat conduction, for example, this connection is given by Fourier's law of heat conduction. The generalized thermodynamic forces I^α linearly depend on the generalized thermodynamic forces. Therefore, one can write

$$I^\alpha = L^{\alpha\beta} F_\beta \quad \text{and} \quad \Phi = L^{\alpha\beta} F_\alpha F_\beta\,. \tag{5.90}$$

Therefore, the dissipation function Φ is a quadratic form of the generalized thermodynamic forces F_α. The theory that is based on the assumption that the dissipation function is a quadratic form of the generalized thermodynamic forces provides the content of Onsager's theory. According to Onsager's theorem, the matrix of the quadratic forms $L^{\alpha\beta}$ is symmetric, that is, $L^{\alpha\beta} = L^{\beta\alpha}$.

6
Problems Posed in Continuum Mechanics

In the previous chapters, the universal equations of mechanics and thermodynamics were obtained and described. The equations are the main relationships that provide the foundations for creating concrete models of the continuum.

For continuous movement, these are equations that describe the differential form of the fundamental laws of the mechanics and thermodynamics. The integral formulations of the laws were also given. For a continuous and sufficiently smooth movement, both formulations are equivalent. For some problems, however, one has to consider the non-continuous space or time characteristics of the movement. In this case, the differential formulation is still used in that part of the movement which is continuous, though on discontinuity surfaces or curves one requires additional relationships that can only be obtained using the integral formulation.

The study of the continuum movement begins with the choice of the continuum model. For this, one needs to introduce additional hypotheses. After choosing the model, one has to select within the context of this model a particular phenomenon or a class of phenomena. This also applies to the additional conditions. For example, if the model of an ideal fluid has been chosen, one can select the incompressible or compressible case. In this context, one can also consider various fluids, that is, water, oil, gas, and so on, movements in various areas, for example, in tubes, canals, or on the surfaces of the fluid and so on. In all cases, the system of equations is the same, but the corresponding solutions have different shapes. This means that besides the system of equations, the additional conditions have to be formulated. In the next section, we consider these conditions.

6.1
Initial Conditions and Boundary Conditions

The solution of the mathematical problem is a function of the points in the continuum volume and of time. The time interval of the movement can start at $t = t_0$ and can be finite or infinite. The domain of the continuum volume can be given or, in some cases, not known a priori. For example, when a fluid flows through a totally filled tube, that area is known. However, if the tube is not totally filled, the

Hydromechanics. Theory and Fundamentals. Emmanuil G. Sinaiski

ISBN: 978-3-527-41026-2

area of the fluid is not known a priori and has to be determined during the solution with the help of additional conditions on the free surface Σ of the volume.

The domain V can consist of parts that are known a priori, that is, the bottom surfaces, the walls of the containers, the moving surfaces of the bodies that are moving within the fluid, and of those parts that are not known a priori, for example, the free surface of the fluid. The domain V can be finite or infinite. The last case is an approximation for some problems for which the dimensions of the body are small in comparison the domain that is occupied by the fluid.

In this case one needs to define conditions at infinity. Sometimes, one chooses as the condition the assumption that the movement of the body produces small disturbances in the undisturbed fluid flow that vanish at infinity, or as it is normally expressed, that the disturbances evaporate at infinity. This means that at infinity, the disturbed state approaches the undisturbed state.

Within the domain V, one sometimes introduces so-called singular points: point sources, point sinks, dipoles and multipoles to model concentrated flows of the fluid, charges, external forces, transfer of energy and absorption of energy. Such points can also model the presence or effect of bodies which are far away from the domain under consideration.

In the theory of differential equations, the Cauchy problem is of great importance. To solve it, one requires the initial conditions whose form and number depends on the order of the system of differential equations. Considering the problem of the non-stationary movement of an ideal, homogeneous and incompressible fluid, it is sufficient to provide the Velocity distribution for the full domain V at the initial point in time.

If the domain V has a boundary Σ, it is necessary to formulate special conditions on Σ. These conditions are called boundary conditions. These are formulated based on the physical principles, taking into account the properties of the continuum model and the interactions with external bodies.

We assume that the position and movement of the boundary Σ or of one of its sections Σ_1 is known. Then, if one approaches the boundary Σ from inside the medium under consideration, a contact between the medium and Σ takes place. Therefore, the displacement of the individual points of the medium as well as those on the boundary have to be connected via the preservation of contact. When there is no slip of the continuum points along the tangent to Σ, the following conditions for the displacement vector $\boldsymbol{W}$ and the velocity $\boldsymbol{V}$ must be fulfilled:

$$\boldsymbol{W}_{\text{medium}} = \boldsymbol{W}_{\text{boundary}}\,, \quad \boldsymbol{V}_{\text{medium}} = \boldsymbol{V}_{\text{boundary}}\,. \tag{6.1}$$

The conditions (6.1) are called non-slip-conditions and are used in the theory of viscous fluids. In the theory of elasticity, the first condition is of importance, while the second condition is used in the theory of the viscous fluid.

The number of boundary conditions depends on the order of the equations. Therefore, the boundary conditions are different for different models. For example, the equations for the ideal incompressible fluid, the Euler equations, only contain the first derivatives of the velocities and three non-slip conditions are too many for an ideal fluid. Therefore, it is assumed that continuum particles can slip on the boundary and the boundary conditions (6.1) are replaced by the condition that the particles cannot flow through each other, that is,

$$V_{n,\text{medium}} = V_{n,\text{boundary}} , \tag{6.2}$$

if the boundary cannot be penetrated. Here, V_n is the normal component of the velocity V on the boundary Σ. Since there is a slip of the continuum particles at the boundary in the case of the ideal fluid, the tangential components on medium and boundary are different

$$V_{t,\text{medium}} \neq V_{t,\text{boundary}} . \tag{6.3}$$

The Navier–Stokes equations for a viscous fluid contain the second derivatives of the velocity with respect to the coordinates. Therefore, the boundary conditions contain both the normal as well as the tangential components V_t, that is,

$$V_{n,\text{medium}} = V_{n,\text{boundary}} , \quad V_{t,\text{medium}} = V_{t,\text{boundary}} . \tag{6.4}$$

A large class of movements is of a potential nature for which $V = \nabla\varphi$. For such movements, the boundary condition (6.2) has the form

$$\left(\frac{\partial\varphi}{\partial n}\right)_{\text{Medium}} = \left(\frac{\partial\varphi}{\partial n}\right)_{\text{Boundary}} . \tag{6.5}$$

In the case of a fixed boundary, we have

$$\left(\frac{\partial\varphi}{\partial n}\right)_{\text{Boundary}} = 0 . \tag{6.6}$$

For the equation of heat inflow, one either provides the temperature or the heat flow, or a linear combination thereof.

For some problems, the boundary Σ or some part of it is not known a priori, and has to be determined as part of the solution. To determine the boundary in this case, the boundary conditions for the velocity (6.1) are insufficient.

In the theory of elasticity, one normally provides the external load on the boundary leading to the following boundary condition:

$$\boldsymbol{p}_n = \boldsymbol{p}_{nn} + \boldsymbol{p}_{nt} = \boldsymbol{f}(M, t). \tag{6.7}$$

Here, $\boldsymbol{p}$ is the density of surface forces and M is a point of the boundary Σ. The fluid surface that borders on gas or air is called the free boundary. On such a boundary, the surface stress is given by the pressure p_0. Therefore, we have on the free boundary:

$$\boldsymbol{p}_{nn} = -p_0\boldsymbol{n}, \boldsymbol{p}_{nt} = 0 . \tag{6.8}$$

For an ideal fluid, the boundary condition (6.8) is reduced to

$$p = p_0 . \tag{6.9}$$

If the boundary is a surface where viscous fluids with different properties are in contact, the additional boundary condition means that both the normal and tangential components of the surface stresses have be the same in both media, that is,

$$(p_{nn})^{(1)} = (p_{nn})^{(2)} , \quad (p_{nt})^{(1)} = (p_{nt})^{(2)} . \tag{6.10}$$

One often deals with problems having mixed boundaries, where one part of the boundary consists of fixed walls and the other of free or moving surfaces. Then, the boundary condition are given by a corresponding combination of the above boundary conditions.

At the boundary between two different media that are in different states of matter, one sometimes has to take into account the surface tension that is generated due to the deformation of the boundary surface (boundary between different states of matter). Such a problem occurs, for example, for the movement of a thin fluid layer on the surface of a body or another fluid. In order to take the surface tension into account, one has to adjust the boundary conditions (6.10). We now discuss this problem further.

We consider the boundary between a fluid and another medium. This boundary is called a separation surface for equal states of matter and a phase boundary for different states of matter. Depending on the media on both sides of the boundary, we can have a fluid–solid, a gas–solid, a fluid–gas or a fluid–fluid interface. If both media at the surface are in a thermodynamic and dynamic equilibrium, temperatures, velocities and surface forces must be equal on both sides of the boundary. Consider the force due to surface tension on the boundary fluid–gas or fluid–fluid. It is known that the bent material surface will experience a stress in the direction of the normal to the surface. We construct the tangential plane through a point O of the surface and choose a Cartesian system of coordinates such that the origin coincides with O, the X and Y axes lie in the plane and the Z-axes lies along the normal on the surface in O. Then, the equation of this surface can be written as

$$Z - \varsigma(X, Y) = 0 ,$$

where ς and its first derivative vanish in O. In the vicinity of the point O, the unit normal vector on the surface is given by

$$\boldsymbol{n}\left(-\frac{\partial \varsigma}{\partial X}, -\frac{\partial \varsigma}{\partial Y}, 1\right),$$

which is accurate to the first order relative to the first partial derivatives. The force that acts on the surface in the vicinity of O is equal to

$$\boldsymbol{F}_{\text{surf}} = -\sigma \oint_L \boldsymbol{n} \times d\boldsymbol{s} , \tag{6.11}$$

where $\boldsymbol{s}$ is the linear vector-element of the closed curve L that encloses the surface σ in the coefficient of surface tension. The coefficient of surface tension has the following meaning: The work needed to increase the area of a separation surface by dS is given by σdS. Therefore, one can consider σ as the free energy per unit surface element. The term surface tension means that every line on the separation surface experiences a force per length unit which is directed along the tangent of the separation surface. For a plane surface, the resulting force vanishes since the unit normal vector $\boldsymbol{n}$ is the same everywhere. However, for a bent surface, the force is acting along the surface normal in O, that is, along the Z-axis. Therefore, one can transform the curve integral (6.11) with the help of Stokes' theorem (see Appendix A.5) into a surface integral and then apply the mean value theorem for a small surface element to obtain:

$$\begin{aligned} \boldsymbol{F}_{fl} &= -\sigma \oint_L \boldsymbol{n} \times d\boldsymbol{s} = -\sigma \oint_L \left(-\frac{\partial \varsigma}{\partial X} dY + \frac{\partial \varsigma}{\partial Y} dX \right) \\ &= \sigma \left(\frac{\partial^2 \varsigma}{\partial X^2} + \frac{\partial^2 \varsigma}{\partial Y^2} \right)_O dS \,. \end{aligned}$$

From this, we conclude that the surface tension is equivalent to an effective pressure, the capillary pressure, defined as

$$p_{\text{cap}} \left(\frac{\partial^2 \varsigma}{\partial X^2} + \frac{\partial^2 \varsigma}{\partial Y^2} \right)_O = \sigma \left(\frac{1}{R_1} + \frac{1}{R_2} \right) , \tag{6.12}$$

where R_1 and R_2 denote the main radii of curvature at the point O. The separation surface can be in a dynamical equilibrium if the capillary pressure compensates for the pressure difference Δp between both sides of the separation surface. Therefore, for every point of the surface, the jump in pressure from one side to the other side where the center of curvature of the surface is must be given by

$$\Delta p = p_{\text{kap}} = \sigma \left(\frac{1}{R_1} + \frac{1}{R_2} \right) . \tag{6.13}$$

Now, it is possible to formulate the conditions of equilibrium on the separation surface. The first condition is the kinematic condition which depends on the continuity of the separation surface. If the boundary does not have a discontinuity, it remains a fluid surface for both media. Therefore, the velocity components must be continuous while crossing the surface (see condition (6.4)). The second condition means that the stress difference between the two media has to be equal to the capillary pressure. By projecting the stress in the normal and tangential directions, one obtains

$$(p_{nn})^{(1)} = (p_{nn})^{(2)} + \sigma \left(\frac{1}{R_1} + \frac{1}{R_2} \right) , \quad (p_{nt})^{(1)} = (p_{nt})^{(2)} \,. \tag{6.14}$$

6.2 Typical Simplifications for Some Problems

The mathematical task of solving the system of equations for the continuum movement in the Eulerian perspective consists of determining the unknown functions such as pressure, velocity, mass density, temperatures and so on of four variables, that is, the coordinates X^i and the time t. This task is normally very difficult and sometimes insoluble. Therefore, when solving concrete tasks, one has to use additional assumptions that are based on the physical circumstances of the problem and apply simplifications that are motivated by valid physical arguments. These simplifications consist mostly of a reduction of the number of independent variables and the assumption of certain relationships between the desired functions of a few variables. We now consider some of the most commonly used simplifications.

Stationary Movement

In a few cases that are important in practice, it is possible to assume that the movement is stationary. Then, one can eliminate the time t from the independent variables using the Eulerian perspective. This results in dropping the partial derivatives with respect to t and initial conditions are not needed, making the task much easier.

Coplanar Movement

A movement is called coplanar if a Cartesian system of coordinates can be chosen such that the velocities of all continuum particles are parallel to the (X, Y)-plane. Then, all characteristics of the movement will only depend on (X, Y) and possibly also on t. The mathematical theory of the coplanar movement is currently well developed and is the basis for the approximate methods of solution for spatial problems.

Potential Movement of the Incompressible Fluid

The considerable success of the methods applied for the coplanar movement of an incompressible fluid is due to the fact that the velocity potential $\Phi(X, Y)$ for a potential movement is a harmonic function. This means that ϕ satisfies the Laplace differential equation

$$\Delta\varphi = \frac{\partial^2\varphi}{\partial X^2} + \frac{\partial^2\varphi}{\partial Y^2} = 0 . \tag{6.15}$$

The harmonic function $\phi(X, Y)$ is connected to the conjugate function $\psi(X, Y)$ according to the Cauchy–Riemann equations

$$\frac{\partial\psi}{\partial X} = -\frac{\partial\varphi}{\partial Y} , \quad \frac{\partial\psi}{\partial Y} = \frac{\partial\varphi}{\partial X} . \tag{6.16}$$

The function ψ is called flow function and we have $\psi = \text{const}$ along each flow line. The conditions (6.16) allow one to introduce the analytical function

$$W(Z) = \phi + \psi i , \quad Z = X + Yi , \tag{6.17}$$

which is referred to as the characteristic function. Finding the velocity potential $\phi(X, Y)$, therefore, is equivalent to the task of determining the function $W(Z)$, which can be accomplished with the help of the theory of complex functions. Therefore, there is only one independent variable Z which simplifies the task considerably.

Axially Symmetric Movement

A large class of tasks are those with axial symmetry. For those tasks, one assumes that an axially symmetric system of coordinates can be chosen in which all functions depend on r, Z and t, and the angular coordinate is not important. All equations and solutions are invariant relative to rotations by any angle around the Z-axis. Axially symmetric tasks include the problems of strength and movement of axially symmetric bodies (tubes and shells etc.), and the problems of fluid or gas flows around axially symmetric bodies and many more.

One-Dimensional Non-stationary Movements

In these processes, only one spatial coordinate and the time are of importance. The one-dimensional non-stationary movements of a fluid only consist of the following three cases:

- The plane wave movements, that is, those movements for which all characteristics are the same on the planes $X = \text{const}$, the wave phase plane. The displacements of particles happen in those directions which are orthogonal to the plane $X = \text{const}$.
- The cylindrical wave movements for which all characteristics only depend on ρ and t, and that are the same on the cylindrical surfaces $\rho = \text{const}$. The displacements happen in the direction of the ρ-coordinate.
- The spherical wave moments for which all characteristics only depend on r and t, and are the same on the spherical surfaces $r = \text{const}$. The displacements happen in the direction of the r-coordinate.

In the framework of one-dimensional movements, many problems of fluid and gas flow in tubes and the propagation of sound and explosion waves are solved.

Similar Solutions

The methods discussed above to simplify the equations eliminate one to three independent variables. However, this is very seldom possible. In some cases, one manages to reduce the number of variables not by the elimination of variables, but by creating new variables as combinations of the old ones, and of which there are fewer.

As one example of this method, we consider the so called similar solutions that allow one to replace the four variables X, Y, Z and t by the three variables X/t^{α}, Y/t^{α} and Z/t^{α} with a constant α. Using this method for some one-dimensional non-stationary movements whose characteristics depend on the arguments X and t, allows one to have only one variable $\eta = X/t^{\delta}$, and thus to transform the corresponding partial differential equation into a ordinary differential equation for η.

Examples of tasks where this method is used are the problem of heat propagation from a point source and the problems of point explosion, that have spherical symmetry, and for which all functions (velocity, pressure and temperature and so on) only depend on the similar variable $\eta = r/t^{\delta}$. Normally, there are no characteristic time and space dimensions, and therefore, the time and space intervals are either infinite or half-infinite. In the following, it will be shown that the similarity coefficient can sometimes be established with the help of the theory of dimensionality.

Linearization of Equations

The problems of continuum mechanics are generally nonlinear. This nonlinearity is due to the nonlinearity of the equations as well as the nonlinearity of the boundary conditions. The latter type of nonlinearity is especially evident for the tasks with an unknown boundary. The nonlinearity makes the tasks very complicated. One method of simplifying the nonlinear equations is linearization. The main idea consists of replacing the nonlinear equations and boundary conditions by linear approximate equations and approximate boundary conditions. A very popular method is the method of small perturbations.

For some problems of continuum mechanics, the movements and corresponding processes can be considered as consisting of a small perturbation of the equilibrium state and the main movement. For example, the deformations of an elastic body are small. Therefore, in the theory of elasticity, the assumption of infinitesimal deformations is widespread, for which the products of small quantities can be neglected. In wave theory, one frequently considers small perturbations of the free surface, which is not much different from a horizontal plane. In this case, the values of speed and the corresponding displacements are small. In aerodynamics, one often considers the movement of slender bodies (projectiles, wings and so on) under a small angle. In this case, the velocities that the bodies induce in the surrounding air are small and the assumption that the desired functions are small can be acceptable.

The linearization consists of the following steps.

- The linearization of the equations. One assumes for the unknown functions:

$$f = f_0 + f_1 \,, \quad f_1 \ll f_0 \,, \tag{6.18}$$

  where f_0 is the value of the function f in the undisturbed movement and f_1 is a small perturbation. Here, f are the components of velocity, pressure, mass-density and temperature. Then, one substitutes the quantities defined in (6.18) into the corresponding equations, additional conditions, initial conditions and boundary conditions, keeps the terms linear in the perturbations and drops the terms of higher order in the perturbations. As a result, one obtains the linear approximate equations.

- The linearization of the boundary form under the condition of a small deformation of the boundary surface. The boundary conditions on the deformed boundary Σ that surround the deformed volume V are transfered to the non-perturbed boundary Σ_0 of the non-perturbed volume V_0 along the normal vector.

Then, the solution of the nonlinear differential equations with nonlinear boundary conditions is replaced by the solution of linear differential equations in the non-perturbed volume V_0 with linear boundary conditions on the non-perturbed boundary Σ_0. In some cases, it is possible to change nonlinear equations to linear ones by a special transformation of the variables.

6.3 Conditions on the Discontinuity Surfaces

During the introduction of the main concepts and for the derivation of the equations of continuum mechanics, we have, up to now, assumed that the given and unknown functions are continuous and possess the required number of continuous derivatives within the volume V. However, these limitations restrict problems of practical importance that can be solved. During the movement of continua with different physical properties (mass densities, viscosities), discontinuities of these parameters exist at the contact boundaries of these media.

During the movement of a homogeneous gas, discontinuities of the velocity, pressure, mass density, temperature and the entropy can develop. It is impossible to solve the problems of such movements within the framework of continuous functions.

The equations of motion for the continuum that have been introduced above are only valid in domains without surfaces of discontinuity, though the integral forms of the laws of continuum mechanics are also valid for piecewise continuous functions. Therefore, the conditions on the discontinuity surfaces can be obtained with the help of these equations.

There are two types of discontinuity surfaces, that is, the weak and strong discontinuity surfaces. These are surfaces where the functions are continuous, though some of the spatial or time derivatives are discontinuous and are referred to as weak discontinuity surfaces. Those surfaces where the functions are discontinuous are called strong discontinuity surfaces. For some problems, the discontinuity surfaces can be specified a-priori, while in other cases, they are obtained only as part of the solution. We now consider strong discontinuity surfaces and formulate the necessary conditions for them.

We first determine the velocity with which the surface is moving. We consider a surface S whose equation at time t is given by

$$f(X, Y, Z, t) = 0 . \tag{6.19}$$

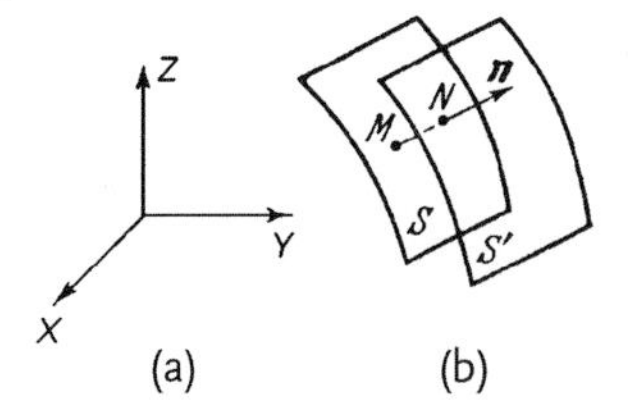

Figure 6.1 Determination of the velocity of a moving surface.

At time $t + \Delta t$, the surface S becomes the surface S_1 (see Figure 6.1). We attach the unit normal vector in point M of the surface in the direction of MN, with $F(M, t) = 0$ and $f(N, t) > 0$. Then, we have

$$\boldsymbol{n} = \frac{\nabla f}{|\nabla f|} . \tag{6.20}$$

The vector $\boldsymbol{D}$ is called the velocity of displacement of the surface S in point M if it is orthogonal to M and given by the following limit:

$$\boldsymbol{D} = \boldsymbol{n} \lim_{\Delta t \to 0} \frac{|\boldsymbol{MN}|}{\Delta t} . \tag{6.21}$$

The components of the vector **n** are

$$n_x = \frac{1}{|\nabla f|} \frac{\partial f}{\partial X} , \quad n_y = \frac{1}{|\nabla f|} \frac{\partial f}{\partial Y} , \quad n_z = \frac{1}{|\nabla f|} \frac{\partial f}{\partial Z} .$$

The equation of the surface S at time $t + \Delta t$ can be written as follows:

$$f(X + MNn_x, Y + MNn_y, Z + MNn_z) = 0 .$$

From this, we obtain that

$$MN \left(n_x \frac{\partial f}{\partial X} + n_y \frac{\partial f}{\partial Y} + n_z \frac{\partial f}{\partial Z} \right) + \frac{\partial f}{\partial t} \Delta t = 0$$

or

$$MN\boldsymbol{n} \cdot \nabla f + \frac{\partial f}{\partial t} \Delta t = MN |\nabla f| + \frac{\partial f}{\partial t} \Delta t = 0 .$$

Using this equation, one obtains from (6.21)

$$\boldsymbol{D} = -\frac{1}{|\nabla f|} \frac{\partial f}{\partial t} \boldsymbol{n} . \tag{6.22}$$

Since all equations that correspond to the laws of physics conserve their form in every inertial system of coordinates, we choose a special system of coordinates with respect to which the velocity $\boldsymbol{D}$ vanishes at time t at the point M.

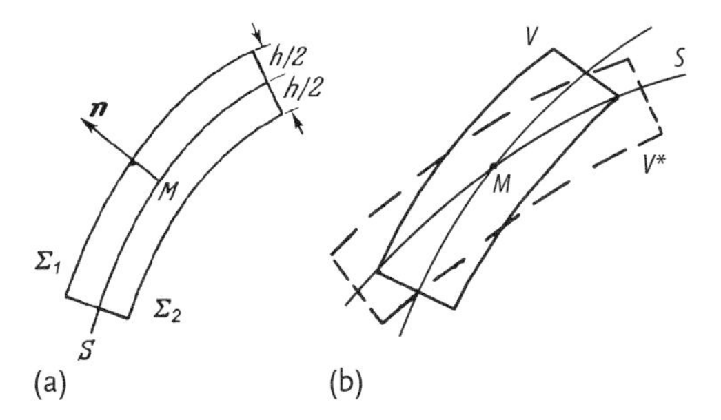

Figure 6.2 Sketch of discontinuity surface S, closed surface Σ (a) and volumes V and V^* (b).

We attach the normal $\boldsymbol{n}$ at point M and mark two lines of length $h/2$ along $\boldsymbol{n}$ on both sides of the discontinuity surface (see Figure 6.2). At time t, the volume V is bounded by the surface Σ, and at time $t + \delta t$, we have the volume V^* which is displaced relative to the volume V (see Figure 6.2). In other words, in the chosen system of coordinates K^*, we have two volumes: a static volume V and a moving volume V^*. Both volumes are bound to the material points of the continuum medium and coincide at time t. However, at time $t + \Delta t$, the volume V^* is displaced relative to volume V. The discontinuity surface is also displaced, but the point M does not move since the velocity relative to the system of coordinates K^* vanishes, that is, $\boldsymbol{D} = 0$. Since the volume V does not move relative to the system of coordinates K^* and the volume V^* does move, we obtain from (2.92)

$$\frac{d}{dt}\int_{V^*} F d\tau = \frac{d}{dt}\int_{V} F d\tau + \int_{\Sigma} F V_n d\sigma \,, \tag{6.23}$$

where V_n is the projection of the velocity at the continuum points relative to the system of coordinates K^* on the outer unit normal of Σ. Consider the first term on the right-hand side of (6.23). Since the discontinuity surface does not move inside of V, according to the mean value theorem, one can write:

$$I = \frac{d}{dt}\int_{V} F d\tau = h\int_{S} F^* d\sigma \,, \tag{6.24}$$

where F^* is the mean value of the function dF/dt along the line orthogonal to S of length h. If S does not move and F is explicitly independent of time, then we have $F^* \to 0$ for $h \to 0$.

On the other hand, in the case of a non-stationary movement and if F is finite and continuous together with its spatial and time derivatives on both sides of the non-moving surface S, the quantity I is a continuous function of t and also goes to 0 for $h \to 0$. For points on the surface S in the vicinity of M, we have $\boldsymbol{D} \neq 0$, and thus S is moving inside of V. We consider an infinitesimal element $d\sigma$ of the surface S in the vicinity of the point M. Since the velocities of the points in this surface are also infinitesimal, we have

$$\lim_{\Delta\sigma \to 0, h \to 0} \frac{I}{\Delta\sigma}\frac{d}{dt}\int_{V} F d\tau = \frac{h}{\Delta\sigma}\int_{S} F^* d\sigma = 0 \,. \tag{6.25}$$

We denote the characteristics of movement on both sides of the discontinuity with indices one and two so that the transition from side two to side one corresponds to the direction of the normal. By taking the limit to $h \to 0$ of (6.23) and using (6.24), one obtains

$$\lim_{\Delta\sigma\to 0, h\to 0} \frac{I}{\Delta\sigma} \frac{d}{dt} \int_{V^*} F d\tau$$

$$= \lim_{\Delta\sigma\to 0, h\to 0} \frac{I}{\Delta\sigma} \left(\int_{\Sigma_1} F V_n d\sigma - \int_{\Sigma_2} F V_n d\sigma + O(h) \right)$$

$$= F_1 V_{n1} - F_2 V_{n2} \,, \tag{6.26}$$

where V_{n1} and V_{n2} are the projections of the velocity at the continuum points on both sides of the discontinuity surface with respect to the normal vector $\boldsymbol{n}$ to the surface S. The negative sign in front of the integral over Σ_2 is due to the direction of the normal vector $\boldsymbol{n}$ being opposite to that of the normal vector to the surface Σ_2. Equation 6.26 allows one to formulate all dynamic and thermodynamic conditions on the discontinuity surface which use the following basic equations of continuum mechanics in the integral form, that is,

- the equation of continuity (3.4)

$$\frac{d}{dt} \int_{V^*} \rho d\tau = 0 \,, \tag{6.27}$$

- the equation of motion (3.24)

$$\frac{d}{dt} \int_{V^*} \rho \boldsymbol{V} d\tau = \int_{V} \rho \boldsymbol{f}_V d\tau + \int_{\Sigma} \boldsymbol{p}_n d\Sigma \,, \tag{6.28}$$

- the equation of angular momentum (3.45)

$$\frac{d}{dt} \left(\int_{V^*} (\boldsymbol{r} \times \mathbf{V}) \rho d\tau + \int_{V^*} \boldsymbol{k} \rho d\tau \right) = \int_{V} (\boldsymbol{r} \times \boldsymbol{f}_V) \rho d\tau + \int_{\Sigma} (\boldsymbol{r} \times \boldsymbol{p}_n) d\Sigma + \int_{V} \boldsymbol{h} \rho d\tau + \int_{\Sigma} \boldsymbol{q}_n d\Sigma \,, \tag{6.29}$$

- the energy equation (5.80)

$$\frac{d}{dt} \int_{V^*} \rho \left(\frac{\mathbf{V}^2}{\rho} + U \right) d\tau = \int_{V} \boldsymbol{f} \cdot \boldsymbol{V} d\tau + \int_{\Sigma} \boldsymbol{p}_n \cdot \boldsymbol{V} d\sigma - \int_{\Sigma} q_n^* d\sigma + \int_{V} \frac{dq_{\text{mas}}^*}{dt} \rho d\tau \tag{6.30}$$

and the entropy equation (5.81)

$$\frac{dS}{dt} = \frac{d}{dt}\int_{V^*} \rho s d\tau = \int_V \frac{1}{T}\left(\frac{dq^{(e)}}{dt} + \frac{dq'}{dt}\right) d\tau . \tag{6.31}$$

Before deriving the conditions at discontinuity surfaces, we need to make the following necessary assumptions:

- All integrands in the surface integrals in (6.27)–(6.30) have for $\Sigma \to S$ finite but different limits on both sides of S.
- For $h \to 0$, we have the following limiting relations:

$$\lim_{h\to 0}\int_V \boldsymbol{f}\rho d\tau = \int_S \boldsymbol{R} d\sigma = 0 , \quad \lim_{h\to 0}\int_V \boldsymbol{h}\rho d\tau = \int_S \boldsymbol{M} d\sigma ,$$

$$\lim_{h\to 0}\int_V \left(\boldsymbol{f}\cdot \boldsymbol{V} + \frac{dq^*_{\text{mass}}}{dt}\right)\rho d\tau = \int_S W d\sigma = 0 ,$$

$$\lim_{h\to 0}\int_V \frac{1}{T}\left(\frac{dq^{(e)}}{dt} + \frac{dq'}{dt}\right)\rho d\tau = \int_S \Omega\, d\sigma . \tag{6.32}$$

These equations mean that according to (6.25), the volume integrals become surface integrals for $h \to 0$ and, if the density of the corresponding characteristic, that is, F^* in (6.25) is inversely proportional to h, these limits of the form (6.32) have finite values. In (6.32), **R**, **M** and W are surface densities of the corresponding forces, torques and energy flows and Ω is the surface density of the entropy change on S due to external heat inflows and irreversible entropy change. If $\rho \boldsymbol{f}$, $\rho \boldsymbol{h}$ and $\rho(\boldsymbol{f}\cdot \boldsymbol{V} + dq^*_{\text{mass}}/dt)$ are finite in the volume V, $\boldsymbol{R}$, W and $\boldsymbol{M}$ are zero. This is, for example, true in the case of external forces, gravity, inertial forces, pondomotoric forces and generally any continuously distributed mass forces, external moments and energy inflows due to electro magnetic fields that are continuous on S. In some cases, the value of quantities $\boldsymbol{R}$, $\boldsymbol{M}$, W and Ω can be nonzero. This is the case for the modeling of surfaces that have a big influence on the medium. If the supporting effect of a wing is modeled, we have $R = M = W = 0$, though $\Omega \neq 0$. From the law of mass conservation (6.27) and the kinematic condition (6.26), it follows that

$$\rho_1 V_{n1} = \rho_2 V_{n2} . \tag{6.33}$$

From the equation of motion (6.28) and (6.32), one obtains

$$\boldsymbol{R} + \boldsymbol{p}_{n1} - \rho_1 V_{n1}\boldsymbol{V}_1 = \boldsymbol{p}_{n2} - \rho_2 V_{n2}\boldsymbol{V}_2 . \tag{6.34}$$

From the angular momentum equation (6.29), one obtains, using (6.32),

$$\boldsymbol{M} + \boldsymbol{Q}_{n1} - \rho_1 V_{n1}\boldsymbol{k}_1 = \boldsymbol{Q}_{n2} - \rho_2 V_{n2}\boldsymbol{k}_2 . \tag{6.35}$$

For the derivation of (6.35), it is taken into account that the projections of the terms with $\boldsymbol{r} \times \boldsymbol{V}$ and $\boldsymbol{r} \times \boldsymbol{p}_n$ do not contribute any components normal to S.

From the energy equation (6.30) and the conditions (6.32), it follows that

$$W + \boldsymbol{p}_{n1} \cdot \boldsymbol{V}_1 - \rho_1 \left(\frac{\boldsymbol{V}_1^2}{2} + U_1 \right) V_{n1} - q_{n1}^* = \boldsymbol{p}_{n2} \cdot \boldsymbol{V}_2 - \rho_2 \left(\frac{\boldsymbol{V}_2^2}{2} + U_2 \right) V_{n2} - q_{n2}^* \,. \tag{6.36}$$

Finally, one obtains from the entropy equation (6.31), the condition (6.32) and (6.33)

$$\rho_1 V_{n1} (s_1 - s_2) = \Omega \,. \tag{6.37}$$

The conditions (6.33)–(6.37) are obtained in the system of coordinates K^* in which the discontinuity surface S does not move, that is, in the system of coordinates that is fixed to the discontinuity surface. Therefore, these conditions are applicable for stationary continuum movements if the system of coordinates K^* and the general system of coordinates which is not fixed to the discontinuity surface coincide. In the case of non-stationary movements, the discontinuity surfaces have velocities of displacement $\boldsymbol{D}_i$ that are different both in direction and absolute value. To obtain conditions on the discontinuity surfaces that are valid in every system of coordinates, one has to replace $\boldsymbol{V}$ relative to the moving system of coordinates K^* by $\boldsymbol{V}^* = \boldsymbol{V} - \boldsymbol{D}$ relative to the observers system of coordinates. By replacing the velocity $\boldsymbol{V}$ by $\boldsymbol{V} - \boldsymbol{D}$ in (6.33)–(6.37), one obtains

$$\rho_1 (D - V_{n1}) = \rho_2 (D - V_{n2}) \,, \tag{6.38}$$

$$\boldsymbol{R} + \boldsymbol{p}_{n1} - \rho_1 (D - V_{n1}) \boldsymbol{V}_1 = \boldsymbol{p}_{n2} - \rho_2 (D - V_{n2}) \boldsymbol{V}_2 \,, \tag{6.39}$$

$$\boldsymbol{W}(\boldsymbol{D}) + \boldsymbol{p}_{n1} \cdot \boldsymbol{V}_1 - \rho_1 \left(\frac{\boldsymbol{V}_1^2}{2} + U_1 \right) (D - V_{n1}) - q_{n1}^* = \boldsymbol{p}_{n2} \cdot \boldsymbol{V}_2 - \rho_2 \left(\frac{\boldsymbol{V}_2^2}{2} + U_2 \right) (D - V_{n2}) - q_{n2}^* \,, \tag{6.40}$$

$$\rho_1 (D - V_{n1})(s_1 - s_2) = \Omega \,, \tag{6.41}$$

which are valid in every system of coordinates and in all points of the discontinuity surface. The quantity $\boldsymbol{W}(\boldsymbol{D})$ in (6.40) is given by $\boldsymbol{W} + \boldsymbol{R} \cdot \boldsymbol{D}$. The conditions for the torques have not been written down in full since we will restrict ourselves in the following to the case $\boldsymbol{M} = \boldsymbol{Q} = \boldsymbol{k} = 0$.

Tangential Discontinuity Surface

The velocities $D - V_{n1}$ and $D - V_{n2}$ can be considered as the velocities of the discontinuity surfaces relative to the corresponding medium. Therefore, if $D - V_{n1} = 0$ and $D - V_{n2}$ or $V_{n1} = V_{n2}$, the particles do not cross through the discontinuity surface and there is no discontinuity in the normal component of the velocity.

However, a discontinuity in the tangential component of the velocity $V_{t1} \neq V_{t2}$ and the mass density $\rho_1 \neq \rho_2$ is generally possible. Such a discontinuity is called a tangential discontinuity. In this case, the conditions (6.39)–(6.41) become

$$R = \boldsymbol{p}_{n2} - \boldsymbol{p}_{n1} , \tag{6.42}$$

$$W = \boldsymbol{p}_{n1} \cdot \boldsymbol{V}_1 - \boldsymbol{p}_{n2} \cdot \boldsymbol{V}_2 + q^*_{n2} - q^*_{n1} , \tag{6.43}$$

$$\Omega = 0 . \tag{6.44}$$

For $\boldsymbol{R} = 0$, the stresses on the discontinuity surfaces are continuous ($\boldsymbol{p}_{n2} = \boldsymbol{p}_{n1}$). Though, for $W = 0$, the work performed by the stress forces due to the difference of the tangential velocity is equal to the difference of energy flow through the discontinuity surface. In the case of an ideal fluid, the conditions (6.42)–(6.44) for $\boldsymbol{R} = 0$ and $\boldsymbol{W} = 0$ result in the continuity of the normal components of energy flow and pressure on a tangential discontinuity surface.

Compaction Jump and Dilution Jump

We now take an in depth look at the case when $V_{n2} \neq V_{n1}$. In this case, the continuum particles move from one side of the discontinuity surface to the other side and as a result, the characteristics undergo a sudden change. We assign indices to the two sides of the discontinuity surface in such a way that the medium moves from side one to side two. We take a system of coordinates such that $\boldsymbol{V}_1 = 0$. This means that this system of coordinates moves with the velocity of the medium on side one, $D_n > 0$ and the discontinuity surface moves relative to the medium one at rest. For $V_{n2} > V_{n1}$, we have $V_{n2} > 0$ and the medium behind the jump moves in the direction of the medium before the jump. From (6.38), it follows that for such jumps, we have $\rho_2 > \rho_1$, that is, the mass density of the medium suddenly increases after the jump. Such a jump is referred to as compactification jump.

In the case $V_{n2} < V_{n1}$, the normal component of the velocity of the medium behind the jump is directed toward the jump propagation velocity in the medium at rest. Therefore, we have $\rho_2 < \rho_1$ in the medium behind the jump and a dilution takes place. Such jumps are called dilution jumps.

The jump conditions formulated above can serve as a foundation for the determination of the boundary conditions for differential equations in the domain of continuous movements. Using these conditions, one can obtain the boundary conditions on the free surfaces of the fluid, on the solid body boundaries and so on.

6.4 Discontinuity Surfaces in Ideal Compressible Media

As an example of the application of discontinuity surfaces, we consider these with respect to an ideal compressible gas. In this case, the internal stress is given by $\boldsymbol{p}_n = -p\boldsymbol{n}$ where p is the pressure. We consider a discontinuity surface with

$\boldsymbol{R} = 0$, $W = 0$, $q_n^* = 0$. This means that there are no external surface effects on the medium and the heat conduction is not affected by the jump. In the system of coordinates K^*, relative to which the discontinuity surface is at rest ($D = 0$), condition (6.38) has the form

$$\rho_1 V_{n1} = \rho_2 V_{n2} \,. \tag{6.45}$$

We assume that $V_{n1} \neq 0$, that is, the continuum particles cross through the discontinuity surface. By projecting (6.39) first on the tangential direction and then on the direction normal to the discontinuity surface and using (6.45), one obtains

$$V_{t1} = V_{t2} \,, \tag{6.46}$$

$$p_1 + \rho_1 V_{n1}^2 = p_2 + \rho_2 V_{n2}^2 \,. \tag{6.47}$$

According to (6.45) and (6.46), the energy equation (6.40) is transformed to

$$U_1 + \frac{V_{n1}^2}{2} + \frac{p_1}{\rho_1} = U_2 + \frac{V_{n2}^2}{2} + \frac{p_2}{\rho_2} \,. \tag{6.48}$$

The last condition (6.41) yields

$$\rho_1 V_{n1}(s_1 - s_2) = \Omega \,. \tag{6.49}$$

Thus, (6.45)–(6.49) give the conditions for a stationary jump in an ideal compressible medium (gas). In the system of coordinates K in which the velocity before the jump vanishes and the jump velocity D is nonzero, one can consider $\boldsymbol{D}$ as the jump propagation velocity relative to the continuum particles on side one. Therefore, in (6.45)–(6.49), one has to substitute $V_{n1} = -D$, $V_{n1} = V_n - D$, $V_{t1} = V_{t2} = 0$. Here, $V_n = V_{n2} - V_{n1}$ is the normal component of the continuum velocity relative to K on S on side two. By introducing the specific volume $v = 1/\rho$ instead of ρ, one obtains from (6.45) and (6.47)

$$-V_{n2} = v_2 \left(\frac{p_2 - p_1}{v_2 - v_1} \right)^{1/2} , \qquad -V_{n1} = D = v_1 \left(\frac{p_2 - p_1}{v_2 - v_1} \right)^{1/2} \tag{6.50}$$

$$V_{n2} - V_{n1} = V_n = D \left(1 - \frac{\rho_1}{\rho_2} \right) = \pm \sqrt{(p_2 - p_1)(v_1 - v_2)} \,. \tag{6.51}$$

Since $D > 0$, one has to take the positive sign for $\rho_2 > \rho_1$ and the negative sign for $\rho_2 < \rho_1$ on the right-hand side of (6.51). Then, from (6.48), one obtains

$$U_2 - U_1 = \frac{1}{2}(p_2 + p_1)(v_1 - v_2) \,. \tag{6.52}$$

From (6.50), it follows that if $v_1 > v_2 (\rho_1 < \rho_2)$, we have $p_2 > p_1$, and if $v_1 < v_2 (\rho_1 > \rho_2)$, we have $p_2 < p_1$.

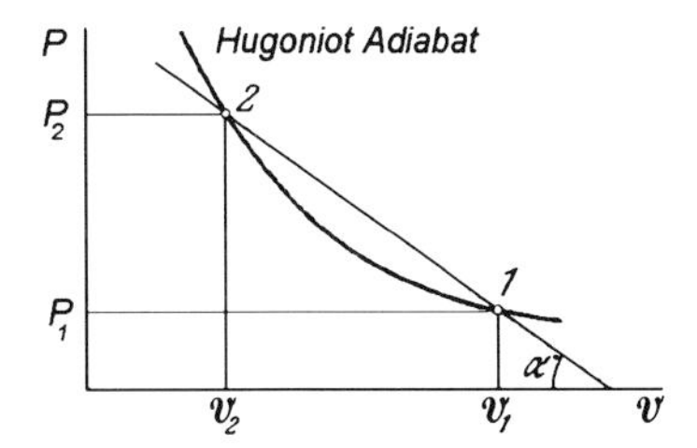

Figure 6.3 Adiabate of Hugoniot.

The internal energy U in (6.52) is a function of the specific volume v, the pressure p and some additional parameters that define the physical and chemical properties. For an ideal gas, we have the following relationship

$$U = c_v T + U_0 = \frac{c_p}{(c_p - c_v)} \frac{p}{\rho} + U_0 \, .$$

For a mixture of ideal gases, the corresponding equation reads

$$U = \sum_i \frac{\rho_i}{\rho} \left[U_{0i} + \int_{T_0}^{T} c_{vi}(T) dT \right] .$$

In general, the parameters c_p, c_v, U_0 and ρ_i/ρ, that is, the gas composition, can change when crossing the discontinuity. Therefore, one needs relationships that follow from physical and chemical laws in addition to (6.52). If the physical and chemical properties do not change during the transition of the particle through the jump and only ρ and p are changed, (6.52) yields the relationship between p_2 and ρ_2 behind the jump, provided that p_1 and ρ_1 before the jump are known.

Consider the state plane $(p, v = 1/\rho)$ (see Figure 6.3). For given values of p_1 and ρ_1, (6.52) corresponds to a curve on the state plane, which is called the adiabate of Hugoniot. The equation of this curve has the following form:

$$U_2(p, v) - U_1(p_1, v_1) = \frac{1}{2} (p + p_1)(v_1 - v) \, . \tag{6.53}$$

Since $U_2(p_1, v_1) - U_1(p_1, v_1) = 0$, the point p_1, v_1 lies on the adiabate of Hugoniot if the functions $U_1(p, v)$ and $U_2(p, v)$ are identically equal. From (6.50), it follows that

$$\tan \alpha = \frac{p_2 - p_1}{v_1 - v_2} = \frac{D^2}{v_1^2} \, .$$

Therefore, the angle α characterizes the jump propagation velocity relative to the particles of medium one.

We now calculate the entropy change along the Hugoniot adiabate. To do this, we consider a reversible process with heat inflow between a fixed $p_1 v_1$ and another arbitrary p, v that belongs to the Hugoniot adiabate. From the entropy equation

$$T ds = dU + p dv \, ,$$

and taking into account (6.52), one obtains for fixed state parameters $p_1 v_1$,

$$\begin{aligned} T ds &= \frac{1}{2}(v_1 - v)\, d(p - p_1) - \frac{1}{2}(p - p_1)\, d(v_1 - v) \\ &= \frac{1}{2}(v_1 - v)^2\, d\left(\frac{p - p_1}{v - v_1}\right) = \frac{1}{2}(v_1 - v)^2\, d(\tan\alpha) \\ &= \frac{1}{2}(p - p_1)^2\, d\left(\frac{1}{\tan\alpha}\right). \end{aligned}$$

On the other hand, for small values of $p - p_1$, we have

$$\begin{aligned} d\left(\frac{1}{\tan\alpha}\right) &= -d\left(\frac{v - v_1}{p - p_1}\right) \\ &= -d\left[\left(\frac{dv}{dp}\right)_{p=p_1} + \frac{1}{2}\left(\frac{d^2 v}{dp^2}\right)_{p=p_1}(p - p_1) + \ldots\right] \\ &= -\frac{1}{2}\left(\frac{d^2 v}{dp^2}\right)_{p=p_1} dp + O(p - p_1)dp\,. \end{aligned}$$

Therefore, it follows that

$$T ds = -\frac{1}{2}\left(\frac{d^2 v}{dp^2}\right)_{p=p_1}(p - p_1)^3 + O(p - p_1)^3\,.$$

Therefore, the entropy change for a small jump in pressure when crossing the discontinuity is a small quantity of order $(p - p_1)^3$. Those jumps for which pressure and mass density change only by a small amount are called weak jumps. For weak jumps, that is, for $p \to p_1$ and $v \to v_1$, it follows from (6.50) that

$$D^2 = -v^2 \frac{dp}{dv} = \left(\frac{dp}{d\rho}\right)_H\,.$$

Here, the subscript H indicates that the derivative is taken along the Hugoniot adiabate. Since $ds \sim (p - p_1)^3$ and $s \to$ const for $p_2 \to p_1$, the process can be kept isentropic. Therefore, we have

$$D^2 = \left(\frac{dp}{d\rho}\right)_H = \left(\frac{dp}{d\rho}\right)_S\,.$$

It is known that the velocity of sound is given by

$$a^2 = \left(\frac{dp}{d\rho}\right)_S\,. \tag{6.54}$$

Therefore, we have $D = a$. Thus, infinitesimal perturbations of the pressure of the mass density are weak jumps that propagate through the particles of the continuum medium with the velocity of sound.

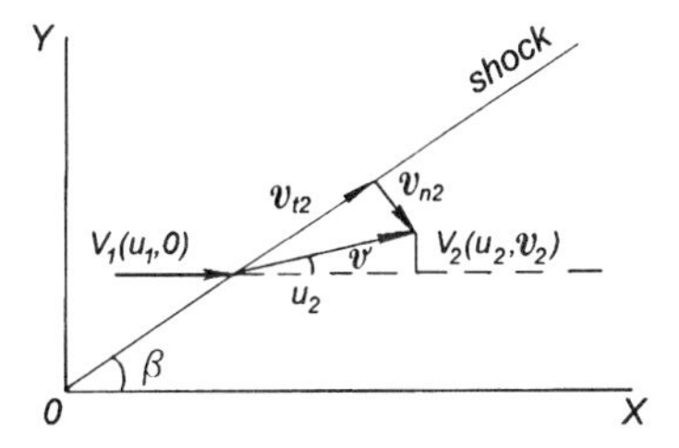

Figure 6.4 Transition of an ideal gas through the shock wave.

In continuous adiabatic flows, the entropy is conserved during the a change of the particle state. Therefore, we have

$$s_2(p, v) - s_1(p_1, v_1) = 0 . \tag{6.55}$$

Equation 6.55 yields the relationship between p and v on both sides of the jump. The corresponding curve on the state plane (p, v) is called Poisson's adiabate.

In many applications, especially in aerodynamics when dealing with the movement of bodies at a constant hypersonic speed, one considers the stationary movement of the ideal gas with a static compactification jump in the system of coordinates that is fixed to the moving body. These jumps are called shock waves. We now consider the conditions in shock waves more closely. For an ideal gas, we have

$$U = \frac{1}{\gamma - 1}\frac{p}{\rho} + \text{const} , \quad a^2 = \frac{\gamma p}{\rho} .$$

If the jump is not moving, the conditions at the jump have the form of (6.45)–(6.49). Therefore, the quantities V_{t2}, V_{n2}, ρ_2 and p_2 behind the jump can be expressed in terms of the quantities V_{t1}, V_{n1}, ρ_1 and p_1 before the jump, that is,

$$V_{t2} = V_{t1}, \quad V_{n2} = V_{n1}\left(\frac{\gamma - 1}{\gamma + 1} + \frac{1}{\gamma + 1}\frac{a_1^2}{V_{n1}^2}\right),$$

$$\rho_2 = \frac{\gamma + 1}{\gamma - 1}\frac{\rho_1}{1 + \frac{2}{\gamma - 1}\frac{a_1^2}{V_{n1}^2}}, \quad p_2 = \frac{2}{\gamma - 1}\rho_1 V_{n2}^2\left(1 + \frac{\gamma - 1}{2\gamma}\frac{a_1^2}{V_{n1}^2}\right). \tag{6.56}$$

We consider the plane spanned by the vectors $\mathbf{V}_1$ and $\mathbf{V}_2$ (see Figure 6.4), align the X-axis with the direction of the velocity $\mathbf{V}_1$ before the pressure wave, denote the inclination angle before the shock wave by β and the inclination angle after the shock wave as ϑ.

We denote the X and Y components of the velocity by u and v. The tangential and normal components of the velocity with respect to the shock wave are then given by

$$V_t = u \cos\beta + v \sin\beta , \quad V_n = u \sin\beta - v \cos\beta . \tag{6.57}$$

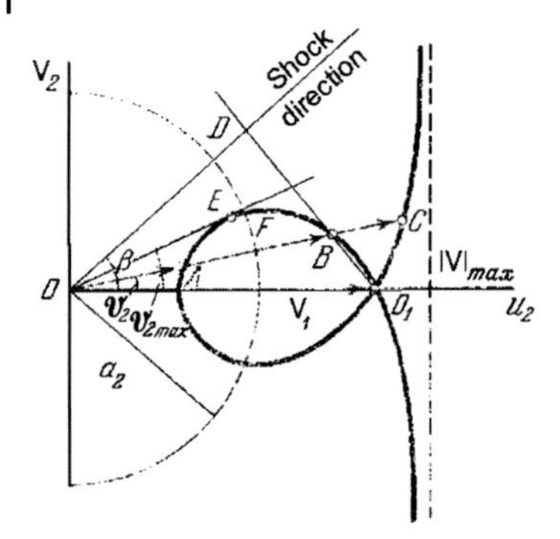

Figure 6.5 Shock Polar.

By using (6.57) to express the quantities V_{t1}, V_{t2}, V_{n1}, V_{n2} in terms of u_1, v_1, u_2, v_2, substituting the resulting expressions in (6.56) and eliminating β, one obtains:

$$v_2^2 = (u_1 - u_2)^2 \frac{\frac{2}{\gamma+1}\left(u_1 - \frac{a_1^2}{u_1}\right) - (u_1 - u_2)}{u_1 - u_2 + \frac{2}{\gamma+1}\frac{a_1^2}{u_1}}. \tag{6.58}$$

We consider the velocity plane (u_2, v_2) that is referred to as a hodograph plane (see Figure 6.5). In the figure, we denote $\boldsymbol{OO}_1 = \boldsymbol{V}_1$ by the velocity vector before the jump, the velocity vector after the jump by $\boldsymbol{OB} = \boldsymbol{V}_2$, and the direction of the shock wave by $\boldsymbol{OD}$. Equation 6.58 corresponds to a curve on the hodograph plane that is described as shock polar. Each inclination angle ϑ_2 of the velocity behind the shock wave corresponds to three values of the velocity, that is, $\boldsymbol{OA}$, $\boldsymbol{OB}$ and $\boldsymbol{OC}$. Since we have $V_{t1} = V_{t2}$, the line that corresponds to the direction of the shock propagation must be orthogonal to the line $\boldsymbol{O}_1\boldsymbol{B}$ that connects the ends of the vectors $\boldsymbol{V}_1$ and $\boldsymbol{V}_2$.

Of the three points, A, B and C, only the two points A and B can correspond to the end of the vector $\boldsymbol{V}_2$, that is, $\boldsymbol{V}_2 = \boldsymbol{OA}$ or $\boldsymbol{V}_2 = \boldsymbol{OB}$ since the jumps that correspond to the branches of the shock polar that go to infinity cannot be physically realized. Indeed, if we choose the vector $\boldsymbol{OC}$ for $\boldsymbol{V}_2$, we have $V_n = V_{n2} - V_{n1} < 0$. However, this condition does not correspond to a compaction jump, but a dilution jump. In the figure, the vector $\boldsymbol{OB}$ is chosen as $\boldsymbol{V}_2$. By comparing the values $|\boldsymbol{OB}|$ and $|\boldsymbol{OA}|$ with the sound velocity a_2 after the jump, one concludes that if the point B is to the right of F, as indicated on Figure 6.5, the velocity after the jump $\boldsymbol{V}_2$ is supersonic. If one would take the point B instead of the point A, the velocity behind the jump would be subsonic. During the transition through the jump, the velocity vector undergoes a rotation by an angle which can be determined via the following equation:

$$\tan(\beta - \vartheta_2) = \tan\beta\left[\frac{\gamma - 1}{\gamma + 1} + \frac{1}{(\gamma + 1)M_1^2(1 - \cos^2\beta)}\right]. \tag{6.59}$$

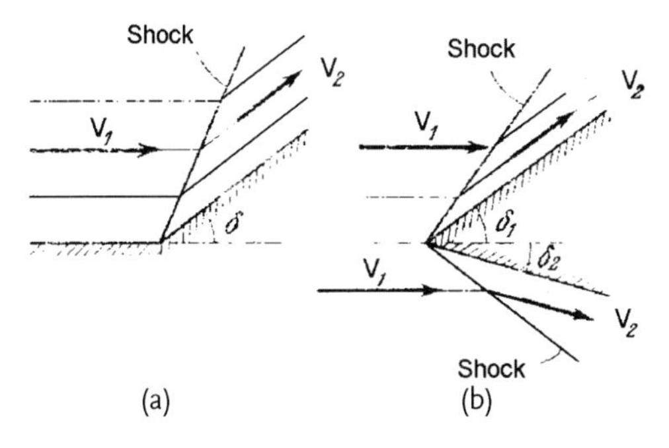

Figure 6.6 Flow around angle (a) and wedge (b).

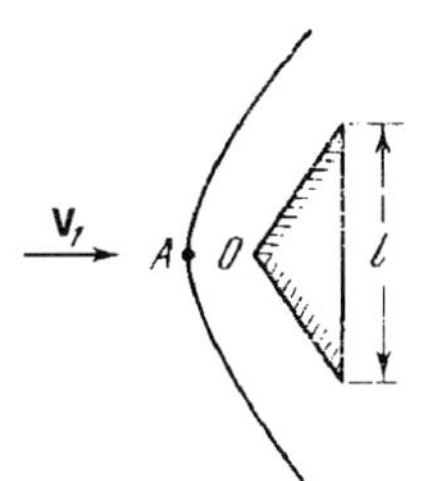

Figure 6.7 Flow around a wedge with large aperture.

Here, $M_1 = V_1/a_1$ is the Mach number before the jump. The maximal possible turning angle is equal to $\vartheta_{2\,\max}$, which corresponds to the tangent DE on the shock polar. The point E corresponds to a supersonic velocity close to the velocity of sound. The quantity $\vartheta_{2\,\max}$ depends on the angle β and the Mach number M_1 in the incoming gas flow. For $M_1 > 1$, the flow is supersonic and for $M_1 < 1$, the flow is subsonic. The compaction jumps can be of two types, even and skewed. A jump for which $V_1 R$ is orthogonal to the jump is called an even jump or a skewed jump. The above conditions for the jump and especially for the compaction jump allow one to obtain solutions for the problems of the supersonic flow around an angle and a wedge (see Figure 6.6).

From the results obtained above, it follows that a flow of the types shown in Figure 6.6 is possible if, for a given Mach number of the incoming gas flow, the angles δ, δ_1 and δ_2 are smaller than or equal to $\vartheta_{2\,\max}$. For each of these problems, two solutions exist for $\delta < \delta_{\max}$ with different directions of the shock wave, and for each of them, the velocity undergoes a rotation. On Figure 6.5, these solutions correspond to the points A and B. For the the slender body, there is, in practice, the movement with an inclined shock wave behind the jump which corresponds to the point B on the shock polar.

For $\delta > \delta_{\max}$, the flow in in the forms of Figure 6.6 is impossible. In this case, a curved shock wave arises in front of the body (see Figure 6.7). The distance $A0$ is proportional to the linear dimension of the body and depends on the aperture of wedge and the Mach number of the incoming gas flow.

6.5 Dimensions of Physical Quantities

To determine the velocities of continuum movements, different systems of coordinates are used that are chosen depending on considerations of simplicity and convenience of the solution for the concrete tasks. The arbitrary choice of the system of coordinates has nothing do to with the phenomenon as such. Therefore, the equations have to be invariant with respect to the system of coordinates. Since the scalars, vectors and tensors possess invariance properties, the above mentioned equations and additional conditions have the form of scalar, vector or tensor equations. On the other hand, when determining or providing quantities, characteristics and medium properties, they have to be expressed as numbers using certain units. The quantities whose numerical values depend on the choice of units are called dimensional quantities. For example, the length is measured in meter, the mass in kilogram, the time in seconds, the acceleration in m/s^2 and so on. The dimensional quantities can be classified as primary or basic quantities, and secondary or derived quantities, that is, dependent quantities. For example, the time T, the length L, and the mass M are primary quantities, while the energy E, the velocity V, the acceleration a, the force F and so on are derived quantities. In mechanics, the quantities T, L and M are used as primary quantities. Sometimes, these are supplemented by the degree Kelvin as a dimensional quantity of the temperature. If the quantities T, L, M are used as primary units, every derived quantity can be expressed as follows:

$$L^l M^m T^t \,. \tag{6.60}$$

Equation 6.60 is referred to as the dimension formula. For example, for the force, we have the dimension formula $F = LMT^{-2}$. The numerical value of the dimensional quantities depends on the system of units. At the same time, one can introduce dimensionless quantities, that is, quantities whose values are independent of the chosen system of units.

As primary quantities, one can take the corresponding quantities in the different systems of units, that is, the mks system or cgs system. Just as the equations that show the physical laws have to be independent of the system of coordinates, those equations also have to be independent of the system of units. This latter property is the condition of invariance of the functional dependencies with respect to the group of similarity transformations. This means that during a similarity transformation, all equations and additional conditions stay unchanged.

We will now carefully consider the structure of the functional dependencies between the dimensional quantities that express the physical laws which are invariant relative to the choice of the system of units. We take a dimensional quantity that is a function of the independent dimensional quantities $a_1, a_2, \ldots, a_n$, that is,

$$a = f(a_1, a_2, \ldots, a_k, a_{k+1}, \ldots, a_n) \,. \tag{6.61}$$

We assume that in (6.61), the first k quantities have independent dimensions, that is, the dimensions of each of these quantities cannot be expressed in terms of others by a monomial, or rather, a product of powers of other units.

The last $n - k$ quantities have dependent dimensions, that is, each of the dimensions of these quantities can be expressed via other ones as a monomial. For example, the length, the mass and the time are independent. However, the acceleration, velocity and the energy are dependent. Among mechanical quantities, there are normally not more than three quantities with independent dimensions.

Now, one can express the dimensions of the quantities $a, a_{k+1}, \dots, a_n$ via monomials in $a_1, a_2, \dots, a_k$. We take $a_1, a_2, \dots, a_k$ as basic quantities and denote their dimensions by

$$[a_1] = A_1\,, \quad [a_2] = A_2, \dots, [a_k] = A_k\,.$$

Using the dimension formula (6.60), the dimensions of the other quantities can be expressed via the dimensions of the basic quantities as

$$[a] = A_1^{m_1} A_2^{m_2} \dots A_k^{m_k}\,,$$
$$[a_{k+1}] = A_1^{p_1} A_2^{p_2} \dots A_k^{p_k}, \dots, [a_n] = A_1^{q_1} A_2^{q_2} \dots A_k^{q_k}\,.$$

If one changes from the chosen system of units to a another one, such that the basic quantities are changed according to $a_1' = \alpha_1 a_1, a' = \alpha_2 a_2, \dots, a_k' = \alpha_k a_k$, then the independent quantities change depending on $\alpha_1 a_1, \alpha_2 a_2, \dots, \alpha_k a_k$, and are given by

$$\begin{aligned} a' &= \alpha_1^{m_1} \alpha_2^{m_2}, \dots, \alpha_k^{m_k} a, \dots, a_{k+1}' = \alpha_1^{p_1} \alpha_2^{p_2}, \dots, \alpha_k^{p_k} a_{k+1}, \dots, a_n' \\ &= \alpha_1^{q_1} \alpha_2^{q_2}, \dots, \alpha_k^{q_k} a_n'\,. \end{aligned}$$

In the new system of units, (6.61) takes on the following form:

$$\begin{aligned} a' &= \alpha_1^{m_1} \alpha_2^{m_2}, \dots, \alpha_k^{m_k} a \\ &= \alpha_1^{m_1} \alpha_2^{m_2}, \dots, \alpha_k^{m_k} f(a_1, a_2, \dots, a_k, a_{k+1}, \dots, a_n) \\ &= f(a_1', a_2', \dots, a_k', a_{k+1}', \dots, a_n') \\ &= f(\alpha_1 a_1, \alpha_2 a_2, \dots, \alpha_k a_k, \alpha_1^{p_1} \alpha_2^{p_2}, \dots, \alpha_k^{p_k} a_{k+1}, \dots, \alpha_1^{q_1} \alpha_2^{q_2}, \dots, \alpha_k^{q_k} a_n)\,. \end{aligned} \tag{6.62}$$

The form of the function F means that it is a homogeneous function relative to the quantities $a_1, a_2, \dots, a_k$.

By choosing $\alpha_1, \alpha_2, \dots, \alpha_k$ as

$$\alpha_1 = \frac{1}{a_1}\,, \quad \alpha_2 = \frac{1}{a_2}, \dots, \alpha_k = \frac{1}{a_k}$$

and introducing the following relationships

$$\Pi = \frac{a}{a_1^{m_1} a_2^{m_2}, \dots, a_k^{m_k}}\,, \quad \Pi_1 = \frac{a_{k+1}}{a_1^{p_1} a_2^{p_2}, \dots, a_k^{p_k}}, \dots,$$
$$\Pi_{n-k} = \frac{a_n}{a_1^{q_1} a_2^{q_2}, \dots, a_k^{q_k}}\,,$$

(6.61) is transformed into

$$\Pi = f(\underbrace{1, 1, \ldots, 1}_{k}, \Pi_1, \Pi_2, \ldots, \Pi_{n-k}) \, . \tag{6.63}$$

Therefore, the functional dependency (6.61) between $n + 1$ dimensional quantities $a, a_1, a_2, \ldots, a_n$, whose values depend on the choice of the system of units, can be represented in the form of a functional relationship between $n - k + 1$ dimensionless quantities $\Pi, \Pi_1, \Pi_2, \ldots, \Pi_{n-k}$, whose values are independent of the choice of the system of units. Equation 6.63 is known as the Π-theorem.

From the Π-theorem, it follows that, given n different parameters of which not more than k have independent dimensions, one cannot create more than $n - k$ independent dimensionless combinations. In addition, every physical functional dependence between dimensional quantities can be formulated as a functional dependence between dimensionless quantities. Since the number of dimensionless parameters is, as a rule, smaller then the number of the dimensional parameters, the number of parameters of the system considered is reduced and the solution method becomes much simpler. If the number of basic units is the same as the number of determining parameters, that is, $n = k$, then it is impossible to create a dimensionless combination from the parameters $a_1, a_2, \ldots, a_k$. Therefore, in this case, the functional dependence (6.63) becomes

$$a = C a_1^{m_1} a_2^{m_2} \ldots, a_n^{m_n} \, . \tag{6.64}$$

6.6
Parameters that Determine the Class of the Phenomenon

To use the main result of the Π-theorem in the form (6.63), it is necessary to first find the arguments of the function $a(a_1, a_2, \ldots, a_n)$, that is, to find the determining parameters of the problem. The first step of setting up the problem consists of choosing the continuum model and the classification of the properties of the required solutions. For example, one takes into account the symmetry as well as the dimension of the problem (one-dimensional or two-dimensional problem, cylindrical or spherical symmetry and so on), and based on this, a suitable system of coordinates is chosen. Thereafter, the system of equations, the properties of the required functions (continuous, continuously differentiable, piecewise continuous) and the independent variables are fixed. The independent variables consist of the coordinates X, Y, Z and the time t, the physical constants, that is, the coefficients of viscosity and heat conduction, the elasticity modulus, the gravitational acceleration, the gas constant, the Boltzmann-constant and so on, the characteristics of the continuum volume V and those quantities that determine the given functions for the formulation of the initial and boundary conditions, or rather, the energy release at the initial time, the heat transfer number on the contact surface with an external body.

Here, we differentiate between two cases. In the first case, the problem under consideration has been formulated in a mathematically complete manner, that is, all equations, additional conditions, initial conditions and boundary conditions have been completely specified. Then, it is easy to find the complete set of arguments in the function given by (6.60). The arguments $a, a_1, a_2, \ldots, a_n$ are all the quantities that have to be specified to calculate the required functions (velocity, pressure, mass density, temperature entropy and so on). In the second case, the mathematical formulation of the problem is missing. This can happen if the model of the continuum to be researched or its physical properties are unknown. In this case, it is sufficient to make use of provisional experimental data and of physically motivated hypotheses. The system of parameters $a, a_1, \ldots, a_n$ that can completely determine the phenomenon or the class of phenomenon must possess the completeness property, that is, the determining parameters must have dimensions by which the dimension of each required quantity can be expressed. Of course, one should only take determining parameters that characterize the phenomenon. For example, the gravitational acceleration must only be included in the list of determining parameters if the gravity plays a role in the problem under consideration. Here, it should be noted that in spite of the considerable advantage that the usage of the dimension theory delivers for the solution of physical problems, this method is limited insofar as the problem cannot be solved by this method alone. The usage of the dimension theory only allows one to determine a common functional dependence of the form (6.63) or (6.64), but the concrete form of the function (6.63) or the constant in (6.64) can only be found from experiments or from the solution of the corresponding mathematical tasks.

Therefore, the main use of the dimension theory for theoretical and experimental research consists of writing the equations that govern the physical processes in a dimensionless form, that is, invariant with respect to the choice of the system of units and then to investigate the physical process further.

Problem of Point Explosion in the Atmosphere

We consider a small volume in which a large amount of energy is initially released and transferred to the air. As a result, a quickly expanding spherically symmetric volume of air is created with a large pressure and mass density disturbance. The problem is formulated as follows. At the initial time, a given energy is released suddenly at a given point into an ideal gas at rest with pressure p_1 and mass density ρ_1. As a result of the explosion, a quickly propagating shockwave is created that separates the air at rest from the moving air behind the shock wave. It is obvious that all characteristics of the movement only depend on the distance r from point of explosion and on the time t. The movement of the ideal compressible gas is described by the following equations (see (4.41) and (4.50)) with

$$U_r = v(r,t)\,, \quad p_r = p(r,t)\,, \quad \rho = \rho(r,t)\,, \quad U_\vartheta = U_\varphi = 0\,,$$

$$\frac{\partial \rho}{\partial t} + \frac{\partial(\rho v)}{\partial r} + \frac{2\rho v}{r} = 0\,, \qquad \frac{\partial v}{\partial t} + v\frac{\partial v}{\partial r} + \frac{1}{\rho}\frac{\partial p}{\partial r} = 0\,,$$

$$\frac{\partial}{\partial t}\left(\frac{p}{\rho^\gamma}\right) + v\frac{\partial}{\partial r}\left(\frac{p}{\rho^\gamma}\right) = 0\,. \tag{6.65}$$

Here, the first equation is the equation of continuity, the second one is the equation of motion, the third one is the adiabatic condition, that is, $ds/dt = 0$ and, according to Eqs. (6.63 and 6.64), though (6.64) follows from (6.63) the entropy is equal to $s = c_v \ln(p/\rho^\gamma)$, $c_v =$ const. To solve this problem, it is necessary to specify the initial conditions, the boundary conditions on the shock wave and the condition that the total energy of gas movement within the spherical domain S is equal to the sum of the energy of the explosion and the initial internal energy of the gas within that volume. The system of determining parameters can be found from the equations (6.65) and the initial conditions. Those parameters are: $p_1, \rho_1, r, t, \gamma$ and E. Since γ is a dimensionless parameter, we have $n = 5$ and $k = 3$. According to the Π-theorem, one can create two dimensionless parameters from the five determining parameters $p_1, \rho_1, r, t, \gamma, E$. The dimensions of these parameters are

$$[\rho_1] = \frac{M}{L^3}, \quad [r] = L, \quad [t] = T, \quad [p_1] = \frac{M}{T^2 L}, \quad [E] = \frac{M L^3}{T^2}.$$

We now look for the dimensionless parameters in the form of monomials according to the dimension formula (6.60)

$$\begin{aligned}&(\rho_1)^\alpha (r)^\beta (t)^\tau (p_1)^\delta (E)^\varepsilon \\ &= \left[\frac{M}{L^3}\right]^\alpha [L]^\beta [T]^\tau \left[\frac{M}{T^2 L}\right]^\delta \left[\frac{M L^3}{T^2}\right]^\varepsilon \\ &= M^{\alpha+\delta+\varepsilon} L^{-3\alpha+\beta-\delta+2\varepsilon} T^{\tau-2\delta-2\varepsilon}.\end{aligned}$$

Since these quantities are supposed to be dimensionless, the powers have to add up to zero. As a result, one obtains three equations for five unknowns. The two linear independent solutions for $\beta = 0$, $\tau = 1$ and $\beta = 1$, $\tau = -2/5$ are $\alpha = -1/2$, $\varepsilon = -1/3$, $\delta = 5/6$ and $\alpha = 1/5$, $\varepsilon = -1/5$, $\delta = 0$. At the end, there are three dimensionless parameters, that is,

$$\gamma; \lambda = \rho_1^{1/5} E^{-1/5} t^{-2/5} r; \theta = p_1^{5/6} E^{-1/3} \rho_1^{1/2} t.$$

To obtain the functional dependence (6.60), one has to use the Π-theorem. As an example, we obtain a formula for the pressure behind the shockwave. First, we create the combination E/r^3 with the help of the dimension formula from the determining parameters. This combination has the dimension of pressure. Then, one obtains

$$p = \frac{E}{r^3} f(\gamma, \lambda, \theta). \tag{6.66}$$

For the determination of the function f, it is necessary to solve the system of equations (6.65) with the corresponding initial and boundary conditions. We consider the boundary conditions on the domain boundary of the disturbance, that is, on the shock wave. These conditions connect the parameter values before and after the shock wave according to (6.56). Since these conditions were written in a system of coordinates relative to which the shock wave was at rest, we assume

$V_{t1} = V_{t2} = 0$, $V_{n1} = -V_2$, $V_{n2} = -D$ in (6.56). By taking into account the formula $a_1 = \sqrt{\gamma p_1/\rho_1}$ for the velocity of sound, one obtains

$$v_2 = D\left(\frac{2}{\gamma+1} - \frac{2\gamma}{(\gamma+1)}\frac{p_1}{\rho_1 D^2}\right), \quad \rho_2 = \frac{(\gamma+1)}{(\gamma-1)}\frac{\rho_1}{\left[1 + \frac{2\gamma p_1}{(\gamma-1)\rho_1 D^2}\right]},$$

$$p_2 = \frac{2}{\gamma+1}\rho_1 D^2\left[1 - \frac{(\gamma-1)p_1}{2\gamma\rho_1 D^2}\right]. \tag{6.67}$$

Due to the strong explosion, the pressure p_2 behind the shock wave is initially much larger than before the shock wave. Therefore, we have $\gamma p_1/\rho_1 D^2 \ll \gamma p_2/\rho_1 D^2$ and $D^2 \gg a_1^2$. Accurate to a_1^2/D^2, one obtains the simplified boundary conditions

$$v_2 = \frac{2}{\gamma+1}D, \quad \rho_2 = \frac{(\gamma+1)}{(\gamma-1)}\rho_1, \quad p_2 = \frac{2}{\gamma+1}\rho_1 D^2. \tag{6.68}$$

If, instead of (6.67), one uses (6.68), the pressure p_1 is dropped from the list of determination parameters. Therefore, there are only two dimensionless parameters, γ and λ. Thus, the solution behind the shock wave for a strong explosion can be written as

$$v = v_2 V(\gamma, \lambda); \quad p = p_2 P(\gamma, \lambda); \quad \rho = \rho_2 R(\gamma, \lambda). \tag{6.69}$$

We denote the radius of the shock wave as r_2. Then, $D = dr_2/dt$, where r_2 and D depend on the determination parameters t, ρ_1, E and γ. Since one cannot create a dimensionless parameter from the dimensional parameters t, ρ_1 and E, every dimensional parameter can be expressed according to (6.64) as monomial in terms of t, ρ_1, E. Using the dimension formula (6.60), one obtains

$$r_2 = k\left(\frac{E}{\rho_1}\right)^{1/5} t^{2/5}, \quad D = \frac{2}{5}k\left(\frac{E}{\rho_1}\right)^{1/5} t^{-3/5} = \frac{2}{5}k^{5/2}\left(\frac{E}{\rho_1}\right)^{1/2} t^{-3/2}. \tag{6.70}$$

The constant k can be obtained from the condition of energy conservation

$$E = \int_0^{r_2}\left(\frac{v^2}{2} + \frac{1}{(\gamma-1)}\frac{p}{\rho}\right)4\pi r^2 \rho dr. \tag{6.71}$$

The integrand consists of kinetic and internal energy.

Problem of the Movement of a Body in a Fluid or Gas

We consider one of the main problems of continuum mechanics, the problem of the body movement in a fluid or a gas. There are two different problem scenarios.

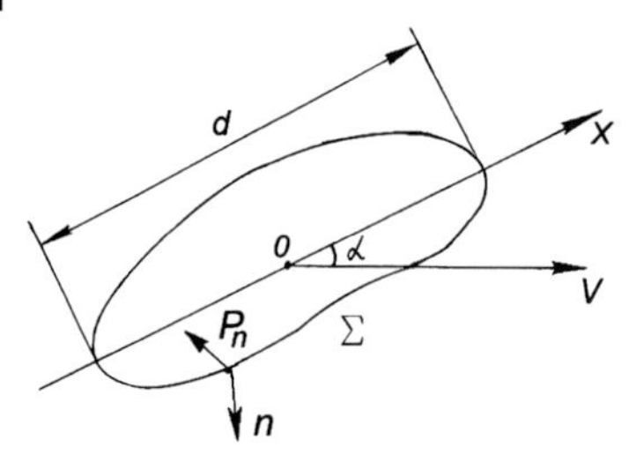

Figure 6.8 Movement of a body in a viscous, incompressible fluid.

The first one relates to the absolute movement of the body, where one assumes that the fluid or gas is at rest far away from the body, but the body undergoes translatory movement with a constant velocity V. The second scenario corresponds to the situation where the body is at rest and the fluid and gas flows around it with constant velocity V at infinity. The only difference between the two problems posed is the system of coordinates, relative to which the movement is observed. In the first scenario, the system of coordinates is fixed to the fluid while it is fixed to the body in the second one. Since both system of coordinates are inertial, the systems of the equation of motion for the fluid coincide in both cases. The interactions connected to forces are also identical. Therefore, one can, when examining the movement of a body in the fluid or gas, consider the equivalent problem of the body at rest around which the fluid is flowing, when dealing with the determination of the force acting from the body on the fluid or from the fluid on the body. We consider these problems and apply the method of dimensional analysis.

We consider the movement of a body in a viscous, incompressible and homogeneous fluid (see Figure 6.8). The mechanical properties of the fluid are determined by two constants: the mass density ρ and viscosity coefficient μ. In the system of coordinates fixed to the body, the stationary velocity field of the fluid movement, the distribution of the pressure and of the internal viscous stresses are described by the Navier–Stokes equations. The corresponding determining parameters of the problem are the mass density ρ, the viscosity coefficient μ, a characteristic linear dimension of the body d, the velocity of the body or the fluid at infinity V, the angles α and β, describing the orientation of the X-axis relative to the vector $\boldsymbol{V}$ (in the figure, we only consider a two-dimensional problem for simplicity) and the pressure at infinity p_∞.

On every surface element of the body $d\sigma$, the fluid exerts the force $\boldsymbol{p}_n d\sigma$ such that the resulting force $\boldsymbol{P}$, the main vector, and the resulting torque $\boldsymbol{M}$, the main torque, are given by the equations:

$$\boldsymbol{P} = \int_\Sigma \boldsymbol{p}_n d\sigma\,; \quad \boldsymbol{M} = \int_\Sigma \left(\boldsymbol{r} \times \boldsymbol{p}_n\right) d\sigma\,. \tag{6.72}$$

We write the main vector as the sum $\boldsymbol{P} = \boldsymbol{W} + \boldsymbol{A}$, $\boldsymbol{P} = \boldsymbol{W} + \boldsymbol{A}$, where $\boldsymbol{W}$ is the component of the force in the direction of the velocity $\boldsymbol{V}$ and is called the resistance force of the body, and $\mathbf{A}$ is the component of the force perpendicular to the direction of the velocity and is called the lift. At first sight, the determining parameters are ρ, μ, d, V, α, β and p_∞, but the last parameter can be eliminated from the determining parameters since the pressure only appears through its derivatives in the Navier–Stokes equations. Therefore, it is always possible to introduce a new pressure $p_1 = p - p_\infty$ without affecting the Navier–Stokes equations. However, replacing p by p_1 could influence the main vector and main torque. However, since we have for a closed surface $\int_\Sigma \boldsymbol{n} d\sigma = 0$ and $\int_\Sigma (\boldsymbol{r} \times \boldsymbol{n})\, d\sigma = 0$, one obtains

$$\boldsymbol{P}_1 = \int_\Sigma (\boldsymbol{p}_n - p_\infty \boldsymbol{n}) d\sigma = \boldsymbol{P}\ ; \quad \boldsymbol{M}_1 = \int_\Sigma \left[\boldsymbol{r} \times (\boldsymbol{p}_n - p_\infty \boldsymbol{n})\right] d\sigma = \boldsymbol{M}\ .$$

Therefore, the force components W and A only depend on ρ, μ, d, V, α and β. It must be noted that this procedure is only feasible for an incompressible fluid. Out of the four remaining dimensional parameters ρ, μ, d, V, one can only construct one dimensionless combination, namely,

$$\mathrm{Re} = \frac{\rho V d}{\mu}\ . \tag{6.73}$$

It is called Reynolds number and plays a fundamental role for problems concerning the movement of viscous fluids. Thus, in the problem at hand, there are three dimensionless parameters α, β and Re. Since the combinations $\rho d^2 V^2$ and $\rho d^3 V^2$ have the dimensions of force and torque, we have, due to the Π-theorem,

$$W = \rho d^2 V^2 C_W(\alpha, \beta, \mathrm{Re})\ ; \quad A = \rho d^2 V^2 C_A(\alpha, \beta, \mathrm{Re})\ ;$$
$$M = \rho d^3 V^2 C_M(\alpha, \beta, \mathrm{Re})\ . \tag{6.74}$$

The influence of the viscosity of the fluid on the movement and the resistance of the body happens via Reynolds number. From the form of Reynolds number, if follows that $\mathrm{Re} \ll 1$ for a highly viscous fluid and $\mathrm{Re} \gg 1$ for a fluid with small viscosity. It should be noted that such large or small values of Reynolds number can be obtained via changing the dimensions of the body or the velocity. We consider these two cases more closely since they cover two large areas of fluid and gas mechanics.

Movement of a Body in a Viscous Fluid for Re ≪ 1

A reduction of Reynolds number is equivalent to an increase of the viscosity of the fluid or a decrease of the density of the fluid, or of the velocity of the fluid V or of the linear dimension of the body. For a small Reynolds number, it follows from the Navier–Stokes equations that the viscous force plays the main role in comparison to the inertial force. Therefore, the Navier–Stokes equations become equivalent to the Stokes equations for $\mathrm{Re} \ll 1$. In the absence of mass forces, they have the following form:

$$\nabla \cdot V = 0\ ; \quad -\nabla p + \mu \Delta V = 0\ . \tag{6.75}$$

The resistance force of the body is determined by the parameters μ, d, V, α and β. Since one cannot form a dimensionless quantity from the quantities μ, d and V, we have

$$W = \mu d V C(\alpha, \beta) = \rho d^2 V^2 \frac{C(\alpha, \beta)}{\text{Re}} . \tag{6.76}$$

Therefore, the resistance force of the body is proportional to V for small Reynolds numbers. The constant C can be determined by the solution of (6.75). In the case of the flow around a sphere, we have $C = 3\pi$.

Movement of a Body in a Viscous Fluid for Re ≫ 1

For an ideal fluid, we have $\mu = 0$ and $\text{Re} = \infty$. Large Reynolds numbers for $\mu \neq 0$ correspond to $Vd \to \infty$. Thus, larger Reynolds numbers are realized for the flow of a fluid with a small viscosity, for the movement of large bodies and for a movement with large velocity.

Then, the viscous force is small in comparison to the inertial force and thus Navier–Stokes equations become equivalent to Euler's equations (4.5). For $\text{Re} \gg 1$, the resistance force and the buoyancy depend on ρ, d, V, α and β.

Since it is impossible to form a dimensionless quantity from ρ, d, V, we have

$$W = \rho d^2 V^2 C_W(\alpha, \beta) ; \quad A = \rho d^2 V^2 C_A(\alpha, \beta) . \tag{6.77}$$

From (6.77), it follows that for $\text{Re} \gg 1$, the resistance force is, unlike the case $\text{Re} \ll 1$, proportional to V^2. The mass forces have not yet been taken into account. However, for some problems, the mass forces, that is, gravity, can play an important role. When considering the movement of a ship on the surface of the water, a part of the surface boundary is free surface. The disturbed movement of the wave depends on its specific weight and therefore the water surface is covered by a system of waves. Therefore, the resistance force and the buoyancy of the ship depend on the specific weight of the fluid. Thus, the gravitational acceleration g has to be included in the list of the determining parameters.

As an example, consider the problem of the movement of a body on the surface of a fluid. The determining parameters are ρ, g, μ, d, V, from which one can form two dimensionless parameters $\text{Re} = \rho d V/\mu$ and $\text{Fr} = V/\sqrt{gd}$. Thus, adding the gravitation leads to a new dimensionless parameter which is called Froud's number. The resistance force can then be written as

$$W = \rho d^2 V^2 C_W(\text{Re}, \text{Fr}) . \tag{6.78}$$

Movement of a Body in an Ideal Gas

We now consider the movement of body in an ideal compressible gas. We replace this problem by the equivalent one of the gas flow around a body at rest under the conditions that at infinity $p_\infty \rho_\infty$ and V_∞ are given. The determining parameters are $p_\infty, \rho_\infty, V_\infty, \gamma, \alpha, \beta, d, X, Y, Z$. The coordinates X, Y, Z are defined in the system of coordinates that is fixed to the body. From ten parameters, one can form seven dimensionless parameters, namely, γ, $M_\infty = V_\infty/a_\infty, \alpha, \beta, X/d, Y/d, Z/d$, where M_∞ is Mach's number at infinity and $a_\infty = \sqrt{\gamma p_\infty/\rho_\infty}$ is the speed of sound at infinity. Thus, the resistance force and the buoyancy have the following forms:

$$W = \rho d^2 V_\infty^2 C_W(\alpha, \beta, M_\infty) ; \quad A = \rho d^2 V_\infty^2 C_A(\alpha, \beta, M_\infty) . \tag{6.79}$$

If, for this problem, the gas flow around an infinite body (wedge or cone) is considered, the linear dimension of the body falls away. Then, the determining parameters are γ, M, α, β and the similarity variables X/Y and Z/Y.

Dimensional Analysis of the Navier–Stokes Equations

Until now, we have not applied the dimensional analysis to processes. It is shown in the following example how the dimensional analysis can be applied to the equations given for a problem. The movement of a viscous, incompressible and homogeneous fluid is described via the Navier–Stokes equations (4.32) and (4.33),

$$\begin{aligned} &\nabla \cdot V = 0 , \\ &\frac{\partial V}{\partial t} + (V \cdot \nabla) V = f - \frac{1}{\rho} \nabla p + \frac{\mu}{\rho} \nu \Delta V . \end{aligned} \tag{6.80}$$

We assume that there are the following characteristic parameters: length d, time τ, force F, velocity u, pressure q, mass density ρ, the coefficient of viscosity μ, the coordinates X_k and the progressing time t. Thus, determining parameters are d, τ, F, u, q, ρ, μ. X_k and t. We introduce the following dimensionless variables: dimensionless time $t' = t/\tau$, dimensionless coordinates $X_k' = X_k/d$, dimensionless density of mass forces $f' = f\rho/F$, dimensionless pressure $p' = p/q$ and dimensionless velocity $V' = V/u$. By substituting these variables into (6.80), one obtains

$$\begin{aligned} &\nabla \cdot V' = 0 , \\ &\mathrm{Sh} \frac{\partial V'}{\partial t'} + (V' \cdot \nabla) V' = \frac{1}{\mathrm{Fr}} f' - \mathrm{Eu} \nabla p' + \frac{1}{\mathrm{Re}} \Delta V' . \end{aligned} \tag{6.81}$$

The dimensionless parameters that appear here are: Strouchal's number $\mathrm{Sh} = d/(u\tau)$, Euler's number $\mathrm{Eu} = q/(\rho u^2)$, Reynolds number $\mathrm{Re} = \rho u d/\mu$ and Froud's number $\mathrm{Fr} = \rho u^2 d^2/F$.

6.7 Similarity and Modeling of Phenomena

Modeling of a certain phenomenon or of a process means the examination of a natural phenomenon by the examination of a similar phenomenon using a smaller or larger model in special laboratory conditions.

The modeling is mostly based on the observation of similar phenomena. The mechanical and physical similarity can be considered as a generalization of the geometric similarity. Two physical phenomena are similar if from the given characteristics, one can easily calculate the characteristics of the other by a simple calculation for the conversion of quantities from one system of units to the other. Thus, the numerical characteristics of two different but physically similar phenomena can be considered as the numerical characteristics of the same phenomenon in different systems of units.

The physically similar phenomena are described by the same system of equations that shows the physical laws of the process. Therefore, its solutions are independent of the system of units if the these equations are written in a dimensionless form. Thus, the main problem consists of determining the system of dimensionless parameter, that completely describes the considered phenomena.

If n determination parameters are chosen, it is possible to form $n - k$ dimensionless combinations where k is the number independent units. Then, it follows from the Π-theorem that all dimensionless characteristics are functions of these $n - k$ parameter, that is, used as a basis.

In order to select from a certain class of phenomena those that are similar, one has to use the following definition. Two phenomena are similar if the numerical values of the corresponding dimensionless parameters that form the basis for these phenomena are identical. The conditions of the invariance of the dimensionless parameters of the basis are called similarity criteria.

Above, we determined the the dimensionless combinations for some problems of fluid and gas flow. We now consider the dimensional analysis introduced above from the perspective of similarity theory.

Similarity for the Flow of a Viscous, Incompressible Fluid around a Body

Since the basis of this problem consists of α, β and Re, the the conditions of the similarity criterion means that the following must be satisfied: $\alpha = \text{const}$, $\beta = \text{const}$, $\text{Re} = \text{const}$.

These conditions mean that when modeling the phenomena, the experimental results can only be transfered to nature for equal values of α, β and Re in the model and in nature. The first two conditions can be easily realized in practice. The third condition, however, is very difficult to satisfy, especially if the body around which the fluid flows has large dimensions.

Since the model is normally smaller than in nature, one either has to increase the velocity of flow around the body or change the mass density and viscosity of the fluid drastically. In the problems of aerodynamics, the condition $\text{Re} = \text{const}$ leads to the necessity to construct large wind channels, in which the planes can be tested in their natural size.

Similarity for the Flow of a Compressible Gas around a Body

The compressibility of the gas only becomes evident at relatively large velocities when $\text{Re} \gg 1$. Then, the viscosity is not important and the following similarity criteria apply to the ideal gas: $\alpha = \text{const}$, $\beta = \text{const}$, $\gamma = \text{const}$, $M_\infty = \text{const}$.

The third condition means that the same gas as in nature must be used for the model. The condition M_∞ has to be satisfied by the experimental conditions. It must be noted that instead of an apparent simplicity for the process of modeling, considerable difficulties exist that are caused by the fact that the attempt to keep the dimensionless parameters constant can change the parameters in such a way as to violate the limits of the class of the phenomena under consideration.

For example, the compressibility of the air is not important for the stationary flow of air around a wing for $M \ll 1$, and the similarity criterion is $\mathrm{Re} = \mathrm{const}$. When reducing the size of the model, this condition is equivalent to a considerable increase of the flow velocity around the wing, which in turn results in an increase of Mach's number for which the compressibility of the gas has to be taken into account and the viscosity does not play a role. Thus, one moves from the class of flows for an ideal incompressible fluid to the class of flows for an ideal compressible gas.

As another example, we can consider modeling the process of a movement of a body on the surface of a fluid. In Section 6.6, we have shown that the main similarity criteria are $\mathrm{Re} = \mathrm{const}$ and $\mathrm{Fr} = \mathrm{const}$. For a model that is of smaller dimensions in nature, the first condition requires an increase of the fluid velocity, while the second requires, on the contrary, a decrease of the fluid velocity. Therefore, it is not possible to fully model the problems of this kind as they, for example, appear in maritime transport.

Similarity for the Time Dependent Flow of a Viscous, Incompressible Fluid around a Body

The flow of a viscous, incompressible fluid is described by the Navier–Stokes equations. In Section 6.6 above, the dimensionless combinations Re, Fr, Eu and Sh were found and the Navier–Stokes equations were brought into a dimensionless form. Therefore, the conditions

$$\mathrm{Re} = \mathrm{const}\,, \quad \mathrm{Fr} = \mathrm{const}\,, \quad \mathrm{Eu} = \mathrm{const}\,, \quad \mathrm{Sh} = \mathrm{const} \tag{6.82}$$

serve as the similarity criteria for modeling the time-dependent flow of viscous, incompressible fluid.

However, these parameters are not determining parameters for all problems. We consider some of the common cases:

- For $\mathrm{Fr} \gg 1$ the mass forces have a small influence on the pattern of the hydrodynamic flow and both this parameter as well as the corresponding term in (6.81) can be neglected. For $\mathrm{Fr} \ll 1$, the mass forces are, on the contrary, a determining parameter and one has to take them into account. If these mass forces have a potential such as the force of gravitation, one can combine them with the pressure gradient in (6.81).
- Euler's number Eu is important for those problems where one has to calculate the pressure gradient. In the opposite case, the pressure gradient can be eliminated from the consideration by applying the curl on both sides of (6.81)

- For the examination of a fast moving process, the characteristic time τ is small and the corresponding Strouchal's number is large. In this case, that local acceleration has a large influence on the flow patterns. Examples of such flows are those that arise because of the collision of a body with the water surface or because of the sudden acceleration of a body in the fluid.
- Reynolds number Re is of particular importance. Since Re is indirectly proportional to the viscosity, we have $\text{Re} \gg 1$ for fluids of low viscosity and one can neglect the viscous force in (6.81). As the final result, one obtains Euler's equations. It must be noted that when considering stationary movement in the potential field of the mass forces and if one is not interested in the pressure distribution, Reynolds number is the only one of the parameters mentioned above that have an influence on the hydrodynamic flow pattern.

Finally, it must be noted that the parameters of all possible dimensionless combinations mentioned above are not complete for certain situations. Depending on the character of the problem, the number of these parameters can decrease or increase because additional dimensional determining parameters can arise. For example, when considering problems of the surfaces of drops and bubbles, a new parameter enters the picture, namely, the coefficient of surface tension. For the hydrodynamic problems of heat and mass transport, the coefficients of diffusion and heat conduction appear, and for the flows of chemically active substances, the constants describing the chemical reactions are added and so on.

7
Hydrostatics

Hydrostatics refers to the theory of the equilibrium of a fluid or gas relative to a chosen system of coordinates.

7.1
Equilibrium Equations

For the equilibrium, we have $V = 0$ and from the equation of continuity (3.6), it follows that $\partial\rho/\partial t = 0$. Therefore, $\rho = \rho(X, Y, Z)$. Then, one obtains, from Euler's equations (4.5) or Navier–Stokes equations (4.31), that

$$\nabla p = \rho \boldsymbol{f} \,. \tag{7.1}$$

If the external forces are not taken into account, that is, $\boldsymbol{f} = 0$, (7.1) becomes

$$\nabla p = 0 \,.$$

Therefore, the pressure p is the same at every point of the fluid. This consequence is called Pascal's law. We now consider the case $\boldsymbol{f} \neq 0$. From (7.1), one obtains

$$\nabla \times \boldsymbol{f} = \nabla \times \left(\frac{1}{\rho}\nabla p\right) = \nabla\left(\frac{1}{\rho}\right) \times \nabla p = \rho\nabla\left(\frac{1}{\rho}\right) \times \boldsymbol{f} \,. \tag{7.2}$$

For the derivation of the above, the following properties of vector fields have been used:

$$\nabla \times (c\boldsymbol{a}) = \nabla c \times \boldsymbol{a} + c\nabla \times \boldsymbol{a} \,, \quad \nabla \times \nabla p = 0 \,.$$

By evaluating the scalar product with $\boldsymbol{f}$ on both sides of (7.2), one obtains

$$\boldsymbol{f} \cdot (\nabla \times \boldsymbol{f}) = 0 \,. \tag{7.3}$$

Thus, a fluid under the influence of an external force $\boldsymbol{f}$ can only be in mechanical equilibrium if (7.3) is satisfied.

For a homogeneous fluid ($\rho = \text{const}$), from (7.2), one obtains

$$\nabla \times \boldsymbol{f} = 0 \quad \text{or} \quad \boldsymbol{f} = \nabla \Phi \,. \tag{7.4}$$

Hydromechanics. Theory and Fundamentals. Emmanuil G. Sinaiski

ISBN: 978-3-527-41026-2

From this, we conclude that the homogeneous, incompressible fluid can only be in equilibrium if the external mass force is (derived from a) potential.

We now consider the case of a compressible fluid in the field of a potential mass force. From (7.1), we conclude that $\nabla p = \rho \nabla \Phi$ and

$$dp = \nabla p \cdot d\boldsymbol{r} = \rho \nabla \Phi \cdot d\boldsymbol{r} = \rho d\Phi \ . \tag{7.5}$$

For $\Phi = \text{const}$, we have $\rho = \text{const}$ so that $p = p(\Phi)$. From the equation $dp/d\Phi = \rho$, one obtains $\rho = \rho(\Phi)$. Therefore, the pressure and mass density in the potential field of an external force only depend on the potential Φ.

From the conditions on discontinuity surfaces (6.33)–(6.38), it follows that the only quantity that can be discontinuous in a fluid at rest is the mass density. Therefore, the pressure p and the force f are continuous on the discontinuity surfaces. If the force is derived from a potential, Φ will be continuous on the discontinuity surfaces. From (7.4), it follows that the density jump on the discontinuity surface can only happen for $dp = d\Phi = 0$. Therefore, a discontinuity surface must coincide with an equipotential surface in a fluid at rest.

7.2 Equilibrium in the Gravitational Field

We now consider the equilibrium of a fluid or gas in the gravitational field. We align the Z-axis opposite to gravity. Then, we have $f_x = f_y = 0, \quad f_z = -g$ and from (7.4) and (7.5), one obtains $\Phi = -gZ + \text{const}$, $p = p(Z)$, $\rho = \rho(Z)$. From the last two equations, it follows that the surfaces of constant pressure (isobars) and constant mass density (isochores) are the horizontal planes $Z = \text{const}$. Using the equation of state $f(p, \rho, T) = 0$, one concludes that for a heavy fluid at rest, the temperature is only a function of the vertical coordinate. From (7.5), it follows that $dp/dZ = \rho d\Phi/dZ = -\rho g$, that is, the pressure decreases with height. The pressure difference between two levels Z and Z_0 is given by

$$p - p_0 = -g \int_{Z_0}^{Z} \rho dZ \ , \tag{7.6}$$

that is, the pressure difference between two points at different heights is equal to the weight of the fluid column between the two heights. Equation 7.6 does not depend on the form of the gas or fluid area or on the physical properties of the gas or fluid. For a homogeneous fluid ($\rho = \text{const}$), from (7.6), one obtains

$$p - p_0 = \rho g(Z - Z_0) \ . \tag{7.7}$$

For $Z_0 = 0$ and $Z = h$, we have

$$p - p_0 = \rho g h \ . \tag{7.8}$$

Using (7.8), one can calculate the pressure on the bottom of container that is filled with a fluid.

For an ideal gas, we have $p = \rho RT$ and from the equation $dp = \rho g dZ$, we obtain

$$p = p_0 \exp\left(-\int_{Z_0}^{Z} \frac{g dZ}{RT(Z)} \rho dZ\right). \tag{7.9}$$

Equation 7.9 is referred to as the barometric height formula. For $T = \text{const}$, for the pressure as a function of height, one obtains

$$p = p_0 \exp\left(-\frac{g}{RT}(Z - Z_0)\right). \tag{7.10}$$

7.3
Force and Moment that Act on a Body from the Surrounding Fluid

We consider a solid body that is submerged in a fluid. For a fluid at rest, the force and the torque that the surrounding ideal or viscous fluid exerts on the body is determined by (6.72), that is,

$$\begin{aligned}
\boldsymbol{A} &= \int_{\Sigma} \boldsymbol{p}_n d\sigma = -\int_{\Sigma} p \boldsymbol{n} d\sigma ; \\
\boldsymbol{M} &= -\int_{\Sigma} (\boldsymbol{r} \times \boldsymbol{p}_n)\, d\sigma = \int_{\Sigma} p\, (\boldsymbol{r} \times \boldsymbol{n})\, d\sigma .
\end{aligned} \tag{7.11}$$

The equilibrium of the surrounding fluid and the body submerged in it will not be disturbed if one replaces the volume of the body in the experiment by an equivalent fluid volume, and the force **A** will also not be changed. By aligning the Z-axis against gravity and applying Gauss theorem to (7.11), one obtains

$$\begin{aligned}
\boldsymbol{A} &= -\int_{\Sigma} p \boldsymbol{n} d\sigma = -\int_{\Sigma} p\, (\cos(n, X)\boldsymbol{i} + \cos(n, Y)\boldsymbol{j} + \cos(n, Z)\boldsymbol{k})\, d\sigma \\
&= -\boldsymbol{i}\int_{\Sigma} p \cos(n, X) d\sigma - \boldsymbol{j}\int_{\Sigma} p \cos(n, Y) d\sigma - \boldsymbol{k}\int_{\Sigma} p \cos(n, Z) d\sigma \\
&= -\boldsymbol{i}\int_{V} \frac{\partial p}{\partial X} d\tau - \boldsymbol{j}\int_{V} \frac{\partial p}{\partial Y} d\tau - \boldsymbol{k}\int_{V} \frac{\partial p}{\partial Z} d\tau = -\int_{V} \nabla p\, d\tau \\
&= \int_{V} \rho \boldsymbol{f} d\tau = \int_{V} \rho g \boldsymbol{k} d\tau = -\boldsymbol{G} ,
\end{aligned} \tag{7.12}$$

where $\boldsymbol{F}$ is the gravitational force due to fluid in the volume of the solid body. Therefore, the buoyancy **G** that acts on a body submerged in the fluid is equal to the gravitational force of the fluid in the volume of the solid body or, as it is said, equal to the gravitational force experienced by the fluid volume that has been displaced by the submerged body. Equation 7.11 contains Archimedes' law. The force of Archimedes' law acts on the center of mass of the displaced fluid. If the body is only partially submerged, one has to use as Σ in (7.11) the surface Σ_{wetted} of the body that is wetted and the cross sectional surface Σ_1 of the body that is at the same level as the surface of the fluid at rest in order to calculate the force $\boldsymbol{A}$. The pressure on this cross sectional surface has to be the same as on the free surface of the fluid, that is, p_0. If one repeats the derivation leading to (7.11) with the displaced fluid volume modified to include only that part of the body that is bounded by $\Sigma_{\text{wetted}} + \Sigma_1$, one obtains Archimedes' law.

The necessary condition for Archimedes' law is that the surface on which the body is in contact with fluid, that is, the osculation surface, is closed. If the osculation surface is not closed, that is, if the body lies on the bottom, Archimedes' law is not valid. In this case, the force of buoyancy does not act and the forces exerted by the fluid on the body push the body to the bottom. Until now, we have only dealt with the body at rest. If the body or the fluid are moving, the force exerted on the body is not only determined by the hydrostatic pressure. The hydrostatic pressure only causes part of the force.

In connection with the phenomenon of buoyancy, it is appropriate to treat the question of stability of an incompressible fluid. If there is another fluid within a container with water, that is, a layer of mercury, there can be two different equilibrium states, namely, when the mercury is either below or on top of the water layer. The the question arises as to whether these equilibrium state are stable. Before giving the answer, one first has to define stability. An equilibrium is called stable if the system strives to return to its initial state after an arbitrary small displacement away from the initial state. An equilibrium is unstable if such a small displacement (disturbance) of the system results in the system moving further away from the initial state. An equilibrium is called indifferent if any small displacement does not disturb the equilibrium.

From these definitions, it follows that if the heavy fluid (mercury) lies on top of the light fluid (water), the equilibrium is unstable. Indeed, one can consider a mercury particle that is accidentally displaced downwards from the separation surfaces as a particle, that is, submerged in water. Then, two forces will act on it: the gravitational force that is directed downwards, and Archimedes' force that is directed upwards. Since the mass density of mercury is larger than the mass density of water, the mercury particle will move further away from the initial state. Therefore, the state when the heavy fluid lies on top of the light fluid is unstable.

By a similar train of thought, for the case when the light fluid lies on top of the heavy fluid, one can easily convince oneself of the stability of this state.

The situation for the stability of the equilibrium state of a gas is more complicated. The reflections above cannot be used for a gas since the mass density of the gas particle that moves from one layer to the other changes as a function of time. However, similar qualitative reflections can be made nonetheless. For example, convection in the atmosphere is often the consequence of a density change caused by heating lower layers of the atmosphere.

7.4
Equilibrium of a Fluid Relative to a Moving System of Coordinates

We consider the equilibrium of a fluid in a container whose wall is turning with a constant angular velocity ω around the vertical Z-axis. We fix the system of coordinates on the bottom of the container and align the Z-axis along the axis of rotation (see Figure 7.1). We assume that the fluid is at rest with respect to the wall. Then, one has to add the centrifugal force density in addition to the gravitational force density on the right-hand side of the equilibrium equations. After projection on the coordinate axes, one obtains

$$\frac{\partial p}{\partial X} = \rho\omega^2 X\,, \quad \frac{\partial p}{\partial Y} = \rho\omega^2 Y\,, \quad \frac{\partial p}{\partial Z} = -\rho g\,. \tag{7.13}$$

The solution of the system of equations has the form

$$p = -\rho g Z + \frac{\rho\omega^2 r^2}{2} + C\,, \quad r^2 = X^2 + Y^2\,.$$

To determine C, we take $p = p_0$ for $r = 0$ and $Z = Z_0$, corresponding to the free surface. Then, one obtains

$$p = p_0 - \rho g(Z - Z_0) + \frac{\rho\omega^2 r^2}{2}\,.$$

Assuming $p = p_0$ in the last equation, one obtains the equation of the free surface in the turning container, that is,

$$Z - Z_0 = \frac{\omega^2 r^2}{2g}\,.$$

Thus, the free surface is paraboloid. All other isobaric surfaces have a similar form. The force vector is directed along the normal to the corresponding fluid. We now

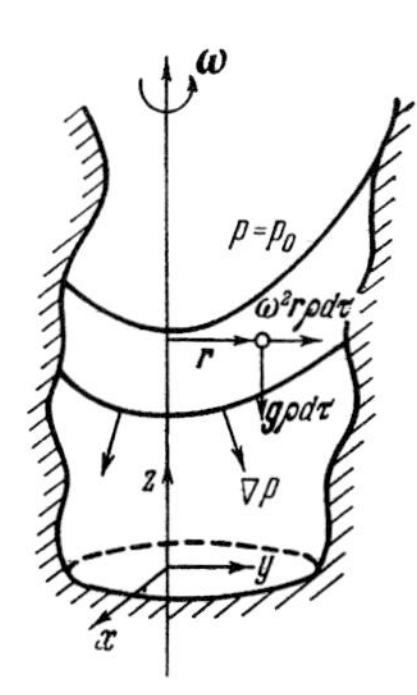

Figure 7.1 Equilibrium of fluid in a container rotating with constant angular velocity.

consider the force that acts on a body that is submerged in the rotating fluid. In preparation, it must be noted that the inertial force in (7.13) corresponds to the potential $\Phi_{\text{in}} = \omega^2 r^2/2$, that is, $\boldsymbol{f}_{\text{in}} = \nabla \Phi_{\text{in}}$. Then, one can write (7.13) in the following form:

$$\nabla p = \rho\omega^2 \boldsymbol{r} + \rho \boldsymbol{g} \,. \tag{7.14}$$

By substituting the last equation into (7.11), one obtains

$$\boldsymbol{A} = \int_V \rho \boldsymbol{g} d\tau - \int_V \rho\omega^2 \boldsymbol{r} d\tau = -\boldsymbol{G} - \rho\omega^2 V \boldsymbol{r}' \,. \tag{7.15}$$

Here, we introduce the notation $\boldsymbol{r}' = \frac{1}{V}\int_V \boldsymbol{r} d\tau$. $\boldsymbol{f}'$ is the vector from the rotation axis to the center of mass of the displaced fluid volume which is orthogonal to the rotation axis.

To summarize the above, one can say that for a fluid in constant rotation, the main vector $\boldsymbol{A}$ of the pressure forces applied by the fluid on the totally submerged body is equal to the sum of Archimedes' buoyancy $\boldsymbol{G}$ and the additional force $\boldsymbol{G}' = -m\omega^2 \boldsymbol{r}$ which is equivalent to a centripetal force directed toward the rotation axis. Now, one can consider all forces that act on the body which is rotating with the fluid around the Z-axis with the angular velocity ω. We assume that the mass of the body is given by $m_B = \rho_B V$. Then, the following forces act on the body: gravity $\boldsymbol{G}_B = m_B \boldsymbol{g}$, centrifugal force $\boldsymbol{f}_C$ and the additional centripetal force $\boldsymbol{G}' = \rho\omega^2 \boldsymbol{r}$. Thus, the resulting force is equal to

$$\boldsymbol{G}'_B - \boldsymbol{G} + (\rho_B - \rho)\omega^2 V \boldsymbol{r}' = (\rho_B - \rho)(\boldsymbol{g} + \omega^2 \boldsymbol{r}')V \,. \tag{7.16}$$

From (7.16), it follows that the bodies will sink in the fluid and be thrown to the periphery for $\rho_B > \rho$. For $\rho_B < \rho$, the bodies emerge from the fluid and are collected at the rotation axis. This effect forms the foundation for the centrifuge process.

8
Stationary Continuum Movement of an Ideal Fluid

8.1
Bernoulli's Integral

The movement of an ideal fluid is described by Euler's equations (4.5). In the Gromeka–Lambs form (4.9), these equations are

$$\frac{\partial V}{\partial t} + \frac{1}{2}\nabla V^2 + 2\,(\boldsymbol{\omega} \times V) = f - \frac{1}{\rho}\nabla p \,. \tag{8.1}$$

We assume that the external mass forces have a potential Φ, that is, $f = \nabla\Phi$. From the stationarity of the movement, it follows that $\partial V/\partial t = 0$. We consider an arbitrary curve L of length l in the fluid flow (see Figure 8.1), which starts in point O. We denote an element of the tangent of L as $d\boldsymbol{l}$ and project (8.1) on the direction of the tangent in M. As final result, one obtains

$$\frac{\partial}{\partial l}\left(\frac{V^2}{2}\right) + \frac{1}{\rho}\frac{\partial p}{\partial l} - \frac{\partial \Phi}{\partial l} = -2\,(\boldsymbol{\omega} \times V)_l \,. \tag{8.2}$$

Along L, mass density and pressure are a function of the length of the curve l. These functions are different for different curves L. Therefore, we have $\rho = \rho(l, L)$ and $p = p(l, L)$. On the other hand, one can consider the mass density along the curve L as a function of the pressure. We define the pressure function P via the equation

$$P = \int\limits_{p_1}^{p} \frac{dp}{\rho(p, L)}\,; \quad \frac{\partial P}{\partial l} = \frac{1}{\rho}\frac{\partial p}{\partial l}\,, \tag{8.3}$$

where $p_1 = \text{const}$. Here, we note that the pressure function is determined up to an additive constant that depends on the choice of the pressure p_1 and will generally depend on L.

However, for some processes, the pressure function can be independent of L. Let us list some examples. In the case of a barotropic change of state, for which the equation $p = p(\rho)$ is given and p_1 does not depend on L, the pressure function is independent of L and can easily be calculated from (8.3). For a homogeneous fluid,

Hydromechanics. Theory and Fundamentals. Emmanuil G. Sinaiski

ISBN: 978-3-527-41026-2

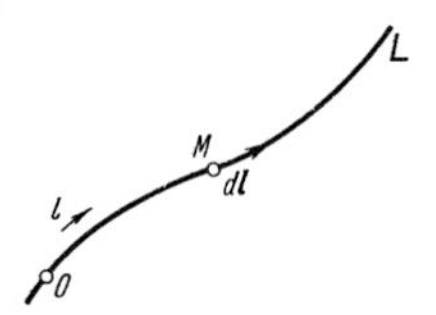

Figure 8.1 Derivation of Bernoulli's integral.

we have ρ = const and $P = p/\rho + \text{const}$. For the case of an isothermal change, that is, T = const in the ideal gas ($p = \rho R T$), one has $P = R T \ln p + \text{const}$.

We now consider the adiabatic reversible flow of an ideal gas. For such a flow, we have $sq^{(e)} = T ds = 0$ and s = const for every fixed particle. However, different particles have different entropy, and therefore the process is not barotropic. This means that $\rho = \rho(p, L)$. Since the flow is stationary, the stream lines and the trajectories of the particles coincide. Then, the particles that are moving along the same stream line have the same entropy, though with different entropy on different stream lines.

Since the equation of state for an ideal gas has, according to (5.37), the following form:

$$\rho = \rho_0 \left(\frac{p}{p_0}\right)^{1/\gamma} e^{(s - s_0)/c_p} = \rho(p, s) ,$$

the pressure function along the stream line L is equal to

$$P(p, L) = \int \frac{dp}{\frac{\rho_0 p^{1/\gamma}}{p_0^{1/\gamma}} \exp\left[\frac{s_0 - s(L)}{c_p}\right]} = \frac{\gamma p_0^{1/\gamma}}{(\gamma - 1)\rho_0} \exp\left[\frac{s_0 - s(L)}{c_p}\right] p^{(\gamma - 1)/\gamma} + \text{const}(L) . \tag{8.4}$$

Using the equation of state once more, one obtains $P(p, L)$ as

$$P(p, L) = \frac{\gamma}{(\gamma - 1)} \frac{p}{\rho} + \text{const}(L) . \tag{8.5}$$

Now, (8.3) is rewritten as

$$\frac{\partial}{\partial l}\left[\frac{V^2}{2} + P(p, L) - \Phi\right] = -2\,(\boldsymbol{\omega} \times \mathbf{V})_l . \tag{8.6}$$

If L is a stream line, we have $(\boldsymbol{\omega} \times \mathbf{V})_l = 0$ since the vector $\boldsymbol{\omega} \times \mathbf{V}$ is orthogonal to the stream line. It must be noted that this result also applies to a vortex line, though the pressure functions are different for the two curves. Therefore, we have along a stream line and a vortex line

$$\frac{\partial}{\partial l}\left[\frac{V^2}{2} + P(p, L) - \Phi\right] = 0 . \tag{8.7}$$

From this, it follows that

$$\frac{V^2}{2} + P(p, L) - \Phi = i^*(L) . \tag{8.8}$$

For a known function $P(p, L)$, (8.8) is the first integral of the equation of motion of the ideal fluid and is called Bernoulli's integral. In the case of a barotropic system, we have $p = p(\rho)$. The constant i^* is the same for the whole mass of the fluid and does not depend on the stream line or vortex line if the condition $\boldsymbol{\omega} \times \mathbf{V} = 0$ is satisfied. This condition is satisfied in three cases: V=0, the fluid is at rest; $\boldsymbol{\omega} = 0$, the case of a potential flow, or the vectors $\boldsymbol{\omega}$ and $\mathbf{V}$ are collinear. The last case is not valid for parallel flows of the fluid since one has $\boldsymbol{\omega} \cdot V = 0$ there. In all cases mentioned above, it is sufficient to know the characteristic of motion on the left-hand side of (8.8). The value of i^* is constant along the stream line if the characteristics V, P and Φ are continuous along it.

It turns out that for an adiabatic flow in an ideal gas, the value of i^* is constant along a stream line, provided that the parameters on all stream lines are continuous without taking the mass forces into account, and even if there are compaction shocks. Indeed, in the present case, (8.5) is applicable and the constant does not depend on L. Therefore, one can assume

$$P = \frac{\gamma}{(\gamma - 1)} \frac{p}{\rho} .$$

From the conditions for jumps that are not moving, it follows that the quantity

$$\frac{V^2}{2} + \frac{\gamma}{(\gamma - 1)} \frac{p}{\rho}$$

is conserved along any stream line that crosses the jump. Therefore, the constant i^* is continuous across the jump while s, P and V can be discontinuous. From this, it follows that the presence of compaction shocks in the flow of an ideal gas does not alter the value of i^*, but the value of the entropy is changed on the stream lines that cross the jump. In this case, there is no barotropy in the gas flow. If the movement is continuous and the quantity i^* and entropy are the same on all stream lines, the equation $\boldsymbol{\omega} \times V = 0$ is valid. That means that either we have a potential flow or the stream lines and vortex lines coincide. For a plane parallel flow, we have $\boldsymbol{\omega}$ and thus the flow is a potential flow.

8.2 Examples of the Application of Bernoulli's Integral

We first consider flows of an incompressible fluid. We assume that fluid which is moving in the gravitational field is homogeneous, incompressible and its weight is not neglected. The potential of the external force Φ and the pressure function P are given by $\Phi = -gZ$, $P = p/\rho + \text{const}$. Using (8.8), we obtain that

$$\frac{V^2}{2} + \frac{\gamma}{(\gamma - 1)} \frac{p}{\rho} + gZ = i^* . \tag{8.9}$$

If we know the quantities p_1 and V_1 at any point of the stream line with $Z = Z_1$, one can determine the constant i^*, and obtains, instead of (8.9),

$$\frac{V^2}{2} + \frac{\gamma}{(\gamma-1)}\frac{p}{\rho} + gZ = \frac{V_1^2}{2} + \frac{\gamma}{(\gamma-1)}\frac{p_1}{\rho} + gZ_1 \,. \tag{8.10}$$

We now consider some examples of making use of (8.10).

Flow of Fluid from a Container

We consider the task of determining the velocity of the stationary outflow of fluid from a large container with a small hole (see Figure 8.2). In general, this problem is non-stationary, but when we consider a small time interval and the level of the fluid in the container is lowered sufficiently slowly, one can consider this process as stationary. We now consider a stream line that starts on the free surface and is terminated at the outflow cross section. Using the given values of the pressure and velocity at the initial and final points, one obtains

$$\frac{V^2}{2} + \frac{p_a}{\rho} + gZ = \frac{p_1}{\rho} + g(Z+h) \,,$$

from which we obtain

$$V = \sqrt{\frac{2(p_1 - p_a)}{\rho} + 2gh} \,. \tag{8.11}$$

If the container is open on top, one has $p_1 = p_a$ and therefore

$$V = \sqrt{2gh} \,. \tag{8.12}$$

This equation is called Torricelli's formula.

Pitot's Tube

Pitot's tube (see Figure 8.3 serves as a measure of the flow velocity. The tube consists of a long thin body with a rounded, streamlined front part that reduces the distortion of the velocity field. The tube has two holes with internal channels, through

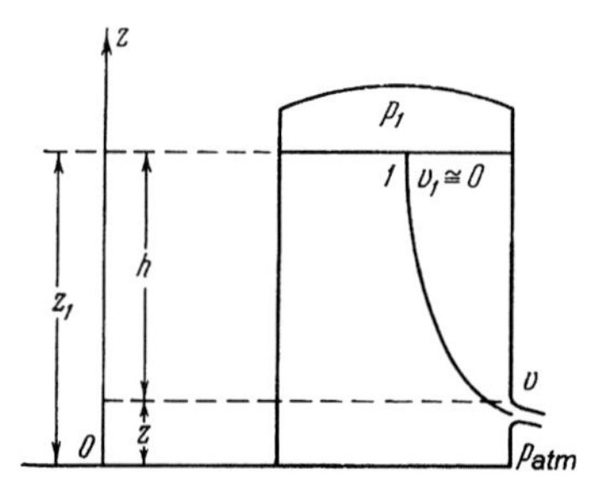

Figure 8.2 Flow of fluid from a container.

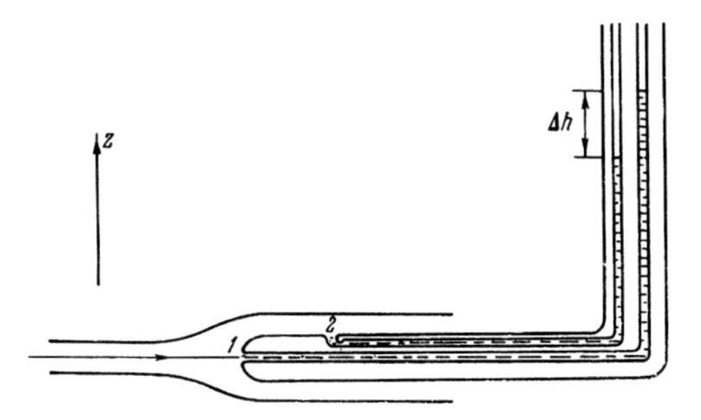

Figure 8.3 Sketch of Pitot's tube.

which the fluid can move. The tube is orientated along the flow. The fluid is lifted up to different heights. Hole one is at the tip of the tube and hole two is on the side farthest from the tip of the tube. Point one is a critical point, for which $V_1 = 0$. We denote the pressures in points one and two by p_1 and p_2. Since the tube is sufficiently thin, one can neglect the height difference $Z_2 - Z_1$. By applying Bernoulli's formula to a stream line that connects points one and two, one obtains

$$\frac{p_1}{\rho} = \frac{V_2^2}{2} + \frac{p_2}{\rho} .$$

From this, it follows that

$$V = \sqrt{\frac{2(p_1 - p_2)}{\rho}} .$$

Since the pressure difference is equal to the difference of the hydrostatic pressures of the fluid column, that is, $p_1 - p_2 = \rho \Delta h$, we obtain

$$V = \sqrt{2g\Delta h} . \tag{8.13}$$

After measuring the difference of the column heights in the channels, one can determine the velocity of the fluid flow.

8.3 Dynamic and Hydrostatic Pressure

According to Bernoulli's equation, the pressure difference between two points on the stream line is given by

$$p - p_1 = \rho g (Z - Z_1) + \frac{\rho V_1^2}{2} - \frac{\rho V^2}{2} . \tag{8.14}$$

We define

$$p_{st} = p_1 + \rho g (Z - Z_1)$$

and call this quantity the hydrostatic pressure. We define the second term on the right-hand side of (8.14) as

$$p_{\text{dyn}} = \frac{\rho V_1^2}{2} - \frac{\rho V^2}{2}$$

and call it dynamic pressure. The total pressure is

$$p = p_{\text{st}} + p_{\text{dyn}} . \qquad (8.15)$$

Therefore, at each point, the pressure is the sum of hydrostatic and dynamic pressures. The contribution of each term to the total pressure depends on the mass density of the fluid, height difference and the velocities of the fluid flow. The forces that act on a body that is immersed in a fluid or gas flow are caused by the inhomogeneous distribution of the hydrostatic (Archimedes') force and the dynamic pressure (driving force). In many cases, the second force is larger then the first one. We consider two cases of the fluid or gas flow around a body. For the case of low gas velocity, one can consider the gas as incompressible and the dynamical pressure as larger than the hydrostatic pressure.

In the case of a horizontal air flow around an asymmetric hydrofoil, the flow velocity is larger on top of the hydrofoil then at the bottom. As a final result, the difference of the dynamic pressures creates the lift. This force is larger, the greater the flight velocity.

For the movement of a body in a heavy fluid, the hydrostatic pressure is larger than the dynamic pressure. Therefore, the buoyancy, Archimedes' force, that acts on the ship is determined by the distribution of the hydrostatic pressure on the wetted surface of the body.

8.4 Flow of an Incompressible Fluid in a Tube of Varying Cross Section

We consider the movement of an incompressible fluid in a thin tube of variable cross sections (see Figure 8.4). We assume that the movement is one-dimensional and stationary. From the equation of continuity, we conclude that $VS = \text{const}$, that is, the same amount of fluid flows per unit time through every cross section. Is is evident that a decrease of S increases the velocity V. The maximum value of the velocity is assumed for the minimum cross section S_{min}. From Bernouli's integral,

$$\frac{p}{\rho} + \frac{V^2}{2} = \text{const} ,$$

we obtain the pressure that will also be a minimum at this cross section. Thus, the reduction of the cross section S leads to an increase of the velocity and a decrease in the pressure. This property is used in the water jet pump (see Figure 8.5).

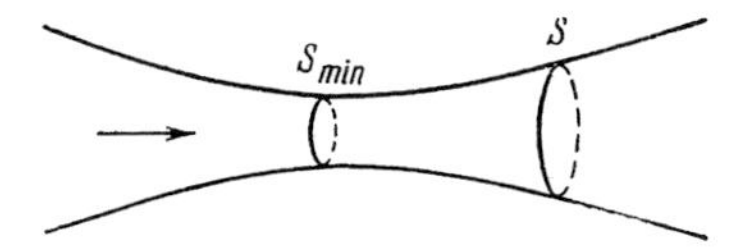

Figure 8.4 Flow of an incompressible fluid in a tube of varying cross sections.

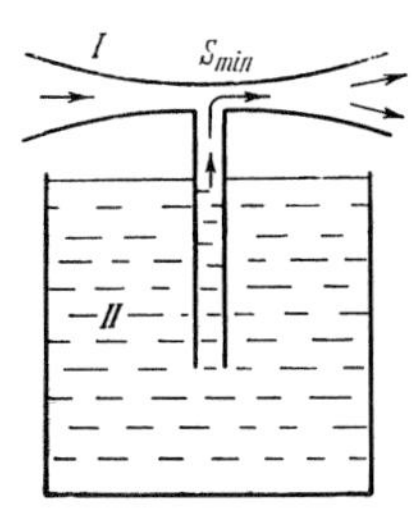

Figure 8.5 Sketch of a water jet pump.

8.5 The Phenomenon of Cavitation

It is a consequence of Bernoulli's integral, that the pressure distribution depends on the velocity distribution for the stationary movement of a gas or an incompressible fluid. When solving the corresponding mathematical tasks, the pressure will become negative or even equal to $-\infty$ if there are points in the flow of the fluid where the velocity goes to infinity.

Real fluids contain hovering particles and dissolved gases. Normally, such fluids are not able to withstand the tractions that result from negative pressures. Normally, the pressure cannot become smaller than a certain positive number p_d approaching zero at 20 °C. In a flow where the pressure decreases to this value, the continuity of the fluid is violated and a domain is created which is filled by gas or steam bubbles. This phenomenon is called cavitation.

The quantity p_d is a physical characteristic which has no effect on the fluid flow for $p > p_d$. For $p = p_d$, cavitation of the fluid can arise that has an important influence on the flow. Cavitation happens in areas with reduced pressures and high velocities, that is, in areas where the diameter of a tube is drastically reduced, in piston pumps behind the piston where the pressure can fall to zero as well as when a fluid flows around a body. From Bernoulli's integral for a stationary movement of a heavy fluid, it follows that

$$p = p_{\text{st}} + \frac{\rho V_\infty^2}{2} - \frac{\rho V^2}{2} ,$$

where V_∞ is the velocity at infinity, that is, far away from the body. This equation can be rewritten as

$$\frac{2(p_{\text{st}} - p)}{\rho V_\infty^2} = \frac{V^2}{V_\infty^2} - 1 . \tag{8.16}$$

The maximum fluid velocity $V_{\max}$ in the flow corresponds to the minimum pressure $p_{\min}$. The dimensionless quantity

$$C_p = \frac{2(p_{st} - p)}{\rho V_\infty^2} \tag{8.17}$$

on the points of body surface is called the pressure coefficient. At the point of minimum pressure, the pressure coefficient is given by

$$C_{p,\min} = \frac{2(p_{st} - p_{\min})}{\rho V_\infty^2} = \frac{V_{\max}^2}{V_\infty^2} - 1 .$$

Cavitation starts if $C_{p,\min} = \chi$ where

$$\chi = \frac{2(p_{st} - p_d)}{\rho V_\infty^2} . \tag{8.18}$$

The dimensionless quantity χ is called the cavitation number. It depends on the pressure at infinity, that is, the immersion depth $Z - Z_1$ via the parameter $p_{st} = p_1 + \rho g(Z - Z_1)$ and especially on the velocity of the fluid flow V_∞. Thus, if χ is equal to $C_{p,\min}$, cavitation will arise in the fluid flow at those points where $V = V_{\max}$. For the dimensional analysis of flows with cavitation, one has to add another parameter, the cavitation number χ to Reynolds number and Froude's number. These parameters can be used as similarity parameters during the modeling of the corresponding hydrodynamic flows.

When cavitation arises on the surfaces of ship elements, submarines and hydraulic machines bubbles with internal pressure close to zero are created. During the displacement of the bubbles together with the fluid into an area of higher pressure, the fluid plunges into the bubbles at high speed. As a result, the bubbles collapse very quickly with a steep increase in local pressure. Cavitation close to the surface of a body destroys the surface of the body around which the fluid flows. This effect is called cavitation erosion. The cavitation is accompanied by a number of undesired phenomena, that is, noise and vibration. The development of cavitation also produces large gas caverns through the combination of bubbles.

8.6
Bernoulli's Integral for Adiabatic Flows of an Ideal Gas

We consider Bernoulli's integral for the reversible adiabatic flow of a weightless ideal gas. The equation of state $p = p(\rho, s)$ of such a gas (see Section 5.4) has the form $p/\rho^\gamma = \text{const}$ or $p/p_1 = (\rho/\rho_1)^\gamma$, with the adiabatic exponent γ.

From (8.3), one obtains the pressure function along a stream line as

$$P = \int_{p_1}^{p} \frac{dp}{\rho(p)} = \frac{\gamma}{\gamma - 1} \frac{p}{\rho} + \text{const} = c_p T + \text{const} .$$

For the stationary adiabatic movement of a two parameter ideal medium, the heat inflow equation (5.27) has the form

$$dU = -p\,d\left(\frac{1}{\rho}\right).$$

According to (5.48), we have

$$di = dU + d\left(\frac{p}{\rho}\right) = -p\,d\left(\frac{1}{\rho}\right) + d\left(\frac{p}{\rho}\right) = \frac{1}{\rho}dp\,,$$

such that the enthalpy is

$$i = i_1 + \int\limits_{p_1}^{p} \frac{dp}{\rho(p)} = i_1 + P\,.$$

Therefore, the change in enthalpy is equal to the pressure function along the stream line.

Thus, Bernoulli's (8.8) is rewritten as

$$\frac{V^2}{2} + i = i^*\,.$$

Since we have $i = c_p T$, for an ideal gas, one obtains $\frac{V^2}{2} + c_p T = c_p T^*$ or

$$\frac{V^2}{2} + \frac{\gamma}{\gamma - 1}\frac{p}{\rho} = c_p T^*\,. \tag{8.19}$$

Here, T^* is the temperature in the point of stagnation, that is, the point where $V = 0$. By denoting the values of the pressure and mass density by p^* and ρ^*, we have

$$i^* = c_p T^* = \frac{\gamma}{\gamma - 1}\frac{p^*}{\rho^*}\,.$$

From (8.19), it follows that pressure and temperature decrease as the velocity increases along the stream line. Since the flow velocity of the gas increases with decreasing pressure, the velocity reaches its maximum value for $p = 0$ which corresponds to the fluid streaming out into vacuum. From (8.19), one obtains

$$V_{\max} = \sqrt{2c_p T^*}\,. \tag{8.20}$$

Thus, we see that in the case of an adiabatic gas flow, the velocity cannot exceed $V_{\max}$, which is determined by (8.20). We note, however, that the flow velocity of the gas can exceed $V_{\max}$ for a non-stationary gas flow.

We introduce the velocity of sound as

$$a = \sqrt{\left(\frac{\partial p}{\partial \rho}\right)_s} = \sqrt{\frac{\gamma p}{\rho}} = \sqrt{\gamma R T}\,. \tag{8.21}$$

Then, Bernoulli's integral (8.19) can be rewritten as

$$\frac{V^2}{2} + \frac{a^2}{\gamma - 1} = \frac{V_{\max}^2}{2} \,. \tag{8.22}$$

Therefore, the velocity of sound changes while the velocity of the gas changes along the stream line. It attains its maximum value at the point of stagnation. By denoting this maximum value by a^*, one can express the constant in Bernoulli's integral as

$$i^* = c_p T^* = \frac{(a^*)^2}{\gamma - 1} = \frac{V_{\max}^2}{2} \,.$$

From this, it follows that

$$a^* = \sqrt{\gamma R T^*}$$

and

$$V_{\max} = \sqrt{\frac{2}{\gamma - 1}} a^* \,.$$

The velocity which is equal to the local velocity of sound is defined by $V_{cr} = a$. From (8.22), one obtains

$$\frac{V_{cr}^2}{2} + \frac{V_{cr}^2}{\gamma - 1} = \frac{V_{\max}^2}{2} = \frac{(a^*)^2}{\gamma - 1}$$

and

$$V_{cr} = \sqrt{\frac{2}{\gamma + 1}} a^* = \sqrt{\frac{\gamma - 1}{\gamma + 1}} V_{\max} \,.$$

We now introduce Mach's number $\mathrm{M} = V/a$. For $\mathrm{M} < 1$, a movement is called a subsonic flow, $\mathrm{M} > 1$ for a supersonic flow and $\mathrm{M} \gg 1$ for a hypersonic flow.

By solving Bernoulli's equation with respect to pressure, mass density and temperature, one obtains

$$p = p^* \left(1 + \frac{\gamma - 1}{2}\mathrm{M}^2\right)^{-\frac{\gamma}{\gamma-1}} \,; \quad \rho = \rho^* \left(1 + \frac{\gamma - 1}{2}\mathrm{M}^2\right)^{-\frac{\gamma}{\gamma-1}} \,;$$
$$T = T^* \left(1 + \frac{\gamma - 1}{2}\mathrm{M}^2\right)^{-1} \,. \tag{8.23}$$

We will now consider the last of these equations. If the T is the temperature far from the point of stagnation, this equation gives

$$T^* = T\left(1 + \frac{\gamma - 1}{2}\mathrm{M}^2\right),$$

that is, the temperature increases at the point of stagnation.

As an example, look at the flow of air around a body. For air, we have $\gamma = 1.4$. Thus, $T^* = (1 + 0.2\mathrm{M}^2)$. If the temperature far from the stagnation point of the flow is $T = 250$ K, one has $T^* = 290$ K $= 17\,^\circ$C for M $= 1$, $T^* = 700$ K for M $= 3$ and $T^* = 1500$ K for M $= 5$. Therefore, when gas at a high supersonic velocity flows around a body, it will heat up very strongly so that it can melt or disintegrate.

We now consider the influence of the gas velocity on the mass density of the gas. This dependence is given by the second equation in (8.23), that is,

$$\rho = \rho^* \left(1 + \frac{\gamma - 1}{2}\mathrm{M}^2\right)^{-\frac{\gamma}{\gamma-1}} = \rho^* \left(1 - \frac{V^2}{V_{\max}^2}\right)^{\frac{1}{\gamma-1}} .$$

We see that one can consider the gas as incompressible for $V/V_{\max} \ll 1$. However, for a large velocity, the compressibility has to be taken into account. To estimate the error during a calculation that treats the gas as incompressible, one needs expand the equations in (8.23) into a Taylor series. For the mass density, one obtains

$$\frac{\rho}{\rho^*} = 1 - \frac{1}{\gamma - 1}\frac{V^2}{V_{\max}^2} + \ldots$$

For air at $V < a^*/5 = 68$ m/s, the density ρ will differ by less than 1% from ρ^*. Thus, the gas flow must be calculated, taking into account the compressibility for velocities close to the sound velocity and even more so for velocities larger than the sound velocity.

8.7
Bernoulli's Integral for the Flow of a Compressible Gas

We consider the stationary movement of a compressible gas in a stream tube. As shown above, it follows from the equation of continuity that the cross section decreases during the movement of an incompressible gas if the velocity increases, that is,

$$S = \frac{\text{const}}{V} .$$

The equation of continuity for a compressible gas can be written in the form

$$\rho V S = \text{const} \quad \text{or} \quad S = \frac{\text{const}}{\rho V} , \tag{8.24}$$

which provides for the conservation of mass flow.

Since the mass density for adiabatic reversible flows depends on the velocity according to the relationship

$$\rho = \rho^* \left(1 - \frac{V^2}{V_{\max}^2}\right)^{\frac{1}{\gamma-1}} , \tag{8.25}$$

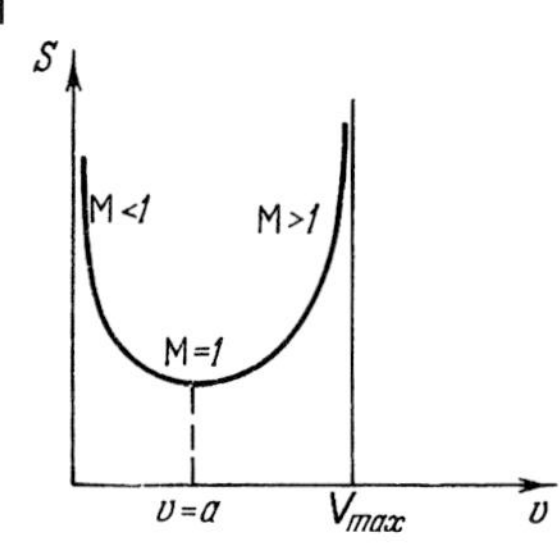

Figure 8.6 Dependence of the cross section of the stream tube on the velocity for an adiabatic and reversible flow of an ideal gas.

one can obtain the dependence $S(V)$ from (8.24). This dependence

$$S = \frac{\text{const}}{\rho^* V \left(1 - \frac{V^2}{V_{\max}^2}\right)^{\frac{1}{\gamma-1}}} \tag{8.26}$$

is plotted in Figure 8.6.

We want to find such a dependency for an arbitrary, but not necessarily adiabatic flow. In order to do so, one has to consider Euler's equation for the stationary one-dimensional flow of a weightless gas. By projecting this equation in the direction of the stream line, one obtains

$$V dV = -\frac{dp}{\rho} . \tag{8.27}$$

Since $a^2 = dp/d\rho$, we have $-dp/d\rho = -a^2 d\rho/\rho$ and one can rewrite (8.27) as

$$V d\rho = -\mathrm{M}^2 \rho \, dV \; ; \quad \mathrm{M} = \frac{V}{a} = \frac{V}{\sqrt{\frac{dp}{d\rho}}} .$$

This leads to

$$d(\rho V) = \rho dV + V d\rho = \rho(1 - \mathrm{M}^2) dV . \tag{8.28}$$

It should be noted that Mach's number, as here defined for non-adiabatic flows, is not equal the normally determined Mach's number $\mathrm{M} = V/\sqrt{dp/d\rho}$ since the entropy is not preserved in a non-adiabatic process.

From (8.28), it follows that for a subsonic flow $\mathrm{M} < 1$, an increase in the velocity $dV > 0$ leads to an increase of ρV ($d(\rho V) > 0$) and thus to a decrease in S. For a supersonic flow $\mathrm{M} > 1$, the situation is reversed, that is, an increase in the velocity leads to an increase in the cross section of the stream tube S.

Since an increase in the velocity leads to a decrease in the cross section for a subsonic flow, the cross section for $\mathrm{M} = 1$ will reach a certain value which is called the critical cross section. In order to increase the velocity even further, it is necessary to increase the cross section S.

From the above, we can deduce that in order to accelerate a subsonic flow into a supersonic flow, one first has to decrease the cross section to a critical value $S_{\min}$, at which the flow velocity is equal to the velocity of sound, and increase the cross

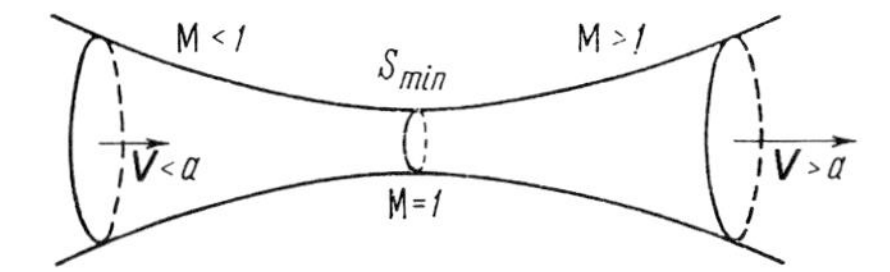

Figure 8.7 Stream tube in a compressible gas.

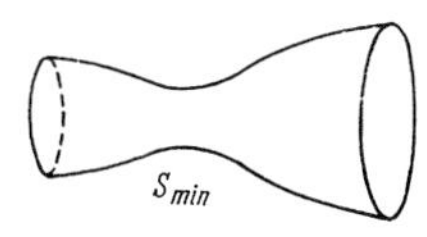

Figure 8.8 The de Laval nozzle.

section after that (see Figure 8.7). A device which realizes the adiabatic transition from subsonic velocity to supersonic velocity is called a de Laval nozzle (see Figure 8.8). A device which only consists of a contraction part is called a simple nozzle or a tapered nozzle. The largest velocity that can be reached with a simple nozzle is the sound velocity, and it is reached at the end of the nozzle. De Laval nozzles are necessary construction elements of rocket engines and wind tunnels.

9
Application of the Integral Relations on Finite Volumes within the Continuum for a Stationary Movement

9.1
Integral Relations

In Chapter 3, the main integral relationships for mechanical and thermodynamic quantities with respect to a finite volume of the continuum have been formulated. Further to this, in Chapter 6, these relationships were applied to determine the conditions of strong discontinuity surfaces. In the present chapter, how one applies the integral relationships to the problem of determining the forces acting on a body immersed in a fluid without solving the corresponding boundary value problem is shown.

We consider a static volume V which is bounded by a closed surface Σ. We denote a finite moving volume by V^* that consists of individual continuum particles. We apply the integral relationships to the volume V^* that coincides at time t with the volume V and whose moving surface coincides with static control surface Σ. For stationary movements, the following integral relationships apply:

- the equation of mass conservation

$$\int_{\Sigma} \rho V_n d\sigma = 0\,; \tag{9.1}$$

- the equation of motion

$$\int_{\Sigma} \rho \boldsymbol{V} V_n d\sigma = \int_{v} \rho \boldsymbol{f} d\tau + \int_{\Sigma} \boldsymbol{p}_n d\sigma\,; \tag{9.2}$$

- the angular momentum equation

$$\int_{\Sigma} \rho(\boldsymbol{r} \times \boldsymbol{V} + \boldsymbol{k}) V_n d\sigma = \int_{v} \rho(\boldsymbol{r} \times \boldsymbol{f} + \boldsymbol{h}) d\tau + \int_{\Sigma} (\boldsymbol{r} \times \boldsymbol{p}_n + \boldsymbol{Q}_n) d\sigma\,; \tag{9.3}$$

Hydromechanics. Theory and Fundamentals. Emmanuil G. Sinaiski

ISBN: 978-3-527-41026-2

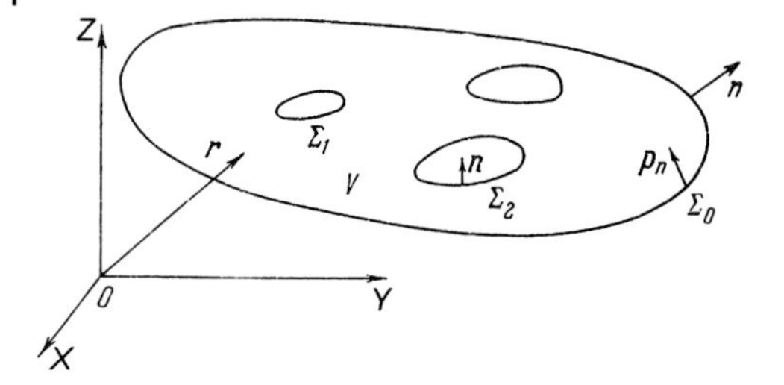

Figure 9.1 Sketch of the control surface.

- and the energy equation

$$\int_{\Sigma} \rho V_n \left(\frac{V^2}{2} + U \right) d\sigma = \int_{v} \rho \left(\boldsymbol{f} \cdot \mathbf{V} + \frac{d q^*_{\text{mas}}}{dt} \right) d\tau + \int_{\Sigma} (\boldsymbol{p}_n \cdot \mathbf{V} - q^*_n) d\sigma \,. \tag{9.4}$$

The main idea of applying (9.1)–(9.4) is to obtain the surface integral accurately or approximately by a suitable choice of the control surface Σ in terms of known or unknown quantities. Normally, the control surface Σ consists of a few closed surfaces Σ_i (see Figure 9.1), such that $\Sigma = \Sigma_0 + \Sigma_1 + \cdots + \Sigma_n$ (see Figure 9.1). The surfaces of bodies which are immersed in the fluid can serve as such control surfaces.

As an example, we consider the stationary movement of a fluid or gas in a beam that flows onto a solid wall and is dispersed on it.

We assume that within the beam of cross section S, far away from the wall, the pressure, mass density and velocity of the ideal weight less fluid are p_0, σ_0 and V_0. At the free surface of the beam, the pressure is p_0 where p_0 is the pressure of the surrounding medium. Then, for every closed surface Σ that surrounds the volume V, the following equation is satisfied

$$\int_{\Sigma} p_0 \boldsymbol{n} d\sigma = \int_{v} \left(\frac{\partial p_0}{\partial X} \boldsymbol{i} + \frac{\partial p_0}{\partial Y} \boldsymbol{j} + \frac{\partial p_0}{\partial Z} \boldsymbol{k} \right) d\tau \,. \tag{9.5}$$

The force that is exerted on the wall by the fluid is given by

$$\boldsymbol{F} = \int_{\Sigma^*} \boldsymbol{p}_n d\sigma \,,$$

where Σ^* is the surface of the wall that is wetted by the fluid. If the external surface Σ^{**}, which is not wetted by the fluid, is subject to the pressure p_0, the resulting force is equal to

$$P = \int_{\Sigma^*} p d\sigma - \int_{\Sigma^{**}} p_0 d\sigma = \int_{\Sigma^* + \Sigma^{**}} (p - p_0) d\sigma \,. \tag{9.6}$$

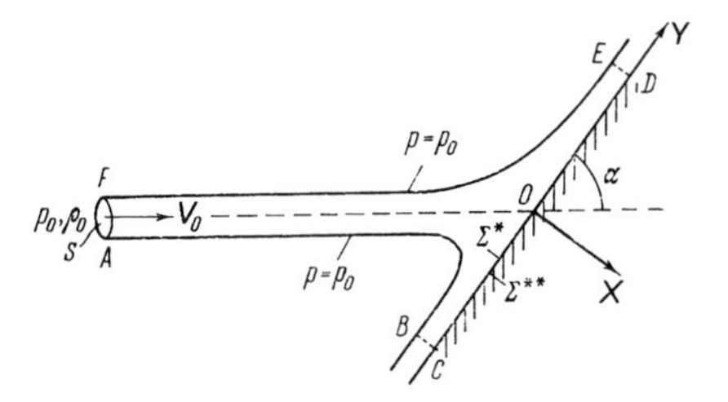

Figure 9.2 Beam hitting the wall.

Using the equation of motion (9.2), we calculate P. We use the control surface Σ which is indicated by the letters $ABCDEF$ in Figure 9.2. From (9.2), we obtain, taking into account the identity $\boldsymbol{p}_n = -p\,\boldsymbol{n}$ where $\mathbf{n}$ is the external unit normal on the area under consideration,

$$\int_{\Sigma} \rho \boldsymbol{V} V_n \, d\sigma = -\int_{\Sigma} p \boldsymbol{n} \, d\sigma \, . \tag{9.7}$$

In the beam that is flowing into the wall, the velocity is parallel to the Y-axis far from the inflow point and the pressure is p_0. Therefore, the pressure on Σ is p_0 except on Σ^*. Since we have $\int_{\Sigma} \boldsymbol{n} \, d\sigma = 0$, for every closed surface, one obtains

$$\int_{\Sigma} \rho \boldsymbol{V} V_n \, d\sigma = -\int_{\Sigma} (p - p_0) \boldsymbol{n} \, d\sigma = -\int_{\Sigma^*} (p - p_0) \boldsymbol{n} \, d\sigma \, .$$

We now determine V_n. We have $V_n = -V_0$, $V_x = V_0 \sin\alpha$ within the cross section S; $V_x = 0$ on BC, and ED and $V_n = 0$ on Σ^*. Therefore, one obtains

$$\left(\int_{\Sigma} \rho \boldsymbol{V} V_n \, d\sigma \right)_x = \int_{S} \rho V_x V_n \, d\sigma = -\rho_0 V_0^2 S \sin\alpha = -G V_0 \sin\alpha \, ,$$

where $G = \rho_0 V_0 S$ is the mass flow of the fluid within the beam.

By substituting the results obtained above into (9.6), one obtains for the force that the flow exerts on the wall

$$P = G V_0 \sin\alpha \, . \tag{9.8}$$

Equation 9.8 is valid for a beam of an ideal fluid of arbitrary cross sectional shape.

9.2
Interaction of Fluids and Gases with Bodies Immersed in the Flow

We consider the stationary movement of a fluid that flows around a body or a system of bodies (see Figure 9.3). We use the control surface Σ which consists of the

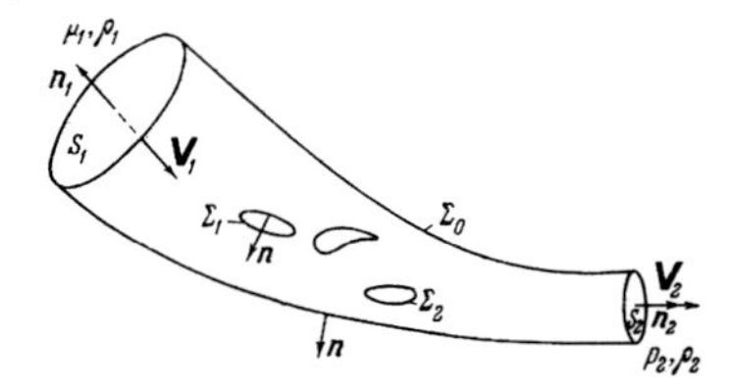

Figure 9.3 Stationary movement of a fluid in a tube around obstacles.

surface of the stream tube or the wall Σ_0, the tube cross sections S_1 and S_2, and the surfaces $\Sigma_1, \Sigma_2, \cdots, \Sigma_n$ of the bodies that are immersed in the fluid. Since we assume that the cross sections S_1 and S_2 are far from the surfaces $\Sigma_1, \Sigma_2, \cdots, \Sigma_n$, we may also assume that the flow parameters in those cross sections are homogeneous. We assume that the internal stresses are given by the pressures in the cross sections S_1 and S_2, and the tangential stresses and energy flows on Σ_0 can be exchanged with the surrounding medium since we do not require the assumptions of an ideal fluid.

The main vector of the surface forces $-\boldsymbol{R}$ that are exerted by the wall Σ_0 and the internal body surfaces $\Sigma_1, \Sigma_2, \ldots, \Sigma_n$ on the fluid, the total torque $-\boldsymbol{M}$ and the total influx of mechanical and other forms energy $-W$ are given by

$$-\boldsymbol{R} = \int\limits_{\Sigma_0+\Sigma_1+\ldots+\Sigma_n} \boldsymbol{p}_n d\sigma \,, \tag{9.9}$$

$$-\boldsymbol{M} = \int\limits_{\Sigma_0+\Sigma_1+\ldots+\Sigma_n} (\boldsymbol{r} \times \boldsymbol{p}_n + \mathcal{Q}_n) d\sigma \,, \tag{9.10}$$

$$-W = \int\limits_v \rho \left(\boldsymbol{f} \cdot \boldsymbol{V} + \frac{dq^*_{\text{mas}}}{dt} \right) d\tau + \int\limits_{\Sigma_0+\Sigma_1+\ldots+\Sigma_n} (\boldsymbol{p}_n \cdot \boldsymbol{V} - q^*_n) d\sigma \,. \tag{9.11}$$

The specific total heat content of the fluid i^* consists of the internal heat content and kinetic energy, that is,

$$i^* = \frac{V^2}{2} + U + \frac{p}{\rho} \,. \tag{9.12}$$

By taking gravity $\boldsymbol{f} = \boldsymbol{g}$ into account, we have

$$\int\limits_v \rho \boldsymbol{f} d\tau = \int\limits_v \rho \boldsymbol{g} d\tau = M \boldsymbol{g} \,, \tag{9.13}$$

$$\int\limits_v \rho (\boldsymbol{r} \times \boldsymbol{f}) d\tau = \int\limits_v \rho (\boldsymbol{r} \times \boldsymbol{g}) d\tau = \int\limits_v \rho \boldsymbol{r} d\tau \times \boldsymbol{g} = M \boldsymbol{r}^* \times \boldsymbol{g} \,, \tag{9.14}$$

$$\int\limits_v \rho \boldsymbol{f} \cdot \boldsymbol{V} d\tau = \boldsymbol{g} \cdot \int\limits_v \rho \boldsymbol{V} d\tau = M \boldsymbol{V}^* \cdot \boldsymbol{g} \,. \tag{9.15}$$

Here, M is the mass of the fluid inside of the volume,and $\boldsymbol{r}^*$ and $\boldsymbol{V}^*$ are the radius vector and velocity of the center of mass of the volume. In most applications, the volume moments $\boldsymbol{h}$, the energy influx due to mass transport dq^*_{mas}/dt

and the mass force f will not be taken into account. From the equation of mass conservation (9.1), it follows that

$$\rho_1 V_1 S_1 = \rho_2 V_2 S_2 = G \tag{9.16}$$

since $V_n \neq 0$ only in the cross sections S_1 and S_2 where we have $V_n = -V_1$ at S_1 and $V_n = V_2$ at S_2. Here, G is the mass flow of the fluid per unit time through the flow tube. Thus, (9.16) provides for the conservation of mass flow.

We now consider the equation of motion (9.2). Since the normal components of the velocity V_n vanish on the surfaces $\Sigma_0, \Sigma_1, \Sigma_2, \ldots, \Sigma_n$, one obtains

$$\int_{S_1+S_2} \rho \boldsymbol{V} V_n d\sigma = -\int_{S_1+S_2} p\boldsymbol{n} d\sigma + \int_{\Sigma_0+\Sigma_1+\ldots+\Sigma_n} \boldsymbol{p}_n d\sigma .$$

In addition, since $V_n = -V_1$ and $\boldsymbol{V}_1 = -V_1 \boldsymbol{n}_1$ on S_1 , $V_n = V_2$ and $\boldsymbol{V}_n = V_2 \boldsymbol{n}_2$ on S_2, and since the last integral is, according to (9.9), equal to $-\boldsymbol{R}$, one obtains

$$\rho_1 V_1^2 S_1 \boldsymbol{n}_1 + \rho_2 V_2^2 S_2 \boldsymbol{n}_2 = -p_1 S_1 \boldsymbol{n}_1 - p_2 S_2 \boldsymbol{n}_2 - \boldsymbol{R} .$$

Using the obvious relationships $\boldsymbol{n}_1 = -\boldsymbol{V}_1/V_1, \boldsymbol{n}_2 = \boldsymbol{V}_2/V_2$, one obtains for the main vector of the surface forces

$$\boldsymbol{R} = \left(p_1 + \rho_1 V_1^2\right) S_1 \frac{\boldsymbol{V}_1}{V_1} - \left(p_2 + \rho_2 V_2^2\right) S_2 \frac{\boldsymbol{V}_2}{V_2} . \tag{9.17}$$

We now consider the angular momentum equation (9.3). We assume that $\boldsymbol{k} = \boldsymbol{h} = \boldsymbol{Q}_n = 0$. Then, one obtains

$$\int_{\Sigma} \rho(\boldsymbol{r} \times \boldsymbol{V}) V_n d\sigma = \int_{\Sigma} (\boldsymbol{r} \times \boldsymbol{p}_n) d\sigma . \tag{9.18}$$

In the first integral, only the integrals over S_1 and S_2 do not vanish. Therefore, we have

$$\int_{\Sigma} \rho(\boldsymbol{r} \times \boldsymbol{V}) V_n d\sigma = \rho_1 V_1^2 \int_{S_1} (\boldsymbol{r} \times \boldsymbol{n}_1) d\sigma + \rho_2 V_2^2 \int_{S_2} (\boldsymbol{r} \times \boldsymbol{n}_2) d\sigma .$$

We denote the radius vectors of the centers of mass of the cross sections S_1 and S_2 by $\boldsymbol{r}_1^*$ and $\boldsymbol{r}_2^*$. Then, the integral on the left-hand side of (9.18) can be written as

$$\int_{\Sigma} \rho(\boldsymbol{r} \times \boldsymbol{V}) V_n d\sigma = \rho_1 V_1^2 S_1 (\boldsymbol{r}_1^* \times \boldsymbol{n}_1) + \rho_2 V_2^2 S_2 (\boldsymbol{r}_2^* \times \boldsymbol{n}_2) .$$

The integral on the right-hand side of (9.18) is equal to

$$\int_{\Sigma} (\boldsymbol{r} \times \boldsymbol{p}_n) d\sigma = \int_{S_1+S_2} (\boldsymbol{r} \times \boldsymbol{p}_n) d\sigma = -\boldsymbol{M} .$$

Since

$$\int\limits_{S_1+S_2} (\boldsymbol{r} \times \boldsymbol{p}_n) d\sigma = -p_1 \int\limits_{S_1} (\boldsymbol{r} \times \boldsymbol{n}_1) d\sigma - p_2 \int\limits_{S_2} (\boldsymbol{r} \times \boldsymbol{n}_2) d\sigma$$

$$= -p_1(\boldsymbol{r}_1^* \times \boldsymbol{n}_1) - p_1(\boldsymbol{r}_2^* \times \boldsymbol{n}_2)$$

and $\boldsymbol{n}_1 = -\mathbf{V}_1/V_1, \boldsymbol{n}_2 = \mathbf{V}_2/V_2$, one obtains

$$\boldsymbol{M} = \left(p_1 + \rho_1 V_1^2\right) S_1 \frac{\boldsymbol{r}_1^* \times \mathbf{V}_1}{V_1} + \left(p_2 + \rho_2 V_2^2\right) S_2 \frac{\boldsymbol{r}_2^* \times \mathbf{V}_2}{V_2} . \tag{9.19}$$

For $q_n^* = 0$, $\boldsymbol{f} = 0$ and $dq_{\text{mas}}^*/dt = 0$, the energy equation changes to

$$\int\limits_{S_1+S_2} \rho V_n \left(\frac{V^2}{2} + U\right) d\sigma = \int\limits_{S_1+S_2} \boldsymbol{p}_n \cdot \boldsymbol{V} d\sigma - W ,$$

which gives

$$W = \rho_1 V_1 S_1 \left(\frac{p_1}{\rho_1} + \frac{V_1^2}{2} + U_1\right) - \rho_2 V_2 S_2 \left(\frac{p_2}{\rho_2} + \frac{V_2^2}{2} + U_2\right) = G(i_1^* - i_2^*) . \tag{9.20}$$

In the absence of energy inflow, we have $W = 0$, and it follows from (9.20) that $i_1^* = i_2^*$, that is, the enthalpy is conserved.

As an example, for using the integral relationships, we consider the movement of a fluid in a curved tube (see Figure 9.4).

According to (9.17), the coercive force of the fluid on the wall of the tube is the vector sum of two vectors. This force is applied at O and shown in the figure. Of special interest, is the problem of determining the resisting force of the body that is moving through a fluid. Therefore, we consider the translational movement of a body within an infinite cylindrical tube. The perturbations of the fluid depend on the form of the body and of the tube, the position of the body relative to the tube, the velocity of the body, the properties of the fluid and the initial state of the undisturbed fluid.

To solve the problem, one has to make simplifying assumptions. For a body whose dimensions are small relative to the width of the tube, one can assume that the disturbances that are created by the movement of the body decay far from the

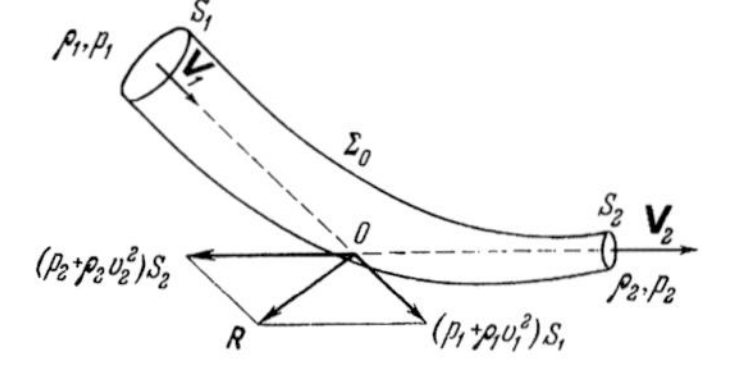

Figure 9.4 Coercive force of the fluid on the wall of the tube.

body, that is, at infinity. Therefore, one can assume that the fluid rests in front of the body at infinity. One can also assume that the movement is stationary in the system of coordinates that is fixed to the body.

The question concerning the properties of the fluid movement far behind the body is more involved, and we therefore consider it more closely. On first inspection, it seems as if, far behind the body and just as far ahead of the body, the movement of the fluid can be considered as undisturbed. However, such an assumption leads to a paradoxical result.

Firstly, we note that instead of considering the translational movement of the body in the fluid, one can consider the opposite movement, that is, the fluid movement relative to the body. This opposite perspective corresponds to a system of coordinates which is at relative to the moving body. Since both moving systems of coordinates and those at rest are inertial, this inversion of the movement under conservation of the interaction forces is allowed for every fluid model.

In this regard, one has to take into account that the inversion of the movement is possible for a viscous fluid if the tube wall can be forced to also move with the same velocity that the fluid has at infinity since the no-slip condition has to be satisfied on the wall.

The no-slip condition of the viscous fluid on the wall only influences a small area next to the wall of the tube. Therefore, the viscosity does not have an important influence on a small body that moves in the fluid flow at the axis of the tube. Thus, the problem of determining the resistance of the body during the movement in the fluid can be replaced by the equivalent problem of a body at rest, around which the fluid flows with a velocity opposite to the velocity of the body in the original problem.

However, in practice and in the experiments, the measurement of the resistance of the body do not show a complete equivalence of both movements. This result is known as the paradox of Du Bois. This can be explained by the fact that in the experiments, the influence of other bodies, that is, the tube wall, is present, which is not taken into account for the opposite movement.

It follows from (9.17) that if the cross sections S_1 and S_2 are take far away from the body and are constant ($S_1 = S_2$), and the pressures at both cross sections are also equal, that is, if the flow parameters far before and far after the body are equal, we have $\boldsymbol{R} = 0$, that is, the total forces that acts on the body and the tube is equal to zero. If the mass forces vanish, we have

$$\boldsymbol{R} = \boldsymbol{R}_1 + \boldsymbol{R}_2 = 0 \,. \tag{9.21}$$

Here, $\boldsymbol{R}_2$ is the force that acts on the tube wall and is perpendicular to the wall and, thus, also perpendicular to the direction of the velocity $\boldsymbol{V}_\infty$ at infinity. Therefore, the resistance force of the body vanishes. This is known as D'Alembert's paradox.

The main reason for D'Alembert's paradox consists of the assumption that the parameters of the fluid are the same far before and far after the body. In reality, there is no such symmetry of the flow. For the fluid flow around the body, one has to take into account the flow separation and the properties of the vortex row behind

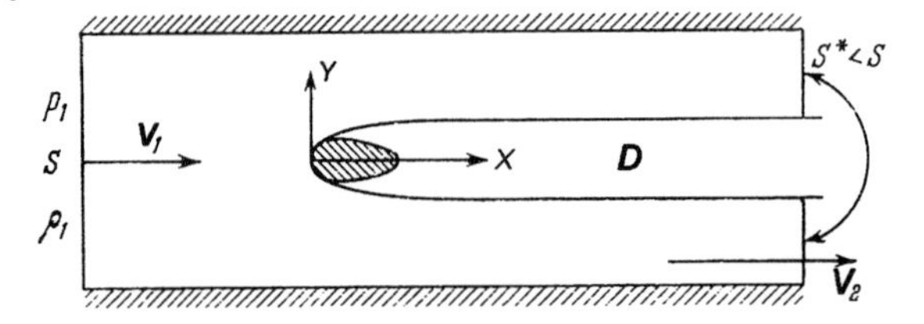

Figure 9.5 Sketch of the flow around a body with flow separation.

the body. This results in the parameter being identical to those in cross section S_1 and only in a part of cross section S_2 (see Figure 9.5).

In Figure 9.5, the flow around the body with flow separation is shown. In the region after the body, we see the trailing flow in which the parameters are different from the corresponding values in S_1 at a large distance from the body. If the body is given a streamlined shape, the trailing flow and the resistance of the body can be substantially reduced. However, it is impossible to totally eliminate the resistance.

In Section 10.4, it will be shown that a similar D'Alembert's paradox arises for the movement of a sphere in the infinite volume of an ideal incompressible fluid and the resistance force also is obtained as zero. The reason for this paradox is the assumed symmetry of the pressure distribution over the surface of the body. For a real fluid, there is no such symmetry and the resistance force is not zero.

10
Potential Flows for Incompressible Fluids

The research into the more than one-dimensional flow of an ideal fluid is a very challenging mathematical task. The assumption that the velocity of a moving fluid is irrotational, that is, $\boldsymbol{\omega} = \frac{1}{2}\nabla \times \boldsymbol{V} = \boldsymbol{0}$, plays a major role when one wants to apply theoretical hydromechanics in concrete technical applications. This assumption means that a function $\varphi(x, y, z)$ exists such that the velocity can be expressed as the gradient of a potential, that is, $\boldsymbol{V} = \nabla\varphi$. This function is called the velocity potential and such a flow is referred to as a potential flow.

The existence of such flows follows two theorems.

Theorem 10.1 Thompson's Theorem

During the barotropic flow of an ideal fluid under the influence of a volume force field with unique potential Φ, the circulation of the velocity vector along a closed contour C remains constant (see Section 2.7), that is,

$$\oint_C \boldsymbol{V} \cdot d\boldsymbol{r} = 2\int_\Sigma \boldsymbol{\omega} \cdot \boldsymbol{n} \, dA = \text{const} ,$$

where Σ is the surface bounded by C.

This theorem leads to

Theorem 10.2 Lagrange's Theorem

If the curl of the velocity vanishes in all points of an ideal barotropic fluid at a certain initial time, then the flow will also remain irrotational for all later times under the influence of volume forces with a unique potential. On the contrary, if the curl of the velocity does not vanish at the beginning, it will remain non-vanishing at all later times.

Let us assume that a solid body starts moving while submerged in a fluid at rest. Because the velocity field in the fluid at rest is initially irrotational, the curl of the velocity field will remain vanishing under the conditions of Lagrange's theorem. However, if the curl of the velocity field becomes non-vanishing due to the

Hydromechanics. Theory and Fundamentals. Emmanuil G. Sinaiski

ISBN: 978-3-527-41026-2

conditions of Lagrange's theorem being violated, the velocity field will have a non-vanishing curl at all later times.

In reality, however, the irrotational character of the velocity field is an idealization of the real flow, just as the ideal fluid is an idealization of the real fluid. In reality, any fluid has some viscosity and thus, internal friction is present. This viscosity leads to a non-vanishing curl of the flow. In the following, it will be shown that the viscosity plays a role especially in the boundary layer to a solid body. While the fluid streams around the body with high velocity, the boundary layer can detach itself from the body and the turbulence can enter the free flow. In spite of the many factors that violate the preconditions for irrotational flow, this theory still delivers a very good approximation in many practical cases.

10.1
The Cauchy–Lagrange Integral

Both for stationary as well as non-stationary potential flows, the first integral of the Euler equations can be obtained. This integral is called the Cauchy–Lagrange integral.

We consider the flow of an ideal fluid and use the equations of motion in the Gromeka–Lambs form, (4.9), that is,

$$\frac{\partial \boldsymbol{V}}{\partial t} + \nabla\left(\frac{V^2}{2}\right) + 2\,(\boldsymbol{\omega} \times \boldsymbol{V}) = \boldsymbol{f} - \frac{1}{\rho}\nabla p\,. \tag{10.1}$$

We now assume that we are dealing with a potential flow, that is, $\boldsymbol{\omega} = 0$, $\boldsymbol{V} = \nabla\varphi$ with the velocity potential φ and that this flow is also barotropic. The latter assumption means that one can introduce the pressure function

$$P = \int \frac{dp}{\rho(p)}\,, \quad \frac{1}{\rho}\nabla p = \nabla P$$

for the entire flow. With these assumptions, (10.1) takes on the following form:

$$\nabla\left(\frac{\partial\varphi}{\partial t} + \frac{V^2}{2} + P\right) = \boldsymbol{f}\,.$$

This shows that the external forces must be potential forces, that is, they must satisfy $\boldsymbol{f} = \nabla\Phi$, yielding

$$\nabla\left(\frac{\partial\varphi}{\partial t} + \frac{V^2}{2} + P - \Phi\right) = 0$$

or

$$\frac{\partial\varphi}{\partial t} + \frac{V^2}{2} + P - \Phi = f(t)\,, \tag{10.2}$$

where $f(t)$ is an arbitrary function of time.

Equation 10.2 is the Cauchy–Lagrange integral. In order to determine $f(t)$, the left-hand side of (10.2) at some point of the flow can be used. In the case of a limited domain, a point on the boundary can be used. However, for an unlimited domain, the point at infinity can be used. It must be noted that introducing the new variable $\varphi_1 = \varphi + \int^t f(t')dt'$ instead of φ, (10.2) is transformed into

$$\frac{\partial \varphi_1}{\partial t} + \frac{V^2}{2} + P - \Phi = 0\,, \tag{10.3}$$

where the introduction of the additional term in φ_1 relative to φ does not affect the velocity field since $\boldsymbol{V} = \nabla\varphi_1 = \nabla\varphi$. In the special case of a stationary flow, Cauchy–Lagrange's integral becomes identical to Bernoulli's integral. Cauchy–Lagrange's integral serves the same purpose as Bernoulli's integral, namely, to determine the pressure distribution from the velocity potential φ and the potential Φ of the external force.

10.2
Some Applications for the General Theory of Potential Flows

For an ideal incompressible fluid, we deduce from the equation of continuity $\nabla \cdot \boldsymbol{V} = 0$ and the condition $\boldsymbol{V} = \nabla\varphi$ that the velocity potential satisfies the Laplace equation, that is,

$$\Delta\varphi = 0\,. \tag{10.4}$$

A function that satisfies this equation is called a harmonic function. The solution of the Laplace equation in the region V is uniquely determined by the values of φ on the surface Σ that is the boundary of V. The boundary value problem of determining a harmonic function on V from its values on the boundary Σ is known as the Dirichlet problem. If instead, the normal derivative $\partial\varphi/\partial n$ is known, the boundary value problem is called a Neumann problem.

The key equations for the non-stationary flow of an ideal barotropic fluid are the equation of continuity,

$$\frac{1}{\rho}\frac{dp}{dt} + \nabla \cdot (\nabla\varphi) = 0 \tag{10.5}$$

and Cauchy–Lagrange's integral

$$\frac{\partial \varphi_1}{\partial t} + \frac{1}{2}(\nabla\varphi)^2 + P(\rho) - \Phi = 0\,. \tag{10.6}$$

Since $dP = dp/\rho = a^2 d\rho/\rho$ where $a^2 = dp/d\rho$, (10.5) assumes the following form:

$$\frac{1}{a^2}\frac{dP}{dt} + \Delta\varphi = 0\,. \tag{10.7}$$

From the system of (10.6) and (10.7), two functions φ and P can be determined. In the following, we consider methods of solution for some types of flow, for which these methods have been well established.

Potential Flows of an Incompressible Fluid

In this case, we have $a^2 = dp/d\rho \to \infty$ and (10.7) turns into

$$\Delta\varphi = 0 . \tag{10.8}$$

Equations 10.8 and (10.6) are used for applications involving the movement of bodies in water, waves on the surface of water, stream flows and many others.

Small Disturbances for an Incompressible Fluid

If we consider the movement of a fluid with small disturbances, we assume that velocity, pressure and their derivatives are given as the sum of known functions and small additional terms. Neglecting the terms of higher order results in linear equations. Considering, for example, small disturbances of a fluid at rest, (10.6) and (10.7) become

$$\frac{1}{a_0^2}\frac{dP}{dt} + \Delta\varphi = 0 , \quad \frac{\partial\varphi}{\partial t} + P - \Phi = 0 , \tag{10.9}$$

where a_0 is the speed of sound in the medium at rest.

Under the condition that Φ is independent of t, one obtains from (10.9) only one equation for φ, that is,

$$\Delta\varphi = \frac{1}{a_0^2}\frac{\partial^2\varphi}{\partial t^2} . \tag{10.10}$$

This is the wave equation that describes the change in the disturbance of the velocity potential. For an incompressible fluid, we have $a \to \infty$ and (10.10) becomes the Laplace equation (10.8).

Stationary Movement of an Incompressible Fluid

The theory of parallel flows is most developed when the unknown functions only depend on two variables x and y. By using certain transformations of the variables and unknown functions, it is possible to arrive at a linear equation, that is, the Chaplygin equation. This method is the foundation for many current theories in gas dynamics.

One-dimensional Non-Stationary Flows

In this case, all parameters depend on one spatial coordinate x and the time t. This situation includes plane, cylindrical and spherical waves as well as the displacement of gas by a piston and the explosion at a point, and many other applications.

10.3
Potential Movements for an Incompressible Fluid

Before considering the potential flow for an incompressible fluid, we consider some methods for solving the Laplace equation (10.8).

Method of Integral Equations

Consider two points in three dimensions, that is, M at (X_0, Y_0, Z_0) and P at (X, Y, Z). The distance between these points is

$$r = \sqrt{(X - X_0)^2 + (Y - Y_0)^2(Z - Z_0)^2} .$$

The function

$$\Phi = \frac{m}{r} \tag{10.11}$$

is called Newton's potential of the point of attraction M. The physical meaning of Newton's potential consists of the following. If there is a material point with mass m in point M and another material point with mass one at point P, then the force $\boldsymbol{F}$ with which point M attracts is given by $\boldsymbol{F} = \nabla\Phi = (F_x, F_y, F_z)$, where

$$F_x = \frac{\partial\Phi}{\partial X} = -\frac{m(X - X_0)}{r^3} ,$$
$$F_y = \frac{\partial\Phi}{\partial Y} = -\frac{m(Y - Y_0)}{r^3} ,$$
$$F_z = \frac{\partial\Phi}{\partial Z} = -\frac{m(Z - Z_0)}{r^3} .$$

It is easy to show that Φ satisfies Laplace's equation. It must also be noted that for $m = -Q/4\pi$, Newton's potential coincides with the potential of a point source $(Q > 0)$ or point sink $(Q < 0)$ at the point (X_0, Y_0, Z_0) with fluid volume outflow Q.

We consider a volume V of the continuum with mass density ρ which is a function of the volume points (X_0, Y_0, Z_0). We introduce the points M and P, where P can be inside the volume V or the boundary surface Σ, or also outside the volume V. The function

$$\Phi(X, Y, Z) = \int_V \frac{\rho(X_0, Y_0, Z_0)}{r} d\tau \tag{10.12}$$

is called Newton's potential of a volume mass. In this case, $\boldsymbol{F} = \nabla\Phi$ is the force with which the volume V attracts the material point.

We now consider the case that the mass m that is distributed over the surface Σ that is the boundary of V, with surface mass density Σ, where the point $M(X_0, Y_0, Z_0)$ lies on Σ. The function

$$\Phi(X, Y, Z) = \int_\Sigma \frac{\rho(X_0, Y_0, Z_0)}{r} d\sigma \tag{10.13}$$

is called the single layer potential.

It has the same physical meaning as the potential of volume masses. If the right-hand side of (10.13) contains, instead of ρ, the surface mass density $q(M)$ of the sources and sinks which are distributed over Σ, Φ will be the flow potential at point P. It is clear that the potentials (10.12) and (10.13) satisfy Laplace's equation.

Since Laplace's equation is linear, one can obtain new solutions by addition or differentiation of particular solutions. Therefore, one can obtain particular solutions in the form of the derivative of the fundamental solution $1/r$ in the direction of $\boldsymbol{s}$

$$\begin{aligned}\varphi &= C\frac{\partial}{\partial s}\left(\frac{1}{r}\right) = C\left(\frac{\partial\left(\frac{1}{r}\right)}{\partial X}\cos\alpha + \frac{\partial\left(\frac{1}{r}\right)}{\partial Y}\cos\beta + \frac{\partial\left(\frac{1}{r}\right)}{\partial Z}\cos\gamma\right)\\ &= -C\frac{(X-X_0)\cos\alpha + (Y-Y_0)\cos\beta + (Z-Z_0)\cos\gamma}{r^3}\\ &= -C\frac{\boldsymbol{r}\cdot\boldsymbol{s}^\circ}{r^3}\,, \end{aligned} \tag{10.14}$$

where $\cos\alpha$, $\cos\beta$ and $\cos\gamma$ are the directional cosines of the unit vector $\boldsymbol{s}^\circ$, $\boldsymbol{r}$ is the radius vector from (X_0, Y_0, Z_0) to (X, Y, Z) and C is a constant.

The solution (10.14) is interpreted as follows. We consider a source and a sink of equal flow Q at a distance of Δs from each other. In the limit $\Delta s \to 0$ and $Q \to \infty$, we obtain the flow of a point dipole. The constant C is called the dipole moment and the direction $\boldsymbol{s}$ of the dipole axis. The point (X_0, Y_0, Z_0) where the dipole is located is a singular point because the velocity is infinite there.

Therefore, the solution (10.14) is the potential of a dipole flow in space. This flow is shown in Figure 10.1. One obtains new solutions of Laplace's equation via differentiating $1/r$ in different directions, that is,

$$\varphi_n = C\frac{\partial}{\partial s_1}\frac{\partial}{\partial s_2}\cdots\frac{\partial}{\partial s_n}\left(\frac{1}{r}\right). \tag{10.15}$$

Flows that are obtained in this way are called multipole flows. These solutions are used to obtain a solution called the double layer potential.

We consider the flow potential due to dipoles distributed over the surface Σ (closed or open) whose axes are directed along the unit normal vector of the surface $\boldsymbol{n}$, that is,

$$\varphi = -\int\limits_{\Sigma} \mu(M)\frac{\partial}{\partial n}\left(\frac{1}{r}\right)d\sigma\,. \tag{10.16}$$

This potential is called the double layer potential and the quantity $\mu(M)$ is called the double layer density. Using the properties of the derivative in the $\boldsymbol{n}$-direction as

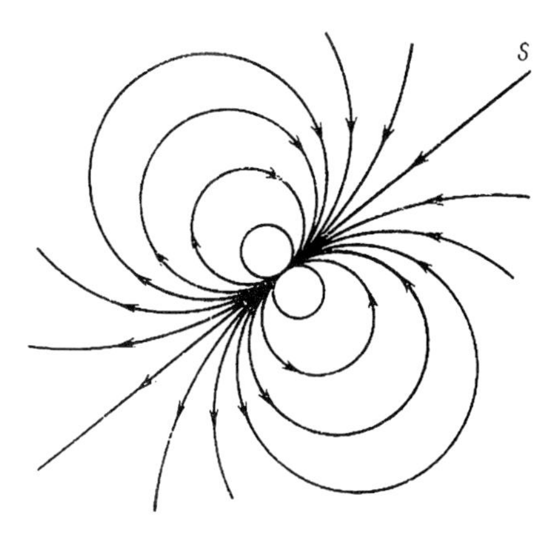

Figure 10.1 Dipole flow.

per (10.14), one obtains

$$\begin{aligned}\varphi &= -\int_{\Sigma} \mu(M)\left[\cos\alpha\frac{\partial}{\partial X_0}\left(\frac{1}{r}\right)+\cos\beta\frac{\partial}{\partial Y_0}\left(\frac{1}{r}\right)+\cos\gamma\frac{\partial}{\partial Z_0}\left(\frac{1}{r}\right)\right]d\sigma \\ &= \int_{\Sigma} \mu(M)\frac{\cos\psi}{r^2}d\sigma\,.\end{aligned} \tag{10.17}$$

Here, ψ is the angle between the external normal and the straight line PM.

We now give the main properties of the single layer and double layer potentials.

Single Layer Potential

The single layer potential is given by (10.13) and solves Laplace's equation. It is a continuous function in the whole space, including the boundary surface Σ. If the point $P(X, Y, Z)$ is removed to infinity, the single layer potential will goto zero as $1/r$, provided, that $\int_{\Sigma} \rho d\sigma \neq 0$. For $\int_{\Sigma} \rho d\sigma = 0$ the single layer potential will decay with a higher order than $1/r$. The normal derivative of the single layer potential with a continuous surface density $\rho(X_0, Y_0, Z_0)$ experiences a discontinuity on the closed surface Σ when moving from the internal (i) to the external (e) domain, such that

$$\frac{\partial\varphi_e}{\partial n} - \frac{\partial\varphi_i}{\partial n} = 4\pi\rho^0\,, \tag{10.18}$$

$$\frac{\partial\varphi_e}{\partial n} + \frac{\partial\varphi_i}{\partial n} = 2\int_{\Sigma} \rho\frac{\cos\psi}{r^2}d\sigma\,. \tag{10.19}$$

The sign O refers to a quantity on the surface Σ and $\boldsymbol{n}$ is the external normal to Σ at the point M.

Double Layer Potential

The double layer potential is given by (10.16) and satisfies Laplace's equation. The double layer potential with continuous density $\mu(X_0, Y_0, Z_0)$ is continuous in the internal (i) and external (e) domains, and undergoes discontinuities while crossing the surface Σ such that

$$\varphi_e = \varphi^0 - 2\pi\mu^0 , \quad \varphi_i = \varphi^0 + 2\pi\mu^0 . \tag{10.20}$$

For the double layer potential with a density $\mu = 1$, Gauss's theorem is satisfied, namely,

$$\int_\Sigma \frac{\cos\psi}{r^2} d\sigma = \begin{cases} 4\pi & \text{for} \quad P \in V_i \\ 2\pi & \text{for} \quad P \in \Sigma \\ 0 & \text{for} \quad P \in V_e \end{cases} . \tag{10.21}$$

Here, V_i and V_e are the internal and external domains.

Sometimes, the problem of determining the potential is reduced to finding the functions $\rho(X_0, Y_0, Z_0)$ and $\mu(X_0, Y_0, Z_0)$. Determining ρ and μ leads to corresponding integral equations. This idea forms the basis of the method of boundary integral equations.

The main problems that are solved by the method of integral equations are the Dirichlet and Neumann problems. Their properties are as follows. We assume it is required to solve Laplace's equation $\Delta\varphi = 0$ in the closed domain V, which is bounded by the surface Σ (see Figure 10.2). If it is required that when approaching the boundary Σ from the inside, the potential φ takes on a given value $f(M^O)$, the problem is called the internal Dirichlet problem. We denote the solution of this problem by $\varphi_i^{(D)}$.

If it is required to solve Laplace's equation $\Delta\varphi = 0$ in the domain external to Σ under the condition that φ takes on a given value $f(M^O)$ when approaching the surface from the outside, this problem is called the external Dirichlet problem. We denote the solution of this problem by $\varphi_e^{(D)}$. In a similar way, the internal and external Neumann problems are formulated. However, on the boundary, one has to give the condition $\varphi/\partial n = f(M^0)$ instead of $\varphi = f(M^0)$. The normal derivative on the surface Σ in point M is defined as a limit of the derivative in the direction MN (N is a point in the external domain) in the internal point P for the internal Neumann problem or in an external point P for the external Neumann problem for

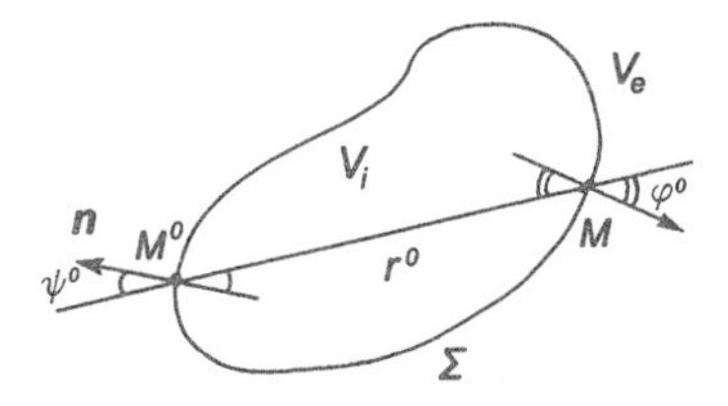

Figure 10.2 Solution of Dirichlet problem.

$P \to M$ along the normal. We denote the solutions of the corresponding problems by $\varphi_i^{(N)}$ and $\varphi_e^{(N)}$.

One looks for a solution of the Dirichlet problem in the form of a double layer potential with unknown density μ. By substituting (10.17) for φ_i into the second condition (10.20), one obtains an integral equation for μ

$$\mu(M_0) + \frac{1}{2\pi} \int\limits_{\Sigma} \mu(M) \frac{\cos\varphi^0}{r_0^2} d\sigma_M = \frac{1}{2\pi} f(M_0) , \tag{10.22}$$

where φ^O is the angle between external normal to Σ in the point M and the line MM_O, the point M_O lies on Σ and $r_0 = |MM_0|$ (see Figure 10.2). After solving (10.22), one obtains $\mu(M)$, from which the solution of the internal Dirichlet problem is obtained via (10.16).

To solve the external Dirichlet problem, one must use the first condition in (10.20) and obtains the following integral equation for μ, that is,

$$\mu(M_0) - \frac{1}{2\pi} \int\limits_{\Sigma} \mu(M) \frac{\cos\varphi^0}{r_0^2} d\sigma = -\frac{1}{2\pi} f(M_0) . \tag{10.23}$$

The boundary condition for the Neumann Problem is $\partial\varphi/\partial n = f(M^0)$. We look for the solution in terms of a single layer potential and use the boundary condition (10.19). As a result, one obtains an integral equation for ρ in the case of the internal Neumann problem

$$\rho(M_0) - \frac{1}{2\pi} \int\limits_{\Sigma} \rho(M) \frac{\cos\psi^0}{r_0^2} d\sigma_M = -\frac{1}{2\pi} F(M_0) \tag{10.24}$$

and for the external Neumann problem

$$\rho(M_0) + \frac{1}{2\pi} \int\limits_{\Sigma} \rho(M) \frac{\cos\psi^0}{r_0^2} d\sigma_M = \frac{1}{2\pi} F(M_0) . \tag{10.25}$$

Among those four problems, only three have unique solutions: both Dirichlet problems and the external Neumann problem. The internal Neumann problem has a solution up to an additive constant under the condition

$$\int\limits_{\Sigma} F(M_0) d\sigma = 0 .$$

The integral equations (10.22)–(10.25) are called Fredholm's integral equations of the second type. They are solved by an iterative approximation method.

Method of Green's Functions

Green's functions are of great importance in the theory of harmonic functions. Green's first formula is a consequence of Gauss' theorem (2.87) where one has to take $U = \psi\partial\varphi/\partial X$, $V = \psi\partial\varphi/\partial Y$, $W = \psi\partial\varphi/\partial Z$. As a result, one obtains Green's first formula

$$\int_V \psi\Delta\varphi d\tau + \int_V \nabla\psi \cdot \nabla\varphi d\tau = \int_\Sigma \psi\frac{\partial\varphi}{\partial n} d\sigma \,. \tag{10.26}$$

The functions φ and ψ are arbitrary continuous functions that are twice continuously differentiable in the volume V. Green's second formula is obtained from the first by swapping the places of the functions φ and ψ, and subtracting the resulting expression from (10.26), that is,

$$\int_V (\psi\Delta\varphi d\tau - \varphi\Delta\psi) d\tau = \int_\Sigma \left(\psi\frac{\partial\varphi}{\partial n} - \varphi\frac{\partial\psi}{\partial n}\right) d\sigma \,. \tag{10.27}$$

We assume that φ is the velocity potential of an ideal fluid. Assuming $\psi = \varphi$ in (10.26), one obtains

$$\frac{1}{\rho}E = \frac{1}{2}\int_V |\Delta\varphi|^2 \, d\tau = \frac{1}{2}\int_\Sigma \varphi\frac{\partial\varphi}{\partial n} d\sigma \,. \tag{10.28}$$

The quantity E is the kinetic energy of the fluid in the volume V and ρ is the mass density of the fluid. It follows from (10.28) that the kinetic energy of a fluid in a closed volume can be expressed through the values of the potential and of the normal derivative of the potential on the boundary surface Σ. If we have $\varphi = \partial\varphi/\partial n = 0$ on the whole boundary surface, or $\varphi = 0$ on parts of the surface Σ and $\partial\varphi/\partial n = 0$ on other parts of the surface, one obtains from (10.28) that inside the volume $|\nabla\varphi| = 0$ and $\varphi = 0$, that is, the fluid is at rest.

Using Green's second formula, a unique harmonic function φ can be expressed inside of V in terms of the values of φ and $\partial\varphi/\partial n$ on the boundary Σ. To do so, we choose the harmonic function of the variables $M(X, Y, Z)$ and $P(X_0, Y_0, Z_0)$ as $\Psi(M, P)$, which has the special form $\psi = 1/r + h$ at $P = M$, where $r = \sqrt{(X - X_0)^2 + (Y - Y_0)^2 + (Z - Z_0)^2}$ and h is a regular harmonic function in the volume V.

We now consider the volume V_1 which is obtained from V by cutting out a small sphere with surface Σ_ε with its center in the point P and radius ε. In this case, from (10.27), one obtains

$$\int_{\Sigma_\varepsilon} \left(\psi\frac{\partial\varphi}{\partial n} - \varphi\frac{\partial\psi}{\partial n}\right) d\sigma + \int_\Sigma \left(\psi\frac{\partial\varphi}{\partial n} - \varphi\frac{\partial\psi}{\partial n}\right) d\sigma \,.$$

Using the condition $\psi \sim 1/r + h$ for $\varepsilon \to 0$, we calculate the integral over Σ_ε as

$$\begin{aligned}\lim_{\varepsilon\to 0}\int_{\Sigma_\varepsilon} \left(\psi\frac{\partial\varphi}{\partial n} - \varphi\frac{\partial\psi}{\partial n}\right) d\sigma &= \lim_{\varepsilon\to 0}\left[\frac{1}{\varepsilon}\int_\Omega \frac{\partial\varphi}{\partial n}\varepsilon^2 d\omega + \int_\Omega \varphi\frac{\partial(1/\varepsilon)}{\partial\varepsilon}\varepsilon^2 d\omega\right] \\ &= -4\pi\varphi(P) \,.\end{aligned}$$

As a result, one obtains

$$\varphi(X,Y,Z) = \frac{1}{4\pi}\int\limits_{\Sigma}\left(\psi\frac{\partial\varphi}{\partial n} - \varphi\frac{\partial\psi}{\partial n}\right)d\sigma\,. \tag{10.29}$$

This formula yields the solution of the internal Dirichlet problem when the boundary condition $\psi = 0$ is satisfied on Σ and of the Neumann problem under the condition $\psi/\partial n = 0$ on Σ.

Two different functions ψ that individually satisfy the conditions $\psi = 0$ and $\partial\psi/\partial n = 0$ on Σ are referred to as the Green's function for the Dirichlet and Neumann problems. The determination of Green's function is equivalent to finding a function ψ, which is harmonic in V with the boundary condition $h = -1/r$ on Σ for the Dirichlet problem (G_D) $\partial h/\partial n = -\partial(1/r)/\partial n$ for the Neumann problem (G_N). Having found G_D, the solution of the Dirichlet problem is

$$\varphi(X,Y,Z) = \frac{1}{4\pi}\int\limits_{\Sigma}\varphi\frac{\partial G_D}{\partial n}d\sigma\,. \tag{10.30}$$

The solution of the Neumann problem is

$$\varphi(X,Y,Z) = -\frac{1}{4\pi}\int\limits_{\Sigma}G_N\frac{\partial\varphi}{\partial n}d\sigma\,. \tag{10.31}$$

By choosing $\psi = 1/r$ in (10.29), that is, $h = 0$, one obtains

$$\varphi(X,Y,Z) = \frac{1}{4\pi}\int\limits_{\Sigma}\left[\frac{1}{r}\frac{\partial\varphi}{\partial n} - \varphi\frac{\partial}{\partial n}\left(\frac{1}{r}\right)\right]d\sigma\,. \tag{10.32}$$

Equation 10.32 represents the potential φ as the sum of a single layer potential and a double layer potential. We now consider a few examples of Green's function.

Green's Function for the Halfspace Bounded by a Plane

We consider the Dirichlet or Neumann problem within a volume V, that is, the upper or lower halfspace. The Green's function of the Dirichlet problem in points of the upper halfspace satisfies the boundary condition $\psi = 0$ at Σ, that is, for $Z = 0$. We consider the following function:

$$\begin{aligned}\psi_1 = &\frac{1}{\sqrt{(X-X_0)^2+(Y-Y_0)^2+(Z-Z_0)^2}}\\ &-\frac{1}{\sqrt{(X-X_0)^2+(Y-Y_0)^2+(Z+Z_0)^2}} = \frac{1}{r_{MP}} - \frac{1}{r_{MP'}}\,.\end{aligned} \tag{10.33}$$

This function satisfies the conditions $\psi_1 = 1/r$ where h is a harmonic function and $\psi_1 = 0$ for $Z = 0$. Therefore, Ψ_1 is the Green's function for the Dirichlet problem of the upper halfspace. The Green's function for the lower halfspace is given by $-\psi_1$ since the solution for $Z < 0$ can be obtained by the mirror image $X = X'$, $Y = Y'$ and $Z = -Z'$. For the Neumann problem, we have $\partial\psi/\partial Z = 0$ for $Z = 0$ and the corresponding Green's function is the same one for $Z > 0$ and $Z < 0$, namely,

$$\psi_2 = \frac{1}{r_{MP}} + \frac{1}{r_{MP'}} . \tag{10.34}$$

Velocity Potential of a System of Singularities in the Halfspace $Z > 0$

We consider the task of determining the velocity potential of the movement of an ideal incompressible fluid in the upper halfspace $Z = 0$ from a system of given singularities: sources, sinks, dipoles and multipoles. We use the method of the mirror image.

To satisfy the boundary conditions of the ideal fluid $V_n = \partial\varphi/\partial n = \partial\varphi/\partial Z = 0$ at $Z = 0$, there has to be a fictitious flow in the halfspace $Z < 0$ consisting of singularities that are mirror images of the singularities in the upper halfspace. For a flow with sources in the points P_k and dipoles in the points Q_j, the velocity potential takes on the following form:

$$\varphi(M) = -\frac{1}{4\pi} \sum_k q_k \left(\frac{1}{r_{MP_k}} + \frac{1}{r_{MP'_k}} \right) + \sum_j m_j \left(\frac{\partial}{\partial s_j} \frac{1}{r_{MQ_j}} + \frac{\partial}{\partial s'_j} \frac{1}{r_{MQ'_j}} \right) . \tag{10.35}$$

Here, q_k and m_j are the given constants which are the source strengths and dipole moments, and P_k, P'_k, Q_j, Q'_j and s_j,s'_j are the pairs of mirror symmetric points and directions relative to the plane $Z = 0$.

Green's Function for the Dirichlet Problem on the Sphere

The method of the mirror image can also be applied to the Dirichlet problem on a sphere. The analogon of the mirror image method for a sphere is the method of the inversion image.

We consider a sphere V with radius R centered at the coordinates origin O. The coordinate transformation

$$\xi = \frac{X R^2}{r^2} , \quad \eta = \frac{Y R^2}{r^2} , \quad \varsigma = \frac{Z R^2}{r^2} , \quad r^2 = X^2 + Y^2 + Z^2 \tag{10.36}$$

moves a point $P(X, Y, Z)$ on the inside of the sphere, which is at a distance $r < R$ from the center to a point $P'(\xi, \eta, \varsigma)$ located outside of the sphere at a distance $r' = R^2/r$. The points P, P' and O lie on a line. The points P and P' are inversion symmetric points relative to a point on the sphere. It is obvious that the inversion image of the surface of the sphere is the surface itself.

We now construct Green's function. To do this, we consider the points $P(X, Y, Z)$ and $P'(\xi, \eta, \varsigma)$. The radius vectors of these points are $\boldsymbol{r}$ and $\boldsymbol{r}'$. By $\boldsymbol{r}'_{PM}$ and $\boldsymbol{r}'_{P'M}$, we denote the vectors between P, P' and the point $M(X_0, Y_0, Z_0)$ with the radius vector $\boldsymbol{r}_0$ and the point M inside of the sphere. Then, the following relationships apply:

$$r_{PM}^2 = (X - X_0)^2 + (Y - Y_0)^2 + (Z - Z_0)^2 = \boldsymbol{r}^2 + \boldsymbol{r}_0^2 - 2\boldsymbol{r} \cdot \boldsymbol{r}_0 ,$$
$$r_{P'M}^2 = (\xi - X_0)^2 + (\eta - Y_0)^2 + (\varsigma - Z_0)^2$$
$$= \boldsymbol{r}'^2 + \boldsymbol{r}_0^2 - 2\boldsymbol{r}' \cdot \boldsymbol{r}_0 = \frac{R^4}{r^2} + r_0^2 - 2\boldsymbol{r}_0 \cdot \frac{R^2}{r^2}\boldsymbol{r} .$$

If the point M is on the sphere, we have $r_0 = R$ and

$$r_{P'M}^2 = \frac{R^2}{r^2} r_{PM}^2 .$$

We now consider a function similar to (10.34):

$$\psi_1 = \frac{1}{r_{PM}} - \frac{R}{r r_{P'M}} . \tag{10.37}$$

This function is symmetric relative to the points (X, Y, Z) and (X_0, Y_0, Z_0), is harmonic on the inside of V, has a singularity of the type $1/r_{PM}$ only inside of V and vanishes on Σ. Then, according to the definition, the function ψ_1 is the Green's function of the internal Dirichlet problem for the sphere.

We now consider another function, namely,

$$\psi_1' = \frac{1}{r_{P'M}} - \frac{r}{R r_{PM}} . \tag{10.38}$$

This function only has one singularity of the type $1/r_{P'M}$ outside of the sphere V and the other properties are the same as for the function ψ_1. Therefore, the function ψ_1' is the Green's function of the external Dirichlet problem for the sphere.

10.4
Movement of a Sphere in the Unlimited Volume of an Ideal, Incompressible Fluid

We consider the movement of a totally rigid sphere in an unlimited ideal and incompressible fluid without mass forces. A sphere of radius a is moving relative to a system of coordinates at rest (X_1, Y_1, Z_1) with the velocity V. The movement of the fluid relative to this system of coordinates is called the absolute movement. We also introduce a moving system of coordinates (X, Y, Z) that is fixed to the sphere and whose origin coincides with the center of the sphere (see Figure 10.3). If the fluid flow due to the movement of the sphere develops from rest and is continuous, this flow is a potential flow and the velocity potential φ satisfies Laplace's equation outside of the sphere, that is,

$$\Delta\varphi = 0 . \tag{10.39}$$

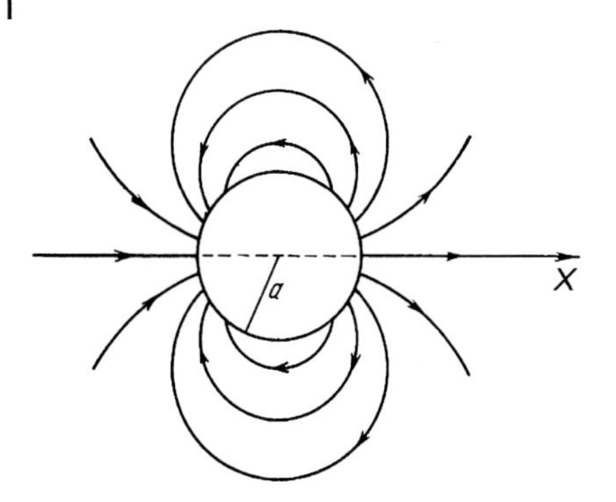

Figure 10.3 Streamlines for the movement of a sphere in an ideal fluid.

The boundary conditions consist firstly of the condition that the fluid is at rest at infinity, that is,

$$\nabla\varphi \to 0 \quad \text{for} \quad r \to \infty \tag{10.40}$$

and the condition that the fluid cannot flow into the sphere on its surface. This condition is equivalent to the equality of the normal components of the velocity of the fluid and the velocity of the sphere. If the sphere is moving along the X-axis with a constant velocity V, the last condition can be written as

$$\left(\frac{\partial\varphi}{\partial r}\right)_{r=a} = V\cos(r, X) = V\cos\vartheta\,. \tag{10.41}$$

The task as formulated is the external Neumann problem for the sphere.

Since the solution of this problem is unique, it can be found with the help of a few solutions that satisfy the boundary conditions (10.40) and (10.41). We now consider two particular solutions that correspond to flows of a point source $-1/r$ or a point dipole centered in the center of the sphere and parallel to the X-axis. The first solution does not satisfy (10.41). Therefore, we must assume

$$\varphi = A\frac{\partial}{\partial X}\left(\frac{1}{r}\right) = -A\frac{X}{r^3} = -A\frac{\cos\vartheta}{r^3}\,. \tag{10.42}$$

From the boundary condition (10.41), it follows that

$$\left(\frac{\partial\varphi}{\partial r}\right)_{r=a} = \left[\frac{\partial}{\partial r}\left(-\frac{A\cos\vartheta}{r^2}\right)\right]_{r=a} = \frac{2A\cos\vartheta}{a^3} = V\cos\vartheta$$

and $A = Va^3/2$ such that

$$\varphi = -\frac{a^3 V\cos\vartheta}{2r^2} \tag{10.43}$$

is the solution of the problem. The corresponding streamlines are shown in Figure 10.3.

If the velocity of the sphere has an arbitrary direction with respect to the streamlines, the velocity potential has the following form:

$$\varphi = -\frac{a^3 V}{2r^3}(V_1 X + V_2 Y + V_3 Z)\,. \tag{10.44}$$

Here, V_1, V_2, V_3 are the components of the velocity that can generally be time-dependent.

To obtain the pressure distribution on the surface of the sphere, one has to use Cauchy–Lagrange's integral (10.2). It must be noted that in (10.2), Cauchy–Lagrange's integral is written for a system of coordinates at rest. To write (10.2) in a moving system of coordinates, one has to use the properties of the substantive derivative (see (2.12)), that is,

$$\frac{d\varphi}{dt} = \frac{\partial\varphi}{\partial t} + \mathbf{V}\cdot\nabla\varphi\,. \tag{10.45}$$

Using (10.45) and the conditions at infinity $\varphi = 0$, $\nabla\varphi = 0$, $p = p_\infty$ in (10.2), one obtains

$$p = p_\infty - \rho\frac{\partial\varphi}{\partial t} + \rho\frac{\partial\varphi}{\partial X}V - \rho\frac{(\nabla\varphi)^2}{2}\,. \tag{10.46}$$

Now, we consider the problem of the flow around a sphere at rest of an ideal and incompressible fluid with constant velocity V at infinity along the X-axis. The potential of such a flow is denoted by φ_{tr} and the flow is called the relative flow. The potential satisfies Laplace's equation, though instead of the boundary conditions (10.40) and (10.41), the potential has to satisfy the following conditions:

$$\left(\frac{\partial\varphi_{tr}}{\partial r}\right)_{r=a} = 0\,, \quad \nabla\varphi \to -V_\infty \quad \text{for} \quad r\to\infty\,. \tag{10.47}$$

In order to make use of the solution (10.43) obtained above, we must imagine that the whole system (fluid plus sphere) is moving with a velocity $-V_\infty$. In this case, the sphere is at rest and the fluid moves with the velocity potential $-V_\infty X$ of the undisturbed flow at infinity. Because of the linearity of the problem, the required potential is the sum of two potentials, that is,

$$\varphi_{tr} = \varphi + \varphi_1 = -\frac{a^3 V_\infty\cos\vartheta}{2r^2} - V_\infty X = -V_\infty\cos\vartheta\left(r + \frac{a^3}{2r^2}\right)\,. \tag{10.48}$$

During the stationary flow, the streamlines flow around a sphere at rest (shown in Figure 10.4).

Using the normal and tangential components of the velocity on the surface of the sphere $V_n = 0$, $V_t \neq 0$, one obtains the following for the distribution of the relative velocities on the surface of the sphere:

$$(V_{tr})_{r=a} = \left(\frac{\partial\varphi_{tr}}{\partial s}\right)_{r=a} = \left(\frac{1}{r}\frac{\partial\varphi_{tr}}{\partial\vartheta}\right)_{r=a} = \frac{3}{2}V_\infty\sin\vartheta\,. \tag{10.49}$$

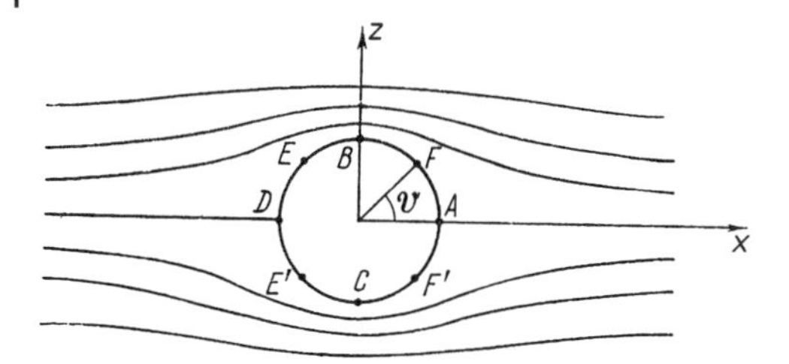

Figure 10.4 Streamlines for the flow of a ideal fluid around a sphere.

From (10.49), it follows that the relative velocity vanishes at the points A and D where $\vartheta = 0$ and $\vartheta = \pi$. These points are called critical points. The maximum value $\frac{3}{2}V_\infty$ for the relative velocity is obtained in the points B and C where $\vartheta = \pi/2$ and $\vartheta = 3\pi/2$. The pressure distribution can be obtained from Cauchy–Lagrange's integral. We restrict ourselves to the stationary case where Cauchy–Lagrange's integral becomes Bernoulli's integral. Thus, we have

$$p + \frac{\rho V_{tr}^2}{2} = p_\infty + \frac{\rho V_{tr\infty}^2}{2} .$$

Using (10.49), one obtains

$$p = p_\infty + \frac{\rho V_\infty^2}{2}\left(1 - \frac{9}{4}\sin^2\vartheta\right) . \tag{10.50}$$

An important dimensionless quantity is the pressure coefficient

$$C_p = \frac{p - p_\infty}{0.5\rho V_\infty^2} = 1 - \frac{9}{4}\sin^2\vartheta . \tag{10.51}$$

The pressure coefficient C_p depends on the position of the point on the body, that is, the angle ϑ, and is independent of the dimension of the body, the velocity and the mass density of the fluid. In critical points, we have $C_p = 1$.

The force that the fluid exerts on the sphere is given by

$$\boldsymbol{R} = -\int_\Sigma p\boldsymbol{n}d\sigma . \tag{10.52}$$

Since (10.50) for the pressure is an even function of ϑ and the pressure force is directed along the surface normal, the resisting force vanishes. The lift is also equal to zero. Therefore, a sphere does not experience any resistance or lift when an ideal fluid flows around it. This effect is known as D'Alembert's paradox. It is valid not only for a sphere, but for any body of finite dimensions and arbitrary shape that is moving at constant velocity without stream separation from the surface.

This can be explained by the fact that a potential flow without stream separation does not really exist in reality. The vortices separate themselves from the surface, the flow becomes asymmetric and the integral (10.52) becomes nonzero.

We now consider the movement of a sphere with variable velocity $V(t)$ along the X-axis. In this case, the movement is non-stationary and it is necessary to use Cauchy–Lagrange's integral (10.46). From (10.43) and (10.46), it follows that during the determination of the force via (10.52), only one term, namely,

$$\frac{\partial \varphi}{\partial t} = -\frac{a \cos\vartheta}{2} \frac{dV}{dt} ,$$

is non-vanishing.

To evaluate the integral (10.52), we decompose the surface of the sphere into infinitesimal stripes (see Figure 10.5) with the surface

$$d\Sigma = 2\pi a^2 \sin\vartheta \, d\vartheta \, .$$

As a result, one obtains

$$R_x = -\int\limits_{\Sigma} p \cos\vartheta \, d\sigma = -\rho\pi a^3 \frac{dV}{dt} \int\limits_{0}^{\pi} \cos^2\vartheta \sin\vartheta = -\frac{2}{3}\pi a^3 \rho \frac{dV}{dt} \, . \tag{10.53}$$

Thus, we write down the equation of motion for a sphere with mass m under the influence of force F_x and the resistance R_x, that is,

$$m\frac{dV}{dt} = F_x - \frac{2}{3}\pi a^3 \rho \frac{dV}{dt} \, .$$

We now rewrite this equation in the form

$$(m + \mu)\frac{dV}{dt} = F_x \, . \tag{10.54}$$

Here, $\mu = 2\pi a^3 \rho/3$ is the virtual mass of the sphere. Therefore, the sphere of mass m moves in the fluid under the influence of a force F_x as if it was moving in the vacuum, and if the mass was increased by the mass of the displaced fluid.

This result has many applications. As an example, we consider the initial acceleration of a spherical balloon that is filled with hydrogen and which is suddenly released from the restraining ropes. We assume that the mass of the balloon is equal to 0.1 times the mass of the displaced air, that is, $m_b = 0,1 m_a$. If one does not take the virtual mass of the balloon into account, the lift is given by

$$m_a g - m_b g = (10 m_b - m_b) g = 9 m_b g \, .$$

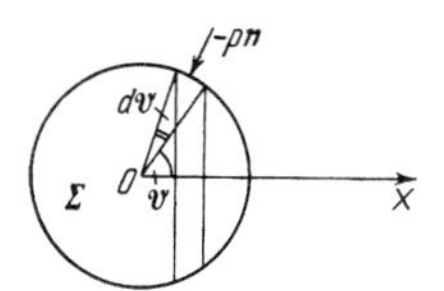

Figure 10.5 Calculation of force on a sphere.

This means that the acceleration of the balloon is equal to 9g. By taking the virtual mass of the balloon into account, the equation of motion for the balloon has the form

$$(m_b + 0.5m_a)a = m_a g - m_b g \,.$$

When the balloon moves in water, we have $m_b = 1000 m_a$ and a similar calculation yields a value of $2g$ when the virtual mass is taken into account while the acceleration without consideration of the virtual mass is $1000g$.

Another example is the pendulum clock. Taking into account that the virtual mass increases the inertia of the physical pendulum by around 0.02%, it is then known that the pendulum watches fall behind by around 10 s per day. The amount by which they fall behind depends on the density of the air, that is, pressure and temperature.

10.5
Kinematic Problem of the Movement of a Solid Body in the Unlimited Volume of an Incompressible Fluid

We consider the movement of a body of finite size and arbitrary form in an infinite volume of an ideal incompressible fluid. We first discuss the task of determining the continuous disturbed movement of the fluid which arises due to the given movement of a solid body from rest.

Together with the non-moving system of coordinates in which the fluid rests at infinity, we introduce a moving system of coordinates X, Y, Z with the basis $\boldsymbol{i}$, $\boldsymbol{j}$, $\boldsymbol{k}$. The velocity of the solid body for an arbitrary movement is given by Euler's formula, that is,

$$\boldsymbol{U} = \boldsymbol{U}_0 + \boldsymbol{\Omega} \times \boldsymbol{r} \,. \tag{10.55}$$

Here, $\boldsymbol{U}_0$ is the translatory velocity of the body point O, $\boldsymbol{\Omega}$ is the instantaneous angular velocity of the body, $\boldsymbol{r}$ is the radius vector from the point O to the point in the body. In the moving system of coordinates, we have

$$\boldsymbol{U}_0 = U^1\boldsymbol{i} + U^2\boldsymbol{j} + U^3\boldsymbol{k}, \boldsymbol{\Omega} = U^4\boldsymbol{i} + U^5\boldsymbol{j} + U^6\boldsymbol{k} \,. \tag{10.56}$$

If the movement of the ideal compressible fluid arises from rest, the movement is continuous and the external mass forces arise from a potential, the fluid flow is a potential flow and the potential satisfies Laplace's equation $\Delta\varphi = 0$. The boundary conditions are $\nabla\varphi \to 0$ for $r \to \infty$ and the impermeability of the body surface Σ, that is,

$$\frac{\partial\varphi}{\partial n} = U_n = \boldsymbol{U}_0 \cdot \boldsymbol{n} + (\boldsymbol{\Omega} \times \boldsymbol{r}) \cdot \boldsymbol{n} = \boldsymbol{U}_0 \cdot \boldsymbol{n} + \boldsymbol{\Omega} \cdot (\boldsymbol{r} \times \boldsymbol{n}) \,. \tag{10.57}$$

Here, $\boldsymbol{n}$ is the external normal relative to V. We look for the solution of the Neumann problem as the following sum:

$$\varphi = U^i\varphi_i = \boldsymbol{U}_0 \cdot \boldsymbol{F}_1 + \boldsymbol{\Omega} \cdot \boldsymbol{F}_2 \,, \tag{10.58}$$

with $F_1 = \varphi_1 i + \varphi_2 j + \varphi_3 k$, $F_2 = \varphi_4 i + \varphi_5 j + \varphi_6 k$.

Thus, we obtain the following external Neumann problems for the φ_i

$$\Delta\varphi_i = 0\,, \tag{10.59}$$

$\nabla\varphi_i \to 0$ at $r \to \infty$, $(\partial\varphi_i/\partial n)_\Sigma = n_i$, $(\partial\varphi_{j+3}/\partial n)_\Sigma = (r \times n)_j$, $(i, j = 1, 2, 3)$.

The boundary conditions on the surface of the body Σ (the last condition in (10.59)) can be written in vector form as

$$\left(\frac{\partial F_1}{\partial n}\right)_\Sigma = n, \left(\frac{\partial F_2}{\partial n}\right)_\Sigma = (r \times n). \tag{10.60}$$

Therefore, the problem of the arbitrary movement of a solid body in an ideal, incompressible fluid is equivalent to six Neumann problems for φ, which are independent of time. The time can only affect the expression for φ by the velocity components U_i. It must be noted that the potentials φ_i only depend on the shape of the body, but not on the kinematics of the movement. Therefore, the φ_i can be determined once for a body of a given shape.

10.6 Energy, Movement Parameters and Moments of Movement Parameters for a Fluid during the Movement of a Solid Body in the Fluid

We now give the dynamical interpretation of the velocity potential for the case of the movement of an ideal, incompressible fluid. We assume that during an infinitesimal time interval τ, an infinite pressure p' is applied to a volume of an ideal, incompressible fluid in such a way that the impulse p_i, defined as

$$p_i = \lim_{\tau\to 0}\int_0^\tau p'\,dt,$$

is finite. The time development of the velocity in an ideal, incompressible fluid is described by Euler's equation as

$$\frac{dV}{dt} = f - \frac{1}{\rho}\nabla p\,.$$

We integrate this equation over the time interval during which the pressure impulse is active and take the limit of τ to zero. Since the external mass force f is finite, the limit of the corresponding integrals is zero. As a result, one obtains

$$\begin{aligned} V' - V &= -\frac{1}{\rho}\lim_{\tau\to 0}\int_0^\tau \nabla p\,dt = -\frac{1}{\rho}\nabla\left(\lim_{\tau\to 0}\int_0^\tau p\,dt\right) \\ &= -\frac{1}{\rho}\nabla p_i = \nabla\left(-\frac{p_i}{\rho}\right). \end{aligned} \tag{10.61}$$

This means that the velocity changes by a finite quantity during an infinitesimal time interval due to a finite pressure impulse and that the velocity change is based on a potential, and that this potential is given by $\varphi = -p\rho$. Therefore, the pressure impulse is given by

$$p_i = -\rho\varphi \,, \tag{10.62}$$

such that $V' = -\nabla\varphi$ for $V = 0$. Therefore, the flow that arises from rest due to a finite pressure pulse is a potential flow.

We now consider a body that is bounded by the surface Σ and immersed in the fluid. We assume that at the initial point in time, the surface Σ exerts a pressure impulse on the fluid. Then, the kinetic energy E, the momentum $\boldsymbol{Q}$ and the angular momentum vector $\boldsymbol{K}$ of the infinite fluid mass can be expressed by the pressure impulse as follows:

$$\begin{aligned}
2E &= \rho \int\limits_{\Sigma} \varphi \frac{\partial \varphi}{\partial n} d\sigma = \rho \int\limits_{\Sigma} \varphi U_n d\sigma = -\int\limits_{\Sigma} p_i U_n d\sigma \,, \\
\boldsymbol{Q} &= -\int\limits_{\Sigma} p_i \boldsymbol{n} d\sigma = \rho \int\limits_{\Sigma} \varphi \boldsymbol{n} d\sigma = \rho \int\limits_{\Sigma} \varphi \frac{\partial \boldsymbol{F}_1}{\partial n} d\sigma \,, \\
\boldsymbol{K} &= -\int\limits_{\Sigma} (\boldsymbol{r} \times p_i \boldsymbol{n}) d\sigma = \rho \int\limits_{\Sigma} \varphi (\boldsymbol{r} \times \boldsymbol{n}) d\sigma = \rho \int\limits_{\Sigma} \varphi \frac{\partial \boldsymbol{F}_2}{\partial n} d\sigma \,,
\end{aligned} \tag{10.63}$$

where $\boldsymbol{n}$ is the external unit normal to Σ and $\boldsymbol{r}$ is the radius vector from the point O to a point on Σ. The vectors $\boldsymbol{Q}$ and $\boldsymbol{K}$ are the total momentum and angular momentum of the fluid due to external forces that the body exerts on the fluid relative to the point O. Since, according to (10.55), we have

$$U_n = \boldsymbol{U}_0 \cdot \boldsymbol{n} + (\boldsymbol{\Omega} \times \boldsymbol{r}) \cdot \boldsymbol{n} \,,$$

where $\boldsymbol{U}_0$ is the translatory velocity of the point O in the body, one obtains, by taking into account the expression expression (10.58), from (10.63) that

$$\begin{aligned}
2E &= \rho \int\limits_{\Sigma} \varphi (\boldsymbol{U}_0 \cdot \boldsymbol{n}) d\sigma + \rho \int\limits_{\Sigma} \varphi \boldsymbol{\Omega} \cdot (\boldsymbol{r} \times \boldsymbol{n}) d\sigma \\
&= \boldsymbol{U}_0 \cdot \boldsymbol{Q} + \boldsymbol{\Omega} \cdot \boldsymbol{K} = \rho \int\limits_{\Sigma} U^i \varphi_i U^k \frac{\partial \varphi_k}{\partial n} d\sigma = \lambda_{ik} U^i U^k \,.
\end{aligned} \tag{10.64}$$

In the last formula, the summation is over i and k runs from one to six, that is, the velocity and angular velocity vectors are combined into one vector. In a similar manner, for the components of the vectors $\boldsymbol{Q}$ and $\boldsymbol{K}$, we have

$$Q_k = \lambda_{ik} U^i \quad (k = 1, 2, 3) \,, \quad K_{k-3} = \lambda_{ik} U^i \quad (k = 4, 5, 6) \,, \tag{10.65}$$

with

$$\lambda_{ik} = \rho \int\limits_{\Sigma} \varphi_i \frac{\partial \varphi_k}{\partial n} d\sigma \,.$$

In the theory of rigid body dynamics, the expression for kinetic energy E_0, momentum Q_0 and angular momentum K_0 form similar to (10.64) and (10.65) in which there are coefficients m_{ik} instead of the λ_{ik}. The matrix $\|m_{ik}\|$ is symmetric and characterizes the inertial properties of the body.

For the fluid, the matrix $\|\lambda_{ik}\|$ is also symmetric, as follows from Green's second formula (10.27), that is,

$$\lambda_{ik} - \lambda_{ki} = \rho \int\limits_{\Sigma} \left(\varphi_i \frac{\partial \varphi_k}{\partial n} - \varphi_k \frac{\partial \varphi_i}{\partial n} \right) d\sigma = 0 \,,$$

because the functions φ_i and φ_k are harmonic.

The total kinetic energy of the system consisting of the fluid and the rigid body is given by

$$2(E + E_0) = (\mu_{ik} + \lambda_{ik}) U^i U^k \,.$$

The coefficients λ_{ik} are called the virtual mass coefficients. The virtual mass matrix $\|\lambda_{ik}\|$ is symmetric and characterizes the inertia of the fluid. In the system of coordinates that is bound to the body, the λ_{ik} do not depend on the time. The quantities λ_{ik} constitute a symmetric tensor of second order.

11
Stationary Potential Flows of an Incompressible Fluid in the Plane

11.1
Method of Complex Variables

We consider a class of stationary potential flows of an incompressible fluid whose velocity potential only depends on the Cartesian coordinates X and Y. For example, one can assume that during the flow around a infinitely long cylinder, the flow has the same shape for each cross section of the cylinder. Therefore, the velocity will only depend on the coordinates X and Y in the plane of the cross section such that the task is equivalent to the solution of Laplace's equation

$$\Delta\varphi = \frac{\partial^2\varphi}{\partial X^2} + \frac{\partial^2\varphi}{\partial Y^2} = 0\,. \tag{11.1}$$

The most important method which is used for the solution of such problems is the method of complex variables. From the equation of continuity

$$\nabla \cdot \mathbf{V} = \frac{\partial U}{\partial X} + \frac{\partial V}{\partial Y} = 0\,, \tag{11.2}$$

it follows that one can introduce a function $\psi(X, Y)$ which is connected as follows with the velocity components U and V

$$U = \frac{\partial\psi}{\partial Y}\,, \quad V = -\frac{\partial\psi}{\partial X}\,, \tag{11.3}$$

and thus identically satisfies (11.2). The function ψ is called the stream function. It has a simple hydrodynamic interpretation. We write down the differential equation for the stream line (2.19)

$$\frac{dX}{U} = \frac{dY}{V}\,. \tag{11.4}$$

By substituting (11.3) into (11.4), one obtains

$$\frac{\partial\psi}{\partial X}dX + \frac{\partial\psi}{\partial y}dY = d\psi = 0\,,$$

Hydromechanics. Theory and Fundamentals. Emmanuil G. Sinaiski

ISBN: 978-3-527-41026-2

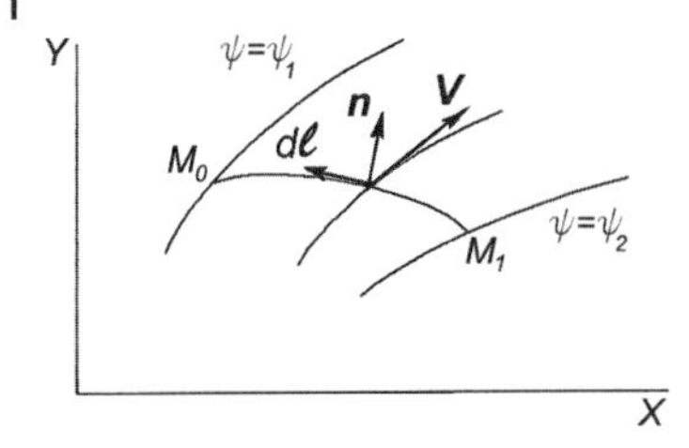

Figure 11.1 Calculation of the fluid flow along an open curve.

which means that $\psi(X, Y) = C = \text{const}$, that is, the stream function preserves its value along the stream line. We now calculate the fluid flow along the open curve $M_0 M_1$ (see Figure 11.1) and obtain

$$\begin{aligned} Q &= \int\limits_{M_0 M_1} V_n \, dl = \int\limits_{M_0 M_1} \boldsymbol{V} \cdot \boldsymbol{n} \, dl = \int\limits_{M_0 M_1} (U n_x dl + V n_y dl) \\ &= \int\limits_{M_0 M_1} (U dY - V dX) = \int\limits_{M_0 M_1} \left(\frac{\partial \psi}{\partial X} dX + \frac{\partial \psi}{\partial Y} dY \right) \\ &= \int\limits_{M_0 M_1} d\psi = \psi_2 - \psi_1 \, . \end{aligned} \tag{11.5}$$

Here, one must take into account that $n_x = dY/dl$ and $n_y = dX/dl$ since $\boldsymbol{n}$ is the unit normal vector. Therefore, the fluid flow along the open curve $M_0 M_1$ which does not coincide with a streamline is given by the difference of the values of the stream functions in the points M_0 and M_1. Since the stream function is defined up to an additive constant, one can define one of the streamlines as the zero streamline with $\psi = 0$. During the fluid flow along a solid surface, one normally defines the contour of this surface as the zero streamline. On the other hand, since the flow is a potential flow, we have $\boldsymbol{V} = \nabla \varphi$ and

$$U = \frac{\partial \varphi}{\partial X} \, , \quad V = \frac{\partial \varphi}{\partial Y} \, . \tag{11.6}$$

By comparing (11.3) and (11.6), one obtains

$$\frac{\partial \varphi}{\partial X} = \frac{\partial \psi}{\partial Y} \, , \quad \frac{\partial \varphi}{\partial Y} = -\frac{\partial \psi}{\partial X} \, . \tag{11.7}$$

Equations 11.7 are called the Cauchy–Riemann conditions. If these are satisfied, the complex quantity

$$w = \varphi + i\psi \tag{11.8}$$

is a function of the complex variable $z = X + iY$. In the following, we will refer to the quantity w as the complex potential.

From the relationships

$$\nabla \varphi \cdot \nabla \psi = \frac{\partial \varphi}{\partial X} \frac{\partial \psi}{\partial X} + \frac{\partial \varphi}{\partial Y} \frac{\partial \psi}{\partial Y} = \frac{\partial \psi}{\partial Y} \frac{\partial \psi}{\partial X} - \frac{\partial \psi}{\partial X} \frac{\partial \psi}{\partial Y} = 0 \, ,$$

we obtain that the lines $\varphi = \text{const}$ and $\psi = \text{const}$, that is, the equipotential lines and streamlines are orthogonal to each other. If we consider the new function $iw = -\psi + \varphi i$ instead of w, ψ and φ exchange their roles. The function $w(z)$ is called the complex potential or characteristic function. φ is called the scalar potential to differentiate it from w. Together with the velocity vector $\boldsymbol{V}(X, Y)$, we introduce the complex velocity v, that is,

$$v = U + iV = |v|\, e^{i\vartheta} , \tag{11.9}$$

with the angle ϑ between $\boldsymbol{V}$ and the X-axis, and the absolute value of the flow velocity $\boldsymbol{v} = |v| = |\boldsymbol{V}| = \sqrt{U^2 + V^2}$.

In addition to the physical plane (X, Y) in which the flow is observed, one introduces a plane (U, V) which is called the hodographic plane. We now consider

$$\begin{aligned} \frac{dw}{dz} &= \frac{dw}{dX} = \frac{\partial(\varphi + i\psi)}{\partial X} = \frac{\partial\varphi}{\partial X} + i\frac{\partial\psi}{\partial X} \\ &= \frac{\partial\varphi}{\partial X} - i\frac{\partial\varphi}{\partial Y} = U - iV = \bar{v} = |v|\, e^{-i\vartheta} , \end{aligned} \tag{11.10}$$

where $\bar{v}$ is the complex conjugate variable (velocity) or, abbreviated, conjugate velocity. For the derivation, the following property of the derivative of complex functions was taken into account:

$$\frac{dw}{dz} = \frac{dw}{dX} = \frac{dw}{d(iY)} .$$

From (11.10), it follows that the derivative of the complex potential is equal to the complex conjugate velocity $\bar{v} = U - iV$ with $U = \text{Re}(\bar{v})$ and $V = -\text{Im}(\bar{v})$.

We consider the integral of the conjugate velocity over a closed contour on the stream plane, that is,

$$\oint_C \bar{v}\, dz = \oint_C \frac{dw}{dz} dz = \oint_C dw = \oint_C (d\varphi + i\, d\psi) .$$

The real part of this integral is the circulation Γ of the velocity around the closed contour C (see (2.82)) and the imaginary part is the flow Q through the contour C (see (11.5))

$$\begin{aligned} \Gamma &= \text{Re} \oint_C \bar{v}\, dz = \oint_C (U dX + V dY) = \oint_C d\varphi , \\ Q &= \text{Im} \oint_C \bar{v}\, dz = \oint_C (U dX - V dY) = \oint_C d\psi . \end{aligned} \tag{11.11}$$

11.2 Examples of Potential Flows in the Plane

In the following, we consider some examples of complex potentials and the corresponding flows.

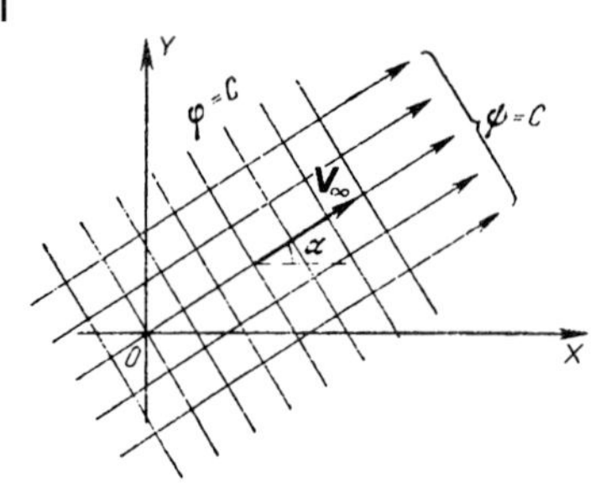

Figure 11.2 The linear complex function $w = az$.

Linear Function $w = az$

We define a to be a constant. To determine the velocity, we use (11.10), that is,

$$\bar{v} = \frac{dw}{dz} = a = |a|\, e^{-i\alpha} . \tag{11.12}$$

From (11.12), it follows that $|v| = |a| = \text{const}$, $\alpha = -\arg(a)$. Therefore, $w = az$ is the complex potential for a homogeneous flow with velocity $|a|$, whose angle to the X-axis is α (see Figure 11.2)

One further obtains

$$w = \varphi + i\psi = |a|\,(\cos\alpha - i\sin\alpha)z = |a|\,(\cos\alpha - i\sin\alpha)(X + iY) .$$

From this, we see that the potential and stream function are given by

$$\varphi = |a|\,(X\cos\alpha + Y\sin\alpha) , \quad \psi = |a|\,(-X\sin\alpha + Y\cos\alpha) .$$

Is the velocity vector equal to $\boldsymbol{V}_\infty = (U_\infty, V_\infty)$, one can rewrite the above as

$$\varphi = (U_\infty X + V_\infty Y) , \quad \psi = (-V_\infty X + U_\infty Y) .$$

Logarithmic Function $w = A \ln z$

We first consider the case of a A being real-valued and write $z = re^{i\varepsilon}$. Then, one obtains

$$w = \varphi + i\psi = A\ln r + A\varepsilon i ,$$

from which it follows that

$$\varphi = A\ln r , \quad \psi = A\varepsilon .$$

The streamlines are the rays $\varepsilon = \text{const}$ and the equipotential lines are the circles $r = \text{const}$ (see Figure 11.3).

Figure 11.3 corresponds to a source (a) or a sink (b). The flow can be obtained from (11.11) as

$$Q = \text{Im}\oint\limits_C \frac{dw}{dz}dz = \text{Im}\oint\limits_C \frac{A}{z}dz = \text{Im}(2\pi i A) = 2\pi A ,$$

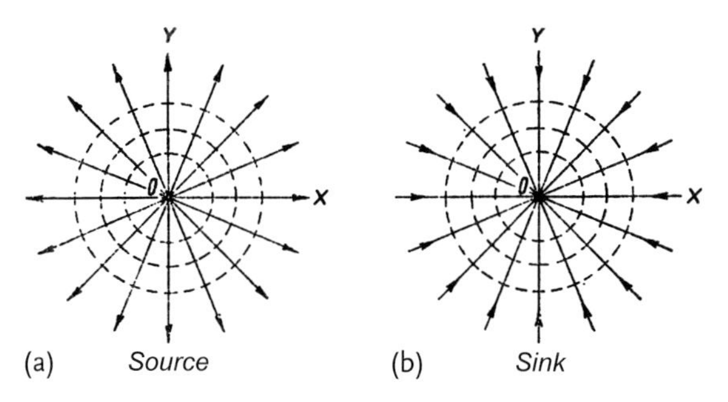

Figure 11.3 Logarithmic complex function $w = A \ln z$ with A real: (a) source, (b) sink.

which means that $A = Q/2\pi$ and

$$w = \frac{Q}{2\pi} \ln z \,, \quad \varphi = \frac{Q}{2\pi} \ln r \,, \quad \psi = \frac{Q}{2\pi} \varepsilon \,. \tag{11.13}$$

For $Q < 0$, we have a sink and for $Q > 0$, a source.

We now consider the case $A = iB$ where B is a real number. We have $w = \varphi + i\psi = Bi \ln z = -B\varepsilon + iB \ln r$ from which it follows that $\varphi = -B\varepsilon$, $\psi = B \ln r$.

Compared to the previous case, one sees that the curves on the z-plane are the same ones as in Figure 11.3, but the roles of the quantities φ and ψ have been swapped around (see Figure 11.3a and b). The velocity circulation of the closed contours is, by (11.11), given by

$$\Gamma = \text{Re} \oint_C \frac{dw}{dz} dz = \text{Re} \oint_C Bi \frac{dz}{z} = \text{Re}(2\pi i Bi) = -2\pi B \,.$$

This means that $B = -\Gamma/2\pi$ and

$$w = -\frac{\Gamma}{2\pi} i \ln z = \frac{\Gamma}{2\pi i} \ln z \,. \tag{11.14}$$

The picture in Figure 11.4 corresponds to a circulatory movement of the fluid around the vortex line that coincides with the Z-axis and is perpendicular to the plane of the figure. For $\Gamma > 0$, the contour is transversed in such a way that the surface bounded by the contour is to the left of the velocity vector. It should be noted that in the case of a source or sink, the absolute value of the fluid velocity is given by $|v| = Q/2\pi r$ and in the case of a vortex by $|v| = \Gamma/2\pi r$. If the source of the vortex is not at the origin though in the point z_0, one has to replace z by $z - z_0$ in (11.14).

We now consider the case if the coefficient in front of $\ln z$ is a complex number $A + Bi$. Then, we have $q = (A + Bi) \ln z$ and the flow consists of the superposition of source or sink and vortex. Such a flow is called a vortex source or a vortex sink (see Figure 11.5).

The complex potential is given by

$$w = \frac{Q - \Gamma i}{2\pi} \ln z \,. \tag{11.15}$$

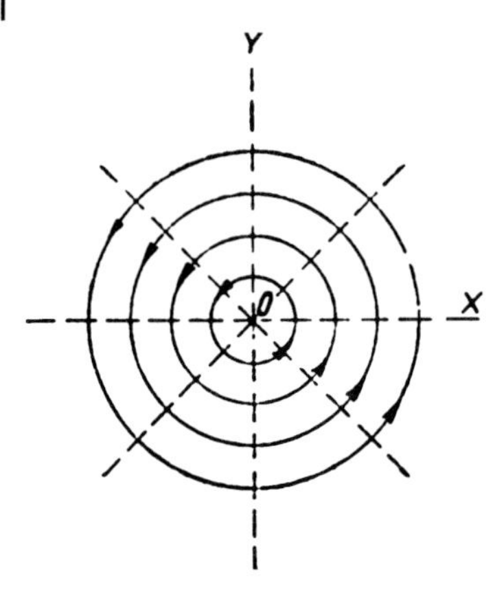

Figure 11.4 Logarithmic complex function $w = A \ln z$ with $A = iB$ where B is a real number.

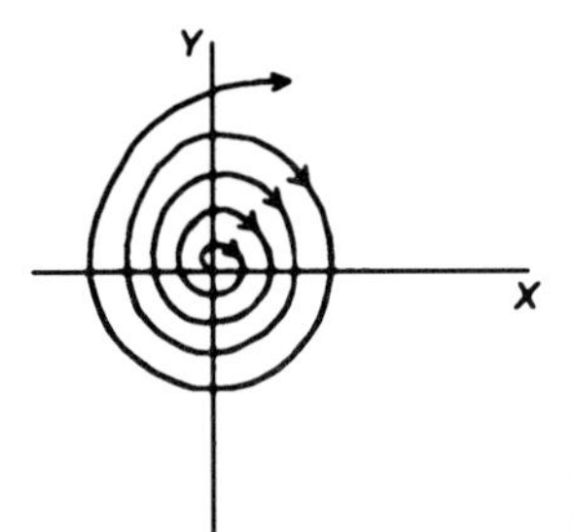

Figure 11.5 Logarithmic complex function $w = A \ln z$ where A is a complex number.

More complicated flows can be obtained by a simple superposition of the above-mentioned flows. This method is often used to determine the complex potentials of complicated flows. For example, for two potential flows w_1 and w_2, the complex potential of the combined flow is given by $w = w_1 + w_2$ and the conjugate velocity is given by

$$\bar{v} = \frac{dw}{dz} = \frac{dw_1}{dz} + \frac{dw_2}{dz} = \bar{v}_1 + \bar{v}_2 \,.$$

The complex velocity is

$$v = v_1 + v_2 \,.$$

Therefore, the complex velocity of a combined flow is the sum of complex velocities for each flow. Maxwell's graphical method for the construction of streamlines of complicated flows from the simple streamlines of the terms in the sum is based on this principle.

Inverse Function $w = 1/z$

By separating the real and imaginary parts, one obtains

$$\varphi = \frac{X}{X^2 + Y^2} \,, \quad \psi = \frac{Y}{X^2 + Y^2} \,.$$

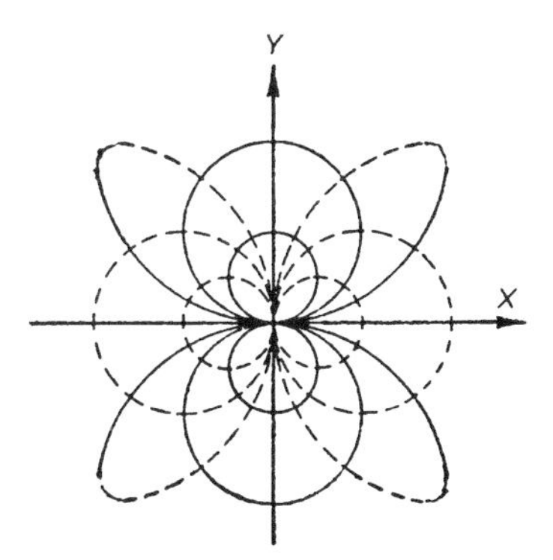

Figure 11.6 The inverse complex function $w = 1/z$.

The streamlines are the curves

$$X^2 + Y^2 = CY .$$

They are circles with their center in the point $(0, C/2)$ and radius $C/2$ (see Figure 11.6). The equipotential lines are also circles, but with their center on the X-axis.

To elucidate the hydrodynamics of the corresponding flow, we put a source of strength $Q > 0$ on the positive side of the X-axis at a distance h from the origin and a sink of the same absolute value, but negative strength, in the same distance h from the origin on the negative side of the X-axis. The complex potential of this flow is

$$W = \frac{|Q|}{2\pi} \ln(z - h) - \frac{|Q|}{2\pi} \ln(z + h) . \tag{11.16}$$

By taking the limits of $h \to 0$ and $Q \to \infty$ under the condition $2Qh \to m = \text{const}$, one obtains

$$\begin{aligned} w &= \lim_{\substack{h \to 0 \\ Q \to \infty}} \left\{ \frac{|Q|}{2\pi} \ln(z - h) - \frac{|Q|}{2\pi} \ln(z + h) \right\} \\ &= \frac{1}{2\pi} \left\{ \frac{\ln(z + h) - \ln(z - h)}{2h} \right\} = -\frac{m}{2\pi} \frac{d}{dz} (\ln z) = -\frac{m}{2\pi z} . \end{aligned} \tag{11.17}$$

Therefore, we have the limit of the flow from a point dipole of moment m, that is,

$$w = \frac{m}{2\pi z} . \tag{11.18}$$

For $m > 0$, the source is on the positive side and the sink is on the negative side of the X-axis, while we have the opposite configuration of sink and source for $m < 0$.

Non-circulatory Flow around a Cylinder of Circular Cross Section

Using the complex potentials of the simple flows mentioned above, one can construct the complex potentials of more complicated flows. We consider such a potential that is obtained by the superposition of a homogeneous flow with velocity $|v_\infty|$ along the X-axis and a dipole flow with the moment m, that is,

$$w = |v_\infty|\, Z + \frac{m}{2\pi z} = \varphi + i\psi \,. \tag{11.19}$$

The streamlines of this flow are given by the equation

$$\psi = |v_\infty|\, Y - \frac{mY}{2\pi(X^2 + Y^2)} = \text{const} \,. \tag{11.20}$$

The zero streamline $\psi = 0$ consists of the circle $X^2 + Y^2 = m/2\pi\,|v_\infty|$ with radius $R = \sqrt{m/2\pi\,|v_\infty|}$ and the X-axis ($Y = 0$) (see Figure 11.7). The other streamlines are obtained from the equation

$$Y\left(1 - \frac{R^2}{X^2 + Y^2}\right) = \text{const} \,.$$

Therefore, the complex potential (11.19) yields the solution of two problems. The first one is the problem of the flow of an ideal, incompressible fluid around a cylinder along the X-axis with velocity $|v_\infty|$ at infinity. The solution of this problem is given by the following complex potential

$$w = |v_\infty|\left(z + \frac{R^2}{z}\right), \quad |z| \geq R \tag{11.21}$$

and $m = 2\pi\,|v_\infty|\,R^2$. The second one is the problem of the flow inside of a cylinder with radius R from a dipole of strength $m = 2\pi\,|v_\infty|\,R^2$. The complex potential of this flow is

$$w = \frac{m}{2\pi R^2}\left(z + \frac{R^2}{z}\right), \quad |z| \geq R \,. \tag{11.22}$$

For the flow around a cylinder, there are two critical points A and B where the velocity vanishes. For the other points, the conjugated velocity is, according to (11.10), given by

$$\bar{v} = U - iV = \frac{dw}{dz} = |v_\infty|\left(1 - \frac{R^2}{z^2}\right).$$

On the cylinder surface, we have $z = \mathrm{Re}^{i\varepsilon}$ and thus

$$\bar{v} = |v_\infty|\left(1 - e^{-2i\varepsilon}\right) = |v_\infty|\, e^{-i\varepsilon}\left(e^{i\varepsilon} - e^{-i\varepsilon}\right) = |v_\infty|\, e^{-i\varepsilon} 2i \sin\varepsilon \,.$$

From this, one obtains

$$|v_R| = 2\,|v_\infty| \sin\varepsilon \,.$$

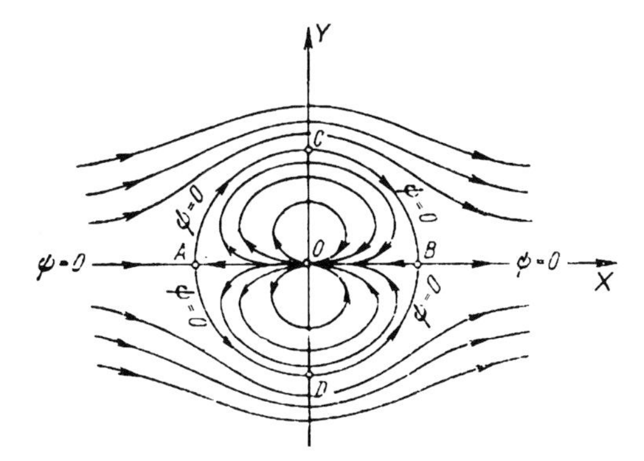

Figure 11.7 Non-circulatory flow around a cylinder of a circular cross section.

The maximum velocity is obtained for $\varepsilon = \pm\pi/2$, that is, in the points C and D. The pressure distribution on the surface can be obtained from Bernoulli's equation, that is,

$$p_R + \rho\frac{|v_R^2|}{2} = p_\infty + \rho\frac{|v_\infty^2|}{2} .$$

By substituting the expression for the velocity, as a pressure coefficient, one obtains

$$c_p = \frac{2(p_R - p_\infty)}{\rho|v_\infty^2|} = 1 - \frac{|v_R^2|}{|v_\infty^2|} = 1 - 4\sin^2\varepsilon . \tag{11.23}$$

This means that the pressure coefficient only depends on the angle ε, but is independent of the mass density of the fluid, the pressure and the radius of the cylinder.

If the flow is at an angle ϑ_∞ with the X-axis, the complex potential is given by:

$$w = \bar{v}_\infty z + v_\infty\frac{R^2}{Z} , \quad |z| \geq R , \tag{11.24}$$

where v_∞ is the complex quantity $v_\infty = |v_\infty|\, e^{i\vartheta_\infty}$. Since the cylinder flow is symmetric, the resistance force vanishes (D'Alembert's paradox).

Circulatory Flow Around a Cylinder of Circular Cross Section

By combining the symmetric flow around the cylinder with the complex potential (11.21) and a circulatory flow (11.14) of a vortex in the origin, as total potential, we obtain

$$w = |v_\infty|\left(z + \frac{R^2}{z}\right) + \frac{\Gamma}{2\pi i}\ln z . \tag{11.25}$$

For $\Gamma > 0$, there is a circulatory flow in a counterclockwise direction. By calculating the complex conjugate velocity, one obtains

$$\bar{v} = \frac{dw}{dz} = |v_\infty|\left(1 - \frac{R^2}{z^2}\right) + \frac{\Gamma}{2\pi i z} .$$

In the critical points, we have

$$z = \frac{\Gamma i}{4\pi\,|v_\infty|} \pm \sqrt{R^2 - \frac{\Gamma^2}{16\pi^2\,|v_\infty|^2}}\,. \tag{11.26}$$

We now consider three possible cases.

1. The circulation is large such that $\Gamma \gg 4\pi v_\infty R$. Both roots z_1 and z_2 of the quadratic equation are imaginary and $|z_1| > R, |z_2| < R$. This means that one critical point is outside of the cylinder and the other one is inside. The flow is shown in Figure 11.8a.
2. For $\Gamma = 4\pi\,|v_\infty|\,R$, the roots of the quadratic equation are $z_1 = z_2 = Ri$. Thus, the critical points A and B coincide, and are on the surface of the cylinder. The streamlines are shown in Figure 11.8b.
3. For the case of small circulation $\Gamma \ll 4\pi\,|v_\infty|\,R$, the roots z_1 and z_2 of the quadratic equation are complex, different and $|z_1| = |z_2| = R$. The critical points A and B lie on the cylinder surface. The streamlines are shown in Figure 11.8c. In the limit $\Gamma \to 0$, the points A and B become the intersections of the circle with the X-axis and the flow becomes symmetric and non-circulatory, as it must be for $\Gamma = 0$.

Let us determine the force on the cylinder. The flow is symmetric with respect to the Y-axis and asymmetric with respect the to Y-axis. Therefore, the resistance force vanishes and the lift is unequal to zero. To determine the lift, we use the expression for the force that the fluid exerts on the body, that is,

$$\boldsymbol{P} = \oint p_n \boldsymbol{n} dl\,.$$

The projection of this force on the coordinate-axis is given by

$$P_y = \oint p_n n_y dl\,.$$

Since $\boldsymbol{n}$ is the external normal of the fluid domain and is directed toward the center, we have

$$P_y = -R\int_0^{2\pi} p_n \sin\varepsilon\, d\varepsilon\,. \tag{11.27}$$

On the cylinder surface, we have $z = \mathrm{Re}^{i\varepsilon}$. Therefore, the conjugate velocity $\bar{v}$ and the velocity $|v_R|$ on the cylinder surface are equal to

$$\begin{aligned}\bar{v} &= |v_\infty|\left(1 - e^{-2i\varepsilon}\right) + \frac{\Gamma}{2\pi i R}e^{-i\varepsilon}\\ &= 2\,|v_\infty|\,ie^{-i\varepsilon}\frac{e^{i\varepsilon} - e^{-i\varepsilon}}{2i} - \frac{\Gamma i}{2\pi R}e^{-i\varepsilon} = ie^{-i\varepsilon}\left(2\,|v_\infty|\sin\varepsilon - \frac{\Gamma}{2\pi R}\right),\\ |v_R| &= 2\,|v_\infty|\sin\varepsilon - \frac{\Gamma}{2\pi R}\,.\end{aligned}$$

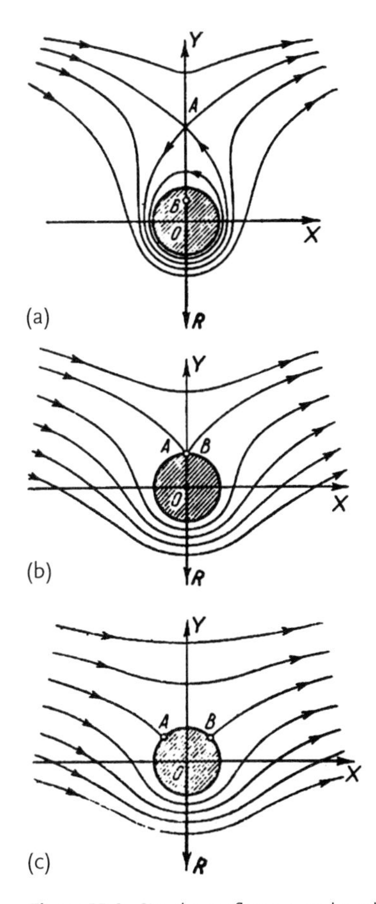

Figure 11.8 Circulatory flow around a cylinder of circular cross section: (a) $\Gamma \gg 4\pi v_\infty R$. (b) $\Gamma = 4\pi |v_\infty| R$ and (c) $\Gamma \ll 4\pi |v_\infty| R$.

The pressure on the cylinder surface can be determined from Bernoulli's equation as

$$p_R = \frac{1}{2}\rho \left|v_\infty^2\right| \left(1 - \frac{v_R^2}{v_\infty^2}\right) = \frac{1}{2}\rho \left|v_\infty^2\right| \left[1 - \left(2\sin\varepsilon - \frac{\Gamma}{2\pi R\,|v_\infty|}\right)^2\right].$$

By substituting the expression p_R into (11.27), one obtains

$$p_Y = -\rho\,|v_\infty|\,\Gamma\,. \tag{11.28}$$

Therefore, during a circulatory flow around a cylinder, the resulting force will be directed downwards for $\Gamma < 0$ and upwards for $\Gamma > 0$. In the latter case, this force is called the lift.

11.3
Application of the Method of Conformal Mapping to the Solution of Potential Flows around a Body

In the previous section, examples of obtaining flows in the plane from a given complex potential were considered. These problems are of an inverse type. However, of greater interest, is the direct problem of determining the complex potential for a given form of a solid body and given conditions of the flow at infinity. We consider the problem of the flow of an ideal, homogeneous and incompressible fluid around a profile, for example, a hydrofoil profile (see Figure 11.9). The incoming flow is characterized by the complex velocity $\bar{v}_\infty$.

The flow area in the physical plane has a hatched cut and therefore it is a doubly connected domain. The circulation Γ along the contour C is also given. The conformal mapping between the z-plane and the ς-plane is realized by the function $\varsigma = f(z)$. Thus, the domain outside of the the contour C is mapped to the domain outside of the circle R, the point $z = \infty$ is mapped to $\varsigma = \infty$, $\bar{v}_\infty$ to $\bar{v}^*_\infty$, where the directions $\bar{v}_\infty$ and $\bar{v}^*_\infty$ coincide. Then, the complex potentials in the physical plane (z) and auxiliary plane (ς) are $w(z)$ and $w*(\varsigma)$, respectively with $w(z) = w*(f(z)) = w*(\varsigma)$. It is obvious that the potential $w*(\varsigma)$ is given by the potential of a circulatory flow of the fluid around a cylinder with angle ϑ_∞ (see (11.24) and (11.25)), that is,

$$w = w* = \bar{v}^*_\infty \varsigma + v^*_\infty \frac{R^2}{\varsigma} + \frac{\Gamma^*}{2\pi i} \ln \varsigma \,, \tag{11.29}$$

where v^*_∞ and Γ^* are the complex velocity in infinity and the circulation along the contour C^* on the ς-plane. Then, we have

$$\frac{dw^*}{d\varsigma} = \frac{dw(f(\varsigma))}{d\varsigma} = \frac{dw}{dz}\frac{dz}{d\varsigma} = \frac{dw}{dz} f'(\varsigma) \,.$$

Since $\bar{v} = dw/dz, \bar{v}^* = dw^*/d\varsigma$, one obtains

$$\bar{v}^* = \bar{v} f'(\varsigma), \bar{v}^*_\infty = \bar{v}_\infty f'(\infty) \,. \tag{11.30}$$

From the condition that the directions of $\bar{v}_\infty$ and $\bar{v}^*_\infty$ coincide, one obtains that $f'(\infty)$ is a real positive number. The function $z = f(\varsigma)$ can be expanded into a

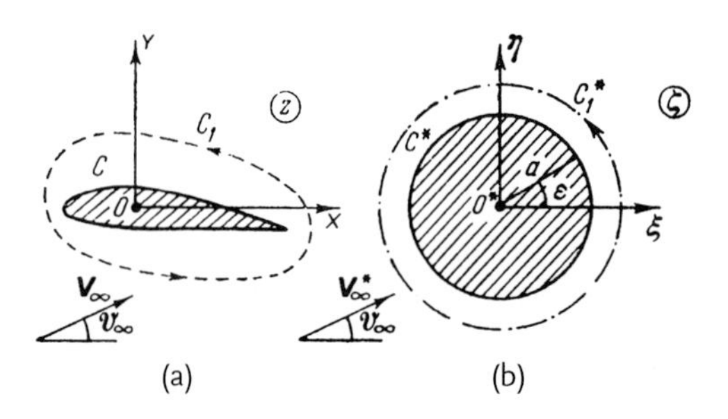

Figure 11.9 Flow around a hydrofoil: (a) physical z-plane (b) z-plane of conformal mapping.

Laurent series, that is,

$$z = m_\infty \varsigma + \sum_{n=0}^{\infty} \frac{m_n}{\varsigma^n} . \tag{11.31}$$

Here, we have $m_\infty = f'(\infty)$ and

$$m_n = \frac{1}{2\pi i} \oint_{C^*} f(\varsigma)\varsigma^{n-1} d\varsigma .$$

On the other hand, the function $f(\varsigma)$ can be expressed by Cauchy's integral as

$$z = f(\varsigma) = \frac{1}{2\pi i} \oint_{C^*} \frac{f'(\varsigma^*)}{\varsigma - \varsigma^*} d\varsigma . \tag{11.32}$$

We now consider the circulation Γ^*. According to (11.11), we have

$$\Gamma^* = \mathrm{Re} \oint_{C^*} \bar{v}^* dz = \mathrm{Re} \oint_{C^*} \bar{v} dz = \mathrm{Re} \oint_{C} \bar{v} dz = \Gamma . \tag{11.33}$$

Therefore, the circulation is conserved during conformal mapping. Thus, by using (11.30)–(11.33), it is possible to obtain $\bar{v}_\infty^*$ and Γ^* from given values of v_∞, Γ, m_∞ and $f(\varsigma)$. If the profile does not have a corner, the conformal mapping can be realized and the solution of the problem can be obtained in parametric form as

$$w(z) = m_\infty \left(\bar{v}_\infty \varsigma + v_\infty^* \frac{R^2}{\varsigma} \right) + \frac{\Gamma^*}{2\pi i} \ln \varsigma, z = f(\varsigma) . \tag{11.34}$$

Only one parameter Γ is unknown. Since this parameter is arbitrary, the solution of the flow around a hydrofoil profile has an infinite number of solutions. For the same profile, the same value and direction of the velocity at infinity and different values of the circulation, there are three different cases of the flow around a hydrofoil (see Figure 11.10), depending on the positions of the critical points that depend on Γ. In the cases *a* and *c*, the fluid flows from one side of the profile to the other side. This means that there are either infinite velocities at the trailing edge which leads to physically impossible negative pressures of infinitely large absolute value, or the flow will be separated from the surface of the profile, leading to the development of vortices.

In case *b*, a separation free flow around the hydrofoil profile with continuous outflow of the streamlines from the trailing edge is realized. This last case only happens for a certain value of Γ, which can be determined by the hypothesis of Joukowski–Chaplygin: among the possible flows around a hydrofoil profile with a given corner, the flow with finite velocity is only realized at this point. The main complication of the conformal mapping of a hydrofoil profile into a circle consists of the fact that the conformity is violated at the trailing edge that is the main property of conformal mappings, namely, the conservation of the angles between

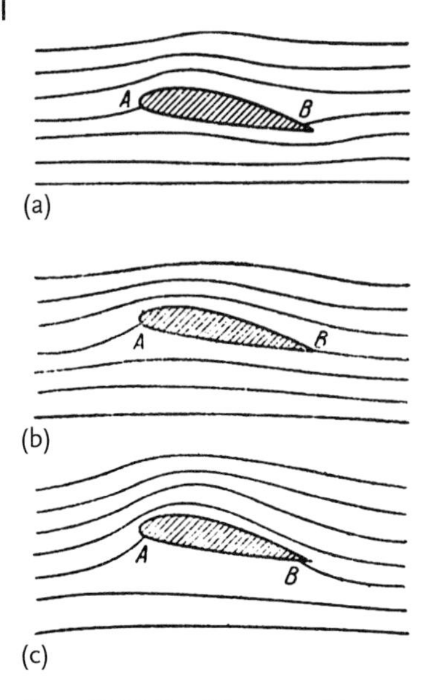

Figure 11.10 Possible cases of the flow around a hydrofoil.

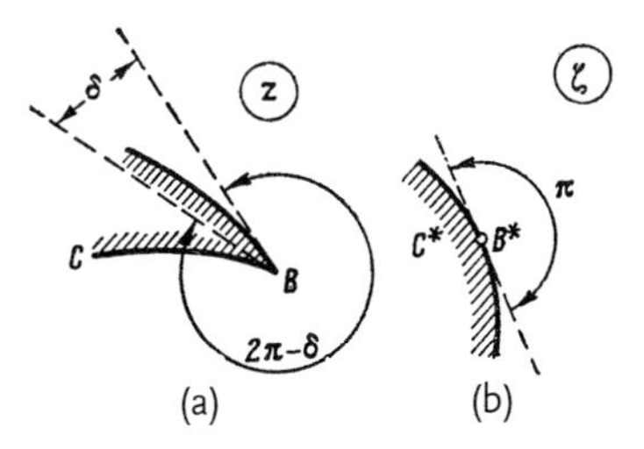

Figure 11.11 Conformal mapping of the hydrofoil profile to a circle if the trailing edge B is mapped to the critical point B^*.

tangents of the mapped profiles is not fulfilled. It can be shown that the hypothesis of Joukowski–Chaplygin leads to the conformal mapping of the hydrofoil profile to a circle mapping the trailing edge point B to the critical point B^* on the circle (see Figure 11.11) where the velocity vanishes.

The conformal mapping

$$z - z_B = M(\varsigma - \varsigma_{B^*})^{\frac{2\pi-\delta}{\pi}} \tag{11.35}$$

maps the external neighborhood of point B on the profile C in the z-plane to the external neighborhood of the point B^* on the circle C^* in such a way that the angle $2\pi - \delta$ is transformed into the angle π. Here, M is a real constant. By using this mapping, one can determine the relationship between the complex velocities $\bar{v}_B$ and $\bar{v}_{B^*}$ in the points B and B^*, that is,

$$\bar{v}_{B^*} = \left(\frac{dw^*}{d\varsigma}\right)_{\varsigma=\varsigma_{B^*}} = \left(\frac{dw}{dz}\right)_{z=z_{B^*}} \left(\frac{dz}{d\varsigma}\right)_{\varsigma=\varsigma_{B^*}} = \bar{v}_B \left(\frac{dz}{d\varsigma}\right)_{\varsigma=\varsigma_{B^*}} .$$

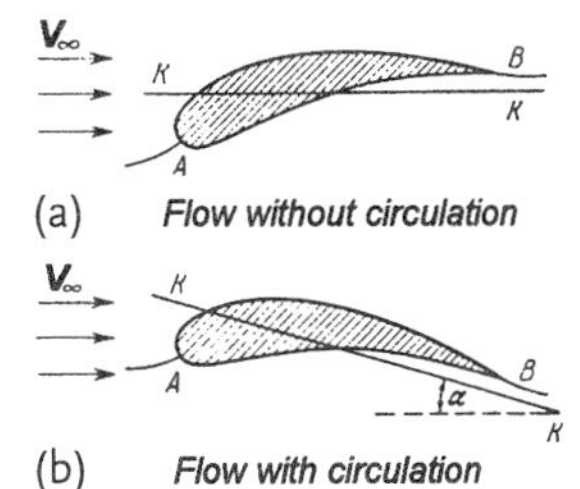

Figure 11.12 Flow around a hydrofoil profile: (a) flow without circulation, (b) flow with circulation.

From (11.35), one obtains

$$\bar{v}_{B^*} = \bar{v}_B \frac{2\pi - \delta}{\pi} M(\varsigma - \varsigma_{B^*})_{\varsigma = \varsigma_{B^*}}^{\frac{2\pi-\delta}{\pi}} .$$

According to the Joukowski–Chaplygin hypothesis, the velocity at point B must be finite. Since $\delta < \pi$, we have $\bar{v}_{B^*} = 0$. Therefore, B^* is the critical point on the circle. From this condition, it follows via (11.34) that

$$\bar{v}_{B^*} = \left(\frac{dw^*}{d\varsigma}\right)_{\varsigma = \varsigma_{B^*}} = m_\infty \bar{v}_\infty - \frac{\bar{v}_\infty R^2}{2\pi i \varsigma} + \frac{\Gamma}{2\pi i} \frac{1}{\varsigma_{B^*}} = 0 . \tag{11.36}$$

At point B^*, we have $\varsigma_{B^*} = \mathrm{Re}^{i\varepsilon_0}$, where ε_0 is the polar angle of the point B^* on the circle C^* with radius R. Let the complex velocity at infinity be $v_\infty = |v_\infty|\, e^{i\vartheta_\infty}$. Then, one obtains from (11.36) the circulation as

$$\Gamma = -4\pi R m_\infty |v_\infty| \sin(\vartheta_\infty - \varepsilon_0) . \tag{11.37}$$

From (11.37), it follows that the flow around the profile is non-circulatory for $\varepsilon_0 = \vartheta_\infty$. This is called the angle of non-circulatory flow. For a circulatory flow, we have $\varepsilon_0 \neq \vartheta_\infty$, and the angle $\alpha = \vartheta_\infty - \varepsilon_0$ is called the angle of attack (see Figure 11.12).

11.4 Examples of the Application of the Method of Conformal Mapping

We consider the following conformal mapping (see Figure 11.13)

$$z = \frac{1}{2}\left(\varsigma + \frac{c^2}{\varsigma}\right), \tag{11.38}$$

or in symmetric form

$$\frac{z - c}{z + c} = \left(\frac{\varsigma - c}{\varsigma + c}\right)^2 . \tag{11.39}$$

This mapping is a special case of the mapping (11.31)

$$m_\infty = 1/2 , \quad m_1 = c^2/2 , \quad m_2 = m_3 = \ldots = 0 .$$

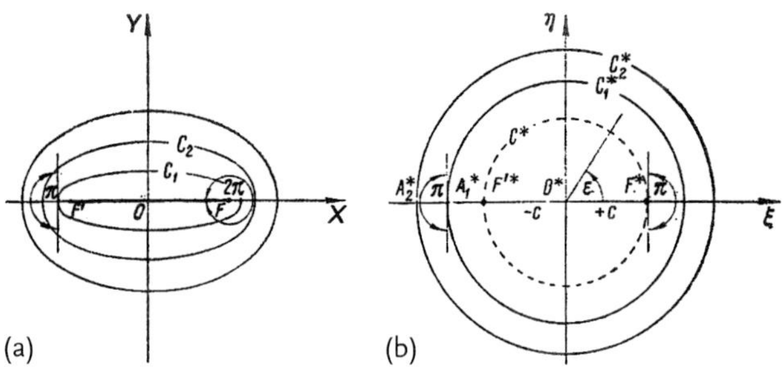

Figure 11.13 Conformal mapping $z = \frac{1}{2}\left(\varsigma + \frac{c^2}{\varsigma}\right)$.

The circle C^* with radius c in the (ς)-plane is mapped to the line FF' on the x-axis in the (z)-plane, whose endpoints are $(-c, 0)$ and $(c, 0)$. The circles $C_1^*, C_2^*, \ldots$ in the (ς)-plane are mapped to confocal ellipses with foci F' and F in the (z)-plane. From the results obtained in the previous section, it follows that the complex potential $w(z)$ of the flow around an elliptical cylinder and in the limit of the plate with velocity at infinity v_∞ at an angle ϑ_0 to the x-axis and with circulation Γ is equal to

$$w = \frac{1}{2}\left(\bar{v}_\infty \varsigma + \frac{v_\infty (a + ib)^2}{\varsigma}\right) + \frac{\Gamma}{2\pi i} \ln \varsigma \,. \tag{11.40}$$

Here, a and b are the half axises of the ellipse such that $r = a + b$ is the radius of the circle to which the ellipse is mapped. By using the formula $\varsigma = z + \sqrt{z^2 - c^2}$ in (11.40) where we take the +-sign because the mapping will then be from the outside of the ellipse to the outside of the circle, one obtains

$$w = \frac{1}{2}\left[\bar{v}_\infty \left(z + \sqrt{z^2 - c^2}\right) + \frac{v_\infty (a + b)^2}{z + \sqrt{z^2 - c^2}}\right] + \frac{\Gamma}{2\pi i} \ln\left(z + \sqrt{z^2 - c^2}\right). \tag{11.41}$$

For $\Gamma = 0$, one obtains a circulation free flow around an elliptical cylinder (see Figure 11.14).

In the special case $c = 0$, $a = b$, one obtains from (11.41) the circulatory flow around a circular cylinder. For $b = 0$, $a = c$, one obtains the flow potential of a plate FF' of length $2c$, that is,

$$w(Z) = U_\infty Z - iV_\infty \sqrt{Z^2 - c^2} + \frac{\Gamma}{2\pi i} \ln\left(Z + \sqrt{Z^2 - c^2}\right), \tag{11.42}$$

where U_∞ and V_∞ are the projections of v_∞ on the coordinate axes. We calculate the velocity using (11.33) and obtain

$$\bar{v} = \frac{dw}{dz} = U_\infty - \frac{iV_\infty z - \Gamma/2\pi i}{\sqrt{z^2 - c^2}} \,. \tag{11.43}$$

From (11.43), it follows that for an arbitrary value of the circulation Γ, the velocity diverges at $z = \pm c$. Using the hypothesis of Joukowski–Chaplygin about the finite

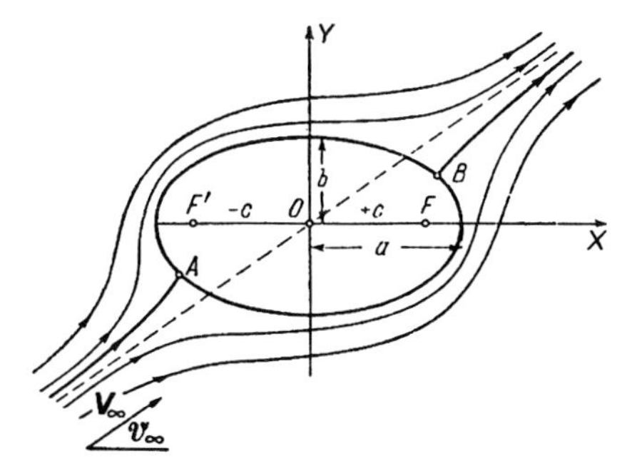

Figure 11.14 Flow around an elliptical cylinder.

value of the velocity at the trailing edge $z = c$, the necessary condition is

$$i V_\infty z - \Gamma / 2\pi i = 0 \,,$$

which yields

$$\Gamma = -2\pi V_\infty c = -2\pi \, |V_\infty| \sin \vartheta_\infty \,. \tag{11.44}$$

The complex velocity is then given by

$$\bar{v} = U_\infty - i V_\infty \sqrt{\frac{z-c}{z+c}} \,. \tag{11.45}$$

It is clear that the velocity at the trailing edge of the plate is finite and equal to U_∞, but diverges at the front edge. Therefore, the velocity cannot become finite at both edges by adjusting the circulation Γ. The streamlines of the corresponding circulatory flow are shown in Figure 11.15.

In (11.41) and (11.42), we have the term

$$w(z) = \frac{\Gamma}{2\pi i} \ln \left(z + \sqrt{z^2 - c^2} \right) \tag{11.46}$$

that corresponds to a purely circulatory flow around an elliptical cylinder or a plate (see Figure 11.16). We restrict ourselves to only consider the flow around a plate.

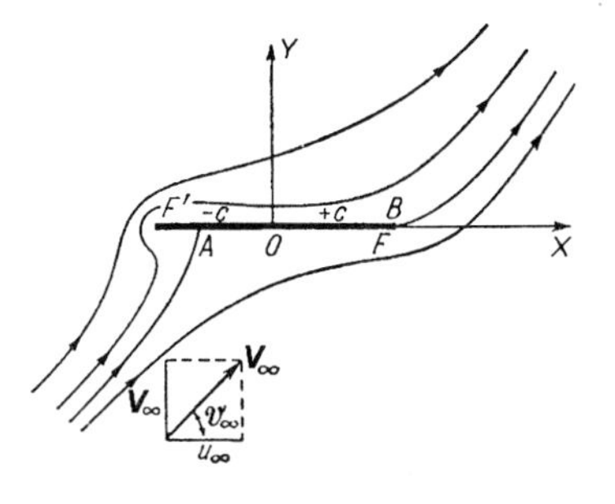

Figure 11.15 Streamlines of circulatory flow.

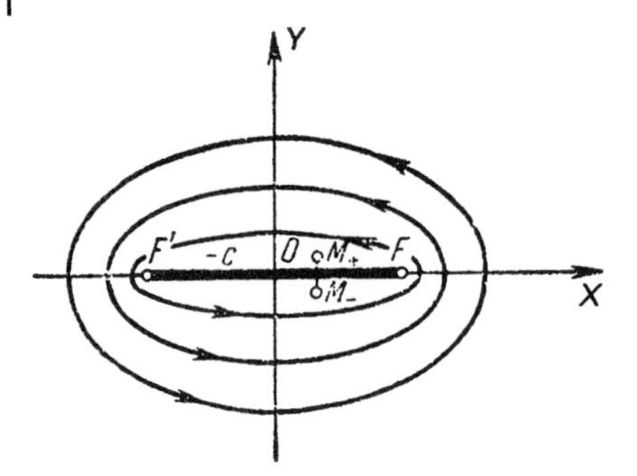

Figure 11.16 Purely circulatory flow around a plate.

The streamlines of this flow are ellipses and the complex conjugate velocity is

$$\bar{v} = -\frac{\Gamma}{2\pi} \frac{1}{\sqrt{c^2 - z^2}} . \tag{11.47}$$

On the upper part of the plate for $Y = 0, -c < X < x$, the velocity U is

$$U_+ = -\frac{\Gamma}{2\pi} \frac{1}{\sqrt{c^2 - X^2}} .$$

On the lower part of the plate, one has to assume the negative sign of the square root in (11.47), resulting in

$$U_- = \frac{\Gamma}{2\pi} \frac{1}{\sqrt{c^2 - X^2}} .$$

If FF' is not considered as a solid plate, but as part of the fluid, then FF' is a line of velocity discontinuity. On FF', the velocity undergoes the jump

$$U_- - U_+ = \frac{\Gamma}{\pi} \frac{1}{\sqrt{c^2 - X^2}} . \tag{11.48}$$

We consider an infinitesimal element dl of the line FF' (see Figure 11.17) and the closed contour $ABCD$. The circulation along this contour is then given by

$$\oint_C \bar{v}\,dl = \int_{AB} \bar{v}\,dl + \int_{Cd} \bar{v}\,dl = U_- dl - U_+ dl = \frac{\Gamma}{\pi} \frac{dl}{\sqrt{c^2 - X^2}} .$$

Therefore, FF' can be considered as a vortex layer. By introducing the density γ of the vortex density per unit length, we have

$$\gamma\,dl = \frac{\Gamma}{\pi} \frac{dl}{\sqrt{c^2 - X^2}} .$$

From this and (11.48), it follows that

$$\gamma = \frac{\Gamma}{\pi} \frac{1}{\sqrt{c^2 - X^2}} = U_- - U_+ . \tag{11.49}$$

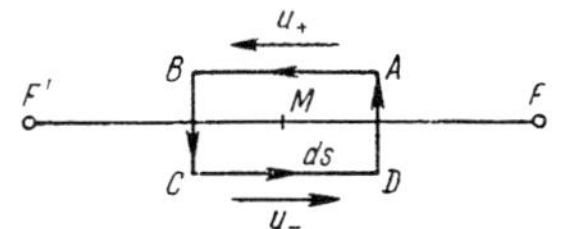

Figure 11.17 Derivation of the total intensity of the vortex layer along FF'.

We see that the purely circulatory flow around an elliptical cylinder is equivalent to a flow that is induced by the vortex layer FF' with the vortex density γ. The total intensity of the vortex layer is

$$\int_{-c}^{c} \gamma \, dX = \int_{-c}^{c} \frac{\Gamma}{\pi} \frac{1}{\sqrt{c^2 - X^2}} dX = \Gamma \,. \tag{11.50}$$

Therefore, the circulation Γ that until now was only introduced as a theoretical concept has a clear physical meaning: it is the total intensity of the vortex layer along the line.

11.5 Main Moment and Main Vector of the Pressure Force Exerted on a Hydrofoil Profile

We now determine the dynamic effect of the flow on a hydrofoil profile (see Figure 11.18). Thus, we determine expressions for the main vector and main moments of the pressure force on the profile C relative to the origin O. According to Bernoulli's theorem, the pressure is given by

$$p = \text{const} - \frac{\rho |v|^2}{2} \,.$$

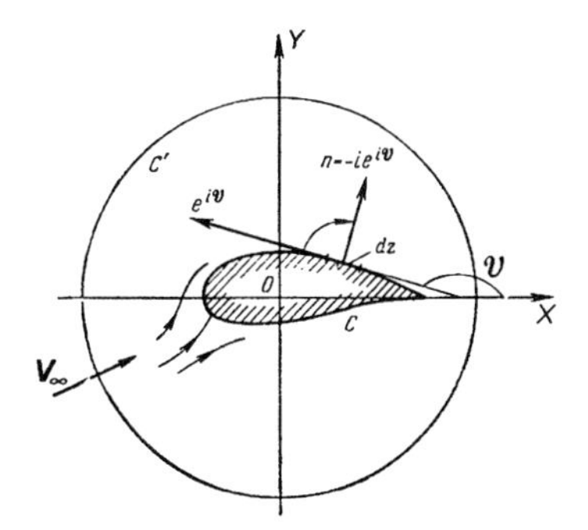

Figure 11.18 Dynamic effect of flow around the hydrofoil profile.

Thus, the main vector $\boldsymbol{R}$ and the main moment $\boldsymbol{M}_0$ of the pressure force are equal to

$$\boldsymbol{R} = -\oint_C p\boldsymbol{n}dl = \frac{\rho}{2}\oint_C |v|^2 \boldsymbol{n}dl ,$$
$$\boldsymbol{M}_0 = -\oint_C (\boldsymbol{r} \times \boldsymbol{n})p\,dl = \frac{\rho}{2}\oint_C (Xn_y - Yn_x)|v|^2 dl . \tag{11.51}$$

By expressing (11.51) by the complex quantities

$$n = -ie^{i\vartheta} , \quad dl = e^{-i\vartheta} , \quad Xn_y - Yn_x = \mathrm{Re}(iz\bar{n})$$

and assuming $v = |v|\, e^{i\vartheta}$ on the contour, one obtains

$$R = R_x + iR_y = -\frac{\rho i}{2}\oint_C |v|^2 dz .$$

$$M_0 = -\frac{\rho}{2}\mathrm{Re}\left(\oint_C |v|^2 e^{-2i\vartheta} z dz, \right) .$$

By replacing in the above formulae $|v|$ by $|v| = \pm v e^{-i\vartheta} = \pm\bar{v}e^{i\vartheta}$, one gets

$$\bar{R} = R_x - iR_y = \frac{\rho i}{2}\oint_C |v|^2 dz = \frac{\rho i}{2}\oint_C |\bar{v}|^2 dz ,$$

$$M_0 = -\frac{\rho}{2}\mathrm{Re}\left(\oint_C |\bar{v}|^2 z dz\right) .$$

Since $\bar{v} = dw/dz$, one obtains

$$\bar{R} = \frac{\rho i}{2}\oint_C \left(\frac{dw}{dz}\right)^2 dz , \quad M_0 = -\frac{\rho}{2}\mathrm{Re}\left(\oint_C \left(\frac{dw}{dz}\right)^2 z dz\right) . \tag{11.52}$$

We now consider the main vector of the pressure force. The complex potential $w(\varsigma)$ given by (11.29),

$$w^*(\varsigma) = \bar{v}_\infty \varsigma + \frac{R^2}{\varsigma} + \frac{\Gamma^*}{2\pi i}\ln\varsigma ,$$

corresponds to the circulatory flow around a circular cylinder. If we consider the flow around the profile in the z-plane, it is mapped to a circle by the conformal mapping $z = f(\varsigma)$. We now expand $f(\varsigma)$ into a Laurent series, that is,

$$z = f(\varsigma) = m_\infty \varsigma + m_0 + \frac{m_1}{\varsigma} + \frac{m^2}{\varsigma^2} + \ldots ,$$

where m_∞ is the real positive number and the other coefficients are complex quantities.

Then, the complex conjugate velocity can be written as

$$|\bar{v}| = \frac{dw}{dz} = \frac{dw^*/d\varsigma}{dz/d\varsigma} = \frac{m_\infty(\bar{v}_\infty - v_\infty R^2/\varsigma^2) + \Gamma/2\pi i\varsigma}{m_\infty - m_1/\varsigma^2 - 2m_2/\varsigma^3}$$
$$= \bar{v}_\infty + \frac{\Gamma}{2\pi i m_\infty}\frac{1}{\varsigma} + \left(\frac{m_1}{m_\infty}\bar{v}_\infty - R^2 v_\infty\right)\frac{1}{\varsigma^2} + \ldots \tag{11.53}$$

Equation 11.53 is the expansion of $\bar{v}$ in the halfplane into a Laurent series. A similar expansion in the physical plane can be written as

$$\bar{v} = a_0 + \frac{a_1}{z} + \frac{a_2}{z^2} + \ldots, \tag{11.54}$$

where

$$a_0 = (\bar{v})_{z=\infty} = \bar{v}_\infty,\ a_1 = \frac{1}{2\pi i}\oint\limits_C \bar{v}\,dz = \frac{\Gamma}{2\pi i},$$
$$a_2 = \frac{1}{2\pi i}\oint\limits_C \bar{v}z\,dz = \frac{m_0\Gamma}{2\pi i} + m_\infty m_1 \bar{v}_\infty - m_\infty^2 R^2 v_\infty\,.$$

By substituting this into (11.52), one obtains

$$\bar{R} = \frac{\rho i}{2}\oint\limits_C\left(a_0 + \frac{a_1}{z} + \frac{a_2}{z^2} + \ldots\right)^2 dz = \frac{\rho i}{2}\oint\limits_C\left(a_0^2 + \frac{2a_0 a_1}{z} + \ldots\right)dz$$
$$= \frac{\rho i}{2}2\pi i \mathrm{Res}\left(a_0^2 + \frac{2a_0 a_1}{z} + \ldots\right) = -2\pi\rho a_0 a_1 = \rho\bar{v}_\infty i\Gamma\,.$$

From this, it follows that

$$R = -i\rho v_\infty \Gamma\,. \tag{11.55}$$

In a similar fashion, one obtains the main moment as

$$M_0 = -\frac{\rho}{2}\mathrm{Re}\left[\oint\limits_C\left(\ldots + \frac{a_1^2 + 2a_0 a_2}{z} + \ldots\right)dz\right]$$
$$= -\pi\rho\mathrm{Re}\left[i(a_1^2 + 2a_0 a_2)\right] = -2\pi\rho\mathrm{Re}(i\bar{v}_\infty a_2)\,.$$

Equation 11.55 is referred to as Joukowski formula for the lift experienced by a hydrofoil in a parallel vortex free flow of an ideal fluid.

An important dimensionless characteristic is the lift coefficient which is defined by

$$c_y = \frac{R}{0.5\rho v_\infty^2 c}, \tag{11.56}$$

where c is the length of the profile tendon, that is, the line which connects the center of the curvature circle for the front and trailing edge.

Lift in an ideal fluid only occurs if a circulation Γ exists. The physical reason for this phenomenon is that the viscosity will induce friction in a real fluid. This force on the surface of the body influences the flow pattern. A thin boundary layer arises on the surface in which the velocity rapidly changes from zero to the velocity of the air flow. This velocity change leads to a considerable vorticity of the flow close to the surface of the body and thus creates a velocity circulation along the contour of the body. In closing, we stress that a body will not experience resistance while moving in an ideal fluid (D'Alembert's paradox), that is, the force that acts on the body along the direction of flight vanishes. Then, the lift exists which acts perpendicular to the flight velocity.

12
Movement of an Ideal Compressible Gas

12.1
Movement of an Ideal Gas Under Small Perturbations

In an incompressible fluid, perturbations propagate with infinite velocity. In a compressible fluid, they propagate with finite velocity, and small perturbations move with the sound velocity defined as $a_0 = \sqrt{(\partial p/\partial \rho)_{\rho=\rho_0}}$.

In Chapter 10, it was shown that determining the velocity potential φ of the barotropic movement of a gas is, for small disturbances, equivalent to solving the wave equation (see (10.10)), that is,

$$\Delta\varphi = \frac{1}{a^2}\frac{\partial^2\varphi}{\partial t^2} . \tag{12.1}$$

When solving (12.1), one has to obtain such solutions that satisfy the corresponding initial and boundary conditions.

Plane Wave Solutions of the Wave Equation

A gas movement for which the potential only depends on one coordinate X and the time t is referred to as a plane wave movement. In this case, (12.1) takes on the following form:

$$\frac{\partial^2\varphi}{\partial X^2} = \frac{1}{a_0^2}\frac{\partial^2\varphi}{\partial t^2} . \tag{12.2}$$

The general solution of (12.2) can be written as

$$\varphi(X,t) = f_1(X - a_0 t) + f_2(X + a_0 t) = f_1(\xi) + f_2(\eta) , \tag{12.3}$$

where $\xi = X - a_0 t$, $\eta = X + a_0 t$.

We consider the solution $\varphi = f_1(\xi)$. At the initial point in time, it will have the shape that is shown in Figure 12.1). At time t, the area of perturbed movement has moved right by a distance $a_0 t$. This movement is called a propagating wave of constant shape which is moving right with the sound velocity a_0. The solution $\varphi = f_2(\eta)$ corresponds to a similar movement to the left with sound velocity a_0.

Hydromechanics. Theory and Fundamentals. Emmanuil G. Sinaiski

ISBN: 978-3-527-41026-2

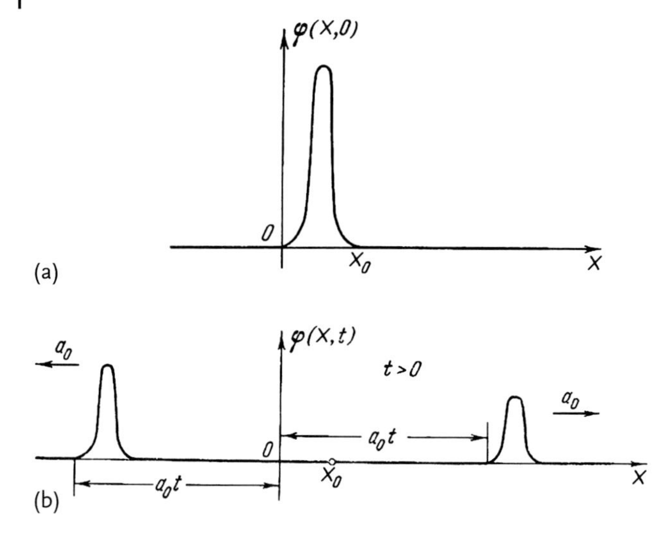

Figure 12.1 Potential describing the one-dimensional propagation of a perturbation; (a) initial disturbance; (b) two propagating waves that travel to the left and the right.

Therefore, the propagation velocity of small perturbation in a gas is equal to the sound velocity a_0.

The solution (12.3) is the sum of two propagating waves that are moving along the X-axis to the right and left. If the interval along the X-axis is infinite, both waves move apart while preserving their initial shapes. For a finite interval on X, this interval must have walls at the boundaries from which the waves are reflected. As a result, reflected waves arise that move from the boundaries inwards.

Solution of the Wave Equation with Spherical Waves

If the movement of the gas has spherical symmetry, ϕ depends on

$$r = \sqrt{X^2 + Y^2 + Z^2}$$

and t. In this case, (12.1) has the form

$$\frac{1}{r^2}\frac{\partial}{\partial r}\left(\frac{\partial \varphi}{\partial r}\right) = \frac{1}{a_0^2}\frac{\partial^2 \varphi}{\partial t^2}\,. \tag{12.4}$$

The solutions of this equation are

$$\varphi(r,t) = \frac{f(r \pm a_0 t)}{r}\,. \tag{12.5}$$

We consider a particular solution in the form of (12.5), that is,

$$\varphi(r,t) = \frac{Q(r - a_0 t)}{4\pi r}\,. \tag{12.6}$$

By expanding Q in a Taylor series in the vicinity of the point $r = 0$, one obtains

$$\varphi(r,t) = -\frac{Q(a_0 t)}{4\pi r} - \frac{Q'(a_0 t)}{4\pi} + o(r)\,.$$

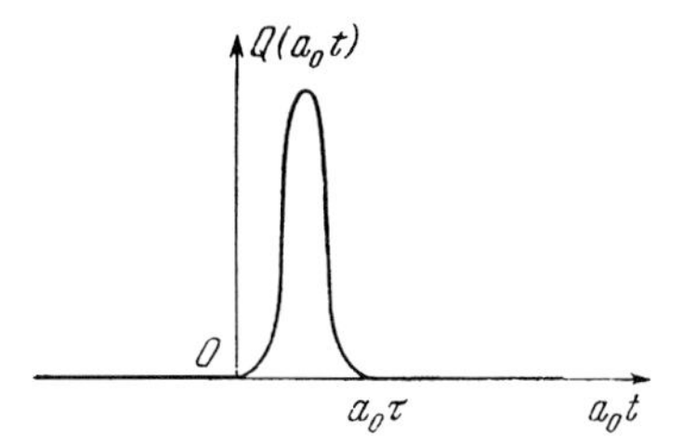

Figure 12.2 Example of the time dependence of a source at the origin.

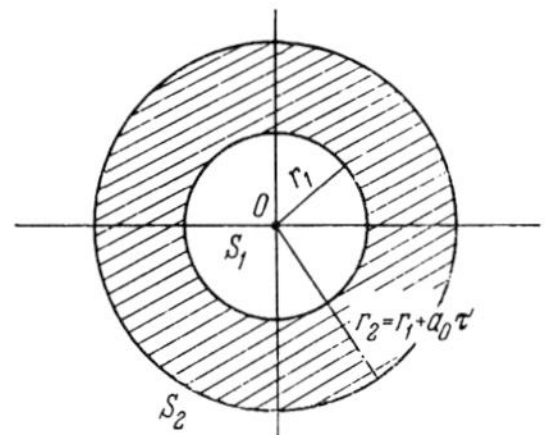

Figure 12.3 Domain of disturbed movement between two spheres.

The main part of the expansion coincides with the expression for the velocity potential of a source at $r = 0$ for a incompressible fluid (see (2.22)). The volume current of this source changes with time and is equal to $Q(a_0 t)$. We now consider the case in which the point $r = 0$ acts for a short time τ as a source with volume current $Q(a_0 t)$ in the infinite volume of the fluid (see Figure 12.2). From (12.6), it follows that for $t > 0$ and $r > 0$, the potential φ is only nonzero in the interval $0 \leqslant a_0 t - r \leqslant a_0 \tau$. Therefore, for a fixed time t, the potential ϕ is unequal to zero only in the interval $a_0(t - \tau) \leqslant r \leqslant a_0 t$. This means that the domain of disturbed movement with $\varphi \neq 0$ lies between the two radii $r_1 = a_0(t - \tau)$ and $r_2 = r_1 + a_0 \tau$, and the center lies in the point $r = 0$ (see Figure 12.3). The velocities with which the circles move through the fluid are identical and given by

$$\frac{dr_1}{dt} = \frac{dr_2}{dt} = a_0 \, .$$

Therefore, the perturbation (12.6) propagates in the form of divergent spherical waves. In contrast to plane waves whose shape and intensity are conserved, the spherical waves only conserve the form, while the amplitude decays like $1/r$ as function of r. In a similar way, one can show that a solution of the form

$$\varphi(r, t) = \frac{Q(a_0 t + r)}{4\pi r} \tag{12.7}$$

corresponds to convergent spherical waves that arrive from a source at infinity and go to the point $r = 0$. The amplitude of such waves grows with decreasing r. From the solutions (12.6) and (12.7), it follows that perturbations originating at $r = 0$ reach a point $r \neq 0$ after a certain time. Therefore, solutions of the form (12.6) and (12.6) are called retarded potentials.

Using the solutions (12.6) and (12.7), one can construct additional solutions of the wave equation. If $\phi(X, Y, Z)$ is a solution of the wave equation, then $\varphi(X - X_0, Y - Y_0, Z - Z_0)$ with arbitrary constants X_0, Y_0, Z_0 is also a solution of the wave equation. Therefore, the function

$$\varphi(X, Y, Z) = -\frac{Q(a_0 t - \sqrt{(X - X_0)^2 + (Y - Y_0)^2 + (Z - Z_0)^2}}{4\pi\sqrt{(X - X_0)^2 + (Y - Y_0)^2 + (Z - Z_0)^2}} \tag{12.8}$$

is also a solution of the wave equation (12.1).

Under the condition that the function $Q(a_0 t)$ looks like Figure 12.2, this solution corresponds to a source that starts, acting at time t_0 in the point (X_0, Y_0, Z_0). Since the wave equation is linear, new solutions can be obtained as superpositions of the solutions of the from (8.8). In this way, a solution can be obtained from all sources that start acting at different times and in different points, and acts for different time intervals with constant or changing intensity.

Perturbations from a Moving Source

The problem of a propagation from a source that moves with constant velocity V in an infinite fluid along a straight line is of large practical interest.

We first consider the case when the source moves with a subsonic velocity $V_0 < a_0$ (see Figure 12.4). At the initial time $t = t_{01}$, the source will be at point M_1 with coordinate X_{01}. We consider another time t_{02}. During the time $t_{02} - t_{01}$, the source moves a distance $V(t_{02} - t_{01})$ and arrives at point M_2. The perturbations from point M_1 propagate during the time $t_{02} - t_{01}$ to the surface of the sphere with radius

$$r_1 = (t_{02} - t_{01})a_0 > (t_{02} - t_{01})V_0$$

and overtake the point M_2 where the source moves in the same time interval. Therefore, the perturbations overtake the source that is moving with subsonic velocity such that the source moves over a medium that has already been perturbed. Thus, the propagation of the perturbation is asymmetric in contrast to a static source. This asymmetry is the reason for the Doppler effect, that is, the fact that observer one, who is in front of the approaching source, will hear sounds of a higher frequency than observer two, who is behind the source that is moving away from him.

We now consider the case when the source is moving with supersonic velocity $V_0 > a_0$ (see Figure 2.5).

Similar arguments to those presented above show that the source moves faster than the spherical perturbation front. Therefore, the medium through which the source is moving is not perturbed. The observer that stands in front of the source moving with supersonic velocity $V_0 > a_0$ does not hear the approaching source, in contrast to a subsonic source. This constitutes the main difference between the propagation of perturbations with subsonic and supersonic velocity.

At time t_{02}, the perturbations from the source moving with supersonic velocity are within a cone with its tip at point M_2 and half angle α at the tip of the cone,

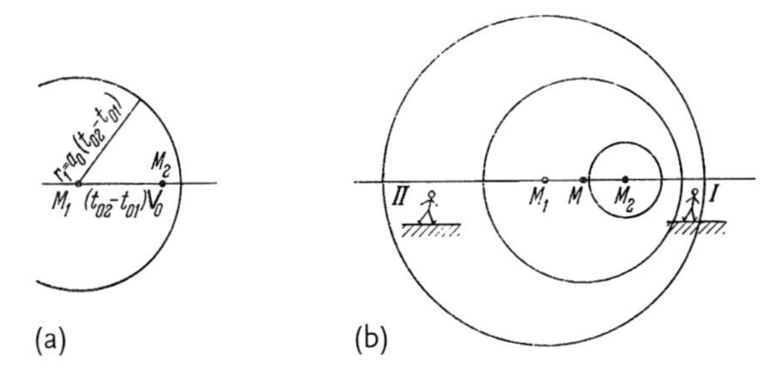

Figure 12.4 Propagation of a perturbation from a source that moves with constant subsonic speed in (a) initial time and (b) ongoing time.

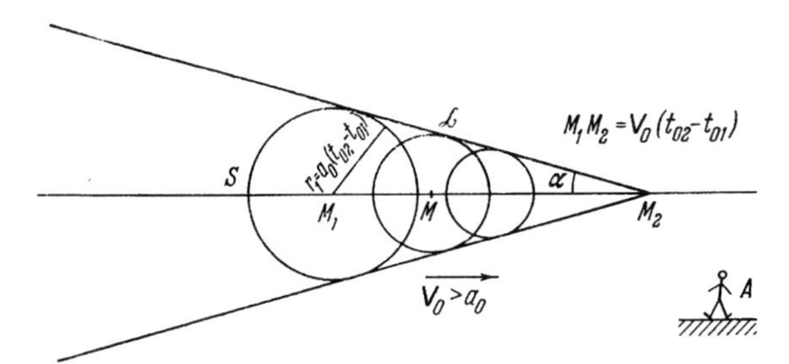

Figure 12.5 Propagation of a perturbation from a source that moves with constant supersonic velocity.

with α defined given by

$$\sin\alpha = \frac{r_1}{M_1 M_2} = \frac{a_0}{V_0} = \frac{1}{M} \, . \tag{12.9}$$

This cone, that separates the perturbed and non-perturbed domains, is called Mach's cone. The side surface of Mach's cone is the characteristic surface of the corresponding partial differential equations. In addition, it is the surface of the perturbation discontinuity.

12.2 Propagation of Waves with Finite Amplitude

We consider one-dimensional movements of an ideal compressible fluid. The movement is described by Euler's equation

$$\frac{\partial U}{\partial t} + U\frac{\partial U}{\partial x} + \frac{1}{\rho}\frac{\partial p}{\partial x} = 0 \, , \tag{12.10}$$

the equation of continuity

$$\frac{\partial \rho}{\partial t} + \rho\frac{\partial U}{\partial x} + U\frac{\partial \rho}{\partial x} = 0 \, , \tag{12.11}$$

and the barotropic equation of state

$$p = p(\rho) . \tag{12.12}$$

In the case of an adiabatic process for an ideal gas, (12.12) becomes

$$p = A\rho^{\gamma} . \tag{12.13}$$

The nonlinear system consisting of (12.10)–(12.12) does not have solutions in the form of plane waves. However, it is possible to construct solutions that are analogous to plane waves and that generalize the solutions of type $f(X \pm a_0 t)$. We want to look for solutions of (12.10)–(12.12) in the form $U = U(\rho)$, $\rho = \rho(X, t)$. Such solutions are called Riemann solutions and the corresponding movements are called Riemann waves. The substitution of $U = U(\rho)$ into (12.10) and (12.11) leads to

$$\begin{aligned} &\frac{dU}{dt}\frac{\partial\rho}{\partial t} + \left(U\frac{dU}{d\rho} + \frac{1}{\rho}\frac{dp}{d\rho}\right)\frac{\partial\rho}{\partial X} = 0 , \\ &\frac{\partial\rho}{\partial t} + \left(\rho\frac{dU}{d\rho} + U\right)\frac{\partial\rho}{\partial X} = 0 . \end{aligned} \tag{12.14}$$

We consider (12.14) as a system of equations for $\partial\rho/\partial t$ and $\partial\rho/\partial X$. In order for non-trivial solutions to exist, the determinant of the system must vanish. This means that

$$\rho\frac{dU}{d\rho} = \frac{\frac{1}{\rho}\frac{dp}{d\rho}}{\frac{dU}{d\rho}} \tag{12.15}$$

and

$$U = \pm\int\sqrt{\frac{1}{\rho^2}\frac{dp}{d\rho}}\,d\rho . \tag{12.16}$$

From this, it follows that if the function $p(\rho)$ is given, $U(\rho)$ can be determined independently from the solution of the system of equations. We define $a(\rho)$ by

$$\frac{dp}{d\rho} = a^2(\rho) . \tag{12.17}$$

Then, (12.14) leads to the following equation for $\rho(X, t)$:

$$\frac{\partial\rho}{\partial t} + (U \pm a)\frac{\partial\rho}{\partial X} = 0 . \tag{12.18}$$

We introduce the velocity c as

$$c = U + a , \tag{12.19}$$

which can be interpreted as the velocity of propagation at constant value of the mass density. Indeed, (12.18) can be rewritten as follows:

$$\frac{d\rho}{dt} = \frac{\partial \rho}{\partial t} + \frac{\partial \rho}{\partial X}\frac{dX}{dt} = 0\,, \qquad \frac{dX}{dt} = c\,.$$

In a similar manner, one can consider the case with velocity $c = U - a$. According to (12.16) and (12.17), we have $c = c(\rho)$ for a barotropic process. Thus, there are two waves of front propagation on constant mass density values. One of them propagates with velocity $c = U + a$ and the other with $c = U - a$ where a is the propagation velocity for perturbations of ρ. For ρ, the following nonlinear equation is valid:

$$\frac{\partial \rho}{\partial t} + c(\rho)\frac{\partial \rho}{\partial X} = 0\,. \tag{12.20}$$

We will now consider the case of the adiabatic movement of an ideal gas and determine the value of $c = U + a$. By using (12.13), one obtains

$$a^2 = \frac{dp}{d\rho} = A\gamma\rho^{\gamma-1}\,,$$
$$U = \pm\int \sqrt{A\gamma}\,\rho^{(\gamma-1)/2-1}\,d\rho = \pm\sqrt{A\gamma}\,\frac{2}{\gamma-1}\rho^{(\gamma-1)/2} + \text{const}\,,$$
$$c(\rho) = U + a = \sqrt{A\gamma}\left(1 + \frac{2}{\gamma-1}\right)\rho^{(\gamma-1)/2} + \text{const}\,. \tag{12.21}$$

Since $\gamma > 1$, it follows from (12.21) that the velocities U and c are monotonically increasing functions of ρ. We consider the movement of the front of constant values of ρ and thus of U. We have

$$\left(\frac{dX}{dt}\right)_{\rho,U} = c(\rho) = U + a\,,$$

from which one obtains

$$X = c(\rho)t + F(\rho)\,. \tag{12.22}$$

Similar solutions exist for $c = U - a$. Equations 12.21 and (12.22) provide exact solutions for the nonlinear (12.10) to (12.12) and are called simple waves.

We now consider the shape change of simple waves as a function of time. The profile $\rho(X)$ at the initial point in time is given by the shape shown in Figure 12.6a. To the left of the point M, ρ grows with increasing X and one has a dilution wave because the mass density drops in this point after passing the fixed point X. To the right of the point M, ρ decreases with increasing X and one has a compression wave. The velocity c with which constant values of ρ are moving depends on ρ. Therefore, the profile of the mass density distribution changes with time. We now consider the case when, as for the adiabatic movement of the ideal gas, the velocity c grows with increasing ρ. This means that the propagation of larger values of ρ

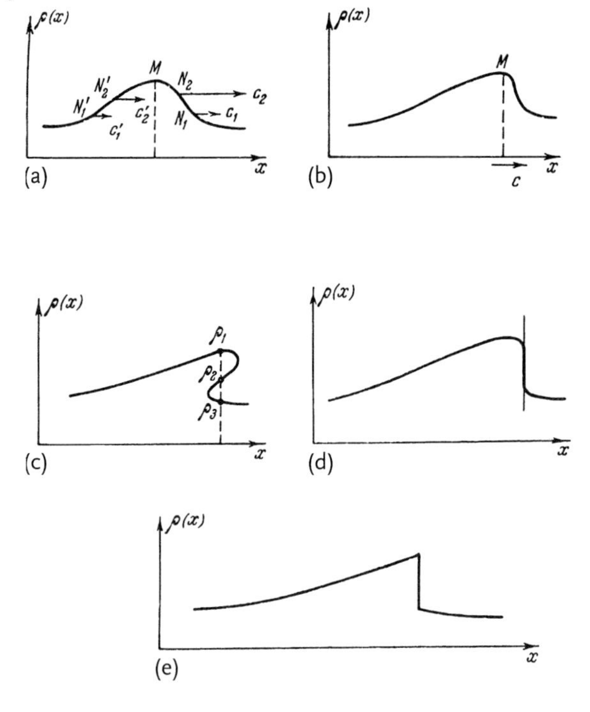

Figure 12.6 Tipping over of the compression wave: (a) distribution at time t, (b) distribution at time $t_1 > t$, (c) breakdown of the unique density distribution, (d) creation of the compression wave, and (e) compression wave.

happens with larger velocity. Therefore, the part of the wave that, to the right of M, corresponds to a compression is shortened because the points N_1 and N_2 approach each other, while the other part that corresponds to a dilution is extended since the points N_1' and N_2' are moving apart (see Figure 12.6b).

From a mathematical point of view, it is possible that more than one value of ρ is observed at some point X at some time (see Figure 12.6c), though this is not allowed from the physical point of view. A unique continuous solution can therefore only exists until the point in time when $\rho(X)$ has a vertical tangent at some point X, that is, $\rho'(X) = \infty$ (see Figure 12.6d). From this point in time, the continuous Riemann solution does not exist and it has to be replaced by a solution of the compression shock type. Therefore, if the Riemann solution has a part that corresponds to a compression wave, compression shocks will be created in the flow of an ideal fluid. If the mass density grows in the propagation direction of the waves, such discontinuities will not arise. Such a situation arises, for example, when a piston is pulled out in a continuous manner from a long tube which is filled with gas. When making use of the Riemann solution with $c = U - a$, all of the above considerations apply, provided that the direction of the X-axis is changed into the opposite direction. We finally note that all results obtained above about the properties of Riemann waves are only valid for a certain dependence $p = f(\rho)$.

12.3
Plane Vortex-Free Flow of an Ideal Compressible Gas

We consider a plane vortex-free flow of an ideal, compressible and barotropic gas. The velocity has two components U and V along the Cartesian coordinate axes X and Y, which only depend on X and Y. The main equations are:

- The equation of continuity

$$\frac{\partial(\rho U)}{\partial X} + \frac{\partial(\rho V)}{\partial Y} = \rho\left(\frac{\partial U}{\partial X} + \frac{\partial V}{\partial Y}\right) + U\frac{\partial \rho}{\partial X} + V\frac{\partial \rho}{\partial Y} = 0\,, \tag{12.23}$$

- The barotropy condition $p = p(\rho)$ or $\rho = \rho(p)$, which leads to

$$\frac{\partial \rho}{\partial X} = \frac{d\rho}{dp}\frac{\partial p}{\partial X} = \frac{1}{a^2}\frac{\partial p}{\partial X}\,, \quad \frac{\partial \rho}{\partial Y} = \frac{d\rho}{dp}\frac{\partial p}{\partial Y} = \frac{1}{a^2}\frac{\partial p}{\partial Y}\,, \tag{12.24}$$

  with the sound velocity a.
- Euler's equation without volume forces

$$\rho\left(U\frac{\partial U}{\partial X} + V\frac{\partial U}{\partial Y}\right) = -\frac{\partial p}{\partial X}\,, \quad \rho\left(U\frac{\partial V}{\partial X} + V\frac{\partial V}{\partial Y}\right) = -\frac{\partial p}{\partial Y}\,. \tag{12.25}$$

  Using (12.24) and (12.25) yields

$$\frac{\partial p}{\partial X} = -\frac{\rho}{a^2}\left(U\frac{\partial U}{\partial X} + V\frac{\partial U}{\partial Y}\right)\,, \quad \frac{\partial p}{\partial Y} = -\frac{\rho}{a^2}\left(U\frac{\partial V}{\partial X} + V\frac{\partial V}{\partial Y}\right)\,.$$

  By substituting these expressions into (12.23), one obtains

$$\left(a^2 - U^2\right)\frac{\partial U}{\partial X} - UV\left(\frac{\partial U}{\partial Y} + \frac{\partial V}{\partial X}\right) + \left(a^2 - V^2\right)\frac{\partial V}{\partial Y} = 0\,. \tag{12.26}$$

The last equation is valid for both vortex-free movements as well as for movements with vortices. For a plane movement, the vorticity vector has only one component $\omega_z = 0.5(\nabla \times \mathbf{V})_z = 0.5(\partial U/\partial Y - \partial V/\partial X)$. Therefore, we have $\partial U/\partial Y - \partial V/\partial X = 0$ for the case of a vortex-free flow. This means that the flow is a potential flow, that is, $\mathbf{V} = \nabla\varphi$ and

$$U = \frac{\partial \varphi}{\partial X}\,, \quad V = \frac{\partial \varphi}{\partial Y}\,. \tag{12.27}$$

Since the flow is stationary and barotropic, Bernoulli's equation applies. In the case of an adiabatic process and an ideal gas, the following equation applies:

$$\frac{1}{2}\left(U^2 + V^2\right) + \frac{a^2}{\gamma - 1} = \frac{1}{2}\left(U_\infty^2 + V_\infty^2\right) + \frac{a_\infty^2}{\gamma - 1}\,. \tag{12.28}$$

Equation 12.23 allows one to introduce the stream function $\psi(X, Y)$ such that

$$\rho U = \rho_\infty\frac{\partial \psi}{\partial Y}\,, \quad \rho V = -\rho_\infty\frac{\partial \psi}{\partial X}\,, \tag{12.29}$$

where ρ_∞ is the value of the mass density at infinity. It is obvious that (12.23) is automatically fulfilled if the equations (12.29) are satisfied. By substituting (12.28) into (12.26), one obtains

$$\left[a^2 - \left(\frac{\partial\varphi}{\partial X}\right)^2\right]\frac{\partial^2\varphi}{\partial X^2} - 2\frac{\partial\varphi}{\partial X}\frac{\partial\varphi}{\partial Y}\frac{\partial^2\varphi}{\partial X\partial Y} + \left[a^2 - \left(\frac{\partial\varphi}{\partial Y}\right)^2\right]\frac{\partial^2\varphi}{\partial Y^2} = 0\,. \tag{12.30}$$

The sound velocity one can obtain from (12.28) as

$$a^2 = a_\infty^2 + \frac{\gamma-1}{2}\left(U_\infty^2 + V_\infty^2\right) - \frac{\gamma-1}{2}\left[\left(\frac{\partial\varphi}{\partial X}\right)^2 + \left(\frac{\partial\varphi}{\partial Y}\right)^2\right]. \tag{12.31}$$

Equations 12.30 and (12.31) describe the vortex-free movement of an ideal compressible gas. To these equations, one has to add the corresponding boundary conditions, for example, the conditions of impermeability on the surface of the body immersed in the flow and the velocity values U_∞ and V_∞ at infinity. Since the solution of this nonlinear system of equations is a difficult mathematical problem, we consider an approximation method that is often used for the problem of the flow around a thin planar profile.

We consider the flow of a homogeneous gas around a thin body which has a small curvature and whose angle of attack is also small. The contour of the body is given by the equation $Y = h(X)$ and $X = X_1$, and $X = X_2$ are the abscissae of the front and trailing edges, respectively. The conditions that the body is thin and has a small curvature can be expressed as the condition of small thickness relative to the length of the body, ($|h(X)| \ll X_2 - X_1$), and of a small angle ϑ of the tangent on the profile relative to the X-axis ($h'(X) = \tan\vartheta \approx \vartheta$). From these conditions, it follows that the perturbations induced in the homogeneous medium are small, and thus (12.30) and (12.31) can be linearized.

We choose the system of coordinates in such a way that the X-axis lies in the direction of the homogeneous flow. By $U_\infty, p_\infty, \rho_\infty$ and a_∞, we denote the corresponding parameters of the homogeneous flow far away from the body at infinity. We denote the perturbations induced by the body by the sign $\sim$ over the corresponding parameter. We assume that the velocity, pressure, mass density and the sound velocity can be written as a sum of the corresponding homogeneous parameter values at infinity and the small perturbations to this parameter, that is,

$$U = U_\infty + \tilde{U}\,, \quad V = V_\infty + \tilde{V}\,, \quad p = p_\infty + \tilde{p}\,,$$
$$\rho = \rho_\infty + \tilde{\rho}\,, \quad a = a_\infty + \tilde{a}\,. \tag{12.32}$$

We assume that $M_\infty = U_\infty/a_\infty \neq 1$ at infinity. By substituting (12.32) into (12.26) and neglecting the terms of second and higher order, one obtains

$$\left(a_\infty^2 - U_\infty^2\right)\frac{\partial\tilde{U}}{\partial X} + a_\infty^2\frac{\partial\tilde{V}}{\partial Y} = 0$$

or

$$\left(1 - M_\infty^2\right)\frac{\partial \tilde{U}}{\partial X} + \frac{\partial \tilde{V}}{\partial Y} = 0\,. \tag{12.33}$$

From the condition that the flow is vortex-free, it follows that

$$\frac{\partial \tilde{U}}{\partial Y} - \frac{\partial \tilde{V}}{\partial X} = 0\,. \tag{12.34}$$

We also assume that the potential ϕ and the stream function ψ can be written as sums of values at infinity and small perturbations, that is,

$$\varphi = \varphi_\infty + \tilde{\varphi}\,, \quad \psi = \psi_\infty + \tilde{\psi}\,. \tag{12.35}$$

By substituting (12.35) into (12.27) and by dropping higher order terms again, one obtains

$$\tilde{U} = \frac{\partial \tilde{\varphi}}{\partial X}\,, \quad \tilde{V} = \frac{\partial \tilde{\varphi}}{\partial Y}\,, \quad \tilde{\rho} U_\infty + \rho_\infty \tilde{U} = \rho_\infty \frac{\partial \tilde{\psi}}{\partial Y}\,, \quad \tilde{V} = -\frac{\partial \tilde{\psi}}{\partial X}\,. \tag{12.36}$$

We write Bernoulli's equation (12.28) in the following form:

$$\frac{1}{2}\left(U^2 + V^2\right) + \frac{a_\infty^2}{\gamma - 1}\left[\left(\frac{\rho}{\rho_\infty}\right)^{\gamma-1} - 1\right] = \frac{U_\infty^2}{2}\,. \tag{12.37}$$

After linearization, one obtains

$$U_\infty \tilde{U} + \frac{a_\infty^2}{\rho_\infty}\tilde{\rho} = 0\,. \tag{12.38}$$

From (12.35) and (12.38), one obtains

$$\tilde{U} = \frac{1}{1 - M_\infty^2}\frac{\partial \tilde{\psi}}{\partial Y}\,, \quad \tilde{V} = -\frac{\partial \tilde{\psi}}{\partial X}\,. \tag{12.39}$$

By substituting those expressions into (12.33), one obtains the equation for the perturbation of the stream function:

$$\left(1 - M_\infty^2\right)\frac{\partial^2 \tilde{\psi}}{\partial X^2} + \frac{\partial^2 \tilde{\psi}}{\partial Y^2} = 0\,. \tag{12.40}$$

A similar equation is obtained for the perturbation of the velocity potential:

$$\left(1 - M_\infty^2\right)\frac{\partial^2 \tilde{\varphi}}{\partial X^2} + \frac{\partial^2 \tilde{\varphi}}{\partial Y^2} = 0\,. \tag{12.41}$$

We now consider the boundary conditions on the surface of the body and at infinity. The conditions at infinity for subsonic flows ($M < 1$) are that the perturbations vanish in the limit of infinite distance from the body, that is,

$$\tilde{\varphi} \to 0\,, \quad \tilde{\psi} \to 0 \quad \text{for} \quad X \to \pm\infty\,. \tag{12.42}$$

It must be noted that at infinity, all parameters are either constant or depend on Y. The corresponding stream function $\Psi_\infty(Y)$ must satisfy the equation $d\psi_\infty/dY = U_\infty$ such that $\psi_\infty = U_\infty Y$.

On the surface of the body, one only has to provide the impermeability condition. For the stream function, this condition is formulated as the condition that the body contour has to be the zero streamline, that is,

$$\psi = \psi_\infty + \tilde{\psi} = U_\infty Y + \tilde{\psi} = 0 \, .$$

From this, it follows that

$$\tilde{\psi} = -U_\infty h(X) \quad \text{for} \quad Y = h(X) \quad \text{and} \quad X_1 < X < X_2 \, . \tag{12.43}$$

The impermeability condition on the body surface is

$$V_n = U_{\infty n} + \tilde{V}_n = 0 \quad \text{for} \quad Y = h(X) \, .$$

Since, on the thin body, we have

$$U_{\infty n} = U_\infty \sin\vartheta \approx U_\infty \tan\vartheta = U_\infty h'(X)$$

and

$$\tilde{V}_n = \tilde{U}\sin\vartheta - \tilde{V}\cos\vartheta \approx -\tilde{V} = -\partial\psi/\partial Y \, ,$$

the boundary condition for $\tilde{\varphi}$ on the body can be written as

$$\frac{\partial\tilde{\varphi}}{\partial Y} = U_\infty h'(X) \quad \text{for} \quad Y = h(X) \, . \tag{12.44}$$

For a thin profile and the case of small perturbations, the conditions (12.43) are not imposed on the profile $h = h(X)$, but on the straight line $Y = 0$ for $X_1 < X < X_2$.

We now obtain the pressure coefficient (see (11.23)) as

$$c_p = \frac{2(p - p_\infty)}{\rho_\infty U_\infty^2} = \frac{2\tilde{p}}{\rho_\infty U_\infty^2} \, . \tag{12.45}$$

From the equation of the adiabatic process, it follows that

$$\frac{p}{\rho^\gamma} = \frac{p_\infty + \tilde{p}}{(\rho_\infty + \tilde{\rho})^\gamma} = \frac{p_\infty}{\rho_\infty^\gamma}$$

and, accurate up to small quantities of second order, we have

$$\frac{\tilde{p}}{p_\infty} = \gamma \frac{\tilde{\rho}}{\rho_\infty} \, .$$

Using this equation and the equation $a_\infty^2 = \gamma p_\infty/\rho_\infty$, we replace $\tilde{\rho}$ in (12.38) by $\tilde{p}$, that is,

$$\tilde{p} = -\rho_\infty U_\infty \tilde{U} \, .$$

By substituting the last equation into (12.45), one obtains

$$c_p = \frac{2\tilde{p}}{\rho_\infty U_\infty^2} = -\frac{2\tilde{U}}{U_\infty} . \quad (12.46)$$

We finally note that (12.40) is a partial differential equation of the elliptic type for a subsonic flow ($M_\infty < 1$), while in the case of a supersonic flow ($M_\infty > 1$), the resulting equations, that is,

$$\left(1 - M_\infty^2\right) \frac{\partial^2 \tilde{\varphi}}{\partial X^2} + \frac{\partial^2 \tilde{\varphi}}{\partial Y^2} = 0 , \quad \left(1 - M_\infty^2\right) \frac{\partial^2 \tilde{\psi}}{\partial X^2} + \frac{\partial^2 \tilde{\psi}}{\partial Y^2} = 0 , \quad (12.47)$$

are of the hyperbolic type. Methods to solve these equations are well developed. In the following two examples, subsonic and supersonic flow around a thin profile are considered.

12.4
Subsonic Flow around a Thin Profile

The thin profile is specified by two contours, h_1 and h_2, for the upper and lower boundary respectively, that is,

$$Y = h_1(X) , \quad Y = h_2(X) , \quad X_1 < X < X_2 .$$

With the definition $\omega^2 = 1 - M_\infty^2$, (12.40) for the perturbation of the stream function can be rewritten as

$$\frac{\partial^2 \tilde{\psi}}{\partial X^2} + \frac{1}{\omega^2} \frac{\partial^2 \tilde{\psi}}{\partial Y^2} = 0 . \quad (12.48)$$

The boundary conditions are (see (12.42) and (12.43))

$$\begin{aligned} &\tilde{\psi} = -U_\infty h_{1,2}(X) , \quad \text{for} \quad X_1 < X < X_2 \\ &\text{and} \quad Y = \pm 0 \quad \tilde{\psi} \to 0 \quad \text{at infinity.} \end{aligned} \quad (12.49)$$

By introducing, instead of X and Y, the new coordinates

$$\xi = X, \quad \eta = \omega Y , \quad (12.50)$$

the boundary value problem becomes

$$\frac{\partial^2 \tilde{\psi}}{\partial \xi^2} + \frac{\partial^2 \tilde{\psi}}{\partial \eta^2} = 0 , \quad (12.51)$$

$$\begin{aligned} &\tilde{\psi} = -U_\infty h_{1,2}(\xi) , \quad \text{for} \quad \xi_1 < \xi < \xi_2 \\ &\text{and} \quad \eta = \pm 0 \quad \tilde{\psi} \to 0 \quad \text{at infinity} . \end{aligned} \quad (12.52)$$

We now compare the problem defined by (12.51) and (12.52) with the problem defined by (12.48) and (12.24) for which we assume $M_\infty = 0$ ($\omega = 1$). We denote the solution of the latter problem by $\tilde{\psi}_0$. Since this problem corresponds to the flow of an incompressible fluid around the same profile, both problems are identical and we have

$$\tilde{\psi}(\xi, \eta) = \tilde{\psi}_0(X, Y) , \quad \frac{\partial \tilde{\psi}}{\partial \xi} = \frac{\partial \tilde{\psi}_0}{\partial X} , \quad \frac{\partial \tilde{\psi}}{\partial \eta} = \frac{\partial \tilde{\psi}_0}{\partial Y} .$$

From (12.38) and (12.26), it follows that

$$\tilde{U} = \frac{1}{1 - M_\infty^2} \frac{\partial \tilde{\psi}}{\partial Y} = \frac{1}{1 - M_\infty^2} \frac{\partial \tilde{\psi}}{\partial \eta} \frac{d\eta}{dY} = \frac{1}{\sqrt{1 - M_\infty^2}} \frac{\partial \tilde{\psi}_0}{\partial Y} = \frac{\tilde{U}_0}{\sqrt{1 - M_\infty^2}} ,$$
$$\tilde{V} = -\frac{\partial \tilde{\psi}}{\partial X} = -\frac{\partial \tilde{\psi}}{\partial \xi} = -\frac{\partial \tilde{\psi}_0}{\partial \xi} = \tilde{V}_0 . \tag{12.53}$$

By substituting the first equation in (12.53) into (12.46), one obtains the pressure function of the gas as

$$c_p = \frac{c_{p0}}{\sqrt{1 - M_\infty^2}} . \tag{12.54}$$

Here, c_{p0} is the pressure coefficient of the flow around the same profile by an incompressible fluid. Equation 12.54 is the Prandtl–Glauert rule: For a given value of $M_\infty < 1$, the distribution of pressure coefficients for a vortex-free linearized subsonic flow can be obtained from the corresponding distribution for the flow of an incompressible fluid by multiplication with $1/\sqrt{1 - M_\infty^2}$.

The lift coefficient is determined as

$$c_y = \frac{R_y}{0.5 \rho_\infty U_\infty^2 c} , \tag{12.55}$$

with the lift R_y and the profile length c. In this case,

$$c_y = \frac{c_{y0}}{\sqrt{1 - M_\infty^2}} , \tag{12.56}$$

where c_{y0} is the lift coefficient in the incompressible fluid ($M_\infty = 0$).

12.5 Supersonic Flow around a Thin Profile

For a supersonic flow, we have $M_\infty > 1$ and (12.39) takes on the form

$$\frac{\partial^2 \tilde{\psi}}{\partial X^2} - \frac{1}{\omega^2} \frac{\partial^2 \tilde{\psi}}{\partial Y^2} = 0 , \tag{12.57}$$

where $\omega^2 = 1 - M_\infty^2$.

This equation has a structure that is different from (12.48) for the subsonic flow. The difference is that (12.57) is a hyperbolic differential equation and thus a wave equation, while (12.48) is an elliptic differential equation. The difference in the types of equations leads to different characteristic physical properties of the phenomena.

The general solution of (12.57) is

$$\tilde{\psi}(X, Y) = \Omega_1(X - \omega Y) + \Omega_2(X + \omega Y) , \tag{12.58}$$

where Ω_1 and Ω_2 are arbitrary functions of $X - \omega Y$ and $X + \omega Y$, whose form is determined from the boundary conditions.

We consider the particular solution $\tilde{\psi}_1(X, Y) = \Omega_1(X - \omega Y)$. which can be interpreted as follows. In the flow plane (X, Y), there exists a class C_1 of straight lines $X - \omega Y = \text{const}$, along which the stream function $\tilde{\psi}$ is constant. These straight lines are the first class of characteristics for the wave equation (12.57). They play the role of perturbation lines in the supersonic flow. They are called Mach's lines or waves.

The second class of characteristics of perturbation lines C_2 are defined by $X + \omega = \text{const}$ which corresponds to the particular solution $\tilde{\psi}_2(X, Y) = \Omega_2(X + \omega Y)$. Along the second class of characteristics, the perturbations of the flow parameters are also conserved.

The angular coefficients of these classes of characteristics are given by $\tan\alpha = \pm 1/\omega = \pm 1/\sqrt{M_\infty^2 - 1}$, where α is the angle between the perturbation lines and the direction of the undisturbed movement, that is, the X-axis, and $\alpha = \arcsin(1/M_\infty)$.

If there is a perturbation source for the supersonic flow along X-axis at any point S in the (X, Y)-plane, one can observe this perturbation only along two rays that originate in S under the angles $\pm\alpha$ to the X-axis. In all other points, the flow is undisturbed and stays homogeneous.

As above, the profile is specified by the equations $h_{1,2}(X)$ for $X_1 < X < X_2$ and the boundary conditions are specified on the profile as

$$\tilde{\psi} = -U_\infty h_{1,2}(X) \quad \text{for} \quad X_1 < X < X_2 \quad \text{and} \quad Y = \pm 0 . \tag{12.59}$$

We fill the flow domain both above and below the profile with the classes of characteristics C_1 and C_2 (see Figure 12.7). The properties of these characteristics and the boundary conditions allow for the solution of (12.57) to be written as

$$\tilde{\psi} = -U_\infty h_{1,2}(X \mp \omega Y) . \tag{12.60}$$

This means that unlike the subsonic flow, the perturbation of the stream function $\tilde{\psi}(X, Y)$ does not vanish at infinity as in the case of the supersonic flow. Inside the upper and lower stripes bounded by the characteristics AA_1, BB_1 and AA_2, BB_2 for $Y \to \pm\infty$, it has the same dependence on X as on the corresponding upper and lower part of the profile.

Outside of the characteristics mentioned above, the flow stays homogeneous and the velocity is equal to U_∞. From the solution (12.60), it also follows that the stream

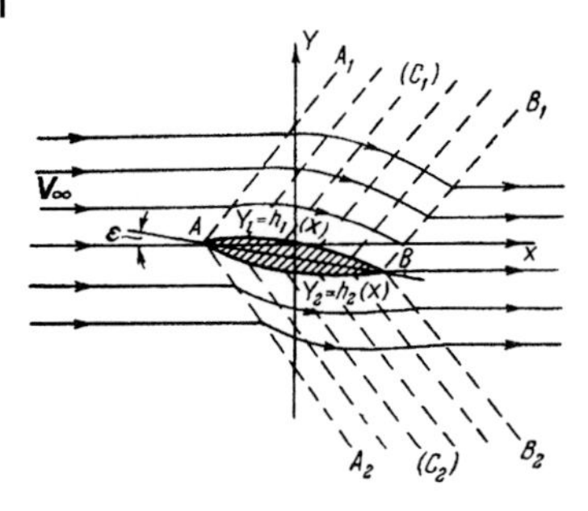

Figure 12.7 Supersonic flow around a thin profile.

lines of the disturbed movement ($\psi = U_\infty Y + \tilde{\psi} = \text{const}$) can be obtained by parallel translation of the upper and lower contour of the profile along the first and second class of characteristics. The streamlines above and below the profile are shown in Figure 12.7.

The perturbations of the velocities are

$$\tilde{U} = -\frac{1}{M_\infty^2 - 1}\frac{\partial \tilde{\psi}}{\partial Y} = \mp \frac{U_\infty}{M_\infty^2 - 1} h'_{1,2}(X \mp \omega Y) ,$$

$$\tilde{V} = -\frac{\partial \tilde{\psi}}{\partial Y} = \mp \tilde{U} h'_{1,2}(X \mp \omega Y) , \tag{12.61}$$

where the $'$ sign indicates the derivative with respect to the whole argument.

The pressure coefficient on the profile can be determined if one makes the approximation $Y = 0$ in (12.46), yielding

$$c_p(X) = \frac{\mp 2h'_{1,2}(X)}{\sqrt{M_\infty^2 - 1}} . \tag{12.62}$$

In conclusion, we give the formula for the lift coefficient

$$c_y = \frac{4\varepsilon}{\sqrt{M_\infty^2 - 1}} , \tag{12.63}$$

where ε is the angle of attack of the profile. This formula is called Ackeret's formula. From (12.63), it follows that the lift coefficient of a thin profile does not depend on the profile form. It only depends on the angle of attack and Mach's number of the incoming flow.

In contrast to the linearized subsonic flow for which the resistance force of the profile vanishes due to the symmetry of the flow with respect to the X-axis, the resistance force of the profile does not vanish for a supersonic flow, which can be explained through the asymmetry of the flow with respect to the X-axis.

13
Dynamics of the Viscous Incompressible Fluid

13.1
Rheological Laws of the Viscous Incompressible Fluid

The previously derived differential equations describing continuity, (3.6), the movement of the fluid (momentum equation), (3.34), the change in angular momentum, (3.47), and the heat inflow, (5.69), can be used for arbitrary media, provided they have the properties of continuity and fluidity. We have considered the following simple models of the continuum: the ideal, that is frictionless, incompressible fluid or gas with small Mach numbers and the ideal compressible gas at subsonic or supersonic speeds when the compressibility plays the largest role such that additional conditions for the properties of the gas have to be assumed, for example, that the gas is ideal and that the process is adiabatic or isothermic.

A gas or fluid is considered as ideal if there is no tangential stress and the internal stress tensor has the form

$$P = -p\,G\,, \quad (p^{ij} = -p g^{ij}) \tag{13.1}$$

with the internal stress tensor P with components p^{ij}, the metric tensor G with components g^{ij} (In a Cartesian system of coordinates, we have $g^{ij} = g^{i}_{j} = \delta^{i}_{j}$.) and the pressure p. Equation 13.1 is the simplest rheological equation of the medium. A rheological equation is a relation that connects the components of the stress tensor $P(p^{ij})$, the strain tensor $E(\varepsilon_{ij})$ and of the strain velocity tensor $\dot{E}(e_{ij})$.

The next level of complexity for a rheological equation is the equation for a Newtonian fluid, that is characterized via the linear dependence of the stress tensor on the strain velocity tensor. In the case of the Navier–Stokes law, this dependence has the general form

$$p^{ij} = A^{ijkl} e_{kl}\,. \tag{13.2}$$

For an isotropic fluid, this dependence becomes

$$p^{ij} = -p g^{ij} + \lambda \nabla \cdot V g^{ij} + 2\mu g^{ik} g^{jl} e_{kl}\,. \tag{13.3}$$

Here, λ and μ are the coefficients of viscosity that generally depend on the temperature, and to a lesser extent, on the pressure. Most of the fluids and solutions with

Hydromechanics. Theory and Fundamentals. Emmanuil G. Sinaiski

ISBN: 978-3-527-41026-2

relatively small molecular mass and all gases obey the rheological law (13.3). However, fluids whose molecules have larger masses, for example, colloidal suspensions, polymer solutions, disperse media such as fluids with different suspended particles, and fluids with a complicated internal structure such as paints and glue, have very unusual properties because these fluids are very different from Newtonian fluids. The rheological laws of these fluids are different from (13.3). Therefore, such fluids are called non-Newtonian. The effective viscosity of these fluids not only depends on the temperature and pressure, but generally also on the strain velocity (for a one-dimensional movement, the strain velocity is equal to $\dot{\varepsilon} = dU/dY$), the deformation, the movement and the time. We now consider a few examples of non-Newtonian fluids.

Of large practical importance are plastic fluids that are characterized by their viscosity and yield stress. If this limit is reached, the fluid becomes very thin. This class of fluids belongs to the viscoplastic (Bingham) fluids. The simplest rheological equation of a viscoplastic fluid for the case of a linear movement along the X-axis with strain velocity $\dot{\varepsilon} = dU/dY$ is given by

$$\tau = \tau_0 + \mu'\dot{\varepsilon} \quad \text{for} \quad \tau > \tau_0 \,, \tag{13.4}$$

with the shear stress τ, the coefficient of structural viscosity μ' and the yield stress τ_0. For $\tau < \tau_0$, there is no fluidity ($\dot{\varepsilon} = 0$), that is, the medium behaves like a solid.

As examples, one can list clay mortar and plaster, oil paint, sludge, and some paste-like substances. The physical explanation of this behavior is that the fluid has, while at rest, an fixed internal spatial structure which resists any external action as long as the shear stress does not exceed the yield stress. Once the shear stress is larger than the yield stress, the structure is dismantled and the fluid begins to behave as a normal Newtonian fluid under the influence of the pseudo stress which is given by the difference of shear stress and yield stress $\tau - \tau_0$. Once the pseudo stress drops to zero, that is, once the actual shear stress returns to the yield stress, the fixed spatial structure is re-established and the flow stops.

Pseudoplastic fluids do not have a constant viscosity, but their pseudo viscosity depends on the strain velocity. Such nonlinear fluids (suspensions of asymmetric particles, solutions of macromolecular polymers) obey the the Ostwald–Reiner equation

$$\tau = k\dot{\varepsilon}^n \,, \tag{13.5}$$

where k and $n < 1$ are constants, and the pseudo viscosity $\tau/\dot{\varepsilon} = k\dot{\varepsilon}^{n-1}$ decreases with increasing $\dot{\varepsilon}$. Dilated fluids ($n > 1$), like pseudoplastic fluids, do not show shear stress, but, in contrast to the pseudoplastic fluids, their pseudo viscosity grows with increasing stress. Examples of dilated fluids are highly concentrated suspensions of solid particles and glue made from starch. Viscoelastic media show both viscous fluidity as well the property of elastic form recovery. Among these media, are some highly viscous fluid and resins. There are two different rheological laws. Voigt's law uses the parallel effect of elasticity and viscosity such that shear stress is the sum of elastic and viscous stresses.

$$\tau = \tau_1 + \tau_2 = G\varepsilon + \mu\dot{\varepsilon} \,, \tag{13.6}$$

with the shear modulus G, the shear strain ε and the coefficient of dynamic viscosity μ. By integrating (13.6) over time for $\tau = \tau_0 = \text{const}$ and $\varepsilon(0) = 0$, one obtains

$$\varepsilon = \frac{\tau_0}{G}\left(1 - e^{-Gt/\mu}\right). \tag{13.7}$$

Equation 13.7 describes the delay of reaching the elastic deformation $\varepsilon = \tau_0/G$ for $t \to \infty$ under the influence of constant stress. The characteristic delay time is given by mu/G. For the situation when the stress is suddenly reduced to zero, one obtains for a given initial deformation from (11.35)

$$\varepsilon = \varepsilon_0 e^{-Gt/\mu} \,. \tag{13.8}$$

This expression shows the delay that affects the reduction of the deformation after the stress has been removed.

In Maxwell's law, the total strain velocity is given by the sums of strain velocities during elastic deformation and viscous movement

$$\dot{\varepsilon} = \dot{\varepsilon}_1 + \dot{\varepsilon}_2 = \frac{\dot{\tau}}{G} + \frac{\tau}{\mu} \,. \tag{13.9}$$

By integrating (13.9) over time with the conditions $\varepsilon = \text{const}$ and $\tau(0) = \tau_0$, one obtains

$$\tau = \tau_0 e^{-Gt/\mu} \,. \tag{13.10}$$

This equation shows the law of stress relaxation with the relaxation time μ/G. The rheological laws mentioned above do not exhaust the whole spectrum of such laws for different media. We have only listed the simplest laws. The rheological laws of many fluids are very complex and cannot always be represented as a combination of the simpler laws. For example, there are fluids whose mechanical properties not only depend on the strain velocities, but also on the duration of the deformation and history of the flows. Such fluids are called thixotropic. Some examples include stiffening cement mortar, molten metal, kefir and so on.

13.2 Equations of the Newtonian Viscous Fluid and Similarity Numbers

The movement of a viscous incompressible fluid is described by the Navier–Stokes system of equations (see (4.30)) as

$$\begin{aligned} &\nabla \cdot V = 0 \,, \\ &\frac{\partial V}{\partial t} + (V \cdot \nabla)V = f - \frac{1}{\rho}\nabla p + \frac{\mu}{\rho}\Delta V \,. \end{aligned} \tag{13.11}$$

We write these equations in a dimensionless form and determine the similarity parameters. We use as the scales of time, length, velocity, pressure and volume force the quantities T, L, v, P and F. Instead of the dimensional quantities, that is, the time t, the coordinates X, Y, Z, the velocity components U, V, W, the pressure p and the force components f_x, f_y, f_z, we employ the following dimensionless quantities

$$t' = \frac{t}{T}, \quad X' = \frac{X}{L}, \quad Y' = \frac{Y}{L}, \quad Z' = \frac{Z}{L}, \quad U' = \frac{U}{v}, \quad V' = \frac{V}{v},$$
$$W' = \frac{W}{v}, \quad p' = \frac{p}{P}, \quad f'_x = \frac{f_x}{F}, \quad f'_y = \frac{f_y}{F}, \quad f'_z = \frac{f_z}{F}.$$

By transforming the (13.11) to new variables, one obtains the system of dimensionless variables

$$\begin{aligned} &\nabla \cdot \boldsymbol{V}' = 0, \\ &\mathrm{Sh}\frac{\partial \boldsymbol{V}'}{\partial t'} + (\boldsymbol{V}' \cdot \nabla)\boldsymbol{V}' = \frac{1}{\mathrm{Fr}}\boldsymbol{f}' - \mathrm{Eu}\nabla p' + \frac{1}{\mathrm{Re}}\Delta \boldsymbol{V}', \end{aligned} \tag{13.12}$$

with Strouchal's number $\mathrm{Sh} = L/vT$, Euler's number $\mathrm{Eu} = P/\rho v^2$, Reynolds number $\mathrm{Re} = \rho v L/\mu$ and Froude's number $\mathrm{Fr} = v^2/FL$.

To solve concrete problems, one has to supplement (13.12) by the corresponding initial and boundary conditions that also have to be written in a dimensionless form.

We consider two flows of a viscous fluid that are described by (13.12) with the same initial and boundary conditions. The similarity conditions for these flows are the following (see Section 6.3): a) geometric similarity, b) kinematic similarity, c) dynamic similarity. To satisfy geometric similarity, it is sufficient that the boundaries of the flow area are similar in a geometric sense. Kinematic similarity means that the ratios of the projections of the velocities on the coordinates axes are the same for both flows. Dynamic similarity refers to the ratios between the projections of the volume force vectors, the pressure and the components of viscous stress to be the same for both flows in every point. The first condition is equivalent to the ordinary geometric similarity. To fulfill the last two conditions, it is necessary that the dimensionless numbers Sh, Eu, Re and Fr coincide for both flows. The number of sufficient conditions depends on the concrete problem, and can be smaller than the above number but not more in any case. The dimensionless numbers mentioned above do not exhaust the problem of similarity since boundary conditions can lead to the need for new dimensionless numbers. For simplicity, we restrict ourselves to the similarity numbers mentioned above. For $\mathrm{Sh} \ll 1$ or when considering a stationary flow, the first term on the left-hand side of the second of the Navier–Stokes equations (13.11) drops out, and we have

$$\begin{aligned} &\nabla \cdot \boldsymbol{V} = 0, \\ &(\boldsymbol{V} \cdot \nabla)\boldsymbol{V} = \boldsymbol{f} - \frac{1}{\rho}\nabla p + \frac{\mu}{\rho}\Delta \boldsymbol{V}. \end{aligned} \tag{13.13}$$

Then, Strouchal's number will not be a similarity number. If there is no characteristic pressure P for the problem, one can use the velocity height ρv^2 in place of P, resulting in $\mathrm{Eu} = 1$ and Euler's number will not be a similarity number as well.

For problems concerning the flow of a heavy, viscous incompressible fluid, Fr and Re are the main similarity numbers, where one has to use g for F in Fr. Froude's number is important for those problems for which gravity has a strong influence on the flow. If gravity does not play a role, Reynolds number is the only similarity number and (13.11) takes on the form

$$\begin{aligned} &\nabla \cdot \boldsymbol{V} = 0\,, \\ &(\boldsymbol{V} \cdot \nabla)\boldsymbol{V} = -\frac{1}{\rho}\nabla p + \frac{\mu}{\rho}\Delta \boldsymbol{V}\,. \end{aligned} \tag{13.14}$$

Since Reynolds number Re is obtained by dividing the characteristic value v^2/L of the inertial term by the characteristic value $\mu v/\rho L$ of the viscous term, Re characterizes the order of magnitude of the ratio of inertial to viscous force. Therefore, for $\mathrm{Re} \ll 1$, the inertial force is small relative to the viscous force and one can neglect the inertial force, that is,

$$\begin{aligned} &\nabla \cdot \boldsymbol{V} = 0\,, \\ &0 = \boldsymbol{f} - \frac{1}{\rho}\nabla p + \frac{\mu}{\rho}\Delta \boldsymbol{V}\,. \end{aligned} \tag{13.15}$$

As a result, the equations are linear and the solution is much simpler. The equations in (13.15) are called Stokes equations. For $\mathrm{Re} \gg 1$, the inertial force is larger than the viscous force such that (13.11) becomes equivalent to Euler's equations, that is,

$$\begin{aligned} &\nabla \cdot \boldsymbol{V} = 0\,, \\ &(\boldsymbol{V} \cdot \nabla)\boldsymbol{V} = \boldsymbol{f} - \frac{1}{\rho}\nabla p\,. \end{aligned} \tag{13.16}$$

13.3 Integral Formulation for the Effect of Viscous Fluids on a Moving Body

The problem of translatory movement of a body in a viscous incompressible fluid is one of the main problems of hydromechanics. As a result, one must obtain the resistance force or drag of the body.

We consider a body that is moving with velocity $\boldsymbol{V}$ in a viscous incompressible fluid (see Figure 13.1). In a Cartesian system of coordinates, the stress acting on a surface element of the body is given by

$$\boldsymbol{p}_n = \boldsymbol{p}_x l + \boldsymbol{p}_y m + \boldsymbol{p}_z n\,, \tag{13.17}$$

where l, m, n are the direction cosines of the normal vector.

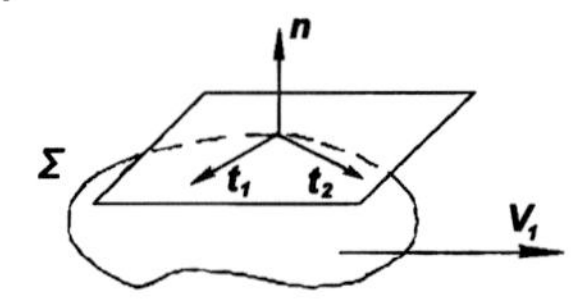

Figure 13.1 Translatory movement of a body in a viscous fluid.

The main vector and main moment of the force exerted on the body by the fluid are

$$\mathbf{R} = \int_{\Sigma} \mathbf{p}_n d\sigma\,, \quad \mathbf{M} = \int_{\Sigma} (\mathbf{r} \times \mathbf{p}_n) d\sigma\,. \tag{13.18}$$

The projection of the vector $\mathbf{p}_n$ on the X-axis is given by

$$p_{nx} = p_{xx} l + p_{yx} m + p_{zx} n\,. \tag{13.19}$$

Using (2.34) for the components of the stress tensor, one obtains

$$p_{xx} = -p + \lambda \nabla \cdot \mathbf{V} + 2\mu \left(\frac{\partial U}{\partial X} \right),$$
$$p_{xy} = p_{yx} = \mu \left(\frac{\partial U}{\partial Y} + \frac{\partial V}{\partial X} \right),$$
$$p_{xz} = p_{zx} = \mu \left(\frac{\partial W}{\partial X} + \frac{\partial U}{\partial Z} \right).$$

By substituting these equations into (13.19), one obtains

$$p_{nx} = (-p + \lambda \nabla \cdot \mathbf{V}) l + \mu \left(\frac{\partial U}{\partial X} l + \frac{\partial U}{\partial Y} m + \frac{\partial U}{\partial Z} n \right) + \mu \left(\frac{\partial U}{\partial X} l + \frac{\partial V}{\partial X} m + \frac{\partial W}{\partial X} n \right). \tag{13.20}$$

Since

$$\frac{\partial U}{\partial X} l + \frac{\partial U}{\partial Y} m + \frac{\partial U}{\partial Z} n = \nabla \cdot \mathbf{n} = \frac{\partial U}{\partial n} \quad \text{and} \quad \frac{\partial U}{\partial X} = \nabla \cdot \mathbf{V} - \frac{\partial V}{\partial Y} - \frac{\partial W}{\partial Z}\,,$$

we have

$$\frac{\partial U}{\partial X} l + \frac{\partial V}{\partial X} m + \frac{\partial W}{\partial Z} n = \nabla \cdot \mathbf{V} l + \frac{\partial V}{\partial X} m - \frac{\partial V}{\partial Y} l + \frac{\partial W}{\partial Z} l\,.$$

We introduce a local system of coordinates with mutually orthogonal basis vectors $(\mathbf{n}, \mathbf{t}_1, \mathbf{t}_2)$, where $\mathbf{n}$ is the unit normal vector and $\mathbf{t}_1, \mathbf{t}_2$ are the tangential unit vectors for the surface Σ at point M in Figure 13.1. The old coordinates X, Y, Z are expressed in terms of the new coordinates (n, τ_1, τ_2) by

$$X(n, \tau_1, \tau_2)\,, \quad Y(n, \tau_1, \tau_2)\,, \quad Z(n, \tau_1, \tau_2)\,.$$

In the new coordinates, one obtains

$$\frac{\partial V}{\partial X} = \frac{\partial V}{\partial n}\frac{\partial n}{\partial X} + \frac{\partial V}{\partial \tau_1}\frac{\partial \tau_1}{\partial X} + \frac{\partial V}{\partial \tau_2}\frac{\partial \tau_2}{\partial X} .$$

Since the body is moving with constant velocity, the velocity components U, V, W on the surface Σ of the body are constant and the derivatives in the directions of the tangent vectors τ_1 and τ_2 are equal to zero. Therefore, we have

$$\frac{\partial V}{\partial X} = \frac{\partial V}{\partial n}\frac{\partial n}{\partial X} = \frac{\partial V}{\partial n} l .$$

The other derivatives can be written down in a similar way, namely,

$$\begin{aligned}
&\frac{\partial U}{\partial X} = \frac{\partial U}{\partial n}\frac{\partial n}{\partial X} = \frac{\partial U}{\partial n} l , \quad \frac{\partial U}{\partial Y} = \frac{\partial U}{\partial n}\frac{\partial n}{\partial Y} = \frac{\partial U}{\partial n} m , \\
&\frac{\partial U}{\partial Z} = \frac{\partial U}{\partial n}\frac{\partial n}{\partial Z} = \frac{\partial U}{\partial n} n , \quad \frac{\partial V}{\partial X} = \frac{\partial V}{\partial n}\frac{\partial n}{\partial Y} = \frac{\partial V}{\partial n} m , \\
&\frac{\partial W}{\partial X} = \frac{\partial W}{\partial n}\frac{\partial n}{\partial X} = \frac{\partial W}{\partial n} l , \quad \frac{\partial W}{\partial Z} = \frac{\partial W}{\partial n}\frac{\partial n}{\partial Z} = \frac{\partial W}{\partial n} n .
\end{aligned}$$

As a final result, one obtains

$$\begin{aligned}
&\frac{\partial V}{\partial X} m - \frac{\partial V}{\partial Y} l = \frac{\partial V}{\partial n} lm - \frac{\partial V}{\partial n} ml = 0 , \\
&\frac{\partial W}{\partial X} n - \frac{\partial W}{\partial Z} l = \frac{\partial W}{\partial n} \ln - \frac{\partial W}{\partial n} nl = 0 , \\
&\frac{\partial U}{\partial X} l + \frac{\partial V}{\partial X} m + \frac{\partial W}{\partial X} n = \frac{\partial U}{\partial X} l + \frac{\partial V}{\partial X}\frac{\partial n}{\partial Y} + \frac{\partial W}{\partial X}\frac{\partial n}{\partial Z} \\
&= \frac{\partial U}{\partial X} l + \frac{\partial V}{\partial n}\frac{\partial n}{\partial Y} l + \frac{\partial W}{\partial n}\frac{\partial n}{\partial Z} l = \left(\frac{\partial U}{\partial X} + \frac{\partial V}{\partial Y} + \frac{\partial W}{\partial Z}\right) l = l \nabla \cdot \mathbf{V} .
\end{aligned}$$

Thus, (13.20) becomes

$$\begin{aligned}
p_{nx} &= (-p + \lambda \nabla \cdot \mathbf{V}) l + \mu \frac{\partial U}{\partial n} + \mu l \nabla \cdot \mathbf{V} \\
&= (-p + (\lambda + \mu) \nabla \cdot \mathbf{V})\, l + \mu \frac{\partial U}{\partial n} .
\end{aligned} \tag{13.21}$$

In a similar way, one can transform the other stress components, yielding

$$p_{ny} = (-p + (\lambda + \mu) \nabla \cdot \mathbf{V})\, m + \mu \frac{\partial V}{\partial n} , \tag{13.22}$$

$$p_{nz} = (-p + (\lambda + \mu) \nabla \cdot \mathbf{V})\, n + \mu \frac{\partial W}{\partial n} . \tag{13.23}$$

From (3.42) and (3.46)–(3.48), it follows that

$$\begin{aligned}
\boldsymbol{p}_n &= p_{nx} \boldsymbol{i} + p_{ny} \boldsymbol{j} + p_{nz} \boldsymbol{k} \\
&= (-p + (\lambda + \mu) \nabla \cdot \mathbf{V})\, (l \boldsymbol{i} + m \boldsymbol{j} + n \boldsymbol{k}) + \mu \frac{\partial \mathbf{V}}{\partial n} .
\end{aligned} \tag{13.24}$$

By substituting (13.24) into (13.18), one obtains the following general formulae for the main vector and main moment of the forces acting on the body:

$$R = \int_{\Sigma} (-p + (\lambda + \mu)\nabla \cdot V)(l i + m j + n k)\, d\sigma + \int_{\Sigma} \mu \frac{\partial V}{\partial n} d\sigma\,, \tag{13.25}$$

$$M = \int_{\Sigma} (-p + (\lambda + \mu)\nabla \cdot V)(r \times (l i + m j + n k))\, d\sigma + \mu \int_{\Sigma} r \times \frac{\partial V}{\partial n} d\sigma\,. \tag{13.26}$$

For incompressible fluids, we have $\nabla \cdot V = 0$ and (13.25) is simplified to

$$R = -\int_{\Sigma} p\,(l i + m j + n k)\, d\sigma + \int_{\Sigma} \mu \frac{\partial V}{\partial n} d\sigma\,. \tag{13.27}$$

The first term is the contribution of the pressure force and the second term is the contribution of the viscous force to the drag.

13.4
Stationary Flow of a Viscous Incompressible Fluid in a Tube

In hydromechanics, very few exact solutions of the Navier–Stokes equations are known. One of the simplest ones is the problem of a laminar (layered) flow in tubes of arbitrary cross section. The streamlines for such a flow are lines that are parallel to the axis of the tube (see Figure 13.2). We assume that the Z-axis lies along the tube axis, that the tube is infinitely long, the fluid flows along the tube axis such that the velocity vector is $V = (0, 0, W)$, the flow is isothermal ($T = \text{const}$) such that the mass density ρ and the viscosity coefficient μ are constant and that there are not volume forces. Under these conditions, one can rewrite (13.11) as follows:

$$\frac{\partial W}{\partial Z} = 0\,, \tag{13.28}$$

$$0 = -\frac{1}{\rho}\frac{\partial p}{\partial X}\,, \quad 0 = -\frac{1}{\rho}\frac{\partial p}{\partial Y}\,, \quad W\frac{\partial W}{\partial Z} = -\frac{1}{\rho}\frac{\partial p}{\partial Z} + \mu \Delta W\,. \tag{13.29}$$

From this, it follows that $p = p(Z)$, $W = W(X, Y)$ and (13.28) and (13.29) result in

$$\mu\left(\frac{\partial^2 W}{\partial X^2} + \frac{\partial^2 W}{\partial Y^2}\right) = \frac{\partial p}{\partial Z}\,. \tag{13.30}$$

Since the left-hand side of (13.30) only depends on X, Y and the right-hand side only on Z, this equality is only possible if both sides are constant, that is,

$$\frac{dp}{dZ} = -\frac{\Delta p}{l} = \text{const}\,, \tag{13.31}$$

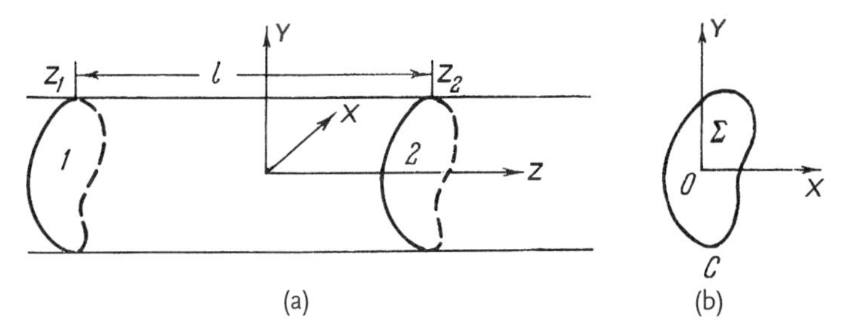

Figure 13.2 Stationary laminar flow of a viscous fluid through a cylindrical tube.

where Δp is the pressure difference per length l. Then, one obtains from (13.30) and (13.31)

$$\frac{\partial^2 W}{\partial X^2} + \frac{\partial^2 W}{\partial Y^2} = -\frac{\Delta p}{\mu l} . \tag{13.32}$$

If the tube wall is impermeable and keeps its form, the condition on the contour of the cross section is $W = 0$. We now consider a few examples.

Flow in a Planar Channel

We consider the flow between two infinite planes $Y = \pm h$ with $h = \text{const}$. This flow can be thought of as a flow in a channel of rectangular cross section with width h and a breadth that tends to infinity, that is, $L \gg 2h$. In this case, W will only depend on Y and (13.32) is transformed into

$$\frac{d^2 W}{d X^2} = -\frac{\Delta p}{\mu l} . \tag{13.33}$$

The boundary condition has the form

$$W = 0 \quad \text{for} \quad Y = \pm h . \tag{13.34}$$

The solution can be easily obtained as

$$W = \frac{\Delta p h^2}{2\mu l}\left(1 - \frac{Y^2}{h^2}\right) . \tag{13.35}$$

The velocity profile is parabolic and the velocity reaches the maximal value such that

$$W_{\text{max}} = \frac{\Delta p h^2}{2\mu l} \tag{13.36}$$

on the tube axis for $Y = 0$.

The linear flow (m^2/s) is

$$Q = \int_{-h}^{h} W d Y = \frac{2\Delta p h^3}{3\mu l} . \tag{13.37}$$

From this, we obtain some important characteristics. The average velocity is given by

$$W_{av} = \frac{1}{2h}\int_{-h}^{h} W\,dY = \frac{1}{3}\frac{\Delta p\,h^2}{\mu l}\,. \tag{13.38}$$

The drag coefficient λ for the flow through a channel of width $2h$ is determined by the equation

$$\Delta p = \frac{\lambda}{2h}\frac{\rho W_{av}^2}{2}\,. \tag{13.39}$$

From (13.38) and (13.39), one obtains

$$\lambda = \frac{4h\Delta p}{\rho W_{av}^2} = \frac{24}{\mathrm{Re}}\,, \tag{13.40}$$

where $\mathrm{Re} = 2h\,W_{av}/\mu$ is Reynolds number.

Flow through Cylindrical Tube of Elliptical Cross Section

We now consider the flow through a straight infinite tube with an elliptical cross section which is defined by

$$\frac{X^2}{a^2} + \frac{Y^2}{b^2} = 1\,. \tag{13.41}$$

We want to look for a solution of (13.32) with the boundary condition $W = 0$ on the elliptical wall of the form

$$W = A\left(1 - \frac{X^2}{a^2} - \frac{Y^2}{b^2}\right). \tag{13.42}$$

By substituting (13.42) into (13.32), one obtains the constant A as

$$A = \frac{\Delta p}{2\mu l}\frac{a^2 b^2}{a^2 + b^2}\,.$$

Thus, the velocity profile is an elliptical paraboloid. The maximum velocity is obtained at the tube axis and is given by

$$W_{max} = \frac{\Delta p}{2\mu l}\frac{a^2 b^2}{a^2 + b^2}\,. \tag{13.43}$$

The volume flow Q (m^3/s) and the average velocity are

$$Q = \int_{-h}^{h} W\,dX\,dY = W_{max}\int_{-h}^{h}\left(1 - \frac{X^2}{a^2} - \frac{Y^2}{b^2}\right)dY$$

$$= \frac{\pi}{2}ab\,W_{max} = \frac{\pi a^3 b^3 \Delta p}{4\mu l\,(a^2 + b^2)}\,, \tag{13.44}$$

$$W_{av} = \frac{Q}{\pi a b} = \frac{1}{2} W_{max} = \frac{\Delta p}{4\mu l} \frac{a^2 b^2}{a^2 + b^2} . \tag{13.45}$$

For $a = b$, one obtains the corresponding formulae for a cylindrical tube of circular cross section,

$$W = W_{max} \left(1 - \frac{r^2}{a^2}\right) , \quad W_{max} = \frac{\Delta p \, a^2}{4\mu l} , \quad Q = \frac{\pi a^4 \Delta p}{8\mu l} . \tag{13.46}$$

The velocity profile is paraboloid of rotation and is referred to as Poiseuille's parabola, and the flow is called Poiseuille's flow.

Flow through Tube of Rectangular Cross Section

We consider the flow through a tube of rectangular cross section. We denote the height of the rectangle parallel to the Y-axis by $2h$ and the base of the rectangle parallel to the X-axis by $2\kappa h$. The Z-axis is chosen through the center of the rectangle in flow direction. We introduce the following dimensionless variables:

$$\xi = \frac{X}{h} , \quad \eta = \frac{Y}{h} , \quad W^* = \frac{W \mu l}{h^2 \Delta p} . \tag{13.47}$$

Equation 13.32 and the boundary conditions at the tube wall assume the form

$$\frac{\partial^2 W^*}{\partial \xi^2} + \frac{\partial^2 W^*}{\partial \eta^2} = -1 , \tag{13.48}$$

$$W^* = 0 \quad \text{for} \quad \xi = \pm\kappa , \quad |\eta| < 1 \quad \text{and for} \quad \eta = \pm 1 , \quad |\xi| < \kappa . \tag{13.49}$$

We want to find a solution for this problem by the method of separation of variables. That means we assume

$$W^* = \sum_{n=0}^{\infty} Y_n(\eta) \cos\left(\frac{2n+1}{2} \frac{\pi}{\kappa} \xi\right) . \tag{13.50}$$

According to a well known relationship for the Fourier series

$$\sum_{n=0}^{\infty} \frac{(-1)^n}{2n+1} \cos\left(\frac{2n+1}{2} \pi \zeta\right) = \begin{cases} \pi/4, |\zeta| < 1 \\ 0, |\zeta| > 1 , \end{cases}$$

the first boundary condition in (13.49) is satisfied and one can rewrite (13.48) as

$$\frac{\partial^2 W^*}{\partial \xi^2} + \frac{\partial^2 W^*}{\partial \eta^2} = -\frac{\pi}{4} \sum_{n=0}^{\infty} \frac{(-1)^n}{2n+1} \cos\left(\frac{2n+1}{2} \frac{\pi}{\kappa} \xi\right) . \tag{13.51}$$

By substituting (13.50) into (13.51), and equating the coefficients of the cosines with equal arguments, one obtains the following system of differential equations and corresponding boundary conditions for the functions $\Upsilon_n(\eta)$:

$$\Upsilon''_n - \left(\frac{2n+1}{2}\frac{\pi}{\kappa}\right)^2 \Upsilon_n = -\frac{4}{\pi}\frac{(-1)^n}{2n+1}, \quad (n = 0, 1, \ldots), \tag{13.52}$$

$$\Upsilon_n = 0 \quad \text{for} \quad \eta = \pm 1. \tag{13.53}$$

The general solution of this equation is

$$\Upsilon_n = A_n + B_n \cosh\left(\frac{2n+1}{2}\frac{\pi}{\kappa}\eta\right) + C_n \sinh\left(\frac{2n+1}{2}\frac{\pi}{\kappa}\eta\right). \tag{13.54}$$

The constants A_n are obtained by substituting (13.54) into (13.52) and the constants B_n and C_n from the boundary conditions of (13.53), yielding

$$A_n = \frac{16\kappa^2}{\pi^3}\frac{(-1)^n}{(2n+1)^3}, \quad B_n = -\frac{A_n}{\cosh\left(\frac{2n+1}{2}\frac{\pi}{\kappa}\right)},$$

$$C_n = 0, (n = 0, 1, 2, \ldots).$$

After substitution of these coefficients into (13.54) and (13.50) and returning to dimensional variables, we obtain the final solution as

$$W^* = \frac{16\kappa^2}{\pi^3}\frac{h^2 \Delta p}{\mu l}\sum_{n=0}^{\infty}\frac{(-1)^n}{(2n+1)^3} \times \left[1 - \frac{\cosh\left(\frac{2n+1}{2}\frac{\pi Y}{\kappa h}\right)}{\cosh\left(\frac{2n+1}{2}\frac{\pi}{\kappa}\right)}\right]\cos\left(\frac{2n+1}{2}\frac{\pi X}{\kappa h}\right). \tag{13.55}$$

Using this solution, one obtains the volume flow of the fluid and the average velocity as

$$Q = \frac{\Delta p}{4\mu l}\kappa h^4 f(\kappa), \quad W_{\text{av}} = \frac{Q}{4\kappa h^2} = \frac{\Delta p h^2}{16\mu l} f(\kappa) \tag{13.56}$$

with $f(\kappa)$ defined by

$$f(\kappa) = \frac{16}{3} - \frac{1024}{\pi^5 \kappa}\left(\tanh\frac{\pi\kappa}{2} + \frac{1}{3^3}\tanh\frac{3\pi\kappa}{2} + \ldots\right). \tag{13.57}$$

Resistance of a Cylindrical Tube

To make an approximate estimate of the resistance of a cylindrical tube with a difficult profile, one applies the procedure of comparing the resistance of this tube with the resistance of an equivalent tube with hydrolic radius $r_h = S/P$ where S and P are the cross section and perimeter of the tube. This procedure can be applied if the tubes to be compared have geometrically similar cross sections.

We consider part of a tube with length L and a pressure difference of Δp between two cross sections. If the movement of the fluid is identical in all cross sections, one can assume that the fluid volume between these cross sections is in a state of equilibrium, and whose condition consists of the equality of longitudinal forces, that is,

$$S\Delta p = L\int_P \tau_w dl = PL\bar{\tau}_w ,$$

where $\bar{(\tau)}_w = \frac{1}{P}\int_P \tau_w dl$ is the average stress on the wetted surface of the tube. From this, one obtains the average frictional stress as

$$\bar{\tau}_w = \frac{\Delta p}{L}\frac{S}{P} = \frac{\Delta p}{L} r_h . \tag{13.58}$$

Therefore, the average frictional tension over the perimeter of a cylindrical tube is given by the pressure difference divided by the length L. For a plane tube with a distance h between parallel walls, we have $r_h = h$. For a tube of circular cross section with radius a, the hydrolic radius is $r_h = a/2$ and the average frictional stress is $\bar{\tau}_w = \frac{\Delta p}{L}\frac{a}{2}$.

13.5 Oscillating Laminar Flow of a Viscous Fluid through a Tube

As an example of such a non-stationary flow, we will consider the flow of a viscous incompressible fluid through a tube of circular cross section. We keep the assumptions of the previous section and only add the condition $W = W(X, Y, t)$. Then, the first two equations in (13.29) are unchanged and the third equation takes on the form

$$\frac{\partial W}{\partial t} - \nu\Delta W = -\frac{1}{\rho}\frac{\partial p}{\partial Z} , \quad \left(\nu = \frac{\mu}{\rho}\right) . \tag{13.59}$$

From (13.29), it follows that $p = p(Z, t)$ and from (13.59), one obtains

$$\frac{\partial p}{\partial Z} = f(t) . \tag{13.60}$$

We now use cylindrical coordinates and take into account that the flow is symmetric with respect to the tube axis, that is, that the flow parameters are independent of the polar angle. Then, (13.59) becomes the following equation:

$$\frac{\partial W}{\partial t} - \nu\left(\frac{\partial^2 W}{\partial r^2} + \frac{1}{r}\frac{\partial W}{\partial r}\right) = \frac{1}{\rho} f(t) . \tag{13.61}$$

The initial and boundary conditions are

$$W = 0 \quad \text{for} \quad r = a , \qquad W = W_0(r) \quad \text{for} \quad t = 0 . \tag{13.62}$$

We assume that the pressure gradient is changing harmonically, that is,

$$f(t) = \rho A \cos \omega t . \tag{13.63}$$

Then, the initial condition is not needed and only the boundary condition remains.

We introduce the new function

$$\psi(r,t) = W(r,t) - \frac{A}{\omega}\sin\omega t \tag{13.64}$$

and instead of t, the new variable $\tau = \nu t$. Then, (13.61) and the boundary condition (13.62) are transformed into

$$\frac{\partial\varphi}{\partial\tau} = \left(\frac{\partial^2\varphi}{\partial r^2} + \frac{1}{r}\frac{\partial\varphi}{\partial r}\right), \quad \varphi(r,\tau) = -\frac{A}{\omega}\sin\frac{\omega\tau}{\nu} \quad \text{for} \quad r = a\,. \tag{13.65}$$

We consider the particular solution of (13.65) given by

$$\varphi(r,\tau) = \mathrm{Re}(R(r)e^{-i\lambda\tau})\,, \tag{13.66}$$

where λ is a real number.

By substituting the expression (13.66) into (13.65), one obtains an ordinary differential equation for $R(r)$, namely,

$$\frac{d^2R}{dr^2} + \frac{1}{r}\frac{dR}{dr} + i\lambda R = 0\,. \tag{13.67}$$

The solution of this equation is provided by the Bessel function of imaginary argument which is non-singular at $r = 0$, that is,

$$R(r) = J_0(r\sqrt{\lambda i}) = (B - iC)\left[\mathrm{Ber}\left(r\sqrt{\lambda}\right) - i\,\mathrm{Bei}\left(r\sqrt{\lambda}\right)\right], \tag{13.68}$$

where $\mathrm{Ber}(z)$ and $\mathrm{Bei}(z)$ are Kelvin's functions.

Coming back to (13.66), one can write

$$\begin{aligned}\varphi(r,\tau) &= \mathrm{Re}\left[(B - iC)\left(\mathrm{Ber}\left(r\sqrt{\lambda}\right) - i\mathrm{Bei}\left(r\sqrt{\lambda}\right)\right)(\cos(\lambda\tau) - i\sin(\lambda\tau))\right]\\ &= B\left(\mathrm{Ber}\left(r\sqrt{\lambda}\right)\cos(\lambda\tau) - \mathrm{Bei}\left(r\sqrt{\lambda}\right)\sin(\lambda\tau)\right)\\ &\quad - C\left(\mathrm{Bei}\left(r\sqrt{\lambda}\right)\cos(\lambda\tau) - \mathrm{Ber}\left(r\sqrt{\lambda}\right)\sin(\lambda\tau)\right).\end{aligned} \tag{13.69}$$

By substituting the expression (13.69) into the boundary condition (13.65) and writing $\lambda = \omega/\nu$, one obtains a system of equations for the unknown coefficients B and C:

$$\begin{aligned}&B\,\mathrm{Ber}\left(a\sqrt{\omega/\nu}\right) - C\,\mathrm{Bei}\left(a\sqrt{\omega/\nu}\right) = 0\,,\\ &B\,\mathrm{Bei}\left(a\sqrt{\omega/\nu}\right) + C\,\mathrm{Ber}\left(a\sqrt{\omega/\nu}\right) = A/\omega\,.\end{aligned}$$

By solving these equations and by the substitution of the resulting coefficients into (13.59) and (13.66), one obtains the final solution as

$$W(r,t) = \frac{A}{\omega}\left[\left(1-\frac{\operatorname{Bei}\left(a\sqrt{\omega/\nu}\right)\operatorname{Bei}\left(r\sqrt{\omega/\nu}\right)+\operatorname{Ber}\left(a\sqrt{\omega/\nu}\right)\operatorname{Ber}\left(r\sqrt{\omega/\nu}\right)}{\operatorname{Ber}^2\left(a\sqrt{\omega/\nu}\right)+\operatorname{Bei}^2\left(a\sqrt{\omega/\nu}\right)}\right)\sin(\omega t)+\frac{\operatorname{Bei}\left(a\sqrt{\omega/\nu}\right)\operatorname{Ber}\left(r\sqrt{\omega/\nu}\right)+\operatorname{Ber}\left(a\sqrt{\omega/\nu}\right)\operatorname{Bei}\left(r\sqrt{\omega/\nu}\right)}{\operatorname{Ber}^2\left(a\sqrt{\omega/\nu}\right)+\operatorname{Bei}^2\left(a\sqrt{\omega/\nu}\right)}\times\cos\omega t\right]. \tag{13.70}$$

The above mentioned method can also be applied to the fluid flow that happens when a constant pressure gradient $\Delta p/l$ is suddenly applied and the fluid starts moving. To solve this equation, one has to assume on the right-hand side of (13.61)

$$f(t) = \frac{\Delta p}{l}\text{const}$$

and the initial and boundary conditions (13.62) become

$$W = 0 \quad \text{for} \quad r = a\,, \qquad W = 0 \quad \text{for} \quad t = 0\,.$$

The solution of this problem is

$$W(r,t) = a^2\frac{\Delta p}{4\pi l}\left[1-\frac{r^2}{a^2}-8\sum_{k=0}^{\infty}\exp\left(-\frac{\nu\lambda_k^2 t}{a^2}\right)\frac{J_0\left(\frac{\lambda_k r}{a}\right)}{\lambda_k^3 J_1(\lambda_k)}\right],$$

where λ_k are the roots of the equation $J_0(\lambda) = 0$ and J_0, and J_1 are the Bessel functions or zeroth and first order.

The volume flow is given by

$$Q = \frac{\pi a^4\Delta p}{8\pi l}\left[1-32\sum_{k=0}^{\infty}\exp\left(-\frac{\nu\lambda_k^2 t}{a^2}\right)\frac{1}{\lambda_k^4}\right].$$

13.6 Simplification of the Navier–Stokes Equations

In Sections 13.4 and 13.5, exact solutions of the Navier–Stokes equations were obtained for very simple fluid flows. The difficulty in solving these equations are due to their nonlinearity. To solve equations in the cases of more complicated problems, it is necessary to simplify them.

We want to restrict this discussion to stationary flows of incompressible fluid in the absence of volume forces. In this case, the Navier–Stokes equations are

$$\nabla \cdot \boldsymbol{V} = 0 \,,$$
$$(\boldsymbol{V} \cdot \nabla)\boldsymbol{V} = -\frac{1}{\rho}\nabla p + \frac{\mu}{\rho}\Delta \boldsymbol{V} \,. \tag{13.71}$$

By denoting the coordinates and velocity components by X_i and V_i, and the corresponding characteristic values by L and V, we can introduce the following dimensionless variables:

$$X_i' = \frac{X_i}{L} \,, \quad V_i' = \frac{V_i}{V} \,, \quad p' = \frac{p}{\rho V^2} \,, \quad \mathrm{Re} = \frac{\rho V L}{\mu} \,.$$

Then, (13.71) can be rewritten as

$$\nabla \cdot \boldsymbol{V}' = 0 \,,$$
$$(\boldsymbol{V}' \cdot \nabla)\boldsymbol{V}' = -\nabla p' + \frac{1}{\mathrm{Re}}\Delta \boldsymbol{V}' \,. \tag{13.72}$$

Equation 13.72 can be simplified for the two limiting cases $\mathrm{Re} \ll 1$ and $\mathrm{Re} \gg 1$. In the first case, the convective term on the left-hand side of the second equation in (13.72) is very small compared to the viscous second term on the right-hand side of (13.72). Therefore, for $\mathrm{Re} \ll 1$, the equations in (13.72) become

$$\nabla \cdot \boldsymbol{V} = 0 \,, \quad \mu \Delta \boldsymbol{V} = \nabla p \,. \tag{13.73}$$

These equations are called Stokes equations or equations of laminar flow. The system of equations (13.73) is linear, which simplifies its solution considerably.

For $\mathrm{Re} \gg 1$, the convective term dominates and (13.72) becomes

$$\nabla \cdot \boldsymbol{V} = 0 \,, \quad (\boldsymbol{V} \cdot \nabla)\boldsymbol{V} = -\frac{1}{\rho}\nabla p \,. \tag{13.74}$$

Equation 13.74 are Euler's equations. They describe the flow of an ideal incompressible fluid. Although the nonlinearity is still there, the order of the equations is reduced.

The reduction of the order of differentiation means that the equations in (13.74) describe the flow far away from the walls with enough accuracy, while these equations can not be used close to the walls or close to bodies around which the fluid flows. The reason is that both the normal and tangential components of the velocity must vanish on the body surface. Since the differential equations (13.74) are of first order, one cannot satisfy both boundary conditions. By neglecting viscous terms with second order derivatives, we therefore restrict the application of these equations to regions that are not close to bodies around which the fluid flows.

In the following, it will be shown that for $\mathrm{Re} \gg 1$, (13.74) can be used far away from bodies around which the fluid flows, while the viscosity of the fluid has a considerable influence close to the walls or the bodies around which the fluid flows.

Therefore, a boundary layer arises on the body surface inside of which the flow is described by (13.71) and outside of which it is described by (13.74).

In Chapters 14 and 15, we will consider laminar flows of a viscous fluid with $\mathrm{Re} \ll 1$ and $\mathrm{Re} \gg 1$. It must be noted that if the Reynolds number exceeds a critical value $\mathrm{Re}_{\mathrm{cr}}$, a laminar flow goes over into a turbulent flow, for which different methods must be used (see Chapter 16).

14
Flow of a Viscous Incompressible Fluid for Small Reynolds Numbers

14.1
General Properties of Stokes Flows

We consider the stationary flow of a viscous incompressible fluid with a small Reynolds number, that is,

$$\mathrm{Re} = \frac{\rho V L}{\mu} \ll 1 . \tag{14.1}$$

The inequality (14.1) is satisfied in the following cases:

1. small flow velocity V
2. flows in an area of small characteristic dimension L, and
3. flows of a highly viscous fluid.

As examples of flows with $\mathrm{Re} \ll 1$, one can consider the following flows: flows that involve the movement of small particles in a fluid; flows of a viscous fluid in thin channels, thin tubes and crevices; slow filtration of a viscous fluid through a porous medium; and flows of a highly viscous fluid through tubes. Before solving any concrete problems, we must consider some general properties regarding Stokes equations.

We recall that in the case of a viscous Newtonian fluid, the stress tensor $\boldsymbol{P}$ with components p_{ij} is in a Cartesian system of coordinates given by

$$p_{ij} = -p\delta_{ij} + \left(\frac{\partial V_i}{\partial X_j} + \frac{\partial V_j}{\partial X_i}\right) = -p\delta_{ij} + \mu e_{ij} , \tag{14.2}$$

where e_{ij} are the components of the strain velocity tensor.

By applying the divergence operator to both sides of the second equation in (13.73) and using the equation of continuity (first equation in (13.74)), one obtains

$$\Delta p = 0 . \tag{14.3}$$

From (14.3), it follows that the pressure is a harmonic function for a Stokes flow. We now apply the Laplace operator on both sides of the second equation in (13.73)

Hydromechanics. Theory and Fundamentals. Emmanuil G. Sinaiski

ISBN: 978-3-527-41026-2

with the result

$$\nabla^4 \boldsymbol{V} = 0 . \tag{14.4}$$

Thus, the velocity in a Stokes flow fulfills the biharmonic equation, with $\Delta^2 \boldsymbol{V} = \Delta(\Delta \boldsymbol{V}) = \nabla^2(\nabla^2 \boldsymbol{V}) = \nabla^4 \boldsymbol{V}$. In a Cartesian system of coordinates and two dimensions, we have

$$\nabla^4 \boldsymbol{V} = \frac{\partial^4 \boldsymbol{V}}{\partial X^4} + 2\frac{\partial^4 \boldsymbol{V}}{\partial X^2 \partial Y^2} + \frac{\partial^4 \boldsymbol{V}}{\partial Y^4} . \tag{14.5}$$

By applying the curl operator on both sides of the second equation in (13.73) and using the equality $\nabla \times (\nabla F) = 0$, one obtains the following equation for the vorticity vector $\boldsymbol{\Omega} = \nabla \times \boldsymbol{V}$:

$$\Delta \boldsymbol{\Omega} = 0 . \tag{14.6}$$

Thus, $\boldsymbol{\Omega}$ is also a harmonic function. From the general properties of harmonic functions, it follows that the maximum vorticity is attained on the boundaries of the flow domain. In the case of a plane (two-dimensional) flow, the stream function Ψ can be introduced by

$$\boldsymbol{V} = \nabla \times (\boldsymbol{k}\Psi) , \tag{14.7}$$

where $\boldsymbol{k}$ is the unit vector that is perpendicular to the (X, Y)-plane. In a Cartesian system of coordinates, (14.7) is written in component form as

$$U = \frac{\partial \Psi}{\partial Y} , \quad V = -\frac{\partial \Psi}{\partial x} .$$

By using (14.7), one can write the vorticity vector as

$$\boldsymbol{\Omega} = \nabla \times \boldsymbol{V} = -\boldsymbol{k}\Delta\Psi .$$

By substituting this equation into (14.6), one obtains the biharmonic equation for the stream function

$$\nabla^4 \Psi = 0 . \tag{14.8}$$

Since $\Delta \boldsymbol{V} = -\nabla \times \boldsymbol{\Omega}$, (13.73) leads to

$$\mu \nabla \times \boldsymbol{\Omega} = -\nabla p ,$$

or in component form

$$\frac{\partial \Omega}{\partial X} = \frac{1}{\mu}\frac{\partial p}{\partial Y} , \quad \frac{\partial \Omega}{\partial Y} = -\frac{1}{\mu}\frac{\partial p}{\partial X} . \tag{14.9}$$

Equations 14.9 are Cauchy–Riemann's equations. Therefore, the function

$$f(Z) = \Omega + i\frac{1}{\mu}p \tag{14.10}$$

is an analytical function of $Z = X + iY$.

The representation (14.10) makes it possible to use the methods of the theory of complex functions to solve problems in the plane. Another method to solve plane problems consists of representing the components of the stress tensor in terms of the Airy function Φ, that is,

$$p_{xx} = \frac{\partial^2 \Phi}{\partial Y^2}\,, \quad p_{xy} = p_{yx} = -\frac{\partial^2 \Phi}{\partial X \partial Y}\,, \quad p_{yy} = \frac{\partial^2 \Phi}{\partial X^2}\,. \tag{14.11}$$

The second equation in (13.73) can be written as $\nabla \cdot \boldsymbol{P} = 0$ (see (3.28)) or $\partial p_{ij}/\partial X_j = 0$, from which it follows that the representation (14.11) satisfies the second equation in (13.73) automatically.

It is easy to show that the Airy function Φ and the stream function ϕ are related, for example,

$$\frac{\partial^2 \Phi}{\partial Y^2} - \frac{\partial^2 \Phi}{\partial X^2} = -4\mu \frac{\partial^2 \Psi}{\partial X \partial Y}\,, \quad \frac{\partial^2 \Psi}{\partial Y^2} - \frac{\partial^2 \Psi}{\partial X^2} = -\frac{1}{\mu}\frac{\partial^2 \Phi}{\partial X \partial Y}\,. \tag{14.12}$$

Using these equations, one can show that the function

$$\chi = \Phi - i2\mu\Psi \tag{14.13}$$

is an analytical function of $Z = X + iY$ and satisfies the equation

$$\left(\frac{\partial^2 \chi}{\partial \bar{Z}^2}\right)_z = 0\,, \tag{14.14}$$

where $\bar{Z} = X - iY$.

The general solution of (14.14) is

$$\chi(\bar{Z}) = \chi_1(Z)\bar{Z} + \chi_2(Z)\,. \tag{14.15}$$

By an appropriate choice of the analytic functions $\chi_1(Z)$ and $\chi_2(Z)$, one can obtain the solution of some problems for plane Stokes flows using the representations in (14.15) and (14.13).

In curvilinear systems of coordinates (cylindrical, spherical), the equation for the stream function defined in (14.8) is similar, but instead of the operator ∇^4, one has to take a different operator E^4,

$$E^4\Psi = 0\,, \tag{14.16}$$

whose form depends on the concrete system of coordinates. We only consider axis-symmetric (two dimensional) flows.

In cylindrical coordinates r, ϕ, Z, the velocity components $\boldsymbol{V}(V_r, 0, V_z)$ are expressed as follows:

$$V_r = \frac{1}{r}\frac{\partial \Psi}{\partial Z}\,, \quad V_z = -\frac{1}{r}\frac{\partial \Psi}{\partial r}$$

and therefore the operator E^2 is given by

$$E^2 = r\frac{\partial}{\partial r}\left(\frac{1}{r}\frac{\partial \Psi}{\partial r}\right) + \frac{\partial^2 \Psi}{\partial Z^2}\,. \tag{14.17}$$

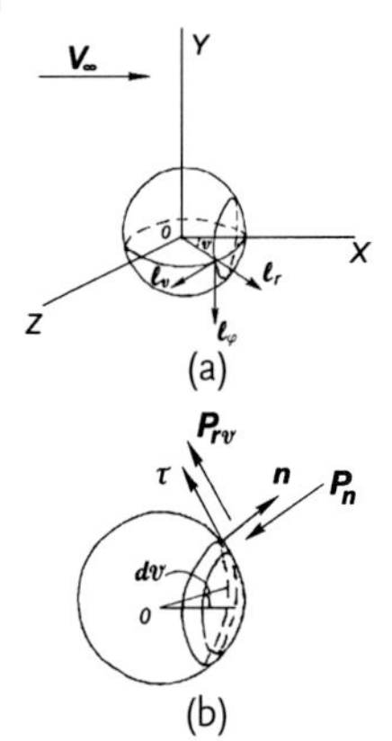

Figure 14.1 Slow flow of viscous incompressible fluid around a sphere.

In spherical coordinates (r, ϑ, ϕ), the velocity components $\boldsymbol{V}(V_r, V_\vartheta, 0)$ and the operator E^2 are given by

$$V_r = -\frac{1}{r^2 \sin\vartheta} \frac{\partial \Psi}{\partial \vartheta}, \quad V_\vartheta = \frac{1}{r \sin\vartheta} \frac{\partial \Psi}{\partial r},$$
$$E^2 \Psi = \frac{\partial^2 \Psi}{\partial r^2} + \frac{\sin\vartheta}{r^2} \frac{\partial}{\partial \vartheta}\left(\frac{1}{\sin\vartheta} \frac{\partial \Psi}{\partial \vartheta}\right). \tag{14.18}$$

14.2
Flow of a Viscous Fluid around a Sphere

We consider the laminar flow of a viscous incompressible fluid around a sphere. We denote the velocity at infinity by V_∞ and the radius of the sphere by a. We align the X-axis parallel to the vector $\boldsymbol{V}_\infty$. The center of the spherical system of coordinates (r, ϑ, ϕ) is taken to coincide with the center of the sphere (see Figure 14.1).

We transform Stokes equations (13.73) into the Gromeka–Lamb form (see (4.9)) and use the following formula from vector analysis, that is,

$$\Delta \boldsymbol{V} = \nabla(\nabla \cdot \boldsymbol{V}) - \nabla \times (\nabla \times \boldsymbol{V}).$$

From the incompressibility of the fluid, it follows that $\nabla \cdot \boldsymbol{V} = 0$. In addition, we have $\nabla \times \boldsymbol{\Omega} = \nabla \times (\nabla \times \boldsymbol{V}) = \nabla(\nabla \cdot \boldsymbol{V}) - \Delta \boldsymbol{V} = -\Delta \boldsymbol{V}$. By substituting this formula into (13.73), one obtains

$$\mu \nabla \times \boldsymbol{\Omega} + \nabla p = 0, \quad \boldsymbol{\Omega} = \nabla \times \boldsymbol{V}, \quad \nabla \cdot \boldsymbol{V} = 0. \tag{14.19}$$

We now apply the curl operator to both sides of the first equation in (14.19). Since $\nabla \times \nabla p = 0$, one obtains

$$\nabla \times (\nabla \times \boldsymbol{\Omega}) = 0. \tag{14.20}$$

Because of the symmetry with respect to the flow axis, the velocity $\boldsymbol{V}$ only has two components $V_r(r, \vartheta)$ and $V_\vartheta(r, \vartheta)$ since $V_\varphi = 0$. Therefore, the vector $\boldsymbol{\Omega}$ only has two non-vanishing components, $\Omega_r \neq 0$, $\Omega_\vartheta \neq 0$ since $\Omega_\varphi = 0$. We denote $|\boldsymbol{\Omega}|$ by Ω.

The components of the vorticity vector in spherical coordinates are

$$\Omega_r = (\nabla \times \mathbf{V})_r = \frac{1}{r \sin\vartheta}\left(\frac{\partial\left(V_\varphi \sin\vartheta\right)}{\partial\vartheta} - \frac{\partial V_\vartheta}{\partial\varphi}\right) = \frac{1}{r\sin\vartheta}\frac{\partial\left(\Omega \sin\vartheta\right)}{\partial\vartheta}\,,$$

$$\Omega_\vartheta = (\nabla \times \mathbf{V})_\vartheta = \frac{1}{r}\left(\frac{1}{\sin\vartheta}\frac{\partial V_r}{\partial\varphi} - \frac{\partial\left(rV_\varphi\right)}{\partial r}\right) = -\frac{1}{r}\frac{\partial\left(r\Omega\right)}{\partial r}\,,$$

$$\Omega_\varphi = (\nabla \times \mathbf{V})_\varphi = \frac{1}{r}\left(\frac{\partial\left(rV_\vartheta\right)}{\partial r} - \frac{\partial V_r}{\partial\vartheta}\right) = 0\,.$$

Thus, we have

$$(\nabla \times (\nabla \times \boldsymbol{\Omega}))_r = 0\,, \quad (\nabla \times (\nabla \times \boldsymbol{\Omega}))_\vartheta = 0\,,$$

$$(\nabla \times (\nabla \times \boldsymbol{\Omega}))_\varphi = \frac{1}{r}\frac{\partial}{\partial r}\left(r(\nabla \times \mathbf{V})_\vartheta\right) - \frac{1}{r}\frac{\partial}{\partial\vartheta}(\nabla \times \mathbf{V})_r$$

$$= \frac{1}{r}\frac{\partial^2(r\Omega)}{\partial r^2} - \frac{1}{r^2}\left(\frac{1}{\sin\vartheta}\frac{\partial(\Omega\sin\vartheta)}{\partial\vartheta}\right).$$

Therefore, (14.20) becomes

$$\frac{1}{r}\frac{\partial^2(r\Omega)}{\partial r^2} - \frac{1}{r^2}\left(\frac{1}{\sin\vartheta}\frac{\partial(\Omega\sin\vartheta)}{\partial\vartheta}\right) = 0\,. \tag{14.21}$$

For $r \to \infty$, we have $\Omega \to 0$ since the flow is homogeneous there. We are looking for a solution for the method of separation of variables in the form $\Omega = R(r)\theta(\vartheta)$. By substituting this ansatz into (14.21), one obtains the following equation:

$$\frac{r}{R(r)}\frac{d^2}{dr^2}(rR(r)) = -\frac{1}{\theta(\vartheta)}\frac{d}{d\vartheta}\left(\frac{1}{\sin\vartheta}\frac{d}{d\vartheta}(\theta(\vartheta)\sin\vartheta)\right). \tag{14.22}$$

Since the coordinates r and ϑ are independent, this means that the left and right-hand sides are equal to a constant. Therefore, one can assume that

$$\frac{r}{R(r)}\frac{d^2}{dr^2}(rR(r)) = -\alpha\,, \quad \frac{1}{\theta(\vartheta)}\frac{d}{d\vartheta}\left(\frac{1}{\sin\vartheta}\frac{d}{d\vartheta}(\theta(\vartheta)\sin\vartheta)\right) = -\alpha\,, \tag{14.23}$$

where α is a constant that has to be determined from the periodicity requirements of the polar angle ϑ. For $\alpha = 2$, the second equation in (14.23) has the solution

$$\theta(\vartheta) = \sin\vartheta \tag{14.24}$$

and the first equation in (14.23) becomes

$$\frac{d^2}{dr^2}(rR(r)) = \frac{2}{r}R(r)\,.$$

The solution of this equation under the condition $R \to 0$ for $r \to \infty$ is given by

$$R(r) = \frac{\text{const}}{r^2} .$$

Thus, the required solution has the form

$$\Omega = \frac{A \sin \vartheta}{r^2} . \tag{14.25}$$

To determine the velocity components V_r and V_ϑ, we have two equations: (14.19) which has, in spherical coordinates, the form

$$\frac{1}{r}\left(\frac{\partial (r V_\vartheta)}{\partial r} - \frac{\partial V_r}{\partial \vartheta}\right) = \frac{A \sin \vartheta}{r^2} \tag{14.26}$$

and the equation of continuity

$$\nabla \cdot \mathbf{V} = \frac{1}{r^2}\frac{\partial (r^2 V_\vartheta)}{\partial r} + \frac{1}{\sin \vartheta}\frac{\partial (V_\vartheta \sin \vartheta)}{\partial \vartheta} = 0 . \tag{14.27}$$

The corresponding boundary conditions are

$$V_r = V_\vartheta = 0 \quad \text{for} \quad r = a , \quad V_r = V_\infty \cos \vartheta \quad \text{and}$$
$$V_\vartheta = -V_\infty \sin \vartheta \quad \text{for} \quad r \to \infty . \tag{14.28}$$

The last two of these boundary conditions are satisfied if we look for solutions in the form

$$V_r = \left(V_\infty + \sum_{k=1}^{\infty} \frac{\lambda_k}{r^k}\right) \cos \vartheta , \quad V_\vartheta = \left(-V_\infty + \sum_{k=1}^{\infty} \frac{\lambda'_k}{r^k}\right) \sin \vartheta . \tag{14.29}$$

By substituting (14.30) into (14.26) and (14.27), and equating the coefficients of the same trigonometric functions, one obtains

$$\sum_{k=1}^{\infty} \left(\lambda_k + (1-k)\lambda'_k\right) \frac{1}{r^{k-1}} = A , \quad \sum_{k=1}^{\infty} \left((2-k)\lambda_k + 2\lambda'_k\right) \frac{1}{r^{k-1}} = 0 .$$

For $k = 1$, we have

$$\lambda_1 = A , \quad \lambda_1 + 2\lambda'_1 = 0 .$$

Since r can be chosen arbitrarily, for $k > 1$, we have

$$\lambda_k + (1-k)\lambda'_k = 0 , \quad (2-k)\lambda_k + 2\lambda'_k = 0 .$$

This system has a non-trivial solution only under the condition that the determinant vanishes, that is, $2 - (1-k)(2-k) = 0$.

Of the roots $k = 0$ and $k = 3$, only $k = 3$ remains since $k \geqslant 1$. Therefore, there are only three non-vanishing quantities: $\lambda_1 = A, \lambda_1' = -A/2$ and $\lambda_3' = \lambda_3/2$. The resulting velocity components are given by

$$V_r = \left(V_\infty + \sum_{k=1}^{\infty} \frac{\lambda_k}{r^k} \right) \cos\vartheta = \left(V_\infty + \frac{A}{r} + \frac{\lambda_3}{r^3} \right) \cos\vartheta \, ,$$

$$V_\vartheta = \left(-V_\infty + \sum_{k=1}^{\infty} \frac{\lambda_k'}{r^k} \right) \sin\vartheta = \left(-V_\infty - \frac{A}{2r} + \frac{\lambda_3}{2r^3} \right) \sin\vartheta \, .$$

The constants A and λ_3 can be determined from the first two conditions in (14.28) as

$$\frac{A}{a} + \frac{\lambda_3}{a^3} = -V_\infty \, , \quad -\frac{A}{2a} + \frac{\lambda_3}{2a^3} = V_\infty \, ,$$

which means that

$$A = -\frac{2}{3} a V_\infty \, , \quad \lambda_3 = \frac{1}{2} a^2 V_\infty$$

and

$$V_r = V_\infty \left(1 - \frac{2a}{3r} + \frac{a^3}{2r^3} \right) \cos\vartheta \, , \quad V_\vartheta = -V_\infty \left(1 - \frac{3a}{4r} - \frac{a^3}{4r^3} \right) \sin\vartheta \, . \tag{14.30}$$

From (14.25), one obtains the vorticity as

$$\Omega = -\frac{3}{2} a V_\infty \frac{\sin\vartheta}{r^2} \, . \tag{14.31}$$

To find the pressure, we take the first equation in (14.19)

$$\nabla p = -\mu \nabla \times \boldsymbol{\Omega} \, ,$$

which has the following form in spherical coordinates:

$$\frac{\partial p}{\partial r} = -\mu \frac{1}{r \sin\vartheta} \frac{\partial(\Omega \sin\vartheta)}{\partial\vartheta} = 3\mu a V_\infty \frac{\cos\vartheta}{r^3} \, ,$$

$$\frac{1}{r} \frac{\partial p}{\partial\vartheta} = \mu \frac{1}{r} \frac{\partial(r\Omega)}{\partial r} = \frac{3}{2} \mu a V_\infty \frac{\sin\vartheta}{r^3} \, .$$

The solution is easily found to be

$$p = p_\infty - \frac{3}{2} \mu a V_\infty \frac{\cos\vartheta}{r^2} \, . \tag{14.32}$$

Now, we obtain the pressure coefficient (see (11.23)) as

$$c_p = \frac{p - p_\infty}{0.5 \rho V_\infty^2} = -\frac{3\mu}{\rho V_\infty a} \frac{\cos\vartheta}{\left(\frac{r}{a}\right)^2} = \frac{6}{\mathrm{Re}} \frac{\cos\vartheta}{\left(\frac{r}{a}\right)^2} \, ,$$

where $\mathrm{Re} = \rho V_\infty d/\mu$ is Reynolds number and $d = 2a$ is the diameter of the sphere.

We now consider the differences for the pressure distribution during the flow of a viscous fluid or an ideal fluid (10.51):

1. The pressure coefficient for a viscous fluid is a function of Reynolds number, that is, it depends on the body dimension, the flow velocity and fluid properties, the mass density and viscosity.
2. The pressure distribution on the surface of the body is asymmetric such that the main vector of the pressure force is unequal to zero.
3. In the critical points $c_p \neq 1$ at the midsection ($\vartheta = \pi/2$), the pressure is equal to the pressure in the undisturbed flow and the maximum dilution of the flow happens at the trailing critical point.

We now consider the components of the stress tensor. Navier–Stokes law gives the following relationship between the stress tensor $\boldsymbol{P}$ and the strain velocity tensor $\boldsymbol{E}$ on the surface of the body with unit normal vector $\boldsymbol{n}$

$$\boldsymbol{P} = -p\boldsymbol{n} + 2\mu\boldsymbol{E} ,$$

or in component form in Cartesian coordinates

$$p_{xx} = -p + 2\mu\frac{\partial U}{\partial X} ,$$

$$p_{yy} = -p + 2\mu\frac{\partial V}{\partial Y} ,$$

$$p_{zz} = -p + 2\mu\frac{\partial W}{\partial Z} ,$$

$$p_{xy} = p_{yx} = \mu\left(\frac{\partial U}{\partial Y} + \frac{\partial V}{\partial X}\right) ,$$

$$p_{xz} = p_{zx} = \mu\left(\frac{\partial W}{\partial X} + \frac{\partial U}{\partial Z}\right) ,$$

$$p_{yz} = p_{zy} = \mu\left(\frac{\partial V}{\partial Z} + \frac{\partial W}{\partial Y}\right)$$

and in spherical coordinates

$$p_{rr} = -p + 2\mu\frac{\partial V_r}{\partial r} , \quad p_{\vartheta\vartheta} = -p + 2\mu\left(\frac{V_r}{r} + \frac{1}{r}\frac{\partial V_\vartheta}{\partial\vartheta}\right) ,$$

$$p_{\varphi\varphi} = -p + 2\mu\left(\frac{V_r}{r} + \frac{1}{r\sin\vartheta}\frac{\partial V_\varphi}{\partial\varphi} + \frac{V_\varphi \cot\vartheta}{r}\right) ,$$

$$p_{r\varphi} = p_{\varphi r} = \mu\left(\frac{\partial V_\varphi}{\partial r} + \frac{1}{r\sin\vartheta}\frac{\partial V_r}{\partial\varphi} - \frac{V_\varphi}{r}\right) ,$$

$$p_{r\vartheta} = p_{\vartheta r} = \mu\left(\frac{1}{r}\frac{\partial V_r}{\partial\vartheta} + \frac{\partial V_\vartheta}{\partial r} - \frac{V_\vartheta}{r}\right) ,$$

$$p_{\vartheta\varphi} = p_{\varphi\vartheta} = \mu\left(\frac{1}{r\sin\vartheta}\frac{\partial V_\vartheta}{\partial\varphi} + \frac{1}{r}\frac{\partial V_\varphi}{\partial\vartheta} - \frac{V_\varphi\cot\vartheta}{r}\right) .$$

From the axial symmetry of the problem, it follows that $V_\varphi = 0$ and that V_r and V_ϑ depend on r and ϑ. The resulting force that acts on the sphere from the fluid is given by

$$\boldsymbol{R} = \int_\Sigma \boldsymbol{p}_n d\sigma \,, \tag{14.33}$$

where $\boldsymbol{p}_n d\sigma$ is the force that acts on the surface of the body with the unit normal vector $\boldsymbol{n}$. The components of the stress tensor on the surface of the sphere are given by

$$\begin{aligned}
(p_{rr})_{r=a} &= (p_{\vartheta\vartheta})_{r=a} = (p_{\varphi\varphi})_{r=a} = -p_a \,, \\
(p_{r\vartheta})_{r=a} &= \mu\left(\frac{\partial V_\vartheta}{\partial r}\right)_{r=a} = -\frac{3}{2}\frac{\mu V_\infty \sin\vartheta}{2} \,, \\
(p_{\vartheta\varphi})_{r=a} &= (p_{\varphi r})_{r=a} = 0 \,.
\end{aligned}$$

Therefore, we have normal forces p_a and tangential forces $p_{r\vartheta}$ acting on the surface of the sphere. On the surface of the sphere, we take an elementary strip of area $d\sigma = 2\pi a^2 \sin\vartheta\, d\vartheta$ (see Figure 14.1), and then multiply it by the frictional tension $p_{r\vartheta}$ and the force p. By projecting these infinitesimal forces on the X-axis and integrating over the angle ϑ from $\vartheta = 0$ to $\vartheta = \pi$, one obtains the resistance force

$$\begin{aligned}
R_x &= \int_0^\pi (-p_{r\vartheta}\sin\vartheta - p\cos\vartheta) 2\pi a^2 \sin\vartheta\, d\vartheta \\
&= \int_0^\pi \left[\frac{3\mu V_\infty \sin\vartheta}{4}\sin\vartheta - \left(p_\infty - \frac{3\mu V_\infty \cos\vartheta}{4}\right)\cos\vartheta\right] \\
&\quad \times 2\pi a^2 \sin\vartheta\, d\vartheta = 6\pi\mu a V_\infty \,.
\end{aligned} \tag{14.34}$$

The expression (14.34) is called Stokes' formula. It gives the resistance force of a sphere, around which a viscous incompressible fluid with small Reynolds number flows. The dimensionless drag coefficient is given by

$$c_x = \frac{R_x}{0.5\rho V_\infty^2 \pi a^2} = \frac{24}{\mathrm{Re}} \,. \tag{14.35}$$

This formula is valid only for $\mathrm{Re} \ll 1$. For a finite Reynolds number, one has to use the full Navier–Stokes equations. The difficulties with their solution lead to having to use approximation methods. One of these methods was used by Oseen in the case of finite, but small $\mathrm{Re} < 1$ Reynolds numbers. He linearized the non-linear convective term $(\boldsymbol{V}\cdot\nabla)\boldsymbol{V}$ by replacing it by $(\boldsymbol{V}_\infty\cdot\nabla)\boldsymbol{V}$. As a result, the drag coefficient is obtained as a power series

$$c_x = \frac{24}{\mathrm{Re}}\left(1 + \frac{3}{16}\mathrm{Re} - \frac{19}{1280}\mathrm{Re}^2 + \ldots\right) \,. \tag{14.36}$$

Until now, we have assumed that V does not depend on the time. If $V = V(t)$, the resistance force must, in a similar manner as for the problem of flow around a sphere considered in Section 10.4, consist of two terms, one of which coincides with (14.34) and the other one corresponds to the virtual mass of the sphere. We now give, without derivation, the formula for the resistance force that acts on a sphere of radius a which moves in a viscous incompressible fluid with velocity $V(t)$ for Re $\ll 1$

$$\begin{aligned} \mathbf{R} = &-6\pi\mu a V(t) \\ &-\frac{2}{3}\pi\rho a^3 \frac{dV}{dt} - 6\pi\mu a \left(\frac{a}{\sqrt{\pi\mu}} \int_0^t \frac{V(\tau)d\tau}{\sqrt{t-\tau}} - 2a^2\sqrt{\frac{\pi\mu\rho}{t}}V(0) \right) . \end{aligned} \tag{14.37}$$

This formula is called the Boussinesq formula.

If the sphere starts to suddenly move with velocity V along a straight line and keeps moving with this velocity, the resistance force changes as a function of time in the following way:

$$R = 6\pi\mu a V \left(1 + \sqrt{\frac{a^2}{\pi\nu t}} \right) . \tag{14.38}$$

For $t \to \infty$, we have $R \to 6\pi\mu a V$, corresponding to Stokes law.

It must be noted that the resistance force during the movement of a sphere in a viscous fluid is not similar to the ideal fluid unequal to zero. The presence of a resistance force can be explained by the fact that the distribution of pressure and frictional force over the surface of the sphere is asymmetric with respect to the midsection ($\vartheta = \pi/2$), while there is not frictional force in the case of the ideal fluid and the pressure distribution over the surface of the sphere is symmetric.

We now ask the question as to which cases the solution can be considered as a sufficiently close approximation to the exact solution. To answer this question, one has to compare the neglected convective terms with the viscous terms in the equations of motion. This comparison shows that this approximation can be made for $r < \nu/V_\infty$, but not in the immediate vicinity of the surface of the body. Far from the sphere, that is, for $r \gg \nu/V_\infty$, the convective terms can be larger than the viscous terms such that the solution obtained above will become invalid in this region. However, in this region, all terms of the equation of motion are small. Therefore, it is not surprising that Stokes law has been confirmed in many experiments also far from the sphere.

The defect of Stokes equation is especially evident for the problem of the flow around an infinite cylinder in the case of Re $\ll 1$. It turns out that in this case, Stokes equations do not have a solution that satisfies the condition $\mathbf{V} \to \mathbf{V}_\infty$ for $r \to \infty$. This phenomenon is also known as Stokes paradox. This contradiction can be resolved by taking the convective term in the equations of motion into account.

14.3
Creeping Spatial Flow of a Viscous Incompressible Fluid

Stokes equations (13.73) can be applied to solving the problems of flows in tight channels, crevices and thin tubes. One also solves the problem of the movement of fluid lubrication oil in the thin gap between the rotary shaft and the bearing support. Therefore, this class of problems is called hydrodynamic lubrication theory.

We consider the movement of a viscous incompressible fluid in a gap between two infinite parallel walls that are a small distance $2h$ apart. We choose the coordinate plane XOY in the middle between the walls and the Z-axis perpendicular to the walls. The movement takes place in planes that are parallel to the walls (see Figure 14.2). The flow is described by

$$\begin{aligned}\mu\left(\frac{\partial^2 U}{\partial X^2}+\frac{\partial^2 U}{\partial Y^2}+\frac{\partial^2 U}{\partial Z^2}\right)&=\frac{\partial p}{\partial X}\,,\\ \mu\left(\frac{\partial^2 V}{\partial X^2}+\frac{\partial^2 V}{\partial Y^2}+\frac{\partial^2 V}{\partial Z^2}\right)&=\frac{\partial p}{\partial Y}\,,\\ 0&=\frac{\partial p}{\partial Z}\end{aligned} \tag{14.39}$$

with the boundary condition $U = V = 0$ at $Z = \pm h$.

From this, it follows that the pressure p is a function of X, Y and satisfies Laplace's equation, that is,

$$\frac{\partial^2 p}{\partial X^2}+\frac{\partial^2 p}{\partial Y^2}=0\,. \tag{14.40}$$

The solution of the above equations is

$$U=-\frac{h^2}{2\mu}\frac{\partial p}{\partial X}\left(1-\frac{Z^2}{h^2}\right),\quad V=-\frac{h^2}{2\mu}\frac{\partial p}{\partial Y}\left(1-\frac{Z^2}{h^2}\right). \tag{14.41}$$

We take the averages of the velocity distributions over the height of the gap and obtain

$$\bar{U}=\frac{1}{2h}\int_{-h}^{h}U dZ=-\frac{h^2}{3\mu}\frac{\partial p}{\partial X}\,,\quad \bar{V}=\frac{1}{2h}\int_{-h}^{h}V dZ=-\frac{h^2}{3\mu}\frac{\partial p}{\partial Y}\,. \tag{14.42}$$

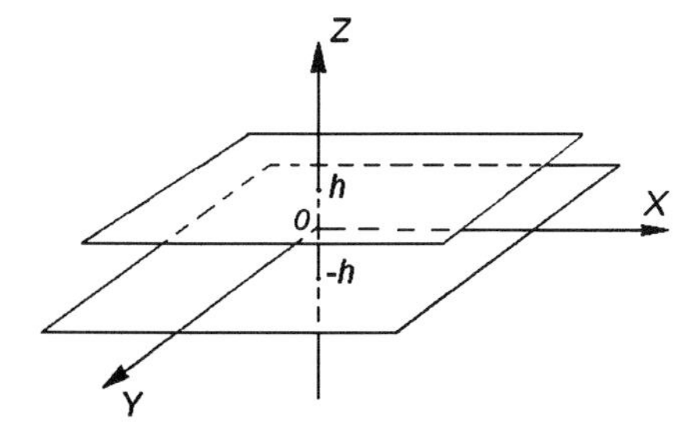

Figure 14.2 Flow of viscous incompressible fluid in a gap.

While the distribution of the true velocities (14.41) is three-dimensional, that is, it depends on X, Y and Z, the distribution of the averages velocities is two-dimensional. Therefore, the distribution of the average velocities $\bar{U}$ and $\bar{V}$ corresponds to an imaginary plane flow similar to the vortex-free flow of an ideal fluid with the potential

$$\bar{\varphi} = -\frac{h^2}{3\mu} p \; . \tag{14.43}$$

Since the pressure p satisfies Laplace's equation, the potential φ also satisfies Laplace's equation and the components of the average velocity satisfy the equation which is equivalent to the equation of continuity of an incompressible fluid, that is,

$$\frac{\partial \bar{U}}{\partial X} + \frac{\partial \bar{V}}{\partial Y} = 0 \; . \tag{14.44}$$

The main feature of the imaginary plane flow is however the proportionality of the velocity vector to the gradient of the pressure, that is,

$$\bar{\boldsymbol{V}} = -\frac{h^2}{3\mu} \nabla p \; . \tag{14.45}$$

A similar flow to the one considered above is the flow of a viscous incompressible fluid through a porous medium. Darcy's law that describes the movement of such a flow is a generalization of (14.45).

To describe the fluid movement in a porous medium one has to choose a model of the medium. By porous medium, one is referring to many particles that are arranged in close proximity. The empty space between the particle is filled by a fluid or gas. If a pressure gradient is applied to a porous medium, the fluid starts moving against the direction of the pressure gradient. The pressure gradient can be created by a pressure difference in two cross sections of the medium or by a height difference between the beginning and end of the area. We now consider the movement that is caused by this pressure gradient.

One can model the movement of the fluid by the flow through a tube of small diameter. The geometric characteristics of this tube depend on the structure of the porous medium. In filtration theory, one uses one parameter, the porosity m which is defined by

$$m = \frac{V_{\text{void}}}{V} \; , \tag{14.46}$$

which characterizes the internal structure of the medium, with the pore volume V_{void} and the medium volume V. These parameters describe the particle geometry and their packing. For real layers, one has $m \sim 0.15 \cdots 0.22$.

The experimentally established filtration law that connects the filtration velocity $\boldsymbol{V}$ and the pressure gradient ∇p by the linear relationship

$$\boldsymbol{V} = -\frac{k}{\mu} \nabla p \; , \tag{14.47}$$

where μ is the viscosity of the fluid and k is the layer permeability, is referred to Darcy's law.

The linear law (14.47) is valid for relatively small velocities if Reynolds number does not exceed a certain critical value. The linear measure L is the characteristic dimension of the considered area.

The filtration process of a viscous incompressible fluid is described by (14.47) and the equation of continuity

$$\nabla \cdot \boldsymbol{V} = 0 . \tag{14.48}$$

If the layer permeability k is constant, these equations mean that

$$\Delta p = 0 . \tag{14.49}$$

By introducing the velocity potential ϕ by

$$\varphi = \frac{k}{\mu} p , \tag{14.50}$$

one obtains from (14.48)

$$\Delta \varphi = 0 . \tag{14.51}$$

Therefore, the filtration flow is a potential flow. Thus, one can obtain the solutions by the method that we have applied to potential flows of ideal incompressible fluids in Chapter 10. For example, one can consider the sources and sinks as creating filtration flows that are caused by the production and injection drills. For a two-dimensional filtration, the method of complex variables is successfully used.

15
The Laminar Boundary Layer

An approximate method for solving the problem of the flow of a viscous fluid for relatively large Reynolds numbers is the foundation for the theory of the laminar boundary layer.

15.1
Equation of Motion for the Fluid in the Boundary Layer

We consider the plane parallel stationary flow of an incompressible fluid along the surface of a solid body without taking into account volume forces (see Figure 15.1).

If the wall is not a plane, we assume that its radius of curvature is sufficiently large compared to the thickness of the boundary layer, such that the flow can be considered as plane parallel. This assumption makes it possible to treat the flow in a non-moving Cartesian system of coordinates that is fixed to the body.

We also assume that the body has an infinite extension in the Z-direction such that the flow parameters only depend on X and Y. For large Reynolds numbers Re, the viscous forces in the equations of motion (13.71) are small relative to the inertial (convective) forces everywhere, except for in a small region close to the surface of the body. Since the no-slip condition $\mathbf{V} = 0$ has to be satisfied on the surface of the body, the viscous terms cannot be smaller than the convective terms.

Therefore, one divides the flow region into two parts. Far from the body, the flow is described by equations of motion without viscous terms, similar to the equations of motion of the ideal fluid, that is, Euler's equations, and in the vicinity of the surface of the body in a layer of limited thickness δ by the Navier–Stokes equations in which both convective and viscous terms are taken into account. The thickness δ is called boundary layer thickness. It is small relative to the characteristic linear dimension L ($\delta \ll L$). There are some problems without a characteristic linear dimension, for example, the problem of the flow around a semi-infinite plate (see Section 15.2).

We now conduct a dimensional analysis of the Navier–Stokes equations in the vicinity of the body. Then, the characteristic dimensional quantities are the velocity components U, V, the linear body dimension L, the specific viscosity $\nu = \mu/\rho$,

Hydromechanics. Theory and Fundamentals. Emmanuil G. Sinaiski

ISBN: 978-3-527-41026-2

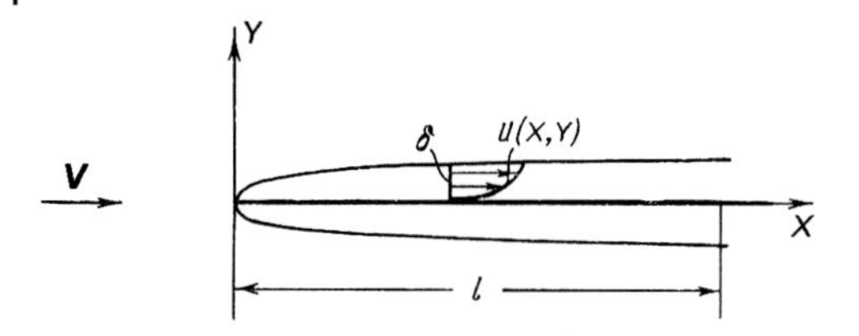

Figure 15.1 Boundary layer for the flow around a plate.

mass density ρ, and we introduce the following dimensionless variables

$$X' = \frac{X}{L}, \quad Y' = \frac{Y}{L}, \quad U' = \frac{V_x}{U}, \quad V' = \frac{V_y}{V}.$$

Thus, the equation of continuity becomes

$$\frac{\partial U'}{\partial X'} + \frac{V}{U}\frac{L}{\delta}\frac{\partial V'}{\partial Y'} = 0. \tag{15.1}$$

In order for the terms in (13.11) to be of the same order, it is necessary that

$$\frac{V}{U}\frac{L}{\delta} \sim 1 \quad \text{or} \quad \frac{V}{U} \sim \frac{\delta}{L} \ll 1. \tag{15.2}$$

Therefore, the problem has a small parameter $\varepsilon = \delta/L$. We also introduce Reynolds number and the dimensionless pressure as

$$\mathrm{Re} = \frac{UL}{\nu}, \quad p' = \frac{\delta^2}{\nu UL} p = \frac{\varepsilon^2 \mathrm{Re}}{\rho U^2} p. \tag{15.3}$$

By transforming the equations of motion

$$\begin{aligned} V_x \frac{\partial V_x}{\partial X} + V_x \frac{\partial V_x}{\partial Y} &= -\frac{1}{\rho}\frac{\partial p}{\partial X} + \nu\left(\frac{\partial^2 V_x}{\partial X^2} + \frac{\partial^2 V_x}{\partial Y^2}\right), \\ V_x \frac{\partial V_y}{\partial X} + V_y \frac{\partial V_y}{\partial Y} &= -\frac{1}{\rho}\frac{\partial p}{\partial Y} + \nu\left(\frac{\partial^2 V_y}{\partial X^2} + \frac{\partial^2 V_y}{\partial Y^2}\right) \end{aligned} \tag{15.4}$$

to dimensionless variables, one obtains

$$\begin{aligned} \frac{U^2}{L} U' \frac{\partial U'}{\partial X'} + \frac{UV}{\delta} V' \frac{\partial U'}{\partial Y'} &= -\frac{U^2}{\varepsilon^2 \mathrm{Re} L}\frac{\partial p'}{\partial X'} + \nu\left(\frac{U}{L^2}\frac{\partial^2 U'}{\partial X'^2} + \frac{U}{\delta^2}\frac{\partial^2 U'}{\partial Y'^2}\right), \\ \frac{UV}{L} U' \frac{\partial V'}{\partial X'} + \frac{V^2}{\delta} V' \frac{\partial V'}{\partial Y'} &= -\frac{U^2}{\varepsilon^2 \mathrm{Re} L}\frac{\partial p'}{\partial Y'} + \nu\left(\frac{V}{L^2}\frac{\partial^2 V'}{\partial X'^2} + \frac{V}{\delta^2}\frac{\partial^2 V'}{\partial Y'^2}\right) \end{aligned}$$

or

$$\begin{aligned} U' \frac{\partial U'}{\partial X'} + \frac{LV}{\delta U} V' \frac{\partial U'}{\partial Y'} &= -\frac{1}{\varepsilon^2 \mathrm{Re}}\frac{\partial p'}{\partial X'} + \frac{1}{\mathrm{Re}}\left(\frac{\partial^2 U'}{\partial X'^2} + \frac{1}{\varepsilon^2}\frac{\partial^2 U'}{\partial Y'^2}\right), \\ U' \frac{\partial V'}{\partial X'} + \frac{LV}{\delta U} V' \frac{\partial V'}{\partial Y'} &= -\frac{1}{\varepsilon^2 \mathrm{Re}}\frac{\partial p'}{\partial Y'} + \frac{1}{\mathrm{Re}}\left(\frac{\partial^2 V'}{\partial X'^2} + \frac{1}{\varepsilon^2}\frac{\partial^2 V'}{\partial Y'^2}\right). \end{aligned} \tag{15.5}$$

Since $LV/\delta U \sim 1$ and $\varepsilon \ll 1$, one can rewrite the first equation in (15.5) as

$$U'\frac{\partial U'}{\partial X'} + V'\frac{\partial U'}{\partial Y'} = -\frac{1}{\varepsilon^2 \text{Re}}\frac{\partial p'}{\partial X'} + \frac{1}{\varepsilon^2 \text{Re}}\frac{\partial^2 U'}{\partial Y'^2} \,. \tag{15.6}$$

One of the main assumptions of the boundary layer theory consists of the hypothesis that the viscous and convective terms in the equations of motion are of the same order of magnitude. This means $\varepsilon^2 \text{Re} \sim 1$. Therefore, one can estimate thickness of the boundary layer as follows

$$\delta \sim \frac{L}{\sqrt{\text{Re}}} \tag{15.7}$$

and rewrite (15.6) as

$$U'\frac{\partial U'}{\partial X'} + V'\frac{\partial U'}{\partial Y'} = -\frac{\partial p'}{\partial X'} + \frac{\partial^2 U'}{\partial Y'^2} \,. \tag{15.8}$$

Since the coefficient of $\partial p'/\partial Y'$ in the second equation of (15.5) is much larger than the viscous and convective terms, one can simplify the second equation in (15.6) to

$$\frac{\partial p'}{\partial Y'} = 0 \,. \tag{15.9}$$

This equation means that the pressure in the boundary layer only changes along the body, that is, $p = p(X)$. Under the conditions (15.2) and (15.7), the boundary layer equations have the form

$$\begin{aligned} \frac{\partial U}{\partial X} + \frac{\partial V}{\partial Y} &= 0 \,, \\ U\frac{\partial U}{\partial X} + V\frac{\partial U}{\partial Y} &= -\frac{1}{\rho}\frac{\partial p}{\partial X} + \nu\frac{\partial^2 U}{\partial Y^2} \,, \\ \frac{\partial p}{\partial Y} &= 0 \,. \end{aligned} \tag{15.10}$$

Since the pressure is constant inside of the boundary layer, the pressure is the same as on the outer boundary of the boundary layer. In addition, since the flow outside of the boundary layer is ideal, the pressure and velocity distributions on the outer boundary of the boundary layer are the same as for the flow of an ideal fluid.

Thus, the problem of the flow of a viscous incompressible fluid around a body is divided into two problems. The first consists of the determination of the velocity and pressure on the surface of the body for the flow of an ideal fluid. To obtain this solution, one has to use Euler's equations. The second problem consists of determining the velocity field by the solution of the boundary layer equations (15.10) subject to the no-slip conditions on the boundary and to the requirement that on the outer boundary of the boundary layer, the velocities and stresses agree with the values obtained as a solution of the first problem. In some cases, one formulates the boundary condition at $Y \to \infty$ during the solution of the second problem. Such a boundary layer is called an asymptotic boundary layer. If the thickness of the boundary layer is finite, the thickness δ is determined during the solution. Therefore, the boundary conditions for the problem at hand can be written as follows:

$$U = V = 0 \quad \text{for} \quad Y = 0 \,,$$

$$U = U_\infty \,, \quad \frac{\partial U_\infty}{\partial Y} = 0 \quad \text{for} \quad Y = \delta \quad \text{or}$$

$$U \to U_\infty(X) \quad \text{for} \quad Y \to \infty \,, \tag{15.11}$$

where U_∞ is the flow velocity just outside of the boundary layer. The condition $\partial U_\infty/\partial Y = 0$ for $Y = \delta$ follows from the equality of tangential tension on the boundary between the boundary layer and the external flow of the ideal fluid. This condition is used to determine the boundary layer thickness δ.

It is known that during the flow around a convex body, the flow (boundary layer) will at some point become detached from the body and a vortex zone will arise behind the body. Therefore, the boundary layer equations are only valid until the point when the boundary layer is detached from the body.

The phenomenon of boundary layer detachment can be explained by the boundary layer theory. From the equations of the plane boundary layer (15.10), it follows that pressure in the boundary layer has the constant value $p = p(X)$ along the thickness of the boundary layer and is equal to the pressure in the flow outside of the boundary layer. According to the boundary layer model, the flow parameters outside of the boundary layer are equal to the corresponding flow parameters of an ideal fluid.

From the solution of the problem of a flow around a cylinder (see (11.23)), it follows that the pressure decreases along the body surface to a minimum value and then grows again. This also applies to the flow around any convex body. In the point M on Figure 15.2 where the pressure reaches its minimum, we have $\partial p/\partial X = 0$. Before the point M, we have $\partial p/\partial X < 0$ and after M, we have $\partial p/\partial X > 0$. On the body, we have $U = V = 0$ and from the second equation in (15.10), it follows that

$$\mu \left(\frac{\partial^2 U}{\partial Y^2}\right)_{y=0} = \frac{\partial p}{\partial X} \,. \tag{15.12}$$

However, at the point M, we have

$$\mu \left(\frac{\partial^2 U}{\partial Y^2}\right)_{y=0} = 0 \,.$$

The quantity

$$\tau = \mu \left(\frac{\partial U}{\partial Y}\right)_{y=0}$$

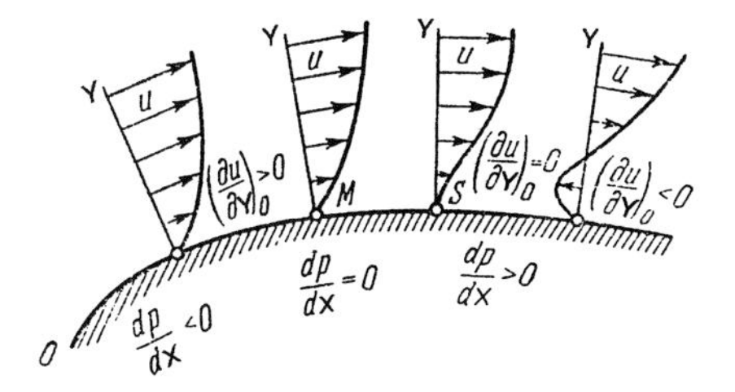

Figure 15.2 Change of tangential tension on the surface of the body and the detachment of the boundary layer.

is equal to the tangential tension on the surface of the body. Therefore, one can rewrite (15.12) as

$$\left(\frac{\partial \tau}{\partial Y}\right)_{y=0} = \frac{\partial p}{\partial X} . \tag{15.13}$$

The expression (15.13) means that the tangential tension on the surface of the body decreases until the point M where it also vanishes. The distribution of longitudinal velocity has a vertical tangent in M. To the left of M, we have $\tau > 0$ and to the right of M, we have $\tau < 0$.

Since the pressure increases to the right of point M, a so-called counter pressure will arise there that forces the fluid on the surface of the body to flow in the opposite direction and thus forces the detachment of the boundary layer. The detachment of the boundary layer is accompanied by a large increase of the boundary layer thickness which leads to strong reconfiguration of the flow, an increase in the resistance force and to vibrations of the body.

15.2 Asymptotic Boundary Layer on a Plate

As an example, we consider the Blasius problem dealing with the flow around a plane plate $X > 0$ of a viscous incompressible fluid with constant speed V_∞ at infinity. Since $V_\infty = \text{const}$, the pressure in the boundary layer is, according to the last equation in (15.10), also constant. Then, the boundary layer equations have the form

$$\frac{\partial U}{\partial X} + \frac{\partial V}{\partial Y} = 0 ,$$
$$U\frac{\partial U}{\partial X} + V\frac{\partial U}{\partial Y} = \nu\frac{\partial^2 U}{\partial Y^2} . \tag{15.14}$$

As boundary conditions, we take

$$U = V = 0 \quad \text{for} \quad Y = 0 , \quad U \to V_\infty \quad \text{for} \quad Y \to \infty . \tag{15.15}$$

This problem only has two characteristic dimensional parameters: the velocity V_∞ and the specific viscosity ν, but no linear dimension. The combination $\sqrt{\nu X/V_\infty}$

has a dimension of length. Therefore, one can introduce a dimensionless coordinate by

$$\eta = Y\sqrt{\frac{V_\infty}{\nu X}}\,. \tag{15.16}$$

Because of the first equation in (15.14), one can introduce the stream function $\Psi(X, Y)$ that satisfies the conditions

$$U = \frac{\partial \psi}{\partial Y}\,, \quad V = -\frac{\partial \psi}{\partial X}\,. \tag{15.17}$$

We are looking for the stream function in the form

$$\psi = \sqrt{\nu V_\infty X}\, f(\eta)\,. \tag{15.18}$$

From (15.17), it follows that

$$U = \sqrt{\nu V_\infty X}\, f'(\eta)\frac{\partial \eta}{\partial Y} = V_\infty f'(\eta)\,, \tag{15.19}$$

$$V = -\frac{1}{2}\sqrt{\frac{\nu}{V_\infty X}}\, f(\eta) - \sqrt{\nu V_\infty X}\, f'(\eta)\frac{\partial \eta}{\partial Y} = \frac{1}{2}\sqrt{\frac{\nu}{V_\infty X}}\,(\eta f'(\eta) - f(\eta))\,, \tag{15.20}$$

and

$$\frac{\partial U}{\partial X} = -\frac{V_\infty}{2X}\eta f''(\eta)\,, \quad \frac{\partial U}{\partial Y} = V_\infty \frac{1}{2}\sqrt{\frac{V_\infty}{\nu X}}\, f''(\eta)\,, \quad \frac{\partial^2 U}{\partial Y^2} = \frac{V_\infty^2}{\nu X} f'''(\eta)\,.$$

By substituting these expressions into (15.14), one obtains the equation for $f(\eta)$

$$2f'''(\eta) + f f'' = 0\,. \tag{15.21}$$

The boundary conditions become

$$f = f' = 0 \quad \text{for} \quad \eta = 0\,, \quad f' \to 1 \quad \text{for} \quad \eta \to \infty\,. \tag{15.22}$$

This boundary value problem does not have an analytic solution. It can be solved with numerical methods. The main problem for the numerical solution is that the boundary conditions are given on both sides. We now show how this problem can be transformed into a Cauchy problem with conditions only on one side of the interval, namely, at $\eta = 0$.

We assume that $f_0(\xi)$ is a solution of (15.21). Then, the function $f(\xi) = \alpha^{1/3} f_0(\alpha^{1/3}\xi)$ for an arbitrary constant α is also a solution of this equation. We define $f_0(\xi)$ as the solution of the Cauchy problem with the conditions

$$f(0) = f_0'(0) = 0\,, \quad f_0''(0) = 1\,. \tag{15.23}$$

From this solution, one can determine $k = \lim\limits_{\xi \to \infty} f(\xi) = (1/0.332)^{2/3}$. We now determine the constant α such that the second boundary condition at ∞ is fulfilled.

Since $f'(\xi) = \alpha^{2/3} f_0'(\eta)$, $f'(\xi) = \alpha f_0''(\eta)$ and $f_0''(0) = \alpha$ where $\eta = \alpha^{1/3}\xi$, we have

$$\lim_{\xi\to\infty} f'(\xi) = \alpha^{2/3} \lim_{\eta\to\infty} f_0'(\eta) = \alpha^{2/3} k \,.$$

If $\alpha^{2/3} k = 1$, we have $f'(\infty) = 1$. This means $\alpha = k^{-3/2} = 0.332$.

Therefore, by solving the Cauchy problem

$$2 f_0'''(\eta) + f_0 f_0'' = 0 \,, \quad f(0) = f_0'(0) = 0 \,, \quad f_0''(0) = 1 \,, \tag{15.24}$$

one obtains the function $f(\xi) = \alpha^{1/3} f_0(\alpha^{1/3}\xi)$, with $\alpha = 0.332$, and by the relationships (15.19) and (15.20), the velocity components U and V. The frictional force on the surface of the plate is given by

$$\tau = \left(\frac{\partial \tau}{\partial Y}\right)_{y=0} = 0.332\sqrt{\frac{\rho\mu V_\infty}{X}} \,. \tag{15.25}$$

Now, one can obtain the resistance R of a rectangular plate of width b and length L along the flow and the drag coefficient c_f as

$$R = b\int_0^L \tau \, dX = 0.664 b\sqrt{\rho\mu L V_\infty^3} \,, \tag{15.26}$$

$$c_f = \frac{R}{\rho b L V_\infty^2} = \frac{1.328}{\sqrt{V_\infty L/\nu}} \,. \tag{15.27}$$

Since there is no exact boundary layer thickness defined for this problem, we take, as δ, the value at which the velocity of the flow has 99% of the value of V_∞. The numerical solution yields

$$\delta = 5\sqrt{\frac{\nu X}{V_\infty}} \,. \tag{15.28}$$

15.3
Problem of the Injected Beam

We consider the broadening of a plane laminar fluid beam which flows through a small gap into a semi-infinite volume of the same fluid that rests at infinity (see Figure 15.3). We choose the system of coordinates in such a way that the X-axis coincides with the beam axis and that the Y-axis is directed along the wall.

The unusual feature of this problem as well as of the problem about the boundary layer on the infinite plate is that there is no characteristic linear dimension. However, in contrast to the previous problem, there is also no characteristic velocity. In addition to the obvious dimensional parameters of mass density ρ and viscosity μ that characterize the properties of the fluid, there is also the parameter

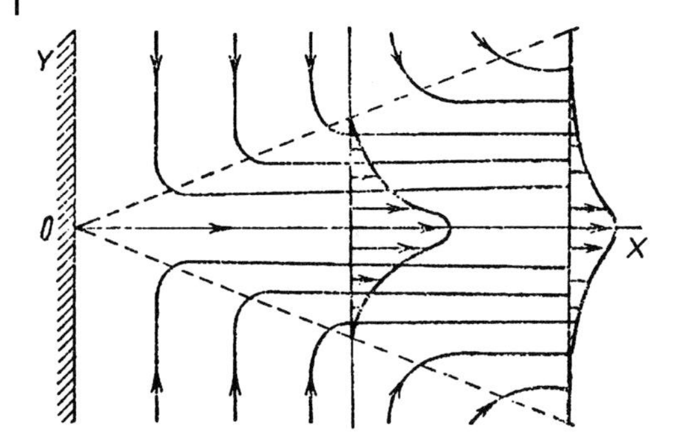

Figure 15.3 Broadening of the injected beam.

of beam momentum

$$J_0 = b \int_{-\infty}^{\infty} \rho U^2 dY \,, \tag{15.29}$$

which characterizes the fluid flow and is constant for all cross sections of the beam.

The experiments show that the amount of movement imparted to the surrounding fluid is greater for larger fluid viscosities, that is for smaller Reynolds numbers. The pressure, however, barely changes in the Y-direction, in full agreement with the boundary layer properties. This means that the flow of the beam can be described by the boundary layer equations. In addition, the flow satisfies (15.29) in every cross section. Therefore, the boundary value problem consists of solving the boundary layer equations

$$\frac{\partial U}{\partial X} + \frac{\partial V}{\partial Y} = 0 \,, \quad U\frac{\partial U}{\partial X} + V\frac{\partial U}{\partial Y} = \nu \frac{\partial^2 U}{\partial Y^2} \,, \tag{15.30}$$

which satisfy the conditions on the beam axis

$$\frac{\partial U}{\partial Y} = 0, \quad V = 0 \quad \text{for} \quad Y = 0 \tag{15.31}$$

and the condition that the fluid is at rest at infinity, that is,

$$U \to 0 \quad \text{for} \quad Y \to \pm\infty \,. \tag{15.32}$$

Using the first equation in (15.30), we rewrite the second equation as

$$\frac{\partial(U^2)}{\partial X} + \frac{\partial(V U)}{\partial Y} = \nu \frac{\partial^2 U}{\partial Y^2}$$

and integrate over Y from $-\infty$ bis $+\infty$:

$$\frac{d}{dX} \int_{-\infty}^{\infty} U^2 dY + (UV)|_{-\infty}^{+\infty} = \nu \left(\frac{\partial U}{\partial Y}\right)_{-\infty}^{+\infty} \,.$$

Since at infinity the condition (15.32) and the obvious condition $\frac{\partial U}{\partial Y} = 0$, which secures the continuous transition $U \to 0$ for $Y \to \pm\infty$ are fulfilled, one obtains

$$\frac{d}{dX} \int_{-\infty}^{\infty} U^2 dY = 0$$

or due to (15.29),

$$\int_{-\infty}^{\infty} U^2 dY = \text{const} = J_0 .$$

Thus, (15.29) has been proven. This relationship means that the momentum transported through any cross section per unit time is constant.

We want to obtain the solution by the method of dimensional analysis (see Section 6.6). The dimensional parameters of the problem are ρ, μ or $\nu = \mu/\rho$, J_0, X, Y, U and V. We denote the hypothetical characteristic parameters by the length L, the velocity V_∞ and the stream function Ψ, and introduce dimensionless variables X', Y' and ψ' via

$$X' = \frac{X}{L} , \quad Y' = \frac{Y}{\frac{L}{\sqrt{\text{Re}}}} , \quad \psi' = \frac{\psi}{\Psi} . \tag{15.33}$$

By transforming (15.29) into the new variables and using the relationship for $\partial\psi/\partial Y$, one obtains

$$\frac{\rho\Psi^2}{L\sqrt{\text{Re}}} \int_{-\infty}^{\infty} \frac{\partial\psi'}{\partial Y'} dY' = J_0 .$$

From this and the dimensions of L and Ψ, it follows that L and Ψ have to be chosen as the following combinations of the characteristic parameters:

$$L = \frac{J_0^2}{\mu\rho V_\infty^3} J_0 , \quad \Psi = \frac{J_0}{\rho V_\infty} . \tag{15.34}$$

Thus, only one undetermined quantity V_∞ remains. Since it does not appear in the problem definition, it cannot appear in the solution. Therefore, the stream function

$$\psi = \frac{J_0}{\rho V_\infty} \psi'(X', Y') \tag{15.35}$$

has to be independent of V_∞. It is easily seen that the dimensionless function

$$\psi' = \sqrt[3]{X'} f\left(\frac{Y'}{X'^{2/3}}\right) , \tag{15.36}$$

or in dimensional form

$$\psi = \sqrt{\frac{\nu J_0 X}{\rho}} f\left(\sqrt[3]{\frac{J_0}{\rho\nu^2}} \frac{Y}{X^{2/3}}\right) , \tag{15.37}$$

satisfies this condition.

Such a solution is called the similarity solution and the corresponding dimensionless variables are referred to as similarity variables.

We introduce the new similarity variable

$$\eta = \frac{Y'}{X'^{2/3}} = \sqrt[3]{\frac{J_0}{\rho \nu^2}}\frac{Y}{X^{2/3}} \tag{15.38}$$

and write the stream function in the form

$$\psi' = \sqrt[3]{X'}\, f(\eta)\,. \tag{15.39}$$

By substituting the stream function by the formulae $U = \partial\psi/\partial Y$, $V = -\partial\psi/\partial X$ into the second equation of (15.30), one obtains

$$\frac{\partial\psi'}{\partial Y'}\frac{\partial^2\psi'}{\partial X'\partial Y'} - \frac{\partial\psi'}{\partial X'}\frac{\partial^2\psi'}{\partial Y'^2} = \frac{\partial^3\psi'}{\partial Y'^3}\,. \tag{15.40}$$

The derivatives are given by

$$\begin{aligned}
&\frac{\partial\psi'}{\partial Y'} = X'^{-1/3} f'(\eta)\,, \quad \frac{\partial^2\psi'}{\partial Y'^2} = X'^{-1} f''(\eta)\,, \quad \frac{\partial^3\psi'}{\partial Y'^3} = X'^{-5/3} f'''(\eta)\,,\\
&\frac{\partial\psi'}{\partial X'} = \frac{1}{3}X'^{-1/3}\left(f(\eta) - 2\eta f'(\eta)\right),\\
&\frac{\partial^2\psi'}{\partial X'\partial Y'} = -\frac{1}{3}X'^{-4/3}\left(f'(\eta) + 2\eta f''(\eta)\right).
\end{aligned}$$

As a result, (15.40) can be transformed into the relationship

$$f''' + \frac{1}{3}\left(f'^2 + f f''\right) = 0\,. \tag{15.41}$$

This equation must be solved subject to the boundary conditions $f = f'' = 0$ at $\eta = 0$, $f' \to 0$ at $\eta \to \pm\infty$ and the integral condition (15.29) which becomes

$$\int_{-\infty}^{\infty} f'^2(\eta)d\eta = 1\,. \tag{15.42}$$

By omitting the details of the solution, we can write the result in dimensional form as

$$\psi = 1.651\sqrt[3]{\frac{\nu J_0 X}{\rho}}\tanh\left(0.2752\sqrt[3]{\frac{J_0}{\rho\nu^2}}\frac{Y}{X^{2/3}}\right). \tag{15.43}$$

The flow Q is given by

$$Q = 2(\psi)_{Y=\infty} = 3.302\sqrt[3]{\frac{\nu J_0 X}{\rho}}\,. \tag{15.44}$$

One can further obtain the distribution of the longitudinal velocity

$$U = \frac{\partial \psi}{\partial Y} = 0.4543 \sqrt[3]{\frac{J_0^2}{\rho^2 \nu X}} \frac{1}{\cosh^2\left(0.2752 \sqrt{\frac{J_0}{\rho \nu^2}} \frac{Y}{X^{2/3}}\right)} \tag{15.45}$$

and the maximum velocity on the beam axis

$$U_m = 0.4543 \sqrt[3]{\frac{J_0^2}{\rho^2 \nu X}} \, . \tag{15.46}$$

Normally, one writes these formulae as

$$\frac{U}{U_m} = \frac{1}{\cosh^2\left(0.2752 \sqrt{\frac{J_0}{\rho \nu^2}} \frac{Y}{X^{2/3}}\right)} \, . \tag{15.47}$$

The beam broadens with the distance from the injection point, but there is no clear boundary between moving fluid and fluid at rest. Therefore, one can only obtain an outer beam boundary $Y = \delta(X)$ if this boundary is the line defined by $U/U_m = \varepsilon \ll 1$. If this small quantity is given a priori, one obtains from (15.47)

$$\delta(X) = \text{const}(\varepsilon) X^{2/3} \, . \tag{15.48}$$

Then, one can write the velocity distribution as follows:

$$\frac{U}{U_m} = F\left(\frac{Y}{\delta}\right) . \tag{15.49}$$

This distribution shows the interesting geometrical property of beam broadening that consists of the affine similarity of the longitudinal velocity profiles at different cross sections.

16
Turbulent Flow of Fluid

16.1
General Information on Laminar and Turbulent Flows

All fluid flows are divided into two distinct categories: smooth layered flows known as laminar flows, and chaotic flows called turbulent flow. The latter term means that the flow in any chosen direction can be described by average values only. Fluid velocity and other parameters such as pressure, temperature, and concentrations of dissolved substances (if the fluid is a multicomponent mixture) fluctuate randomly about their average values, displaying a very irregular spatial and temporal behavior. The fluctuations are characterized by different periods and different amplitudes of both hydrodynamic and non-hydrodynamic parameters. This indicates a complex internal structure of the turbulent flow which distinguishes it from the laminar flow and explains the difference in properties of these two types of flows. As compared to the laminar flow, the turbulent flow has better ability to transport momentum and heat, to spread impurities and products of chemical reactions throughout the volume, to transport particles, to promote interparticle interactions (collisions, coagulation) and so on.

The transport of each quantity (momentum, heat, matter, etc.) is characterized by its own coefficient. Transport coefficients for a turbulent flow are called effective (or turbulent) transport coefficients in order to distinguish them from their ordinary counterparts that describe a laminar flow. Thus, momentum transport is characterized by the coefficient of turbulent kinematical viscosity ν_t, which is defined by a constitutive (phenomenological) proportionality relation between the stress tensor and the strain velocity tensor. Similarly, we can define the effective coefficients of heat conductivity, diffusion, and other coefficients.

The kinetic molecular theory, which focuses on the internal molecular structure of substances, is the proper tool to derive ordinary transport coefficients of fluids and gases. In contrast, turbulent transport coefficients follow from the internal structure of the flow. In a sense, there are similarities between these two derivations as both rely on statistical methods.

Nevertheless, the kinetic molecular theory (i.e., statistical theory of molecular ensembles) is vastly different from statistical hydrodynamic models of viscous fluids. First, under the standard assumptions that the kinetic molecular theory makes

Hydromechanics. Theory and Fundamentals. Emmanuil G. Sinaiski

ISBN: 978-3-527-41026-2

about molecule interactions, the total kinetic energy of a set of moving molecules is constant in time, whereas in hydromechanics, viscous dissipation causes kinetic energy of a real fluid to change as the fluid moves. The second difference is that molecular ensembles are, by their nature, discrete, and their evolution is described by ordinary differential equations (i.e., method of molecular dynamics), while hydromechanics describes flows of continuous media by partial differential equations. Turbulent flow must always obey the basic conservation laws which are implicitly present in the equations of hydro- and thermodynamics. The reader will find the basics of turbulent flow theory and the relevant mathematical techniques explained in [1–11].

16.2
Momentum Equation of a Viscous Incompressible Fluid

An isothermal flow of a viscous incompressible fluid is completely determined by specifying four quantities in each point $\boldsymbol{X}$ at the moment t: the three components u_i of velocity $\boldsymbol{u}(\boldsymbol{X}, t)$ and the pressure $p(\boldsymbol{X}, t)$. These quantities are completely described by the continuity equation

$$\nabla \cdot \boldsymbol{u} = 0 , \tag{16.1}$$

which translates to

$$\frac{\partial U_i}{\partial X_i} = 0 , \tag{16.2}$$

in the component form in a Cartesian coordinate system, and the momentum equation

$$\rho \frac{D\boldsymbol{u}}{Dt} = -\nabla \cdot \boldsymbol{T} + \boldsymbol{F} , \tag{16.3}$$

where $\boldsymbol{T}$ is the stress tensor

$$\boldsymbol{T} = \left(-p + \frac{2\mu}{3} \Delta \cdot \boldsymbol{u} \right) \boldsymbol{I} + 2\mu \boldsymbol{E} , \tag{16.4}$$

$\boldsymbol{I}$ – the unit tensor, and $\boldsymbol{E}$ – the strain velocity tensor with components

$$E_{ij} = \frac{1}{2} \left(\frac{\partial U_i}{\partial X_j} + \frac{\partial U_j}{\partial X_i} \right) . \tag{16.5}$$

Equation 16.3 can be rearranged as follows:

$$\rho_e \frac{D\boldsymbol{u}}{Dt} = -\nabla p + \mu_e \Delta \boldsymbol{u} + \boldsymbol{F} \tag{16.6}$$

or, in the component form in a Cartesian coordinate system,

$$\rho_e \frac{D u_i}{Dt} = -\frac{\partial p}{\partial X_i} + \mu_e \sum_{k=1}^{3} \frac{\partial^2 u_i}{\partial X_k^2} + F_i , \quad (i = 1, 2, 3) . \tag{16.7}$$

Here, ρ_e is the fluid density, $\boldsymbol{F}$ – the density of external forces acting on the unit volume of the fluid, and μ_e – the dynamic viscosity coefficient of the fluid. D/Dt is the substantial derivative

$$\frac{D}{Dt} = \frac{\partial}{\partial t} + u_k \frac{\partial}{\partial X_k} . \tag{16.8}$$

Here, and further on, a repeated index implies summation over all values of the index.

Now, (16.7) can be rewritten as

$$\frac{\partial u_i}{\partial t} + u_k \frac{\partial u_i}{\partial X_k} = -\frac{1}{\rho_e}\frac{\partial p}{\partial X_i} + \nu_e \sum_{k=1}^{3} \frac{\partial^2 u_i}{\partial X_k^2} + f_i , \quad (i = 1, 2, 3) , \tag{16.9}$$

where we have introduced: $\nu_e = \mu/\rho_e$ – the coefficient of kinematics viscosity of the fluid; and $f_i = F_i/\rho_e$ – the density of external forces acting on a unit mass.

Equations 16.1 and (16.9) are called the Navier–Stokes equations. The hydrodynamics of viscous incompressible fluids is based on these equations.

The momentum equation can be written in another form. One must introduce the vorticity vector

$$\omega = \nabla \times \boldsymbol{u} , \tag{16.10}$$

with components

$$\omega_k = \varepsilon_{kji} \frac{\partial u_i}{\partial X_j} , \quad (i, j, k = 1, 2, 3) . \tag{16.11}$$

Here, ε_{kji} are permutation symbols (components of a complete antisymmetric tensor).

If the external force is absent, then by taking the curl $(\nabla\times)$ of both sides of (16.6) and using the component expression for the curl,

$$(\nabla\times)_k = \varepsilon_{kji} \frac{\partial}{\partial X_j} ,$$

we eliminate the pressure from the equations and after some algebra obtain

$$\frac{\partial \omega_k}{\partial t} + u_i \frac{\partial \omega_k}{\partial X_i} - \omega_i \frac{\partial u_k}{\partial X_i} = \nu_e \Delta \omega_k , \quad (k = 1, 2, 3) . \tag{16.12}$$

Equations 16.12 and (16.11) help to determine the components of velocity $\boldsymbol{u}$ and vorticity $\boldsymbol{\omega}$. By calculating the divergence of (16.6) and using (16.1) with $f = 0$, one gets the Poisson equation for pressure

$$\Delta p = -\rho_e \frac{\partial^2 (u_i u_j)}{\partial X_i \partial X_j} , \tag{16.13}$$

whose general solution is

$$p(\boldsymbol{X}) = \frac{\rho_e}{4\pi} \int \frac{\partial^2 [u_i(\boldsymbol{X}')u_j(\boldsymbol{X}')]}{\partial X_i' \partial X_j'} \frac{d\boldsymbol{X}}{|\boldsymbol{X} - \boldsymbol{X}'|} + F(\boldsymbol{X}) , \tag{16.14}$$

where $F(\boldsymbol{X})$ is an arbitrary harmonic function and integration is performed over the whole volume of the fluid. If the flow is taking place in an unbounded fluid, then from the requirement of finiteness of both pressure and velocity at the infinity, $F = \text{const}$ follows. Since pressure is always determined up to the hydrostatic constant, it is natural to take $F = 0$.

Suppose the flow is stationary and the external forces are absent. Then, ratio of the inertia term and the viscous friction term in (16.9) is equal by the order of magnitude to the Reynolds number

$$\text{Re} = \frac{UL}{\nu_e} , \tag{16.15}$$

where U is the characteristic velocity and L is the characteristic linear scale.

The Reynolds number is an important flow parameter that determines the relative role of inertia and friction forces. At $\text{Re} \ll 1$, the inertia force in the momentum equation can be neglected and the Navier–Stokes equations reduce to the linear Stokes equations, which considerably simplifies their solution.

At $\text{Re} \gg 1$, the inertial force exceeds the friction force everywhere with the exception of the thin layers of thickness $\delta \sim L/\text{Re}^{1/2}$ (they are called viscous boundary layers) adjacent to the rigid boundaries of the flow volume. The other peculiarity of a flow with a high Reynolds number is that the effect of inertia forces leads energy to transfer from large-scale components of the flow to small-scale components, causing the formation of sharp local inhomogeneities in the flow.

For the stationary flow of a viscous incompressible fluid in the absence of external forces, the Reynolds number is the only similarity criterion. For non-stationary flows characterized by a time scale T different from L/U and for flows that are considerably affected by external forces such as gravity, there are additional similarity criteria.

In order to solve the Navier–Stokes equations, it is necessary to specify boundary conditions. The zero relative velocity condition holds at rigid boundary surfaces (i.e., flow velocity $\boldsymbol{u}$ must be equal to the velocity of the boundary's motion). For a boundary between two fluid phases, we require the equality of velocities and stresses on both sides of the boundary. The effect exerted by the boundary conditions will be different at low and high Reynolds numbers. At low Reynolds numbers, small changes (perturbations) of boundary conditions (for example, velocity values or shape of the boundary) lead to small changes of flow parameters, whereas at high Reynolds numbers, even small variations of boundary conditions may completely change the character of the flow.

Remember that the term "high Reynolds number" is intrinsically ambiguous. For example, while the transition from a laminar flow to a turbulent one is known to occur at high Reynolds numbers Re_{cr} (they are referred to as critical values), it cannot be characterized by one universal value of $\mathrm{Re}_{\mathrm{cr}}$ because the values of $\mathrm{Re}_{\mathrm{cr}}$ can differ considerably for various flow types.

16.3 Equations of Heat Inflow, Heat Conduction and Diffusion

In problems involving non-isothermal flows, one has to add to the system of equations (16.1)–(16.6) an equation describing the temperature change such as the heat inflow equation, which represents the physical law of energy conservation:

$$\rho_e \frac{D}{Dt}\left(\frac{u^2}{2} + e\right) = \nabla \cdot (\boldsymbol{T} \cdot \boldsymbol{u}) - \nabla \cdot \boldsymbol{q} + \rho \boldsymbol{F} \cdot \boldsymbol{u} + \rho q_\varepsilon , \tag{16.16}$$

where e is the internal energy density, $\boldsymbol{q}$ is the heat flux vector, and q_ε is the density of internal heat sources (Joule heat, heat due to chemical reactions and so on).

If the heat flux $\boldsymbol{q}$ is caused by heat conduction, then by using Fourier's law $\boldsymbol{q} = -\kappa \nabla \vartheta$ and employing the Navier–Stokes equations, we write (16.16) as

$$\rho_e \frac{D}{Dt}\left(\frac{u^2}{2} + e\right) = \boldsymbol{T} : \boldsymbol{E} + \nabla \cdot (\kappa \nabla \vartheta) + \rho q_\varepsilon , \tag{16.17}$$

where κ is thermal conductivity (also known as heat conduction coefficient); $\boldsymbol{T} : \boldsymbol{E}$ is the complete (double-dot) scalar product of stress and strain velocity tensors,

$$\boldsymbol{T} : \boldsymbol{E} = T_{ij} E_{ij} = -p \frac{\partial u_i}{\partial x_i} + \rho_e \varepsilon$$

and ε is the so-called dissipative function characterizing the rate of dissipation of a part of kinetic energy into heat:

$$\varepsilon = \frac{1}{2} \nu_e \sum_{i,j} \left(\frac{\partial u_i}{\partial x_j} + \frac{\partial u_j}{\partial x_i}\right)^2 . \tag{16.18}$$

In other words, ε is the amount of heat released in a unit volume per unit time due to the viscous effect. For incompressible flows, $\boldsymbol{T} : \boldsymbol{E} = T_{ij} E_{ij} = -p \partial u_i / \partial x_i + \rho_e \varepsilon$. The internal energy density of a fluid is equal to $e = c_v \vartheta + e_0$ where c_v is thermal capacity (specific heat at constant volume) and e_0 is an additive constant. If thermal expansion of the fluid is neglected, and in the absence of internal sources of heat ($q_\varepsilon = 0$), the equation of heat inflow becomes

$$\frac{\partial \vartheta}{\partial t} + \boldsymbol{u} \cdot \nabla \vartheta = \chi \Delta \vartheta + \frac{\varepsilon}{c_p} , \tag{16.19}$$

where $\chi = \kappa / \rho_e c_p$ is thermal diffusivity. The derivation of (16.19) hinges upon the thermodynamic relation $p/\rho_e = (c_p - c_v)\vartheta$.

The second term on the right-hand side of (16.19) has the meaning of the amount of heat released in a unit volume per unit time due to internal friction forces. In the majority of hydrodynamic problems, this term is negligibly small and can be dropped. As a result, (16.19) reduces to a well-known convective heat conduction equation

$$\frac{\partial \vartheta}{\partial t} + \boldsymbol{u} \cdot \nabla \vartheta = \chi \Delta T \,. \tag{16.20}$$

Keep in mind that while it is safe to neglect the heat produced as a result of viscous dissipation when calculating temperature distribution, viscous dissipation by itself (or, to be more precise, specific dissipation ε) has proved to be a very important physical parameter of the turbulent flow. It will be shown later that this parameter characterizes the local structure of turbulence. A more casual term, "energy dissipation", is often used for this parameter.

The heat conduction equation (16.20) will have the same look as the diffusion equation if the temperature $\vartheta(\boldsymbol{X}, t)$ is replaced by impurity concentration $C(\boldsymbol{X}, t)$ and thermal diffusivity coefficient by the coefficient of molecular diffusion D_{m}:

$$\frac{\partial C}{\partial t} + \boldsymbol{u} \cdot \nabla C = D_{\mathrm{m}} \Delta C \,. \tag{16.21}$$

If the impurity present in the fluid (it can be a soluble or an insoluble substance) does not exert any influence on the dynamics of the flow and on the rheological properties of the fluid, it is called a passive impurity. So, the assumption that the impurity is passive means that velocity $\boldsymbol{u}$ in the diffusion equation (16.21) can be determined independently from the hydrodynamic equations. A similar assertion is true for the heat conduction equation when the viscosity coefficient does not depend on the temperature.

The form of the heat conduction and diffusion equations suggests the introduction of an additional similarity parameter known as the Peclet number. For the first equation, it is called the thermal Peclet number and for the second the diffusion, the Peclet number

$$\mathrm{Pe}_T = \frac{UL}{\chi} \,, \quad \mathrm{Pe}_D = \frac{UL}{D_{\mathrm{m}}} \,. \tag{16.22}$$

These parameters characterize the ratio of the convective term to the thermal diffusivity or to the diffusion term. Thus, the Peclet number has the same physical meaning as the Reynolds number. The heat conduction or diffusion term dominates at $\mathrm{Pe} \ll 1$, while convective transfer prevails at $\mathrm{Pe} \gg 1$. As in the case of a high Reynolds number, a thin boundary layer called the thermal or diffusion boundary layer forms at the surface of a body placed into the fluid if $\mathrm{Pe} \gg 1$. Two other dimensionless parameters can be introduced in addition to the Peclet number: the Prandtl number $\mathrm{Pr} = \nu_{\mathrm{e}}/\chi$ and the Schmidt number $\mathrm{Sc} = \nu_{\mathrm{e}}/D$. These parameters are related to the Reynolds and Peclet numbers through the expressions

$$\mathrm{Pe}_T = \mathrm{Re}\mathrm{Pr} \,, \quad \mathrm{Pe}_D = \mathrm{Re}\mathrm{Sc}. \tag{16.23}$$

16.4
The Condition for the Beginning of Turbulence

The equations of motion (16.1)–(16.6) and heat inflow (16.21) are employed in calculations of various flows, particularly in flows in pipes and flows around solid objects. However, the solutions obtained show poor agreement with the actual behavior of flows. For example, the flow in a circular pipe known as the Poiseuille flow shows a good match between the theory and the experiment only for sufficiently low Reynolds numbers. The theoretical solution of the problem about the boundary layer on a plate (i.e., on a flat surface) known as a Blasius solution agrees with experimental data only at sufficiently low UX/ν_e where U is the velocity of the bulk flow and X is the distance from the leading edge of the plate. The situation is similar for many other flows.

As a rule, theoretical solutions of hydrodynamic equations can adequately describe the flows observed in experiments only under some special conditions. When these are not met, the flow can change its character. A smooth variation of hydrodynamic parameters, which is suggested by theoretical solution, can turn into disorderly fluctuations (both spatial and temporal) of these parameters as the flow changes from laminar to turbulent.

The general condition for the beginning of turbulence has been established by Reynolds. Namely, the flow remains laminar until the Reynolds number $\mathrm{Re} = UL/\nu_e$ stays below some critical value Re_{cr}, whereas at $\mathrm{Re} > \mathrm{Re}_{cr}$, the flow becomes turbulent.

It is easy to see why the Reynolds number is responsible for the transition from the laminar to the turbulent flow if we recall that this is equal by order of magnitude to the ratio between the inertial to viscous term in the momentum equation (16.6) and that at $\mathrm{Re} \gg 1$, the inertial term becomes dominant. Inertial forces draw together fluid volumes initially located far apart, promoting the formation of sharp inhomogeneities in the flow. Viscous forces, on the other hand, tend to smooth small inhomogeneities. Therefore, at $\mathrm{Re} \ll 1$, when viscous forces prevail over the inertial ones, there are no sharp inhomogeneities and the flow is laminar. As the Reynolds number increases, inertial forces prevail, the smoothing effect of viscosity becomes less pronounced, there appear small-scaled regions with sharp flow inhomogeneities, and disordered fluctuations are observed in the flow. At first, these fluctuations decay shortly after their birth, but starting from some Reynolds number Re_{cr} called the critical number, these fluctuations increase and the flow becomes turbulent.

Experimental research shows that Re_{cr} is not a universal constant that would apply to all kinds of flows. Even repeated measurements for one and the same flow under identical conditions produce different values of Re_{cr}. Such a spread in the values of Re_{cr} is explained by the fact that the critical Reynolds number depends not only on the type of flow, but also on the amount of perturbation ("initial turbulence") to the laminar flow at the tube entrance to the pipe or at the leading edge of the plate. These perturbations decay at $\mathrm{Re} < \mathrm{Re}_{cr}$, but increase at $\mathrm{Re} > \mathrm{Re}_{cr}$. The value of Re_{cr} for a pipe flow is approximately equal to 2800. Perturbations can also be caused by small protuberances and other irregularities on the wall surface.

An increase of Re_{cr} helps to postpone the emergence of turbulence. This can be achieved by special technical procedures, for example, by grinding the wall surface, by optimization of the hydrodynamic conditions at the entrance to the pipe, and so on. Multiple studies of flows in boundary layers around rigid bodies have produced similar conclusions about the conditions for the beginning of turbulence.

16.5 Hydrodynamic Instability

Any theoretical analysis of the condition for the emergence of turbulence should begin with the observation that the velocity and pressure fields in laminar and turbulent flows are solutions of hydrodynamic equations with the corresponding initial and boundary conditions. The laminar flow is described by stationary solutions of these equations, while the turbulent flow should be described by non-stationary equations. Hydrodynamic equations have a stationary solution for any Reynolds number. However, not every solution corresponds to an actually existing flow. This is due to the fact that real flows should also be stable in addition to obeying the hydrodynamic equations. This means that small perturbations appearing in the flow must decay with time without changing the general character of the flow. If perturbations increase with time, then such a flow cannot exist for a long time. At some point, it will evolve into a different flow.

This is precisely what happens when a laminar flow becomes turbulent. Therefore, we should expect the critical Reynolds number to be a criterion of stability, in other words, at $\mathrm{Re} < \mathrm{Re}_{cr}$, the laminar flow will be stable. However, at $\mathrm{Re} > \mathrm{Re}_{cr}$, it will become unstable and small perturbations that always exist in a flow will eventually make it turbulent. Thus, our task is to formulate the mathematical problem of the stability of hydrodynamic equations (16.1) and (16.6) describe the laminar fluid flow in order to obtain the theoretical value of Re_{cr}.

One valuable theoretical method of examining stability of the flow is the method of small perturbations, whose essence consists of the following. Let $U_i(\boldsymbol{x}, t)$ and $P(\boldsymbol{x}, t)$ be particular solutions of the Navier–Stokes equations, and let u'_i and p' be small perturbations of these fields appearing in the flow at the initial moment such that $u'_i \ll U_i$ and $p' \ll P$. The resulting fields of velocity $u_i = U_i + u'_i$ and pressure $p = P + p'$ also obey (16.1) and (16.6). Consider a flow in absence of external forces. By substituting $u_i = U_i + u'_i$ and $p = P + p'$ into equations and neglecting second-order terms, one obtains the following linear equations for perturbations u' and p'_i:

$$\frac{\partial u'_i}{\partial X_i} = 0\,, \tag{16.24}$$

$$\frac{\partial u'_i}{\partial t} + U_k \frac{\partial u'_i}{\partial X_k} + u'_k \frac{\partial U_i}{\partial X_k} = -\frac{1}{\rho_e}\frac{\partial p'}{\partial X_i} + \nu_e \Delta u'_i \tag{16.25}$$

or, in vector form,

$$\nabla \cdot \boldsymbol{u}' = 0 \,, \tag{16.26}$$

$$\frac{\partial \boldsymbol{u}'}{\partial t} + (\boldsymbol{U}\nabla)\boldsymbol{u}' + (\boldsymbol{u}'\nabla)\boldsymbol{U} = -\frac{1}{\rho_e}\nabla p + \nu_e \Delta \boldsymbol{u}' \,. \tag{16.27}$$

The boundary condition on a rigid surface is $\boldsymbol{u}' = 0$.

By differentiating (16.25) with respect to x_i, summarizing the result over i and using the continuity equation (16.24), we obtain (16.13). Therefore, the general solution of equations (16.24) and (16.25) will be determined once we set the initial values of velocity perturbations $u_i'(\boldsymbol{x}, 0)$. After solving equations (16.24) and (16.25), we can find conditions under which perturbations will not decay in time. These are the hydrodynamic instability conditions, that is, the conditions for the transformation of the flow from laminar to turbulent.

If the solutions $\boldsymbol{U}(\boldsymbol{x})$ and $P(\boldsymbol{x})$, whose stability are investigated, are time-independent, the system of equations (16.26) and (16.27) allows a solution in the form

$$\boldsymbol{u}'(\boldsymbol{X}, t) = e^{-i\omega t} f_\omega(\boldsymbol{X}), \quad p'(\boldsymbol{X}, t) = e^{-i\omega t} g_\omega(\boldsymbol{X}) \,. \tag{16.28}$$

Here, ω is the complex frequency, f_ω and g_ω, which are amplitudes that must be found by solving the eigenvalue problem for the system of linear partial differential equations.

If the coefficients of this system do not depend on some spatial coordinates, the number of unknown variables can be reduced by assuming an exponential dependence of f_ω and g_ω on this coordinate. Thus, when an undisturbed flow only depends on one coordinate X_3, we can write

$$f_\omega(\boldsymbol{X}) = e^{i(k_1 X_1 + k_2 X_2)} \tilde{f}(X_3) \,; \quad g_\omega(\boldsymbol{X}) = e^{i(k_1 X_1 + k_2 X_2)} \tilde{g}(X_3) \,.$$

The eigenvalue problem reduces to a system of ordinary differential equations and we can find the eigenfrequencies ω. If the flow region is finite, the eigenfrequencies form a discrete set. For the flow to be stable, it is necessary and sufficient that all eigenfrequencies should satisfy the condition $\mathrm{Im}(\omega) < 0$.

Since the Reynolds number enters the equations (16.26) and (16.27) written in dimensionless form, the eigenfrequencies functionally depend on Re as a parameter. Since at $\mathrm{Re} \to 0$ (the rest state) the flow is stable and at $\mathrm{Re} \to \infty$ it should be unstable, there always exists a $\mathrm{Re}_{\mathrm{cr}}$ at which the stable flow changes to an unstable one. It means that as Re increases, imaginary parts of some eigenfrequencies should increase and become positive. Because different eigenvalues can change their sign at different values of Re, we should take the smallest of these critical Reynolds numbers as $\mathrm{Re}_{\mathrm{cr}}$.

There are papers devoted to the subject of hydrodynamic stability that contain solutions of the stability problem for various flows: flow taking place between two rotating cylinders, convective flow in a fluid layer heated from below, plane-parallel flows, flow in pipes, boundary layer flows, and so on. Comparisons with the corresponding experimental data shows that theoretical values of $\mathrm{Re_{cr}}$ do not always agree with experimental values. For example, the theory of stability of a plane Poiseuille flow gives a noticeably higher value of $\mathrm{Re_{cr}}$ than the one obtained in experiments on turbulent flows in a plane channel. A large discrepancy between theoretical and experimental values of $\mathrm{Re_{cr}}$ for this and other flows shows that the transition from a laminar flow to a turbulent one may not always be described by the linear perturbation theory. In the above-considered formulation of the problem, perturbations of hydrodynamic quantities were assumed to be small. However, in case of an unstable flow, an initially small perturbation can become finite after a while, and then the small perturbation theory will no longer be applicable. Hence, the linear theory of small perturbation is capable of describing the initial stage of turbulence only, and cannot give a complete picture of the process. For finite perturbations, the stability problem reduces to a system of nonlinear equations. This is why the nonlinear theory of the beginning of turbulence is very complex and does not offer a complete solution of the problem.

16.6 The Reynolds Equations

The primary characteristic feature of a turbulent flow is the disordered (random) character of its fluctuating hydrodynamic parameters. As a result, the dependence of these parameters on spatial coordinates at a given time as well as the dependence on time at a given spatial point is highly complex and difficult to handle. Besides, even if we reproduce one and the same flow under the same conditions, hydrodynamic parameters will still assume different values. In practice, one has no choice but to consider a set of similar flows, assuming that hydrodynamic parameters are random variables. This means that in a turbulent flow, an individual (deterministic) description of hydrodynamic fields of velocity, pressure, and so on, is practically impossible, and reliance on statistical methods becomes unavoidable. We can then define the turbulent flow as the one for which there exists a statistical ensemble of similar flows characterized by known probability distributions with continuous probability density functions (PDFs) for the hydrodynamic fields.

In practice, it is quite unnecessary to know all the minutiae of hydrodynamic fields as we are primarily interested in their average characteristics. So we have a powerful reason to employ the averaging methods which will allow us to operate with smooth and reproducable average values of the flow parameters.

As we discussed in Appendix C.7, ensemble averaging can be replaced by averaging over the time by spatial averaging thanks to the ergodic hypothesis. Further on, the validity of the ergodic hypothesis will be assumed by default.

The simplest statistical characteristics of random hydrodynamic fields are their average values $\langle u \rangle$, $\langle p \rangle$ and so on. We shall reserve the term "fluctuations" for the deviations of individual values from their averages, for example, $u' = u - \langle u \rangle$, $p' = p - \langle p \rangle$, and so on. Then, any hydrodynamic field can be expressed as a sum of the average value and the fluctuation:

$$u = \langle u \rangle + u' , \quad p = \langle p \rangle + p' \quad \text{etc} . \tag{16.29}$$

Average values behave in a rather smooth manner, while fluctuations are characterized by intense spatial and temporal "jumps". Fluctuations define turbulent inhomogeneities. Note that the scale and period of inhomogeneities can, generally speaking, be arbitrarily small. However, small-scale inhomogeneities must be accompanied by large velocity gradients, which requires high expenditures of energy to overcome friction forces that become quite considerable on such small scales. Thus, the existence of microflows on very low scales is almost impossible. This is why turbulent motions should be characterized by minimum scales and minimum periods of inhomogeneities.

For many turbulent flows inside pipes, the characteristic minimum scale of fluctuation ranges from 0,1–1 mm. On distances comparable to the minimum scale fluctuations and on time intervals comparable to the minimum period of fluctuations, all hydrodynamic fields vary slowly and can be described by differentiable functions. Therefore, a description of turbulent flows by means of differential equations is quite possible. However, direct application of these equations proves to be very difficult and sometimes outright impossible because hydrodynamic fields in a turbulent flow are non-stationary and depend on initial conditions. Thus, even small perturbations will lead to unstable solutions. Therefore, conventional hydrodynamic equations are all but useless for calculating individual hydrodynamic fields. However, this does not mean that hydrodynamic equations cannot be used at all. Hydrodynamic equations have proved to be extremely useful for obtaining connections between statistical characteristics of turbulent hydrodynamic fields. The simplest of these connections were first established by Reynolds, who averaged the equations of motion of viscous incompressible fluid. The resulting equations are known as the Reynolds equations. Before we proceed to average hydrodynamic equations, we must formulate the basic rules for the averaging of hydrodynamic fields that were first established by Reynolds.

$$\langle f + g \rangle = \langle f \rangle + \langle g \rangle , \tag{16.30}$$

$$\langle a f \rangle = a \langle f \rangle , \quad \text{if} \quad a = \text{const} , \tag{16.31}$$

$$\langle a \rangle = a , \quad \text{if} \quad a = \text{const} , \tag{16.32}$$

$$\left\langle \frac{\partial f}{\partial s} \right\rangle = \frac{\partial \langle f \rangle}{\partial s} , \quad s: \text{coordinate or time} \tag{16.33}$$

$$\langle \langle f \rangle g \rangle = \langle f \rangle \langle g \rangle . \tag{16.34}$$

Equations 16.30–(16.34) are called the Reynolds rules. Taking in consecutive order $g = 1$, $g = \langle u \rangle$, and $g = u' = u - \langle u \rangle$, we derive additional rules from

(16.30)–(16.34):

$$\langle\langle f\rangle\rangle = \langle f\rangle\,, \quad \langle f'\rangle = \langle f - \langle f\rangle\rangle = 0\,, \quad \langle\langle f\rangle\langle u\rangle\rangle = \langle f\rangle\langle u\rangle\,,$$
$$\langle\langle f\rangle u'\rangle = \langle f\rangle\langle u'\rangle = 0\,. \tag{16.35}$$

Let us proceed to average the Navier–Stokes equations (16.2) and (16.7), taking

$$\boldsymbol{u} = \langle\boldsymbol{u}\rangle + \boldsymbol{u}'\,, \quad p = \langle p\rangle + p'\,, \tag{16.36}$$

and using the continuity equation (16.2) to transform the second term in the left-hand side of (16.35) to

$$u_k\frac{\partial u_i}{\partial X_k} = \frac{\partial}{\partial X_k}(u_i u_k)\,.$$

By applying the Reynolds rules, we obtain:

$$\frac{\partial\langle u_i\rangle}{\partial X_i} = 0\,, \tag{16.37}$$

$$\frac{\partial\langle u_i\rangle}{\partial t} + \frac{\partial}{\partial X_k}\left(\langle u_i\rangle\langle u_k\rangle + \langle u_i' u_k'\rangle\right) = -\frac{1}{\rho_e}\frac{\partial\langle p\rangle}{\partial X_i} + \nu_e\Delta\langle u_i\rangle + \langle f_i\rangle\,. \tag{16.38}$$

Equations 16.37 and (16.38) are called Reynolds equations. The advantage of these equations as compared to (16.2) and (16.7) is that they operate on smoothly varying averaged quantities. At the same time, they contain new unknown variables $\langle u_i'\, u_k'\rangle$, characterizing fluctuational components of velocity. One should think of them as components of a second-rank correlation tensor (see Appendix C.10). The appearance of new unknowns is the consequence of nonlinearity of the Navier–Stokes equations.

The physical meaning of the terms with new unknowns becomes apparent if one carries them to the right-hand side of (16.38) and combines them with the viscous term:

$$\frac{\partial\langle u_i\rangle}{\partial t} + \langle u_k\rangle\frac{\partial\langle u_i\rangle}{\partial X_k} = -\frac{1}{\rho_e}\frac{\partial\langle p\rangle}{\partial X_i} + \frac{\partial}{\partial X_k}\left(\nu_e\frac{\partial\langle u_i\rangle}{\partial X_k} - \langle u_i' u_k'\rangle\right) + \langle f_i\rangle\,. \tag{16.39}$$

By comparing the resulting equation with (16.3), one concludes that the stress tensor is not represented by the viscous stress tensor $\sigma_{ij} = 2\mu_e E_{ij}$ as in the laminar flow, but by the tensor

$$\tau_{ij} = 2\mu_e E_{ij} - \rho_e\left\langle u_i' u_j'\right\rangle = \sigma_{ij} + \tau_{ij}^{(1)}\,. \tag{16.40}$$

It follows from (16.40) that turbulence gives rise to additional stresses $\tau_{ij}^{(1)}$, which are induced by turbulent fluctuations. These additional stresses are called the Reynolds stresses. Turbulent stresses are examined using the same method that is commonly applied in hydromechanics. In particular, one can show that the quantities $\rho_e \langle u'_i \, u'_j \rangle$ stand for the normal components of turbulent stresses at $i = j$ and the tangential components at $i \neq j$. In this context, we should mention that the effect of turbulent mixing on the averaged flow is similar to the effect of viscosity because turbulent fluctuations promote additional momentum transfer from one fluid volume to another in the same way as molecular viscosity forces promote the transport of momentum in the kinetic molecular theory.

When considering the Reynolds stress tensor, we are primarily interested in the tensor component describing the transfer of momentum from the flow to the body placed into the flow because momentum transfer characterizes the friction force (the drag force) acting on the body. Let us look at a simple case: a plane wall $X_3 = 0$ is placed into a turbulent flow which is moving along the X_1-axis parallel to the wall. Then, the friction force acting on a unit area of the wall is directed along the X_1-axis and equal to

$$\tau_0 = (\langle \sigma_{13} \rangle - \rho_e \langle u'_1 u'_3 \rangle)_{X_3=0} \,, \quad \langle \sigma_{13} \rangle = \rho_e \nu_e \left(\frac{\partial \langle u_1 \rangle}{\partial X_3} + \frac{\partial \langle u_3 \rangle}{\partial X_1} \right) .$$

The zero relative velocity condition is satisfied at the wall surface. Therefore, u'_i, $\langle u'_i \rangle$ and their derivatives with respect to X_1 are all equal to zero at $X_3 = 0$ and

$$\tau_0 = \rho_e \nu_e \left(\frac{\partial \langle u_1 \rangle}{\partial X_3} \right)_{X_3=0} . \tag{16.41}$$

Near the wall, the average flow velocity is directed parallel to the wall, and the friction stress is equal to

$$\tau = \rho_e \nu_e \frac{\partial \langle u_1 \rangle}{\partial X_3} - \rho_e \langle u'_1 u'_3 \rangle . \tag{16.42}$$

Suppose that

$$-\rho_e \langle u'_1 u'_3 \rangle = \rho_e \nu_t \frac{\partial \langle u_1 \rangle}{\partial X_3} . \tag{16.43}$$

The factor ν_t has the dimensionality of a kinematic viscosity coefficient (m^2/s) and, by analogy, is called the coefficient of turbulent viscosity. In contrast to the ordinary (molecular) viscosity coefficient ν_e, the turbulent viscosity coefficient ν_t characterizes statistical properties of fluctuation motion rather than physical properties of the fluid. In the general case, ν_t does not remain constant but varies in space and in time. The turbulent viscosity coefficient is much larger than the molecular viscosity coefficient since $\nu_t/\nu_e \sim \mathrm{Re} \gg 1$.

By virtue of (16.43), the friction stress near the wall can be written as

$$\tau = \rho_e (\nu_e + \nu_t) \frac{\partial \langle u_1 \rangle}{\partial X_3} , \tag{16.44}$$

while far away from the wall, we have $\nu_t \gg \nu_e$ and $\tau \approx \tau^{(1)} = -\rho_e \langle u'_1 \, u'_3 \rangle$.

Hence, the Reynolds equations (16.39) are essentially the equations of conservation of momentum for a turbulent flow, and the Reynolds stresses describe the turbulent transport of momentum.

Similarly, we can derive the equations of conservation of other substances such as heat and matter. Taking $\vartheta = \langle\vartheta\rangle + \vartheta'$ in (16.20)–(16.21), where ϑ may be temperature or concentration, and averaging these equations, we get

$$\frac{\partial\langle\vartheta\rangle}{\partial t} + \frac{\partial}{\partial X_i}\left(\langle u_i\rangle\langle\vartheta\rangle + \langle u_i'\vartheta'\rangle\right) = \chi\Delta\langle\vartheta\rangle$$

or

$$\frac{\partial\langle\vartheta\rangle}{\partial t} + \langle u_i\rangle\frac{\partial\langle\vartheta\rangle}{\partial X_i} = \frac{\partial}{\partial X_i}\left(\chi\frac{\partial\langle\vartheta\rangle}{\partial X_i} - \langle u_i'\vartheta'\rangle\right). \tag{16.45}$$

The last equation is written in the divergent form (16.39) and ϑ can be either temperature or concentration of the passive impurity. In the former case, χ has the meaning of thermal diffusivity and in the latter, it is the diffusion coefficient.

The equations of heat and mass transfer have the same structure as the Reynolds equation. In these two equations, an additional flux caused by turbulent fluctuations appears: the heat flux $-c_p\rho_e\langle\vartheta' u_i'\rangle$ or the mass flux of passive impurities $-\rho_e\langle\vartheta' u_i'\rangle$. By analogy with (16.43), the additional heat flux can be written as

$$-c_p\rho\langle\vartheta' u_i'\rangle = c_p\rho\chi_t\frac{\partial\langle\vartheta\rangle}{\partial X_i}, \tag{16.46}$$

where ϑ is temperature and χ_t is turbulent thermal diffusivity. By the same token, we write for the mass transfer of passive impurity:

$$-\rho\langle\vartheta' u_i'\rangle = -\rho D_t\frac{\partial\langle\vartheta\rangle}{\partial X_i}, \tag{16.47}$$

where ϑ is the concentration of passive impurities and D_t is the turbulent diffusion coefficient.

We have defined the coefficients of turbulent viscosity ν_t, thermal diffusivity χ_t, and diffusion D_t for the case of one-dimensional flow. For multidimensional flows, these coefficients will be tensor, rather than scalar, quantities.

The appearance of additional terms containing Reynolds stresses $\tau_{ij}^{(1)}$ in the Reynolds equations means that the system of equations will no longer be a closed. In order to obtain a closed system of equations, one should write down additional equations that would describe $\tau_{ij}^{(1)}$.

A general method that aims at deriving the necessary equations for the Reynolds stresses has been proposed by Keller and Friedman. However, in each of the newly derived equations, new unknowns still appear whose determination in its turn requires new equations. The resulting system of equations (the Keller–Friedman chain) becomes infinite because any finite subsystem turns out to be unclosed. Nevertheless, the equations for $\tau_{ij}^{(1)}$ still lead us to some important qualitative conclusions about the properties of turbulent flows.

16.7
The Equation of Turbulent Energy Balance

Kolmogorov was the first to suggest using the energy balance equation in addition to the Reynolds equations. Since the quantity $\langle u'_i u'_j \rangle$ is a second order moment, let us employ the following general method that enables us to compile the equations for moments.

Let $u_1, u_2, \ldots, u_N$ be N hydrodynamic fields of the turbulent flow, and $\boldsymbol{X}_1, \boldsymbol{X}_2, \ldots, \boldsymbol{X}_N - N$ points in the volume filled by the fluid. The fields as well as the points might all be different, or some of them might be the same. Let us consider the N-th order moment

$$B_{u_1 u_2 \ldots u_N}(\boldsymbol{X}_1, \boldsymbol{X}_2, \ldots, \boldsymbol{X}_N, t) = \langle u_1(\boldsymbol{X}_1, t) u_2(\boldsymbol{X}_2, t) \ldots u_N(\boldsymbol{X}_N, t) \rangle .$$

By differentiating this relation with respect to time and using (16.33), one obtains

$$\begin{aligned}
&\frac{\partial}{\partial t} B_{u_1 u_2 \ldots u_N}(\boldsymbol{X}_1, \boldsymbol{X}_2, \ldots, \boldsymbol{X}_N, t) \\
&= \left\langle \frac{\partial u_1(\boldsymbol{X}_1, t)}{\partial t} u_2(\boldsymbol{X}_2, t) \ldots u_N(\boldsymbol{X}_N, t) \right\rangle \\
&\quad + \left\langle u_1(\boldsymbol{X}_1, t) \frac{\partial u_2(\boldsymbol{X}_2, t)}{\partial t} \ldots u_N(\boldsymbol{X}_N, t) \right\rangle \\
&\quad + \ldots + \left\langle u_1(\boldsymbol{X}_1, t)\, u_2(\boldsymbol{X}_2, t) \ldots \frac{\partial u_N(\boldsymbol{X}_N, t)}{\partial t} \right\rangle .
\end{aligned} \tag{16.48}$$

By eliminating the derivatives $\partial u_i(\boldsymbol{X}_j, t)/\partial t$ on the right-hand side with the help of (16.9), we obtain a balance equation for the moment $B_{u_1 u_2 \ldots u_N}$ in the form of a combination of hydrodynamic fields and spatial coordinates. Let us first apply this method to the unaveraged quantities $\rho u_i u_j$. From the self-obvious equality

$$\frac{\partial \rho u_i u_j}{\partial t} = \rho u_i \frac{\partial u_j}{\partial t} + \rho u_j \frac{\partial u_i}{\partial t} ,$$

and from the momentum equation (16.9) expressed in the form

$$\frac{\partial \rho u_i}{\partial t} + \frac{\partial}{\partial X_k}(\rho u_i u_k + p \delta_{ik} - \sigma_{ik}) = F_i ,$$

use of the continuity equation (16.2) yields

$$\begin{aligned}
&\frac{\partial (\rho u_i u_j)}{\partial t} + \frac{\partial}{\partial X_k}(\rho u_i u_j u_k + (p u_i \delta_{jk} + p u_j \delta_{ik}) - (u_i \sigma_{jk} + u_j \sigma_{ik})) \\
&= (\rho u_i F_j + \rho u_j F_i) + p \left(\frac{\partial u_i}{\partial X_j} + \frac{\partial u_j}{\partial X_i} \right) - \left(\sigma_{ik} \frac{\partial u_j}{\partial X_k} + \sigma_{jk} \frac{\partial u_i}{\partial X_k} \right) .
\end{aligned} \tag{16.49}$$

If we introduce density of kinetic energy

$$e_k = \rho u_i \frac{u_i}{2} ,$$

then for $i = j$, the equation (16.49) turns into an equation for kinetic energy

$$\frac{\partial e_k}{\partial t} + \frac{\partial}{\partial X_k}(e_k u_k + p u_k - u_i \sigma_{ki}) = p u_k X_k - \rho\varepsilon \,, \tag{16.50}$$

where

$$\rho\varepsilon = \frac{\mu_e}{2} \sum_{l,m} \left(\frac{\partial \langle u_l \rangle}{\partial X_m} + \frac{\partial \langle u_m \rangle}{\partial X_l} \right)^2$$

is the dissipation of kinetic energy in a unit volume of the fluid in a unit time.

It follows from this equation that the change of kinetic energy occurs due to the following factors: energy transfer by convective flux and the work performed by pressure forces and molecular forces (second term on the left-hand side); work of the body forces (first term on the right-hand side); viscous dissipation of energy where the density ε is given by (16.18).

The averaged continuity equation (16.37) allows one to bring the Reynolds equations into the form

$$\frac{\partial \rho \langle u_i \rangle}{\partial t} + \frac{\partial}{\partial X_k} \left(\rho \langle u_i u_k \rangle + \rho \langle u_i' u_k' \rangle + \langle p \rangle \delta_{ik} - \langle \delta_{ik} \rangle \right) = \langle F_i \rangle \,. \tag{16.51}$$

Application of this method to the moments $\rho \langle u_i u_j \rangle$ yields

$$\begin{aligned}
&\frac{\partial \rho \langle u_i \rangle \langle u_j \rangle}{\partial t} + \frac{\partial}{\partial X_k} \left(\rho \langle u_i \rangle \langle u_j \rangle \langle u_k \rangle + \rho \langle u_i' u_k' \rangle \langle u_j \rangle + \rho \langle u_j' u_k' \rangle \langle u_i \rangle \right. \\
&\quad \left. + \left(\langle p \rangle \langle u_i \rangle \delta_{jk} + \langle p \rangle \langle u_j \rangle \delta_{ik} \right) - \left(\langle u_i \rangle \langle \delta_{jk} \rangle + \langle u_j \rangle \langle \delta_{ik} \rangle \right) \right) \\
&= \left(\rho \langle u_i \rangle \langle F_j \rangle + \rho \langle u_j \rangle \langle F_i \rangle \right) + \langle p \rangle \left(\frac{\partial \langle u_i \rangle}{\partial X_j} + \frac{\partial \langle u_j \rangle}{\partial X_i} \right) \\
&\quad - \left(\langle \delta_{ik} \rangle \frac{\partial \langle u_j \rangle}{\partial X_k} + \langle \delta_{jk} \rangle \frac{\partial \langle u_i \rangle}{\partial X_k} \right) + \left(\rho \langle u_i' u_k' \rangle \frac{\partial \langle u_j \rangle}{\partial X_k} + \rho \langle u_j' u_k' \rangle \frac{\partial \langle u_i \rangle}{\partial X_k} \right) .
\end{aligned} \tag{16.52}$$

If we now introduce the density of kinetic energy of the averaged turbulent flow by the relation

$$e_s = \rho \langle u_i \rangle \langle u_i \rangle / 2 \,,$$

(16.51) turns into an equation for the density of kinetic energy of the averaged turbulent flow:

$$\begin{aligned}
&\frac{\partial e_s}{\partial t} + \frac{\partial}{\partial X_k} \left(e_s \langle u_k \rangle + \rho \langle u_k' u_i' \rangle \langle u_i \rangle + \langle p \rangle \langle u_k \rangle - \langle u_i \rangle \langle \sigma_{ki} \rangle \right) \\
&= \rho \langle u_k \rangle \langle F_k \rangle - \rho\varepsilon_s + \rho \langle u_k' u_i' \rangle \frac{\partial \langle u_i \rangle}{\partial X_k} \,,
\end{aligned} \tag{16.53}$$

where

$$\varepsilon_s = \frac{1}{\rho}\langle\sigma_{lm}\rangle\frac{\partial\langle u_l\rangle}{\partial X_m} = \frac{\nu_e}{2}\sum_{l,m}\left(\frac{\partial\langle u_l\rangle}{\partial X_m} + \frac{\partial\langle u_m\rangle}{\partial X_l}\right)^2$$

is the specific dissipation of energy of the averaged flow that occurs due to the viscous forces. The physical meaning of the terms in (16.53) is the same as the meaning of the terms in (16.50), with the exception of the term $\rho\langle u'_k u'_i\rangle\langle u_i\rangle$ which corresponds to the transport of energy by turbulent viscosity.

The equation for components of the Reynolds stress tensor can be derived by subtracting (16.52) term by term from the averaged equation (16.49):

$$\begin{aligned}
&\frac{\partial\rho\langle u_i\rangle\langle u_j\rangle}{\partial t} + \frac{\partial}{\partial X_k}\Big[\rho\left\langle u'_i u'_j\right\rangle\langle u_k\rangle + \rho\left\langle u'_i u'_k u'_j\right\rangle \\
&\quad + \left(\langle p' u'_i\rangle\delta_{jk} + \left\langle p' u'_j\right\rangle\delta_{ik}\right) - \left(\left\langle u'_i\sigma'_{jk}\right\rangle + \left\langle u'_j\sigma'_{ik}\right\rangle\right)\Big] \\
&= \rho\left\langle u'_i F'_j\right\rangle + \rho\left\langle u'_j F'_i\right\rangle + \left\langle p'\left(\frac{\partial u'_i}{\partial X_j} + \frac{\partial u'_j}{\partial X_i}\right)\right\rangle - \left\langle\sigma'_{ik}\frac{\partial u'_j}{\partial X_k} + \sigma'_{jk}\frac{\partial u'_i}{\partial X_k}\right\rangle \\
&\quad - \left(\rho\langle u'_i u'_k\rangle\frac{\partial\langle u_j\rangle}{\partial X_k} + \rho\left\langle u'_j u'_k\right\rangle\frac{\partial\langle u_i\rangle}{\partial X_k}\right).
\end{aligned} \tag{16.54}$$

We see that in addition to the average velocity $\langle u_i\rangle$ and Reynolds stresses $\rho\langle u'_i u'_j\rangle$, (16.54) contains new unknowns: third order central moments $\rho\langle u'_i u'_k u'_j\rangle$; second order moments of fluctuations of velocity and its spatial derivatives appearing in $\langle u'_i\sigma'_{jk}\rangle$ and $\langle\sigma'_{jk}(\partial u'_i/\partial X_k)\rangle$; and second order mutual moments of pressure and velocity fields $\langle p' u'_i\rangle$ and $\langle p'(\partial u'_i/\partial X_j)\rangle$. The latter moments can be represented as two-point, third order moments of the type $\langle u'_i(\mathbf{X}, t)u'_j(\mathbf{X}', t)u'_k(\mathbf{X}', t)\rangle$ with the help of (16.14).

From (16.54), one can obtain an equation for the average kinetic (turbulent) energy density of fluctuational motion:

$$e'_k = \frac{\rho\langle u'_i\rangle\langle u'_i\rangle}{2}.$$

By putting $i = j$ in (16.54), one gets:

$$\begin{aligned}
&\frac{\partial e'_k}{\partial t} + \frac{\partial}{\partial X_k}\left(e'_k\langle u_k\rangle + \frac{1}{2}\rho\langle u'_i u'_i u'_k\rangle + \langle p' u'_k\rangle - \langle u'_i\sigma'_{ki}\rangle\right) \\
&= \rho\langle u'_k F'_k\rangle - \rho\langle\varepsilon_k\rangle - \rho\langle u'_k u'_i\rangle\frac{\partial\langle u_i\rangle}{\partial X_k},
\end{aligned} \tag{16.55}$$

where

$$\langle\varepsilon_k\rangle = \frac{1}{\rho}\left\langle\sigma'_{lm}\frac{\partial u_l}{\partial X_m}\right\rangle = \frac{\nu_e}{2}\sum_{l,m}\left\langle\left(\frac{\partial u'_l}{\partial X_m} + \frac{\partial u'_m}{\partial X_l}\right)^2\right\rangle$$

is the average specific energy of fluctuational motion under the action of viscous forces.

The terms in (16.55) have the following physical meaning. The second term on the left-hand side expresses the change of turbulent energy flux density. The four summands in this term represent the contribution from energy transfer by the averaged flow, from turbulent viscosity, from pressure fluctuations, and from molecular viscosity, respectively. The term

$$A = -\rho \left\langle u'_k u'_i \right\rangle \frac{\partial \left\langle u_i \right\rangle}{\partial X_k} \tag{16.56}$$

describes energy exchange between the averaged and fluctuational motions.

Thus, (16.55) is the equation of turbulent energy balance. It follows from there that turbulent energy density at a given point inside the flow can change via the following mechanisms: transport of turbulent energy from other regions in the fluid; work performed by external force fluctuations; viscous dissipation of turbulent energy and finally, transformations of energy of the averaged motion into turbulent energy and vice versa.

One characteristic that we shall be using frequently in the subsequent discussion is the average kinetic energy of fluctuational motion per unit mass of the fluid, $e_k = e'_k/\rho = \left\langle u'_i u'_i \right\rangle /2$. The equation for e_k can easily be obtained from (16.55)

$$\begin{aligned} \frac{De_k}{Dt} &= -\left\langle u'_k u'_i \right\rangle \frac{\partial \left\langle u_i \right\rangle}{\partial X_k} - \left\langle \varepsilon_k \right\rangle + \frac{\partial}{\partial X_k} \left(-\frac{1}{2} \left\langle u'_i u'_i u'_k \right\rangle - \frac{1}{\rho} \left\langle p' u'_k \right\rangle \right) \\ &\quad + \nu_e \left\langle u'_i \left(\frac{\partial \left\langle u'_i \right\rangle}{\partial X_k} + \frac{\partial \left\langle u'_k \right\rangle}{\partial X_i} \right) \right\rangle + \left\langle u'_k F'_k \right\rangle . \end{aligned} \tag{16.57}$$

This equation, in its turn, contains new unknown quantities $\left\langle u'_i u'_i u'_k \right\rangle$, $\left\langle p' u'_k \right\rangle$ and $\left\langle \varepsilon_k \right\rangle$. Therefore, a system of equations containing the Reynolds equation (16.51) and either the equation for Reynolds stresses (16.54) or the equation for turbulent energy (16.55) will not be closed. One can always try to construct new equations for the new unknowns, but the derived system will also be unclosed because it will contain unknown moments of higher orders. Hence, construction of additional equations for higher-order moments gets us nowhere as we try to obtain a closed system of equations describing the turbulent flow. The Reynolds equations and the equation of turbulent energy balance only allow us to infer the existence of certain connections between different statistical characteristics of turbulence, but they cannot be solved.

The only way out from this situation is to attempt to close the system of equations by making additional assumptions that are based on a certain physical considerations and justified by their agreement with experimental data. In other words, we aim to specify the missing connections between statistical characteristics of turbulence irrespective of the available equations.

The "easiest" way to close the system is to simply drop the higher-order moments. It turns out, however, that this works only for relatively small Reynolds numbers that do not present any practical interest. We are interested specifically in the case of great Reynolds numbers, or, to use another term, in the case of fully developed turbulence.

In a few cases, the form of additional connections between statistical characteristics can be guessed from dimensionality considerations; the expressions derived in this way are accurate up to a certain small number of empirical constants. However, the dimensionality theory still stops halfway in solving the problem because the resulting relations contain unknown functions and (or) constants which then have to be determined experimentally. The total number of these functions and constants can be large because different functions and constants are needed for different flows (flows in pipes, flows around a solid body, flows in boundary layers, jet flows, etc.).

Yet another closure method uses transport equations to find the characteristics of turbulence such as turbulent energy, turbulent viscosity and the integral scale of turbulence. This method is the most popular as of today and is widely used in numerical calculation for different turbulent flows.

It is quite natural that one would like to make his task simpler by finding the minimum required number of additional relations, functions or constants that would be applicable all at once to many different flows. Unfortunately, as of today, we are still lacking a universal theory that would describe all kinds of turbulent flow. Inevitably, all the existing turbulent flow models are only valid for one type of flow.

Theories of turbulence that encompass relations found empirically or guessed from physical considerations and then proved by experiments (in addition to the available hydrodynamic equations) are referred to as semi-empirical theories.

The existing models of turbulence will be examined in more detail in Section 16.10. However, we must first discuss the internal structure of turbulence, and in particular, the underlying concept of isotropic turbulence.

16.8 Isotropic Turbulence

Turbulence is called homogeneous when all hydrodynamic fields are homogeneous random fields (see Section C.9), and isotropic when all hydrodynamic fields are isotropic random fields (see Section C.10). Isotropic turbulence is mathematical idealization, which is suitable for the approximate description of some special turbulent flows. In fact, turbulence can be isotropic only when the fluid occupies the entire space. Real flows always have boundaries, and this is where the isotropy ends. The case of isotropic turbulence is the simplest one, yet it allows us to establish some distinguishing properties of turbulence. This explains why the concept of isotropic turbulence, which was first introduced by Taylor, has played such a crucial role in the development of the modern theory of statistical turbulence. Later on, Kolmogorov proposed a more general concept of local isotropic turbulence which embraced a greater variety of real flows and has since proved itself to be a powerful tool for the analysis of various turbulent flows.

Consider isotropic turbulence in a viscous incompressible fluid in the absence of external forces. As discussed in Sections 16.5 and 16.7, we are primarily interested in the components of correlation tensors which according to the definition of a homogeneous, isotropic random vector field, depend on $\boldsymbol{r} = \boldsymbol{X}' - \boldsymbol{X}$ and t, where $\boldsymbol{X}'$ and $\boldsymbol{X}$ are two arbitrary points in space. For a homogeneous, isotropic random velocity field, the second order correlation tensor

$$B_{ij}(\boldsymbol{r}, t) = \left\langle u_i(\boldsymbol{X}, t) u_j(\boldsymbol{X} + \boldsymbol{r}, t) \right\rangle$$

is expressed in terms of two scalar functions, $B_{LL}(r, t)$ and $B_{NN}(r, t)$:

$$B_{ij}(\boldsymbol{r}, t) = \left[B_{LL}(r, t) - B_{NN}(r, t) \right] \frac{r_i r_j}{r^2} + B_{NN}(r, t) \delta_{ij} , \tag{16.58}$$

where $B_{LL}(r, t) = \langle u_L(\boldsymbol{X}, t) u_L(\boldsymbol{X} + \boldsymbol{r}, t) \rangle$ and $B_{NN}(r, t) = \langle u_N(\boldsymbol{X}, t) u_N(\boldsymbol{X} + \boldsymbol{r}, t) \rangle$ are the longitudinal and transverse correlation functions; u_L and u_N are projections of the velocity vector $\boldsymbol{u}$ onto the directions parallel and transverse to $\boldsymbol{r}$; and $r = |\boldsymbol{r}|$. By the same token, the third order correlation tensor

$$B_{ij,k}(\boldsymbol{r}, t) = \left\langle u_i(\boldsymbol{x}, t) u_j(\boldsymbol{x}, t) u_k(\boldsymbol{x} + \boldsymbol{r}, t) \right\rangle$$

can be expressed in terms of three scalar functions $B_{LL,L}(r, t)$, $B_{LN,N}(r, t)$ and $B_{NN,L}(r, t)$

$$\begin{aligned} B_{ij,k}(\boldsymbol{r}, t) = {} & \left[B_{LL,L}(r, t) - 2 B_{LN,N}(r, t) - B_{NN,L}(r, t) \right] \frac{r_i r_j r_k}{r^3} \\ & + B_{NN,L}(r, t) \frac{r_k}{r} \delta_{ij} + B_{LN,N}(r, t) \left(\frac{r_i}{r} \delta_{jk} + \frac{r_j}{r} \delta_{ik} \right) . \end{aligned} \tag{16.59}$$

The relations (16.58) and (16.59) can be considerably simplified if the velocity field is solenoidal or potential. The former case is true for an incompressible fluid ($\nabla \cdot \boldsymbol{u} = 0$), and the second – for an ideal fluid ($\nabla \times \boldsymbol{u} = 0$). Since the fluid is assumed to be incompressible, we have a solenoidal velocity field. Then, the continuity equation leads us to

$$B_{NN}(r, t) = B_{LL}(r, t) + \frac{r}{2} \frac{\partial}{\partial r} \left[B_{LL}(r, t) \right] , \tag{16.60}$$

which is the so-called Karman equation. One can see that the second order correlation tensor of an isotropic solenoidal vector field can be expressed in terms of a single scalar function.

The third order correlation tensor of an isotropic solenoidal vector field $B_{ij,k}(\boldsymbol{r}, t)$ is also expressed in terms of a single scalar function because

$$B_{NN,L}(r, t) = -\frac{1}{2} B_{LL,L}(r, t) .$$

$$B_{LN,N}(r, t) = \frac{1}{2} B_{LL,L}(r, t) + \frac{r}{4} \frac{\partial}{\partial r} \left[B_{LL,L}(r, t) \right] . \tag{16.61}$$

We should mention yet another important property of an isotropic turbulent vector field. Any isotropic random vector field $\boldsymbol{u}(\boldsymbol{x})$ can be represented as a sum of two mutually uncorrelated fields, one of which is solenoidal and the other a potential. The corollary is that no scalar isotropic field can correlate with a solenoidal vector field. If we choose pressure to be our scalar field and velocity to be our vector field, then this corollary reduces to the statement that

$$B_{Pi}(\boldsymbol{r}, t) = B_{PL}(\boldsymbol{r}, t)\frac{r_i}{r} = 0 \,, \tag{16.62}$$

where $B_{Pi}(\boldsymbol{r}, t) = \langle p(\boldsymbol{r}, t) u_i(\boldsymbol{r}, t)\rangle$, $B_{PL}(\boldsymbol{r}, t) = \langle p(\boldsymbol{r}, t) u_L(\boldsymbol{r}, t)\rangle$.

It is now easy to derive the dynamical equation for the correlation tensor $B_{ij}(\boldsymbol{r}, t)$. Let us apply the equations (16.9) (with $\boldsymbol{f}$ set to zero) to the i-th component of velocity u_i at the point $\boldsymbol{X}$ and to j-th component of velocity u'_j, at the point $\boldsymbol{X}+\boldsymbol{r} = \boldsymbol{X}'$. Multiply the first equation by u'_j and the second by u_i, add both equations together and average the result. The result is

$$\frac{\partial\left\langle u_i u'_j\right\rangle}{\partial t} + \frac{\partial\left\langle u_i u_k u'_j\right\rangle}{\partial X_k} + \frac{\partial\left\langle u_i u'_j u'_k\right\rangle}{\partial X'_k}$$

$$= -\frac{1}{\rho}\left(\frac{\partial\left\langle p u'_j\right\rangle}{\partial X_i} + \frac{\partial\left\langle p' u'_i\right\rangle}{\partial X'_j}\right) + \nu_e\left(\frac{\partial^2\left\langle u_i u'_j\right\rangle}{\partial X_k \partial X_k} + \frac{\partial^2\left\langle u_i u'_j\right\rangle}{\partial X'_k \partial X'_k}\right). \tag{16.63}$$

It follows from the homogeneity of turbulence that all two-point moments depend on $\boldsymbol{r} = \boldsymbol{X}' - \boldsymbol{X}$. Therefore, $\partial/\partial X_k$ and $\partial/\partial X'_k$ are respectively equal to $-\partial/\partial r_k$ and $\partial/\partial r_k$. As a result, (16.63) reduces to

$$\frac{\partial B_{ij}(\boldsymbol{r}, t)}{\partial t} = \frac{\partial}{\partial r_k}\left[B_{ik,j}(\boldsymbol{r}, t) - B_{i,jk}(\boldsymbol{r}, t)\right]$$

$$+\frac{1}{\rho}\left[\frac{\partial B_{pj}(\boldsymbol{r}, t)}{\partial r_i} - \frac{\partial B_{ip}(\boldsymbol{r}, t)}{\partial r_j}\right] + 2\nu_e\frac{\partial^2 B_{ij}(\boldsymbol{r}, t)}{\partial r_k \partial r_k}. \tag{16.64}$$

From the property of isotropy, one can deduce the relations (16.62), thus establishing that $B_{pj}(\boldsymbol{r}, t) = B_{ip}(\boldsymbol{r}, t) = 0$, and the relations (16.60), (16.61), which mean that tensors $B_{ij}(\boldsymbol{r}, t)$, $B_{ik,j}(\boldsymbol{r}, t)$ and $B_{i,jk}(\boldsymbol{r}, t) = B_{jk,i}(-\boldsymbol{r}, t)$ can be expressed through scalar functions $B_{LL}(\boldsymbol{r}, t)$ and $B_{LL,L}(r, t)$. After some algebra, we arrive at the Karman–Howarth equation

$$\frac{\partial B_{LL}(\boldsymbol{r}, t)}{\partial t} = \left(\frac{\partial}{\partial r} + \frac{4}{r}\right)\left[\partial B_{LL,L}(r, t) + 2\nu_e\frac{\partial B_{LL}(\boldsymbol{r}, t)}{\partial r}\right]. \tag{16.65}$$

Just as the Reynolds equations, (16.65) cannot be solved because it contains two unknowns $B_{LL}(r, t)$ and $B_{LL,L}(r, t)$.

Consider some important corollaries that follow from the derived equations for correlation functions. These are equations for some functions of r and t, from which one can obtain certain numerical characteristics describing turbulence as a whole (in other words, these characteristics are independent of the distance r between the two points under consideration). To this end, it is sufficient to expand the functions that appear in these equations as a Taylor series in the powers of $r$and then equate the terms having the same power.

We begin with the Karman–Howarth equation (16.65). The zeroth term of the expansion (i.e., $r = 0$) gives us

$$\frac{d B_{LL}(0)}{dt} = 10\nu_e \left(\frac{\partial^2 B_{LL}}{\partial r^2}\right)_{r=0} . \tag{16.66}$$

Since $B_{LL}(0) = \langle u^2 \rangle$, (16.66) can be rewritten in the form

$$\frac{d}{dt}\left(\frac{3}{2}\langle u^2 \rangle\right) = 15\nu_e \langle u^2 \rangle f''(0) = -\frac{15\nu_e \langle u^2 \rangle}{\lambda_t^2} , \tag{16.67}$$

where u is the velocity component along the X-axis, $f(r) = B_{LL}(r)/B_{LL}(0)$, and $\lambda_t^2 = -1/f''(0)$. Note that $B_{LL}(0) = \langle (\boldsymbol{u}(\boldsymbol{x}))^2 \rangle /3$.

Equation 16.67 represents the balance of energy for isotropic turbulence. It describes the rate of decrease of the average kinetic energy of turbulence due to the action of viscosity forces. The parameter λ_t has the dimensionality of a length and is called the Taylor microscale. It can be regarded as the smallest size of eddies, which are responsible for energy dissipation. As far as $d\langle u^2 \rangle /dt \sim \langle u^2 \rangle /\tau_t$, where τ_t is the characteristic time of hydrodynamic relaxation (the Taylor time microscale), (16.67) gives $\tau_t \sim \lambda_t^2/10\nu_e$. One can use the Karman equation to express the microscale λ_t through the transverse correlation $\lambda_t^2 = -2/g''(0)$, where $g(r) = B_{NN}(r)/B_{NN}(0)$.

The expressions for Taylor microscales can be brought to the form

$$\left\langle \left(\frac{\partial u}{\partial X}\right)^2 \right\rangle = \frac{\langle u^2 \rangle}{\lambda_t^2} , \quad \left\langle \left(\frac{\partial u}{\partial Y}\right)^2 \right\rangle = 2\frac{\langle u^2 \rangle}{\lambda_t^2} . \tag{16.68}$$

The relations (16.68) make it possible to determine λ_t by finding the values of $\langle (\partial u/\partial X)^2 \rangle$ and $\langle (\partial u/\partial Y)^2 \rangle$ from the experiment. By obtaining λ_t and $\langle u^2 \rangle$ from independent measurements (they can be conducted, for example, behind the gate in a wind tunnel at different distances from the gate), we can prove the relation (16.8.10) and the established attenuation formula for $\langle u^2 \rangle$. The averaged squares of all velocity components decrease with time in accordance with the "5/2 law", namely,

$$\langle u^2 \rangle = \frac{C}{(t - t_0)^{5/2}} , \tag{16.69}$$

where t_0 is some arbitrary chosen initial moment ("initial time reading"), and λ_t^2 increases linearly with time

$$\lambda_t^2 = 4\nu_e(t - t_0) . \tag{16.70}$$

Since $B_{LL}(r)$ is an even function of r, while $B_{LL,L}(r)$ is an odd function of r, both sides of (16.65) only contain even degrees of r. By equating coefficients by r^2 in the Taylor expansion, one gets

$$\frac{1}{2}\frac{d}{dt} B''_{LL}(0) = \frac{7}{6} B'''_{LL,L}(0) + \frac{7}{3}\nu_e B^{IV}_{LL}(0) . \tag{16.71}$$

This equation is interpreted as the balance equation for a vortex since the correlation tensor for a vortex is equal to $B_{\omega_i \omega_i}(\mathbf{r}) = -\Delta B_{ii}(\mathbf{r})$.

If $r \to \infty$, the quantity $B_{LL}(r)$ goes to zero faster than r^{-5}. Then, from (16.65) follows the relation

$$\int_0^\infty r^4 B_{LL}(r) dr = \Lambda = \text{const}\,, \tag{16.72}$$

which has the form of a conservation law. The quantity Λ is called the Loitsyansky integral (or the Loitsyansky invariant).

Similarly, using (16.13) that expresses pressure in terms of velocity, we can study the statistical properties of a scalar hydrodynamic field, that is, pressure, and by using the equations of heat conduction (16.20) and diffusion (16.21), we can study the statistical properties of scalar fields, namely, temperature and concentration (see the end of this section).

In addition to the above-mentioned Taylor microscale, the theory of turbulence introduces four other length scales: longitudinal and transverse differential scales

$$\lambda_1 = \left(-\frac{B_{LL}(0)}{2B''_{LL}(0)}\right)^{1/2}, \quad \lambda_2 = \left(-\frac{B_{NN}(0)}{2B''_{NN}(0)}\right)^{1/2} \tag{16.73}$$

and longitudinal and transverse integral scales

$$L_1 = \frac{1}{B_{LL}(0)} \int_0^\infty B_{LL}(r) dr\,, \quad L_2 = \frac{1}{B_{NN}(0)} \int_0^\infty B_{NN}(r) dr\,. \tag{16.74}$$

By comparing the relations (16.74) with the formulae (C59) and (C102), one can conclude that the integral length scale has the meaning of a characteristic correlation length, that is, the average distance that turbulent perturbations can travel. Since correlation between velocities at two different points decreases with increase of the distance between these points, the integral scale is equal by order of magnitude to the maximum distance between these points at which the velocities still show a noticeable correlation.

A further insight into isotropic turbulence can be gained by examining correlation functions in the wavenumber space. As mentioned in Section C.10, spectral representations of random functions have the meaning of superposition of harmonic oscillations for stationary random processes. For an isotropic turbulent field, the spectral representation looks especially simple. Representations for the components of the correlation tensor $B_{ij}(\mathbf{r}, t)$ and its spectral tensor $F_{ij}(\mathbf{k})$ are found from the definitions (C104) and (C105) as:

$$B_{ij}(\mathbf{r}, t) = 4\pi \int_0^\infty \frac{\sin(k\mathbf{r})}{kr} F_{ij}(k, t) k^2 dk\,, \tag{16.75}$$

$$F_{ij}(\mathbf{k}, t) = \frac{1}{2\pi^2} \int_0^\infty \frac{\sin(k\mathbf{r})}{kr} B_{ij}(r, t) r^2 dr\,. \tag{16.76}$$

The spectrum $F_{ij}(\boldsymbol{k})$ is symmetric and non-negative, and its corresponding quadratic form is positive definite. The isotropy condition implies that $F_{ij}(\boldsymbol{k})$ can be represented in the form (16.58), namely,

$$F_{ij}(\boldsymbol{k}, t) = (F_{LL}(k, t) - F_{NN}(k, t))\frac{k_i k_j}{k^2} + F_{NN}(r, t)\delta_{ij} , \tag{16.77}$$

where $F_{LL}(k, t)$ and $F_{NN}(k, t)$ are the longitudinal and transverse spectra.

The Spectral representation of the average energy is

$$\frac{1}{2}\left\langle \boldsymbol{u}^2(\boldsymbol{X}, t)\right\rangle = \frac{1}{2}B_{ii}(0, t) = \int_0^\infty E(k, t)dk . \tag{16.78}$$

In the isotropic case, the last relation becomes

$$E(k, t) = 4\pi k^2 \frac{F_{ij}(k, t)}{2} = 2\pi k^2 (F_{LL}(k, t) + 2F_{NN}(k, t)) . \tag{16.79}$$

The conditions of solenoidality and potentiality allow us to simplify the expressions for $E(k, t)$ and $F_{ij}(k, t)$ with the result

$$E(k, t) = \begin{cases} 4\pi k^2 F_{NN}(k, t) , & \text{for solenoidal field} , \\ 2\pi k^2 F_{LL}(k, t) , & \text{for potential field} , \end{cases} \tag{16.80}$$

$$F_{ij}(k, t) = \begin{cases} \dfrac{E(k)}{4\pi k^2}\left(\delta_{ij} - \dfrac{k_i k_j}{k^2}\right), & \text{for solenoidal field} , \\ \dfrac{E(k) k_i k_j}{2\pi k^4} , & \text{for potential field} . \end{cases} \tag{16.81}$$

For a solenoidal field, the longitudinal B_{LL} and transverse B_{NN} correlation functions are connected with $E(k)$ through the relations

$$B_{LL}(r) = 2\int_0^\infty \left(-\frac{\cos kr}{(kr)^2} + \frac{\sin kr}{(kr)^2}\right) E(k, t)dk ,$$

$$B_{NN}(r) = 2\int_0^\infty \left(\frac{\sin kr}{kr} + \frac{\cos kr}{(kr)^2} - \frac{\sin kr}{(kr)^3}\right) E(k, t)dk , \tag{16.82}$$

whereas $E(k)$ is expressed through B_{LL} as

$$E(k) = \frac{1}{\pi}\int_0^\infty (kr \sin kr - k^2 r^2 \cos kr) B_{LL}(r)dr . \tag{16.83}$$

Similar relations exist for the spectrum of a third order correlation tensor of an isotropic field $\boldsymbol{u}(\boldsymbol{x})$:

$$F_{ij,k}(k) = iF_{LN,N}(k)\left(\delta_{jk}\frac{k_i}{k} + \delta_{ik}\frac{k_j}{k} - 2\frac{k_i k_j k_k}{k^3}\right),$$

$$F_{LN,N}(k) = \frac{1}{8\pi^2}\int_0^{\infty}\left(\sin kr + \frac{3\cos kr}{kr} - \frac{3\sin kr}{(kr)^2}\right)B_{LL,L}(r)r^2 dr\,.$$

The Karman–Howarth equation (16.65) has the following spectral representation:

$$\frac{\partial F_{NN}(k,t)}{\partial t} = -2kF_{LN,N}(k,t) - 2\nu k^2 F_{NN}(k,t) \tag{16.84}$$

$$\frac{\partial E(k,t)}{\partial t} = -8\pi k^3 F_{LN,N}(k,t) - 2\nu k^2 E(k,t). \tag{16.85}$$

Equations 16.84 and (16.85) describe the rate of time change of the spectral distribution of isotropic turbulence energy. The second term on the right-hand side gives energy dissipation due to viscosity. The viscosity-rated increase in the dissipation of kinetic energy of a perturbation with the wave number k is proportional to the intensity of this perturbation: $2\nu k^2$ is the proportionality coefficient. Hence, the energy of long-wave perturbations (small values of k) decreases under the action of viscosity at much slower rates than the energy of short-wavelength perturbations. The reason for this is that short-wavelength perturbations produce large velocity gradients, and the viscous friction force is proportional to the velocity gradient. The first term on the right-hand side of (16.84) and (16.85) describes the energy change of the spectral component of turbulence with the wave number k due to nonlinear inertial terms of hydrodynamic equations. This change leads to redistribution of energy between spectral components without changing the total energy of turbulence. Hence, any change of the total energy of turbulence is caused exclusively by viscosity forces, that is,

$$\frac{\partial}{\partial t}\frac{\langle u_i u_j\rangle}{2} = \frac{\partial}{\partial t}\int_0^{\infty} E(k,t)dt = -2\nu_e\int_0^{\infty} k^2 E(k,t)dt\,. \tag{16.86}$$

The first term in the right-hand side of (16.85) is negative at small values of k and positive at large values of k. Therefore, turbulent mixing leads to the breakup of turbulent perturbations, that is, to energy transfer from large-scale to small-scale components, with energy being spent to overcome viscous friction. Hence, viscosity becomes a major factor for small-scale components. This fact will be used in the next section as we examine the inner structure of developed turbulence.

When looking at the inner structure of developed turbulence, we are not as much concerned with correlations between components of velocities at different points $\boldsymbol{X}+\boldsymbol{r}$ and $\boldsymbol{X}$ at a given moment of time (i.e., with components of tensor B_{ij}) as we are with correlations between components of velocity differences $\Delta_r = \boldsymbol{u}(\boldsymbol{X}+\boldsymbol{r}) -$

$-\boldsymbol{u}(\boldsymbol{r})$ at these points. For isotropic turbulence, the condition $\langle \Delta_r \boldsymbol{u} \rangle = 0$ must be valid (see Section C.10). The corresponding symmetric tensor has the components

$$b_{ij} = \langle [\boldsymbol{u}_i(\boldsymbol{X} + \boldsymbol{r}) - \boldsymbol{u}_i(\boldsymbol{r})][\boldsymbol{u}_j(\boldsymbol{X} + \boldsymbol{r}) - \boldsymbol{u}_j(\boldsymbol{r})] \rangle , \tag{16.87}$$

known as the structure functions. For simplicity's sake, we are considering stationary processes only, thus explaining the omission of the time t in (16.86).

Structure functions for an isotropic field can be written in time similar to that of (16.58)

$$b_{ij}(\boldsymbol{r}) = \left[b_{LL}(r) - b_{NN}(r) \right] \frac{r_i r_j}{r^2} + b_{NN}(r)\delta_{ij} , \tag{16.88}$$

where $b_{LL}(r, t)$ and $b_{NN}(r, t)$ are the longitudinal and transverse structure functions equal to

$$b_{LL}(r, t) = \left\langle [\boldsymbol{u}_L(\boldsymbol{X} + \boldsymbol{r}) - \boldsymbol{u}_L(\boldsymbol{r})]^2 \right\rangle , \quad b_{NN}(r, t) = \left\langle [\boldsymbol{u}_N(\boldsymbol{X} + \boldsymbol{r}) - \boldsymbol{u}_N(\boldsymbol{r})]^2 \right\rangle .$$

The longitudinal and transverse structure functions are connected with corresponding correlation functions through the relations

$$b_{LL}(r) = 2[B(0) - B_{LL}(r)] , \quad b_{NN}(r) = 2[B(0) - B_{NN}(r)] . \tag{16.89}$$

Here, $B(0) = B_{LL}(0) = B_{NN}(0) = \langle u^2 \rangle /3$.

For a solenoidal field $\boldsymbol{u}(\boldsymbol{x})$ (incompressible fluid), the longitudinal and transverse structure functions b_{LL} and b_{NN} are mutually connected through an equation similar to the Karman equation (16.60)

$$b_{NN}(r, t) = b_{LL}(r, t) + \frac{r}{2} \frac{\partial}{\partial r} [b_{LL}(r, t)] . \tag{16.90}$$

In addition to the two-points second order moments of velocity difference, b_{ij}, one can introduce two-point third order moments of velocity difference,

$$b_{ijk} = \langle [u_i(\boldsymbol{X} + \boldsymbol{r}) - u_i(\boldsymbol{r})][u_j(\boldsymbol{X} + \boldsymbol{r}) - u_j(\boldsymbol{r})][u_k(\boldsymbol{X} + \boldsymbol{r}) - u_k(\boldsymbol{r})] \rangle ,$$

which can be expressed through a single scalar function $b_{LLL}(r)$ by virtue of the isotropy condition:

$$b_{ijk} = \frac{1}{2} \left[b_{LLL}(r) - r \frac{\partial b_{LLL}(r)}{\partial r} \right] \frac{r_i r_j r_k}{r^3} + \frac{1}{6} \left[b_{LLl}(r) + r \frac{\partial b_{LLl}(r)}{\partial r} \right] \left[\frac{r_i}{r} \delta_{jk} + \frac{r_j}{r} \delta_{ik} + \frac{r_k}{r} \delta_{ij} \right] .$$

Once again, the system of hydrodynamic equations for an isotropic turbulent flow turns out not to be closed, which is evident, for example, from (16.90) that contains two unknown functions $b_{NN}(r, t)$ and $b_{LL}(r, t)$. We have to come up with additional hypotheses and relations in order to close this system.

Our previous analysis for the isotropic vector field can be extended to the case of an isotropic scalar random field, for example, the field of passive impurity concentration. Let $C(\boldsymbol{X}, t)$ be the concentration of a substance in the fluid and let $C'(\boldsymbol{X}, t)$ be its fluctuation relative to the average value $\langle C \rangle$. The fields of velocity (a vector field) and concentration (a scalar field) are assumed to be isotropic, so $\langle \boldsymbol{u} \rangle$ and $\langle C \rangle$ are constants. A theoretical examination of the turbulent scalar field can be performed in the same manner as for the vector field in the preceding discussion.

Let $C'_a = C'(\boldsymbol{X}_a, t)$ and $C'_b = C'(\mathbf{X}_b, t)$ denote concentration fluctuations at the points $\mathbf{X}_a$ and $\boldsymbol{X}_b = \boldsymbol{X}_a + \boldsymbol{r}$ at one and the same instant of time. The correlation of these quantities is $B_{ab} = \langle C'_a C'_b \rangle$ and the corresponding correlation coefficient is $\Psi_{cc} = \langle C'_a C'_b \rangle / \langle (C')^2 \rangle$. The quantity $\langle (C'_a)^2 \rangle = \langle (C'_b)^2 \rangle = \langle (C')^2 \rangle$ is called intensity of concentration fluctuations. Just as for the vector field (see (16.73) and (16.74)), we can introduce two length scales: differential scale (microscale) λ_c and integral scale (macroscale) L_c

$$\frac{2}{\lambda_c^2} = -\left(\frac{\partial^2 \psi_{cc}}{\partial r^2}\right)_{r=0} , \quad L_c = \int_0^\infty \psi_{cc}(\boldsymbol{r}) d\boldsymbol{r} . \tag{16.91}$$

Having looked at correlations between the values of one and the same scalar quantity C, we may now ask about correlations between C and components of the velocity vector u_i, either at one and the same point or at different points. It turns out that due to the fact that no scalar isotropic field can correlate with a solenoidal vector field, these correlations are absent, that is, $\langle (u_i)_a C_a \rangle = 0$ and $\langle (u_i)_a C_b \rangle = 0$.

We can also introduce correlations of higher order, for example, third order correlations at two points $\langle (u_i)_a C_a C_b \rangle$ and $\langle (u_i)_a (u_j)_b C_b \rangle$. It is obvious that $\langle (u_i C^2)_a \rangle = 0$ and $\langle (u_i^2)_a C_b^2 \rangle = 0$.

The correlation function $B_{ab}(r, t)$ for an isotropic turbulent field satisfies a dynamic equation of the Karman–Howarth type

$$\frac{\partial B_{ab}}{\partial t} = 2\left(\frac{\partial}{\partial r} + \frac{2}{r}\right)\left(B_{La,b} + D_{\mathrm{m}} \frac{\partial B_{ab}}{\partial r}\right), \tag{16.92}$$

where D_{m} is the coefficient of molecular diffusion; $B_{La,b} = \langle (u_L)_a C_a C_b \rangle$; $(u_L)_a$ is the velocity component along the vector $\boldsymbol{r}$ connecting the points a and b. This equation is called the Corrsin equation.

The Corrsin equation leads to the dynamic equation for the intensity of concentration fluctuations $\langle (C')^2 \rangle$. One can derive it by going to the limit $r \to 0$ and expanding the functions entering in (16.92) in a Taylor series similarly to the derivation of (16.66)

$$\frac{d\langle (C'^2) \rangle}{dt} = -12 \frac{D_{\mathrm{m}}}{\lambda_c^2} \langle (C')^2 \rangle . \tag{16.93}$$

One can see from (16.93) that the intensity of concentration fluctuations $\langle (C')^2 \rangle$ decreases with time, and furthermore, the characteristic time of this decrease is inversely proportional to the coefficient of molecular diffusion. So, in the final analysis, attenuation of intensity of concentration fluctuation is caused solely by the molecular diffusion just as attenuation of turbulence is caused solely by the molecular viscosity. By introducing τ_c, the characteristic relaxation time (also known as time of micromixing) of the scalar field, through the relation $d\langle (C')^2\rangle/dt \sim \langle (C')^2\rangle/\tau_c$, we get $\tau_c \sim \lambda_c^2/12D_m$ from (16.93).

Scalar fields can also be represented in the spectral form. Let us introduce the spectral representations of correlations B_{ij} according to the formulae (16.75) and (16.76) in which i and j should be replaced by a and b:

$$B_{ab}(r,t) = 4\pi \int_0^\infty \frac{\sin kr}{kr} k^2 F_{ab}(k,t)\,dk\,, \tag{16.94}$$

$$F_{ab}(k,t) = \frac{1}{2\pi^2} \int_0^\infty \frac{\sin kr}{kr} k^2 B_{ab}(r,t)\,dr\,. \tag{16.95}$$

At $r \to 0$, one gets the spectral representation of concentration fluctuation intensity $\langle (C')^2\rangle$:

$$\langle (C')^2\rangle = 4\pi \int_0^\infty k^2 F_{ab}(k,t)\,dk\,. \tag{16.96}$$

By analogy with (16.79), we can introduce the function

$$E_c(k,t) = 4\pi k^2 F_{ab}(k,t)\,.$$

Thus, (16.96) takes on the form

$$\langle (C')^2\rangle = \int_0^\infty E_c(k,t)\,dk \tag{16.97}$$

and (16.94) and (16.95) are transformed to

$$B_{ab}(r,t) = \int_0^\infty \frac{\sin kr}{kr} E_c(k,t)\,dk\,, \quad E_c(k,t) = \frac{1}{2\pi} \int_0^\infty kr \sin kr\, B_{ab}(r,t)\,dr\,.$$

The differential and integral length scales of the scalar field are expressed as

$$\frac{2}{\lambda_c^2} = \frac{1}{3}\frac{1}{\langle (C')^2\rangle} \int_0^\infty k^2 E_c(k,t)\,dk\,, \quad L_c = \frac{\pi}{2}\frac{1}{\langle (C')^2\rangle} \int_0^\infty \frac{E_c(k,t)}{k}\,dk\,. \tag{16.98}$$

Spectral representations of the Corrsin equation gives rise to the dynamic equation for the spectrum $E_c(k,t)$

$$\frac{\partial E_c(k,t)}{\partial t} = F_c(k,t) - 2D_m k^2 E_c(k,t)\,, \tag{16.99}$$

where $F_c(k,t) = -8\pi k^2 F_{La,b}(k,t)$; $F_{La,b}(k,t)$ is the spectral representation of $B_{La,b}$ in accordance with (16.94).

16.9
The Local Structure of Fully Developed Turbulence

The concept of isotropic turbulence introduced in the previous section is a mathematical idealization that has little to do with real turbulent flows. Yet, it would be a mistake to think that it has no practical importance. In the present section, we are going to introduce the concept of local isotropic turbulence, making it possible to examine the local structure of the turbulent flow with rather simple methods. This concept has direct applications to real turbulent flows at very high Reynolds numbers, that is, at $\mathrm{Re} \gg \mathrm{Re}_{cr}$ [12–15]. We reserve the term "developed turbulent flow" for this turbulent flow regime. It remains to be seen if such flows present the greatest practical interest in practical applications.

A distinguishing feature of developed turbulence is the presence of fluctuational motions with various amplitudes that get superimposed on the averaged flow described by the velocity U. To describe turbulent fluctuations, one has to specify not only the absolute velocity values, but also the distance at which velocity can noticeably change. Such distances are called "motion scales" or "scales of eddies". The latter notion eludes the precise definition, but for all practical purposes, it is acceptable to imagine a region of size λ, within which the turbulent motion is localized. In the subsequent discussion, the term "motion scale" will be understood to refer to such a region. The most rapid fluctuational motions have the largest motion scales. Their velocities are equal by the order of magnitude to the average flow velocity U and their motion scale which is equal to the characteristic linear scale L of the flow. For example, if the fluid is flowing inside a pipe, then U is the average flow rate velocity and L is the diameter of the pipe. Such fluctuations are called "large-scale". Small-scale fluctuations also exist. Before we define the meaning of "small-scale", let me remind you that in principle, the size of fluctuations can be as small as desired, that is, up to the mean free path of a molecule. However, fluctuations that have a very small scale give rise to extremely large velocity gradients which in their turn invoke strong forces of viscous friction, causing a very rapid decay of such fluctuations. Hence, the size of fluctuations should be bounded from below by some scale λ_0 (we will talk about this motion scale later). Fluctuations with scales $\lambda \ll L$ are defined as small-scale fluctuations. Small-scale fluctuations of the size $\lambda \sim \lambda_0$ are accompanied by considerable energy dissipation with subsequent conversion of energy into heat. Finally, there is an intermediate region with scales $\lambda_0 \ll \lambda \ll L$. Hence, the entire spectrum of motion scales can be divided into three regions: the energy region $\lambda \sim L$, the inertia region $\lambda_0 \ll \lambda \ll L$, and the viscous dissipation region $\lambda \leqslant \lambda_0$. To be sure, this classification is very inexact because it is impossible to establish sharply defined boundaries between these regions.

It turns out that small-scale perturbations in a turbulent flow with a very high Reynolds number can be regarded as isotropic, and it is just this property of developed turbulence that we called local isotropy. This statement is based on the following qualitative model of developed turbulence. According to this scheme, developed turbulence consists of a set of disordered perturbations (eddies) that differ from each other by their scale λ and velocity u_λ. As we gradually increase Re, the fluid flow accomplishes a transition from a laminar to a turbulent flow, and then, at a further increase of Re, to a developed turbulence. Perturbations of different scales do not appear at the same moment. First, when Re becomes larger than Re_{cr}, large scale fluctuations emerge. As Re keeps increasing, these fluctuations give birth to small-scale perturbations, transferring to them a part of their kinetic energy. Those perturbations, in their turn, give birth to even smaller perturbations and so on. Eventually, we get the entire spectrum of fluctuations where each perturbation gets its kinetic energy from its larger-scale "parent". Perturbations can disintegrate because of their instability. Indeed, each perturbation (fluctuation) is characterized by its own Reynolds number $\text{Re}_\lambda = u_\lambda \lambda / \nu_e$. For the largest fluctuations, the Reynolds number is equal by the order of magnitude to the Reynolds number of the bulk flow, and since $\text{Re} \gg \text{Re}_{cr}$, large fluctuations are instable and disintegrate into small-scale fluctuations. The Reynolds number of these newly-generated fluctuations is still too large, so they too disintegrate into smaller fluctuations and so on. The chain of ever-smaller fluctuations continues until the scale of resulting fluctuations approaches λ_0. This scale corresponds to the Reynolds number $\text{Re}_{\lambda_0} \sim 1$ and is called the inner (or Kolmogorow) scale of turbulence. Motions whose scale is λ_0 or less are hydrodynamically stable and do not disintegrate. For such fluctuations, viscous friction forces are essential. The energy of such fluctuations eventually dissipates into heat.

Hence, instability of the averaged motion leads to a continuous flux of energy over the spectrum of fluctuation, that is, from large-scale fluctuations to the fluctuations of minimum scale with subsequent conversion into heat. In order for developed turbulence to be sustainable, one has to continuously supply the averaged motion with energy from an external source. It is easy to see that the average specific dissipation energy $\bar{\varepsilon}$ (average amount of energy per unit mass per unit time) is an important parameter characterizing the intensity of developed turbulence.

The average fluid flow is generally inhomogeneous, anisotropic and non-stationary. Because of the random character of energy transfer from large-scale motions to small-scale ones, the orientating influence of the averaged flow will have less and less effect on statistical characteristics of fluctuations as the scale decreases. It is therefore quite natural to assume that in the case of developed turbulence, all perturbations except the largest ones are isotropic. The change of the average flow velocity $\langle \boldsymbol{u} \rangle = \boldsymbol{U}$ with distance becomes noticeable only for distances of the order L. The distance has to be that large in order for inhomogeneity to affect the average flow velocity. Therefore, inhomogeneity is only important for large-scale fluctuations, but does not affect small-scale fluctuations. Hence, the second assumption boils down to that of statistical homogeneity of small-scale fluctuations. As the fluctuation scale λ decreases, so does its characteristic period $t_\lambda = \lambda / u_\lambda$. For small-scale fluctuations, it becomes much shorter than the characteristic time $t_L = L/U$ during which the averaged flow remains non-stationary. In other words, the change of the average velocity that is responsible for flow non-stationary character of the flow takes much longer than the change of statistical characteristics of small-scale fluctuations. Therefore, small-scale perturbations can be regarded as stationary, or, more precisely, quasi-stationary. Recall that a quasistationary flow is defined as a flow whose parameters do not explicitly depend on time, while the flow itself does change with time because of its dependence on the integral characteristics of the flow.

Hence, the mechanism outlined above for developed turbulence leads us to the logical assumption that the statistical regime of small-scale fluctuations (i.e., the ones with length scale $\lambda \ll L$ and time scale $t_\lambda \ll T_L$) will be stationary, homogeneous and isotropic over sufficiently small spacetime regions. This assumption forms the basis of the theory of local isotropic turbulence, which was first formulated by Kolmogorov. Though this assumption cannot be proven rigorously, many functional dependences that follow from the theory of local isotropic turbulence have been confirmed by numerous experiments.

We shall now determine the general qualitative characteristics of developed turbulent flow, keeping in mind what we have just said about the pattern of developed turbulence and using some dimensionality consideration. We begin with consideration by considering large-scale fluctuations. In accordance with the preceding discussion, large-scale fluctuations are characterized by the following parameters: characteristic external integral length scale, equal by the order of magnitude to the characteristic length scale of the averaged flow L; characteristic velocity change ΔU of the most rapid fluctuations on the distance equal to the scale of fluctuations $\lambda \sim L$ (ΔU has the same order of magnitude as U); specific dissipation of energy $\bar{\varepsilon}$ equal to

$$\bar{\varepsilon} = \frac{\varepsilon}{\rho_e} = \frac{1}{2}\frac{\mu_e}{\rho_e}\sum_{i,j}\left\langle\left(\frac{\partial u_i}{\partial X_j} + \frac{\partial u_j}{\partial X_i}\right)^2\right\rangle , \tag{16.100}$$

and fluid density ρ_e. Since for large-scale fluctuations, the Reynolds number is large, $\mathrm{Re} \gg 1$, the coefficient of molecular viscosityμ_e is not included in the list of characteristic parameters. Nevertheless, energy dissipation does not take place and by analogy with the formula (16.100), it should be characterized by the coefficient of turbulent viscosity μ_t. Since the expression inside the brackets in (16.100) has the same order of magnitude as $(\Delta U)^2/L^2$, we have

$$\varepsilon \sim \mu_t \frac{(\Delta U)^2}{L^2} . \tag{16.101}$$

On the other hand, in view of dimensionality considerations, the quantities ε and $\bar{\varepsilon}$ should be expressed through dimensional parameters L, ΔU and ρ. Therefore,

$$\varepsilon \sim \frac{\rho(\Delta U)^3}{L} , \quad \bar{\varepsilon} \sim \frac{(\Delta U)^3}{L} . \tag{16.102}$$

Thus, (16.101) and (16.102) give us the dynamic and kinematic turbulent viscosities:

$$\mu_t \sim \rho \Delta U L , \tag{16.103}$$

$$\nu_t = \frac{\mu_t}{\rho} = \Delta U L . \tag{16.104}$$

The ratio of molecular and turbulent viscosities is equal to

$$\frac{\nu_e}{\nu_t} \sim \frac{\nu_e}{\Delta U L} \sim \frac{1}{\mathrm{Re}} \ll 1 . \tag{16.105}$$

So, the coefficient of turbulent viscosity is much larger than the coefficient of molecular viscosity.

The pressure change is approximately (i.e., by the order of magnitude) equal to

$$\Delta p \sim \rho(\Delta U)^2 . \tag{16.106}$$

Let us go to small-scale fluctuations with the scale $\lambda \ll L$. We begin with the inertia region $\lambda_0 \ll \lambda \ll L$ where the motion can be considered as non-viscous. The velocity u_λ of fluctuation having the scale λ does not depend on μ_e or on the external parameters L and ΔU because $\lambda \ll L$. Therefore, u_λ can only depend on ρ, λ and $\bar{\varepsilon}$ (or ε). The only combination of these quantities that has the dimensionality of velocity is $(\bar{\varepsilon}\lambda)^{1/3} = (\varepsilon\lambda/\rho)^{1/3}$. By substituting ε from (16.102) into this formula, we get

$$u_\lambda \sim (\bar{\varepsilon}\lambda)^{1/3} = \Delta U \left(\frac{\lambda}{L}\right)^{1/3} . \tag{16.107}$$

We see from (16.107) that the change of fluctuation velocity on a small distance λ is proportional to $\lambda^{1/3}$. This principle is known as the Kolmogorov–Obuchov law. It can be represented in a spectral form ("spectral" here refers to the spatial spectrum) by assigning to each fluctuation its wave number $k \sim 1/\lambda$ instead of λ and the kinetic energy $E(k)dk$ per unit mass contained in fluctuations with wave numbers in the interval $(k, k + dk)$. Since the dimensionality of $E(k)$ is $\mathrm{m}^3/\mathrm{s}^2$, we should compose from the parameters $\bar{\varepsilon}$ and k a combination that would have this dimensionality:

$$E(k) \sim \bar{\varepsilon}^{2/3} k^{-5/3} . \tag{16.108}$$

By integrating (16.108) over k from k up to ∞, one gets the total kinetic energy contained within fluctuations whose scale is $\leqslant \lambda$:

$$\int\limits_k^\infty E(k)dk \sim \frac{\bar{\varepsilon}^{2/3}}{k} \sim (\bar{\varepsilon}\,\lambda)^{2/3} \sim u_\lambda^2 .$$

Then, u_λ^2 is equal by order of magnitude to the total kinetic energy contained within such fluctuations. If we define the coefficient of turbulent viscosity as $\nu_t \sim \lambda\, u_\lambda$ by analogy with the formula (16.104), the relation (16.107) will take on the form

$$\bar{\varepsilon} \sim \frac{u_\lambda^3}{\lambda} \sim \nu_t \left(\frac{u_\lambda}{\lambda}\right)^2 . \tag{16.109}$$

Let us introduce the characteristic period of fluctuations, $t_\lambda = \lambda/u_\lambda$ and determine the order of velocity change Δu_t at a given point in space during the time t_λ

that is small compared to the characteristic external time $t_L \sim L/U$. The presence of the averaged flow leads us to conclude that after the time t_λ, any arbitrary point in space will be filled with the fluid that initially was separated from this point by the distance Ut_λ. Therefore, Δu_t can be derived from the formula (16.107) in which λ should be replaced by Ut_λ:

$$\Delta u_t \sim (\bar{\varepsilon}\, U\, t_\lambda)^{1/3} . \tag{16.110}$$

One should distinguish the quantity Δu_t from the change of velocity $\Delta u'_t$ of a given volume element of the fluid (fluid particle) moving in space. Since the latter only depends on the parameters $\bar{\varepsilon}$ and t_λ, we must find a combination of these parameters that has the dimension of velocity:

$$\Delta u'_t \sim (\bar{\varepsilon} t_\lambda)^{1/2} . \tag{16.111}$$

It is readily seen that the change of velocity of a moving fluid particle is proportional to $\Delta u_t \sim t_\lambda^{1/2}$ while the change of velocity of the fluid at a given point in space obeys another law: $\Delta u_t \sim t_\lambda^{1/3}$, so when $t_\lambda \ll T$, we have $\Delta u'_t \ll \Delta u_t$. Now, (16.107) and (16.110) can be represented as

$$\frac{u_\lambda}{\Delta U} \sim \left(\frac{\lambda}{L}\right)^{1/3} , \quad \frac{u_t}{\Delta U} \sim \left(\frac{t_\lambda}{T}\right)^{1/3} . \tag{16.112}$$

The form of these relations shows that characteristics of small-scale fluctuation in different developed turbulent flows differ from each other only by their length and velocity (or length and time) scales. This statement forms the essence of the self-similarity property of the local isotropic turbulence.

Let us determine the distance λ_0 at which viscous effects become significant. As it was noted earlier, this distance corresponds to the local Reynolds number $\mathrm{Re}_\lambda = u_\lambda \lambda/\nu_e \sim 1$. By substituting the relation (16.107), we write

$$\mathrm{Re}_\lambda = \frac{\Delta U \lambda^{4/3}}{\nu_e L^{1/3}} \sim \mathrm{Re}\left(\frac{\lambda}{L}\right)^{4/3} \sim 1 , \tag{16.113}$$

where Re is the Reynolds number of the average flow. This condition yields λ_0:

$$\lambda_0 \sim \frac{L}{\mathrm{Re}^{3/4}} \sim \left(\frac{\nu_e^3}{\bar{\varepsilon}}\right)^{1/4} . \tag{16.114}$$

The characteristic velocity and characteristic time of such fluctuations are obtained from (16.107):

$$u_{\lambda_0} \sim \frac{\Delta U}{\mathrm{Re}^{1/4}} = \lambda_0 \left(\frac{\bar{\varepsilon}}{\nu_e}\right)^{1/2} , \quad \tau_{\lambda_0} = \frac{\lambda_0}{u_{\lambda_0}} \sim \left(\frac{\nu}{\bar{\varepsilon}}\right)^{1/2} . \tag{16.115}$$

The scales λ_0 and τ_{λ_0} are respectively known as the Kolmogorov (or inner) spatial and temporal microscales. The values of λ_0 and u_{λ_0} decrease with the increase of the Reynolds number of the average flow.

At $\lambda \leqslant \lambda_0$, the motion of the fluid has viscous character. Turbulent fluctuations do not vanish suddenly; instead, they gradually decay, subject to viscous forces. Since velocity changes rather smoothly in this region, it can be expanded in a Taylor series over the powers of λ. Let us only keep the first term of the series $u_\lambda \sim \text{const}\,\lambda$ and determine the constant from the condition $u_\lambda \sim u_{\lambda_0}$ at $\lambda \sim \lambda_0$. Then,

$$u_\lambda \sim \frac{u_{\lambda_0}}{\lambda_0}\lambda \sim \frac{\Delta U}{L}\lambda \mathrm{Re}^{1/2} . \tag{16.116}$$

Scales of turbulent fluctuations are a function of their spatial characteristics. In addition, we may consider the time characteristics of fluctuations, namely, the frequencies ω_λ. The whole frequency spectrum can be divided into three intervals. The lower end of the spectrum, $\omega_L \sim U/L$, corresponds to the energy region; the upper end, $\omega_{\lambda_0} \sim U/\lambda_0 \sim U\mathrm{Re}^{3/4}/L$, corresponds to the dissipation region; and the intermediate interval $\omega_L \ll \omega_\lambda \ll \omega_{\lambda_0}$ corresponds to the inertia region. The inequality $\omega_\lambda \gg \omega_L$ means that the external (average) flow can be considered as stationary with respect to the local properties of small-scale fluctuations.

Energy distribution over the frequency spectrum in the inertia region is derived from (16.108) by replacing k with ω_L/U:

$$E(\omega) \sim (U\bar{\varepsilon})^{2/3}\omega_\lambda^{-5/3} . \tag{16.117}$$

The frequency ω_λ defines the repetition period of velocity at a fixed point in space. Together with ω_λ, we can introduce another frequency ω'_λ which stands for the repetition period velocity of a chosen fluid particle. The distribution of energy over the frequency spectrum for such particles does not depend on U, but only on $\bar{\varepsilon}$ and ω_λ. We conclude from dimensionality considerations that

$$E(\omega'_\lambda) \sim \frac{\bar{\varepsilon}}{\omega'_\lambda} . \tag{16.118}$$

We now apply the results obtained for small-scale fluctuations to estimate the velocity of inertialess particles buoyant in the fluid. Turbulent mixing causes particles to gradually move away from each other. Consider two particles such that the initial interparticle distance does not exceed the size of fluctuations from the inertia region. We make this requirement because otherwise, large fluctuations would just transport the two particles without changing the interparticle distance. Our assumption allows us to find the rate of change of interparticle distance δ from the equation

$$\frac{d\delta}{dt} \sim u_\lambda \sim (\bar{\varepsilon}\delta)^{1/3} . \tag{16.119}$$

By solving this equation at a given initial value of the interparticle distance δ_0, we find the time it takes for the two particles to move away from each other so that the gap between them reaches the value δ_1. In the limiting case $\delta_1 \gg \delta_0$, this time is

$$t \sim \frac{\delta_1^{2/3}}{\bar{\varepsilon}^{1/3}} . \tag{16.120}$$

Now, consider the correlations of velocity differences of two neighboring particles at a fixed instant of time. These correlations were introduced as structure functions in Section 16.8. Even the formula (16.107) gives a qualitative correlation of velocities at two points separated by a distance $\lambda \ll L$. In other words, it provides a connection between velocity values at two neighboring points. Components of the correlation tensor $b_{ik}(\boldsymbol{r}, t)$ serve as quantitative characteristics of this correlation. In an isotropic vector field, these components depend on two scalar functions, namely, longitudinal and transverse functions $b_{LL}(r, t)$ and $b_{NN}(r, t)$, where $r = |\boldsymbol{r}_2 - \boldsymbol{r}_1|$ is the length of the radius vector between the points $\boldsymbol{r}_1$ and $\boldsymbol{r}_2$. In the case of local isotropic turbulence, we have $\lambda_0 \leqslant r \ll L$.

The change of velocity at small distances is caused by small-scale fluctuations and is independent of the average flow. Therefore, our analysis of correlation and structure functions can be simplified if we assume that isotropy and homogeneity take place not only at small scales, but at large scales as well. Then, the average velocity can be taken to be zero (see Section C.10), and we can take advantage of the relations between the functions b_{LL} and b_{NN} that have been established in Section 16.8.

Because of (16.107), the difference of velocities over the distance r in the inertia region is proportional to $r^{1/3}$. Therefore, b_{LL} and b_{NN} are proportional to $r^{2/3}$. In other words, in any turbulent flow with a sufficiently high Reynolds number, the root-mean-square value of the difference of velocities at two points separated by the distance r (where r is neither too small nor too large) should be proportional to $r^{2/3}$. This law, which was established by Kolmogorov, is one of the most important laws describing turbulent flows and is called “the law of two thirds”.

Similar to the spectral form, we can formulate a similar law for the energy spectrum

$$E(k) \sim \bar{\varepsilon}^{2/3} k^{-5/3} \tag{16.121}$$

which is called “the law of five thirds”.

Let us now obtain the connection between b_{LL} and b_{NN} in the inertia region. First, we transform (16.90) to the form

$$b_{NN} = \frac{1}{2r}\frac{d}{dr}(r^2 b_{LL}) . \tag{16.122}$$

Recalling that both b_{NN} and b_{LL} are proportional to $r^{2/3}$, we can write

$$b_{NN} = 4b_{LL}/3 , \quad (\lambda_0 \ll r \ll L) . \tag{16.123}$$

In the dissipation region ($\lambda \leqslant \lambda_0$), the velocity difference at two neighboring points is proportional to r as follows from (16.116). Then, b_{LL} and b_{NN} are proportional to r^2 and the formula (16.122) reduces to

$$b_{NN} = 2b_{LL} , \quad (r \leqslant \lambda_0) . \tag{16.124}$$

The longitudinal and transverse functions b_{LL} and b_{NN} for small-scale fluctuations can be expressed in terms of the specific dissipation of energy

$$b_{NN} \sim \frac{2\bar{\varepsilon}}{15\nu_e} r^2 , \quad b_{LL} \sim \frac{\bar{\varepsilon}}{15\nu_e} r^2 . \tag{16.125}$$

The case considered above corresponds to the situation when the average fluid flow is absent, for example, when the fluid has been subjected to intensive shaking and then left alone. Such motion decays with time, and small-scale fluctuations decay in accordance with the power law

$$u_\lambda \sim t^{-5/4} \,. \tag{16.126}$$

The deduced statistical characteristics have been examined only using consideration of similarity and dimensionality theory, which do not invoke hydrodynamic equations. The main conclusion is that the statistical regime of small-scale components of turbulence at high Reynolds numbers is quite independent from the properties of the macroscopic structure of the flow which can only affect the value of $\bar{\varepsilon}$. Therefore, dynamic equations for the characteristic of locally isotropic turbulence do not depend on the character of large-scale motions, and it is sufficient to consider the case of isotropic turbulence in unbounded space and find the connections between its local characteristics. The obtained characteristics will then be the same for all turbulent flows sharing the same values of $\bar{\varepsilon}$ and ν_e if $\mathrm{Re} \gg 1$. Hence, all the relations given above are universal for any locally isotropic turbulent flow.

Note that in the case of local isotropy, the system of dynamic equations is also not closed, and we need additional hypotheses and relations to close such a system.

Our analysis of the local structure of the velocity vector field can be repeated for a scalar field of passive impurity concentration $C(\boldsymbol{X}, t)$. Consider a developed turbulent flow of some fluid containing passive impurity; the impurity does not influence the turbulent flow of the carrier fluid. An intense mixing of fluid volumes with different impurity concentration occurs in a developed turbulent flow. Under the action of fluctuations with different scales, mixing of both small volumes (microvolumes) and relatively large volumes occurs (macrovolumes or, to use a different term, moles). As shown earlier, small-scale perturbations in a developed turbulent flow can be considered stationary and isotropic, that is, locally isotropic. It is natural to expect that perturbations of the field of concentration in a small regions of space will also be stationary and isotropic, in other words, that the scalar field $C(\boldsymbol{X}, t)$ in such regions will be scalar isotropic.

Gradual disintegration of fluctuations starting from largest-scale ones (whose size is equal by the order of magnitude to the characteristic linear size L of the flow region) all the way down to the Kolmogorov microscale λ_0 is the underlying process responsible for the formation of the velocity fluctuation spectrum. Of all characteristics of large-scale motions, only the specific energy dissipation $\bar{\varepsilon}$ has an effect on small-scale motions. The same reasoning can be applied to the field of concentration by replacing the Reynolds number $\mathrm{Re} = \Delta U L/\nu_e$ with the diffusion Peclet number $\mathrm{Pe}_D = L\delta U/D_\mathrm{m}$, assuming $\mathrm{Pe}_D \gg 1$ and, of course, keeping the condition $\mathrm{Re} \gg 1$. Here, L is the characteristic linear scale of change of average concentration $\langle C(\boldsymbol{X})\rangle$, δU is the change of average velocity on the distance L and D_m is the coefficient of molecular diffusion. If $L > \mathrm{L}$, we should take ΔU as our δU.

Because of their instability, large-scale fluctuations of concentration will give rise to smaller and smaller fluctuations, all the way down to the minimum fluctuation which has the inner concentration scale $\lambda_c^{(0)}$. Using the same reasoning as for the vector field, we arrive at the statement that in a spatial region of scale $\lambda_c \ll L$, the field of concentration will be locally isotropic so that the average concentration $\langle C \rangle$ can be considered constant. The degree of concentration inhomogeneity in these regions is given by the parameter characterizing the change of concentration fluctuations $C' = C - \langle C \rangle$. The meaning of this parameter is analogous to specific energy dissipation $\bar{\varepsilon}$, which is determined by velocity gradient $\nabla \mathbf{u} = \partial u_i / \partial X_j$ (see (16.100) rather than velocity fluctuations. Therefore, it is quite natural to assume that a quantity similar to (16.100), namely,

$$\langle N \rangle = D_{\mathrm{m}} \left\langle (\nabla C')^2 \right\rangle , \qquad (16.127)$$

will serve as a measure of concentration inhomogeneity.

This parameter is called dissipation of concentration inhomogeneity. Since we have $\langle C \rangle = \text{const}$ for small-scale ($\lambda_c \ll L$) fluctuations of concentration, the formula (16.9.28) can be rewritten as

$$\langle N \rangle = D_{\mathrm{m}} \left\langle (\nabla C)^2 \right\rangle . \qquad (16.128)$$

As far as there are two characteristic length scales L and L, let us introduce $L_0 = \min(\mathrm{L}, L)$ and divide the entire spectrum of concentration fluctuations into two intervals: the interval of large-scale fluctuations with $\lambda_c \sim L_0$ and the interval of small-scale fluctuations with $\lambda_c \ll L_0$. For the first interval, the characteristic quantities are the length scale L_0, the change of average velocity $\Delta_{L_0} U$ and the change of average concentration $\Delta_{L_0} \langle C \rangle$. We can build a combination having the dimensionality of $\langle N \rangle$ from these parameters:

$$\langle N \rangle \sim \frac{\Delta_{L_0} U (\Delta_{L_0} \langle C \rangle)}{L_0} . \qquad (16.129)$$

In practice, the characteristic length scales L and L are equal by order of magnitude, that is, $L_0 = L \sim \mathrm{L}$. Therefore, we can introduce the coefficient of turbulent diffusion D_t with the formula (16.101):

$$\langle N \rangle = D_t \left(\frac{\Delta \langle C \rangle}{L} \right)^2 . \qquad (16.130)$$

By comparing (16.129) and (16.130) with (16.101) and (16.104), and taking into account that $L \sim \mathrm{L}$, we get

$$D_t \sim \mathrm{L} \Delta U \sim \nu_t . \qquad (16.131)$$

Thus, the coefficients of turbulent diffusion and turbulent viscosity have the same order of magnitude.

Now, consider the small-scale interval $\lambda_c \ll L_0$. The velocity field of small-scale perturbations is characterized by two dimensional parameters: specific dissipation energy $\bar{\varepsilon}$ and kinematic viscosity ν_e. When examining the concentration field in this region, we should bring in two more parameters, that is, the coefficient of molecular diffusion D_m, which enters the diffusion equation for the impurity,

$$\frac{\partial C}{\partial t} + u_k \frac{\partial C}{\partial X_k} = D_m \Delta C \,, \tag{16.132}$$

and the dissipation of concentration inhomogeneity $\langle N \rangle$. The ratio between the convective term (second summand on the left-hand side) and the diffusion term (right-hand side) in (16.132) is equal by the order of magnitude to the diffusion Peclet number $Pe_D = \Delta_{L_0} L_0 / D_m$. Molecular diffusion plays a considerable role only for $Pe_D < 1$. For the averaged concentration, the Peclet number is usually $Pe \gg 1$, so the effect of molecular diffusion is negligible for large scale perturbations of concentration, and turbulent diffusion emerges as the main mechanism behind macroscopic mixing of regions with different impurity concentrations. Since $Pe_D \sim \lambda_c$, a smaller scale of concentration fluctuations means a smaller Pe_D. When $\lambda_c = \lambda_c^{(0)}$, the Peclet number is $Pe_D = 1$, and when $\lambda_c \leqslant \lambda_c^{(0)}$, the inequality $Pe_D \leqslant 1$ is satisfied. Thus, the interval of small-scale concentration fluctuations contains two sub-intervals: the convective interval $\lambda_c^{(0)} \ll \lambda_c \ll L$, for which $Pe_D \gg 1$, and the dissipation interval $\lambda_c \leqslant \lambda_c^{(0)}$, for which $Pe_D \leqslant 1$.

In the convective interval, molecular diffusion does not play any noticeable role. Therefore, the parameter D_m does not figure among its governing parameters. The governing parameters in this interval are ε, ν_e, and $\langle N \rangle$. In the dissipative interval where molecular diffusion plays a noticeable role, the governing parameters will include $\bar{\varepsilon}$, ν_e, D_m and $\langle N \rangle$. Two combinations with the dimensionality of length can be built from these four parameters: the inner scale of turbulence (Kolmogorov scale)

$$\lambda_0 = \left(\frac{\nu_e^3}{\bar{\varepsilon}} \right)^{1/4} = \frac{L}{Re^{3/4}}$$

and the inner scale of diffusion (Batchelor scale)

$$\lambda_c^{(0)} = \lambda_b = \left(\frac{\lambda_e D_m^2}{\bar{\varepsilon}} \right)^{1/4} = \lambda_0 Sc^{-1/2} \,, \tag{16.133}$$

where $Sc = \lambda_e / D_m$ is the Schmidt number. It should be noted that for fluids $Sc \sim 10$, whereas for infinite dilute solutions, $Sc \sim 10^3$.

By its physical meaning, the convective interval should comprise those concentration scales for which molecular diffusion is negligibly small as compared to convection. However, the presence of two quantities (ν_e and D_m) of the same dimensionality leads to the appearance of a new dimensionless numbers, namely, the Schmidt number Sc and to the dependence of the lower end of the interval on Sc in accordance with (16.133). Therefore, the condition $\lambda \gg \lambda_c^{(0)}$ alone does not guarantee that the fluctuation belongs to the convective interval. To determine where the intervals where convection or molecular diffusion dominates, one should compare the transport coefficients ν_e and D_m. If they are of the same order of magnitude, then $Sc \sim 1$ and the length scales λ_0 and λ_b are roughly the same. The cases of

Sc > 1 or Sc < 1 requires additional study. As long as Sc ≫ 1, we have $\lambda_b \ll \lambda_0$. In other words, there exists within the viscous interval, a visco-diffusional interval where molecular diffusion plays a significant role.

In conclusion, we shall give an approximate expression for the structure function of the concentration field $d_{cc}(r) = \langle [C(\boldsymbol{X} + \boldsymbol{r}) - C(\boldsymbol{X})]^2 \rangle$. For small-scale fluctuations of concentration, the dimensionality theory suggests the expression

$$d_{cc}(r) = \langle N \rangle\, (\bar{\varepsilon})^{-1/2}\, D_{\mathrm{m}}^{1/2}\, F\left(\frac{r}{\lambda_b}, \frac{\nu_e}{D_{\mathrm{m}}}\right). \tag{16.134}$$

In the dissipation region, at $r \ll \lambda_b$, there exists the following representation:

$$d_{cc}(r) \approx \frac{\langle N \rangle}{3\, D_{\mathrm{m}}}\, r^2\,. \tag{16.135}$$

In the convective interval, at $L \gg r \gg \lambda_b$, the structure function is given by

$$d_{cc}(r) \approx \frac{\langle N \rangle}{3(\bar{\varepsilon})^{1/3}}\, r^{2/3}\,. \tag{16.136}$$

The formula (16.136) is called "the law of two thirds" for the concentration field. The spectral "law of five thirds" for the local isotropic concentration field has the form

$$E_c(k) \approx \frac{\langle N \rangle}{(\bar{\varepsilon})^{1/3} k^{5/3}}\,. \tag{16.137}$$

16.10 Models of Turbulent Flow

It was shown in Section 16.6 that the system of Reynolds equations describing a turbulent fluid flow is not closed because the number of unknowns is greater then the number of equations. Attempts to close the system by adding equations for higher-order moments were unsuccessful because those additional equations contain new moments of higher order. Therefore, neither the Reynolds equations on their own, nor a system of Reynolds equations plus equations for higher moments (e.g., the energy equation discussed in Section 16.7), or the simplified equations for isotropic (see Section 16.8) or locally isotropic turbulence (see Section 16.9) can be solved. All they can do for us is to establish certain connections between different statistical characteristics of turbulence.

Several possible ways exist to close the system of Reynolds equations. The first is to use experimental data to determine the functional connections between moments of some definite order and the lower-order moments. A second way is to deduce these connections from simple hypotheses that are well justified on physical grounds and are accurate up to some empirical constants. This method lies at the basis of all semi-empirical theories of turbulence. Finally, the third and currently most widespread method is based on the use of transport equations for some characteristics of turbulence.

It should be emphasized that there are no universal relations that would be applicable to all turbulent flows. Each of the existing approximations is suitable only for some type of flows, for example, flows in tubes, boundary layer flows, jet flows, flows past a body and so forth.

The present section offers a review of several models from which one can derive additional equations and thus close the system.

16.10.1 Semi-empirical Theories of Turbulence

The failure of the Reynolds equations (16.37) and (16.39) to form a closed system is explained by the presence of new unknowns, namely, correlations of velocity fluctuations that appear in the Reynolds stresses $\tau_{ij}^{(1)}$. The simplest way to close the Reynolds equations is to establish connections between the Reynolds stresses and the average hydrodynamic fields. Such methods are called local equilibrium algebraic methods and the corresponding relationships are said to be of the gradient type.

The methods based on approximating the Reynolds stresses with the help of parameters determined by the average velocity profile in the given cross section are well-developed and widely used for calculating different flows. The range of application of these methods is limited to the turbulent flows whose turbulence characteristics in a given cross section do not depend on their distributions in the preceding cross sections. The main difficulty is finding the range of applicability of these methods. Some of the resulting connections are adduced below.

1. *Boussinesq model:*
   For simplicity's sake, we consider a stationary fluid flow in a flat channel in the absence of external forces. The average velocity has one component $\langle u_x \rangle = U$ parallel to the channel wall and depending only on the transverse coordinate Z. Suppose that $u_x = U + u'$, $u_z = w'$, where u' and w' are fluctuations of the longitudinal and the transverse velocity components. Then, (16.39) takes on the form

$$\frac{\partial \tau}{\partial z} = \frac{\partial \langle p \rangle}{\partial X}, \quad \rho_e \frac{\partial \left\langle w'^2 \right\rangle}{\partial Z} = -\frac{\partial \langle p \rangle}{\partial Z}, \tag{16.138}$$

   where the stress τ is equal to

$$\tau = \rho_e \nu_e \frac{d \langle u \rangle}{d Z} - \rho \left\langle u' w' \right\rangle. \tag{16.139}$$

   A new unknown function $\tau' = -\rho_e \langle u' w' \rangle$ appears in the equations. Thus, in order to close the set of equations (16.138) and (16.139), it is sufficient to express τ' through $U(Z)$. The Boussinesq hypothesis states that the following equality

$$-\rho_e \left\langle u' w' \right\rangle = \rho_e \nu_t \frac{d U}{d Z} \tag{16.140}$$

is valid where ν_t is a quantity with the dimension of viscosity; it is called the turbulent viscosity coefficient.

Strictly speaking, (16.140) does not constitute a closure relation because in order to determine the new unknown ν_t, one needs to have experimental data or to formulate a supplementary hypothesis. The simplest way out is to take $\nu_t = \text{const}$. Then, in a notable analogy with (16.44), the introduction of ν_t will be tantamount to replacing the fluid viscosity ν_e with $\nu_e + \nu_t$, and the Reynolds equations will be equivalent to equations for the laminar flow with a new viscosity coefficient. In this case, the obtained velocity profile will be a parabolic Polseuille profile, although it is well known that such turbulent flows have a logarithmic rather than parabolic velocity profile. We conclude that ν_t cannot be a constant; instead, it should be a function of Z.

Let us estimate the form of this function by using dimensional analysis. Consider the flow near a flat wall. Let ΔU be the characteristic variation of velocity at the distance Z from the wall. Since no characteristic linear size has been assigned to our flow, we shall take Z as the characteristic linear size. From the two governing parameters ΔU and Z, one can form the quantity

$$\nu_t = \Delta U Z \tag{16.141}$$

that has the dimensionality of viscosity. On the other hand, in the vicinity of the wall, we can assume $\Delta U \sim Z dU/dZ$. Then, the friction force per unit area of the wall is

$$\tau_\text{f} = \mu_t \frac{dU}{dZ} = \rho_e \nu_t \frac{dU}{dZ} \sim \rho_e Z^2 \left(\frac{dU}{dZ}\right)^2 , \tag{16.142}$$

from which it follows that

$$\frac{dU}{dZ} \sim \left(\frac{\tau_\text{f}}{\rho_e}\right)^{1/2} \frac{1}{Z} .$$

Since the value of τ_f at the wall has to be constant, we have

$$U \sim \left(\frac{\tau_\text{f}}{\rho_e}\right)^{1/2} \ln Z . \tag{16.143}$$

Thus, a simple estimation shows that ν_t decays linearly as we get closer to the wall and that the longitudinal velocity profile has a logarithmic form. Actually, the structure of the flow in the vicinity of the wall is more complex. A detailed analysis of the flow structure, which takes into account the transformation of the turbulent boundary layer into a viscous boundary layer, shows that in fact, in the region adjacent to the wall, ν_t decays much faster (never slower than z^3). In spite of the fact that the assumption $\nu_t = \text{const}$ is inadmissible for turbulent flows inside pipes, there are some flows for which this simple model is acceptable, such as, for example, turbulent jet flows and flows in the open atmosphere. For such flows, ν_t should be considered as a parameter that varies for different flows and is determined from experiments.

We should also mention that a similar model can be applied to the problems involving heat or passive impurity propagation in a turbulent flow once we introduce the coefficient of turbulent thermal diffusion χ_t and the diffusion coefficient D_t (see (16.46) and (16.47)).

2. *The Prandtl model:*

   The model proposed by Prandtl is based on the concept of mixing length. Prandtl attributes a physical meaning to the quantity L in (16.104), taking his hint from the analogy between turbulent flow and the random molecular motion in the molecular kinetic theory of gases. According to this theory, the viscosity ν_e is defined by the same formula where ΔU is the average velocity and L is the mean free path of molecule. The velocity of molecular motion is certainly much greater then the average velocity of a turbulent fluid flow, whereas the mean free path of molecules is much shorter than the scale of fluctuation. Therefore, the product of these quantities gives the difference between ν_e and ν_t in good agreement with the (16.105).

   In a similar manner, as for the exchange of momentum between molecules, in a turbulent exchange, a finite fluid volume leaving the layer separated from the given layer by a certain distance conserves its average momentum until it reaches the given layer. There, it mixes with the ambient fluid, transferring the entire momentum difference to the fluid. The average distance between the initial layer from where the volume has started its journey, and the destination layer, where it mixes with the ambient fluid, is called the mixing length. That is why Prandtl's theory is sometimes called the mixing length theory.

   As in the previous model, we shall consider a plane-parallel flow with the average velocity U along the X-axis. The Z-axis is perpendicular to the X-axis and is pointing in the upward direction. The adopted model says that the volumes coming from the lower layer $Z - l'$ and from the upper layer $Z + l'$ will reach the layer Z. If the mixing of the arriving volumes with the ambient fluid happens instantaneously, the volumes bring to the layer Z the same momentum which they held initially while inside the layers $Z - l'$ and $Z + l'$. Such an exchange will lead to the emergence of fluctuations of the transverse velocity w' which by their order of magnitude are equal to

$$w' \sim U(Z \mp l') - U(Z) \sim \mp l' \frac{d\,U(Z)}{dZ} . \tag{16.144}$$

   We may now determine the friction force per unit area exerted on the layer Z by the upper and lower layers. If we designate the momentum from the upper layer as positive and the one from the lower layer as negative, then

$$\tau_\mathrm{f} = \rho_e \left\langle w'(U(Z \mp l') - U(Z))\right\rangle \sim \mp\rho_e \left\langle w'l'\right\rangle \frac{d\,U(Z)}{dZ} .$$

   Plugging in the expression (16.144) for w' and designating $l^2 = \langle l'^2 \rangle$, we obtain

$$\tau_\mathrm{f} = \mp\rho_e l^2 \left(\frac{d\,U(Z)}{dZ}\right)^2 .$$

Since τ_f should be a positive quantity, the latter formula can be represented as

$$\tau_\text{f} = \rho_e l^2 \left| \frac{dU(Z)}{dZ} \right| \frac{dU(Z)}{dZ} \,. \tag{16.145}$$

Now, similarly to (16.42), if we take $\tau_t = \rho \nu_t dU/dZ$ and use (16.145), we have

$$\nu_t = l^2 \left| \frac{dU(Z)}{dZ} \right| . \tag{16.146}$$

For the problem of a flow in a plane channel near the wall, we can take $l \sim Z$. Then, using (16.145) and (16.146), we obtain a logarithmic velocity profile near the wall but outside the viscous sublayer. However, in the region close to the symmetry axis of the channel, this approach is not acceptable. In the latter case, it is better to take $l = \text{const}$.

In contrast to the Boussinesq model, the unknown parameter in the Prandtl model is the mixing length l, which depends on coordinates and must be obtained experimentally for any specific flow. Prandtl's model as well as Boussinesq's is not applicable to all turbulent flows.

3. *The Taylor model:*

Taylor has suggested his model, known as the theory of eddy transport, in an attempt to properly account for the influence of pressure fluctuations on fluid particles. The theory is similar to Prandtl's in that it also uses the concept of mixing length. However, unlike Prandtl, Taylor considers the mixing layer for the velocity vorticity, and not for the momentum.

Consider a two-dimensional flow with average velocity $\langle \boldsymbol{u} \rangle = (U, W)$ in the (X, Z)-plane. In this flow, the average vorticity only has one component:

$$\langle \omega_y \rangle = \Omega = \frac{\partial U}{\partial Z} - \frac{\partial W}{\partial X} \,.$$

Let the average flow be parallel to the X-axis. Then, $\langle \boldsymbol{u} \rangle = (U, 0)$ and the turbulent component of stress $\tau^{(1)} = \rho_e \langle u'w' \rangle$ obeys the momentum equation

$$\begin{aligned} \frac{\partial \tau}{\partial Z} &= -\frac{\partial}{\partial Z} \langle u'w' \rangle = -\left(\left\langle u' \frac{\partial w'}{\partial Z} \right\rangle + \left\langle w' \frac{\partial u'}{\partial Z} \right\rangle \right) \\ &= -\left\langle w' \left(\frac{\partial u'}{\partial Z} - \frac{\partial w'}{\partial Z} \right) \right\rangle + \frac{1}{2} \frac{\partial}{\partial X} \left(\left\langle u'^2 \right\rangle - \left\langle w'^2 \right\rangle \right) . \end{aligned} \tag{16.147}$$

When deriving this equation, we used the continuity equation for velocity fluctuations

$$\frac{\partial u'}{\partial Z} + \frac{\partial w'}{\partial Z} = 0 \,.$$

Let the flow be uniform along the X-axis. Then, all derivatives with respect to X are equal to zero and (16.147) becomes

$$\frac{\partial \tau}{\partial Z} = -\rho_e \left\langle w' \omega'_y \right\rangle . \tag{16.148}$$

We now introduce the mixing length l_1' for the vorticity vector through a relation similar to (16.144):

$$\omega_y' = l_1' \frac{\partial \Omega}{\partial Z} , \tag{16.149}$$

where $\Omega = \langle \omega_y \rangle = dU/dZ$. For the fluctuation of transverse velocity w', we assume the validity of (16.144), thus

$$w' = l' \frac{\partial U}{\partial Z} . \tag{16.150}$$

Then,

$$\frac{\partial \tau}{\partial Z} = -\rho_e \left\langle w' \omega_y' \right\rangle = -\rho_e \left\langle l_1' l' \right\rangle \frac{\partial U}{\partial Z} \frac{\partial \Omega}{\partial Z} = \rho_e l_1^2 \frac{\partial U}{\partial Z} \frac{\partial^2 U}{\partial Z^2} , \tag{16.151}$$

where $l_1 = (-\langle l_1' l' \rangle)^{1/2}$ is the characteristic length that plays the same role in the Taylor model that l plays in the Prandtl model.
Taking $l_1 = \text{const}$, from (16.151), we get

$$\tau_f = \frac{1}{2} \rho_e l_1^2 \left(\frac{dU}{dZ} \right)^2 . \tag{16.152}$$

This expression coincides with the formula (16.145) in the Prandtl theory if we take $l_1 = \sqrt{2} l$. In all other respects, the Taylor theory is different from the Prandtl theory, and makes different predictions. For example, the velocity profile for the channel flow predicted by the Taylor theory is in good agreement with experimental data all the way to the central axis of the channel, in stark contrast with predictions of the Prandtl theory.
As all other semi-empirical models, the Taylor model does not solve the closure problem entirely because it reduces to a single empirical parameter, that is, the vortex mixing length l_1. The main shortcoming of the Taylor theory is the limited range of application, for example, it is only suitable for two-dimensional problems.
Finally, it is necessary to make the following note. The coefficient of turbulent viscosity ν_t in the Boussinesq model and the mixing lengths l and l_1 in the Prandtl and Taylor models have been introduced formally (albeit with some supporting physical rationalization) for plane-parallel fluid flows in an attempt to describe the simplest fluid flows, that is, in pipes, channels and boundary layers. We still need to show how these parameters can be introduced in the general case of an arbitrary spatial flow.
Let us suppose that turbulence emerges as a result of the transition of a part of the average flow energy into small scale perturbations. Then, according to the energy balance equation (16.54), the inequality should hold where A is the term in the energy equation (see (16.55)) describing the exchange of energy between the average motion and fluctuational motions. Indeed, the condition means that the turbulent energy density e_k at a given point increases at the expense of

energy of the averaged flow. Then, all statistical characteristics of turbulence, including the Reynolds stresses, should depend on the field of average velocity. The Reynolds stresses $\tau_{ij}^{(1)}$ play the same role with respect to the averaged motion as viscous forces with respect to the laminar flow. Therefore, when deformation of fluid particles is not taken into account, the averaged flow is similar to the motion of a rigid body, and the Reynolds stresses are pointing along the normal to any surface element arbitrary selected within the fluid. Then, the tensor $\rho\left\langle u_i'\, u_i'\right\rangle$ is isotropic and can be represented as a spherical component of the strain velocity tensor (see (16.4))

$$\rho\left\langle u_i' u_j'\right\rangle = c\delta_{ij}\,, \quad c = \frac{1}{3}\rho\left\langle u_k' u_k'\right\rangle = \frac{2}{3}\rho e_k\,. \tag{16.153}$$

The turbulent energy ρe_k is similar to $-p$ in the incompressible fluid law (16.4). In the general case that takes into account the deformation of fluid particles, the stresses $\tau_{ij}' = \rho\langle u_i' u_j'\rangle$ depend on the derivatives of the average velocity with respect to coordinates. Since the tensor $\tau_{ij}^{(1)}$ is symmetrical, it depends on the strain velocity tensor E_{ij} (see (16.5)). In the case of small deformations, this dependence is linear and the proportionality coefficient has the meaning of the turbulent viscosity coefficient, analogously to the Navier–Stokes law.
Let us now dwell on the analogy with the kinetic theory of gases. This theory holds that the coefficient of molecular viscosity is equal to $\nu_e \sim u_m l_m$, where value of velocity fluctuation and l_m are, respectively, the average velocity and the mean free path of molecules. Suppose that a similar relation is true for the turbulent motion, with the root-mean-square value of velocity fluctuation functioning as our u_m, and the integral scale of turbulence, namely, as l_m. In the Prandtl theory, this scale is the mixing length which has the order of the integral scale of turbulence and, as we noted in Section 16.8, has the meaning of the average distance that turbulent fluctuations can travel. As we are concerned with spatial motions, the turbulence will be characterized by different scales assigned to different directions. The set of scales l_{ij} will then form a symmetric scale tensor. Now, using this tensor and taking advantage of the symmetry of the tensor $\rho\langle u_i'\, u_j'\rangle$, we can assume

$$\tau_{ij}^{(1)} = \rho\left\langle u_i' u_j'\right\rangle = \frac{2}{3}\rho e_k \delta_{ij} - \rho\sqrt{e_k}(l_{ik}E_{kj} + l_{jk}E_{ki})\,. \tag{16.154}$$

This formula was first suggested by Monin. It can be thought of as a generalization of the Boussinesq and Prandtl models.
Sometimes, in the first approximation, we can take

$$l_{ij} = l\delta_{ij}\,.$$

Then, (16.154) takes on the form

$$\tau_{ij}^{(1)} = \frac{2}{3}\rho e_k \delta_{ij} - \rho\, l\sqrt{e_k}\, E_{ij}\,.$$

Defining the coefficient of turbulent viscosity as

$$\nu_t = l\sqrt{e_k}\,,$$

one gets

$$\tau_{ij}^{(1)} = \frac{2}{3}\rho e_k \delta_{ij} - 2\rho\nu_t E_{ij}\,. \tag{16.155}$$

16.10.2
The Use of Transport Equations

Following the papers [16–18], we offer a brief review of several models of turbulence that are based on transport equations for various statistical characteristics of turbulent flows. In these models, the minimum possible number of parameters that can be used to describe turbulence is equal to three. The turbulent stress, the energy of turbulence and a third parameter which, when combined with the energy of turbulence, would result in a quantity having the dimensionality of length is the most common choice of parameters. These models have been tested for problems that involve flows in channels and boundary layers. The most influential publications that shaped the development of these models are [19–22].

Numerous models tailored for different types of turbulent flows have been proposed as of today. All of them can be classified as one-, two- and three- parametric according to the number of transport equations employed. If a model contains less than three transport equations, it means that the model includes some algebraic relations between various characteristics of turbulence. An increase of the number of transport equations complicates the problem considerably, as we are faced with the necessity to measure the constants involved and determine the range of applicability of the model. The models listed below as well as the corresponding equations are written in the approximation of a stationary plane boundary layer for a homogeneous incompressible fluid (ρ = const). The velocity $\boldsymbol{u}$ has two components, $u_1 = u$, $u_2 = v$ along X and Y coordinate axes.

1. *One-parametric models:*
   These models use one equation for the turbulence energy $e_k = 0.5\sum\langle u_i'^2\rangle$, for the Reynolds stress (shear stress) $\tau' = \tau_{12}^{(1)}/\rho = -\langle u'\,v'\rangle$, or for the turbulent viscosity ν_t.
   a) Kolmogorov [19] was the first to suggest using the equation for turbulence energy. In the stationary plane boundary layer approximation, this equation has the following form:

$$\frac{De_k}{Dt} = \frac{\partial}{\partial Y}\left(D_E\frac{\partial e_k}{\partial Y}\right) + \tau'\frac{\partial\langle u\rangle}{\partial Y} - \bar{\varepsilon}\,, \tag{16.156}$$

   where $D/Dt = \langle u\rangle\,\partial/\partial X + \langle v\rangle\,\partial/\partial Y$, $\bar{\varepsilon} = \nu_e\big\langle(\partial u_i'\partial X_k)(\partial u_i'\partial X_k)\big\rangle$ is the specific dissipation of energy and D_E is the effective diffusion coefficient. The respective terms on the right-hand side of (16.156) describe the processes of diffusion, production, and dissipation of energy. The diffusional term is written in the gradient form with the effective diffusion coefficient D_E. The parameters $\bar{\varepsilon}$ and D_E entering in this equation were given by Kolmogorov based on dimensional considerations:

$$\bar{\varepsilon} = C_E L^{-1} e_k^{3/2} \,, \quad D_E = C_D L \sqrt{e_k} \,. \tag{16.157}$$

Making a correction to account for molecular viscosity, we rewrite (16.157) as

$$\bar{\varepsilon} = C_E L^{-1} e_k^{3/2} + C'_E \nu_e \frac{e_k}{L^2} \,, \quad D_E = C_D L \sqrt{e_k} + \nu_e \,. \tag{16.158}$$

Here, L is the integral scale of turbulence, and C_E, C'_E and C_D are empirical constants.

A shortcoming of this model is the need to specify the integral scale of turbulence L which depends on the flow pattern.

b) A Transport equation for the Reynolds stress $\tau' = -\langle u' \, v' \rangle$ is derived in [23]. For a plane-parallel flow of an incompressible fluid in a boundary layer, it has the following form:

$$\frac{D\tau'}{Dt} = a_1 \tau' \frac{\partial \langle u \rangle}{\partial Y} - a_1 \frac{\tau^{3/2}}{L} - a_1 \sqrt{\tau_m} - \frac{\partial}{\partial Y} (G \tau') \,. \tag{16.159}$$

The presence of the empirical functions L and G together with the empirical constants a_1 and τ_m is the main shortcoming of the model.

It should be noted that, in contrast to the model *a*) where the transport equation (16.156) is an equation of parabolic type owing to the first term on the right-hand side, (16.159) is hyperbolic.

c) Transport equation for turbulent viscosity ν_t was first proposed in [24] and later specified in more detail in [25]. For a plane boundary layer of incompressible fluid, it has the form

$$\frac{D\nu_t}{Dt} = \alpha \nu_t \left| \frac{\partial \langle u \rangle}{\partial Y} \right| - \gamma \frac{\nu_t (\nu + \beta \nu_t)}{s^2} + \frac{\partial}{\partial Y} \left[(\nu_e + \kappa \nu_t) \frac{\partial \nu_t}{\partial Y} \right] , \tag{16.160}$$

where s is the minimal distance from the wall; α, β, γ and κ are empirical constants.

One-parametric models of turbulence use the transport equation to determine only one of the quantities characterizing the turbulent flow. As a rule, this quantity is either the energy of turbulence e_k (see (16.156)) or the turbulent viscosity ν_t (see (16.160)). Somewhat less common, is the one-parametric model that uses the transport equation (16.159) for the Reynolds stress.

A serious disadvantage of one-parametric models is the necessity to specify the scale of turbulence L, which is not known beforehand and cannot be determined without additional hypotheses dependent on the type of the flow. For simple flows, the scale of turbulence can be determined through the governing parameters at a given point inside the flow. Thus, for a flow near a plane wall, L can be taken proportional to the distance between the given point and the wall, whereas for jet flows, L is usually taken proportional to the width of the jet. In complex turbulent flows, it is impossible to express L in terms of the governing parameters of the flow at a given point. Besides, in such flows, the scale of turbulence as well as other governing parameters usually depends not only on their values at a given point, but also on the entire prehistory of the flow, (for example, on the conditions at the channel entrance plus the boundary conditions).

d) A transport equation for the scale of turbulence L was proposed in [22] where it was used to calculate the shape of the turbulent boundary layer on a flat plate. The equation for $F = L^2/2$ has the form

$$u\frac{\partial F}{\partial X} + w\frac{\partial F}{\partial Y} = \nu_e\frac{\partial^2 F}{\partial Y^2} - C_L\frac{\nu_t}{e_k}\left(\frac{\partial u}{\partial Y}\right)^2 F + C_F\left|1 - \frac{2F}{s^2}\varphi\left(\frac{2F}{s^2}\right)\right|\frac{F^2}{e_k}\varepsilon_E\,,$$
$$w = \nu + 0.5\nu_e\frac{\partial \ln F}{\partial Y} - (\nu_e + D_E)\frac{\partial \ln e_k}{\partial Y}\,, \tag{16.161}$$

where e_k is the energy of turbulence, s,that is, the distance from the plate; $\varepsilon = \varphi e_k^{3/2} L$, C_L, C_F, and φ_1, that is, the empirical constants.

2. *Two-parametric models:*

Turbulence models that use two transport equations to determine the characteristics of turbulence are called two-parametric models. The majority of such models involve a transport equation for the energy of turbulence e_k and a transport equation for the specific dissipation energy $\bar{\varepsilon} = \langle(\partial u_i' \partial X_k)(\partial u_i' \partial X_k)\rangle$ or for the function $F = e_k^m L^n$.

The first two-parametric model was proposed in [19]. This work considers transport equations for e_k and the combination $\sqrt{e_k}/L$. The Reynolds stress is determined by the relation

$$\tau' = -\langle u'v'\rangle = C\sqrt{e_k}L\frac{\partial\langle u\rangle}{\partial Y}\,, \tag{16.162}$$

where C is a constant.

Paper [26] was the first to use a transport equation for $\bar{\varepsilon}$ in a two-parametric model. As of today, the most popular two-parametric models are ones that describe turbulent flows by two transport equations for the functions e_k and $\bar{\varepsilon}$. In many publications, the energy of turbulence is denoted by k, hence the commonly used term "k–ε models".

Consider two models: the $e - F$ and the k–ε model.

a) In addition to (16.156) for turbulent energy e_k, the two-parametric $e - F$ model contains the following for the function $F = e_k^m L^n$:

$$\frac{DF}{Dt} = \frac{\partial}{\partial Y}\left(D_F\frac{\partial F}{\partial Y}\right) - (C\sqrt{e_k}L + C_1\nu_e)\frac{e_k}{L^2} + \gamma F\frac{\partial\langle u\rangle}{\partial Y} + \Psi\,, \tag{16.163}$$

where $D_F = a_F\sqrt{e_k}L + \alpha_F\nu$; $L = (e_k^m/F)^{1/n}$; a_F and α_F are empirical constants, γ is a function of τ', e_k, and $\partial\langle u\rangle/\partial Y$; Ψ is a function that depends on

the sign of n: at $n < 0$, it is zero, while at $n > 0$, a special form of Ψ is required. The reason for such behavior of Ψ is that the first term on the right-hand side of (16.163) describes a "diffusion process" characterized by the diffusion coefficient D_F. For positive D_F, the maximal value of F must decrease with time, which happens only at $n < 0$. Therefore, in case $n > 0$ requires the presence of such a function Ψ that D_F would not change its sign.
Accordingly, all e–F models fall into two categories: with $n > 0$ and those with $n < 0$. The models with $n > 0$ use the function $e_k L$ or $\sqrt{e_k} L$ for F. The models with $n < 0$, on the other hand, use one of the functions $\sqrt{e_k}/L$, e_r/L^2 and $e_k^{3/2}/L$.
b) k–ε models use the transport equation for specific energy dissipation $\bar{\varepsilon}$ (see (16.100)). In the plane boundary layer approximation, this equation has the form

$$\frac{D\bar{\varepsilon}}{Dt} = \frac{\partial}{\partial Y}\left(D_\varepsilon \frac{\partial \bar{\varepsilon}}{\partial Y}\right) - C_1 f_1 \frac{\bar{\varepsilon}^2}{e_k} + C_2 f_2 \frac{\bar{\varepsilon}}{e_k}\frac{\partial \langle u \rangle}{\partial Y} + \Psi \,. \tag{16.164}$$

Here, $D_\varepsilon = a_\varepsilon \sqrt{e_k} L + \alpha_\varepsilon \nu_e$; C_1, C_2, a_ε, and α_ε – constants; f_1, f_2, and Ψ are functions that depend on the governing parameters.
Equation 16.164 has some peculiarities, one of which has to do with the behavior of $\bar{\varepsilon}$ near the wall.
Two-parametric models rely on the following relation to determine the Reynolds stress:

$$\tau' = -\langle u'v' \rangle = C_\mu f_\mu(\mathrm{Re}) \sqrt{e_k} L \frac{\partial \langle u \rangle}{\partial Y} \,. \tag{16.165}$$

The scale of turbulence L that appears in this formula is defined differently in different models. The purpose of the function $f_\mu(\mathrm{Re})$ is to describe the effect of viscosity on τ'. This dependence is not universal but varies with the type of flow.

3. *Three-parametric models:*
A distinguishing feature of three-parametric models that sets them apart from other models is that transport equations are written for all characteristics of turbulence that are employed by the model. Instead of introducing the turbulent viscosity to find the Reynolds stresses, these models rely on the corresponding transport equation whose structure is similar to that of the transport equation for the energy of turbulence. Models of this type are sometimes called Reynolds stress models.
Three-parametric models include transport equations for the shear stress $\tau' = -\langle u' v' \rangle$, for the energy of turbulence $e_k = 0.5 \sum \left\langle u_i'^2 \right\rangle$ and for the parameter $F = e_k^m L^n$.
Since the equations for all three characteristics of turbulence have identical structure, they can be written in the general form

$$\frac{d\Phi}{dt} = -D_\Phi \frac{\Phi}{L^2} + \gamma_\Phi \Gamma_\Phi \frac{\partial \langle u \rangle}{\partial Y} + \frac{\partial}{\partial Y}\left(D_\Phi^* \frac{\partial \langle u \rangle}{\partial Y}\right), \tag{16.166}$$

where Φ successively assumes the values e_k, τ' and $F = e_k^m L^n$. The values of D_Φ, γ_Φ, Γ_Φ, and D_Φ^* are different for each of these equations:

$$D_E = \alpha_E \sqrt{e_k} L + \beta_E \nu \, ; \quad \Gamma_E = \tau' \, ; \quad \gamma_E = 1 \, ;$$

$$D_\tau = \alpha_\tau \sqrt{e_k} L + \beta_\tau \nu_e \, ; \quad \Gamma_\tau = e_k \, ; \quad \gamma_\tau = m \frac{\tau}{E_\tau} - n \gamma'_F \text{sign} \left(\frac{\partial \langle u \rangle}{\partial Y} \right) ,$$

where $\alpha_E, \beta_E, \alpha_\tau, \beta_\tau, \gamma_\tau$, and γ'_F are constants.
A number of current publications present theoretical results for stationary turbulent flows obtained from three-parametric models. The experience with three-parametric (as well as other, more sophisticated models) indicates that as we increase the number of differential equations (e.g., by using equations for third- order moments), we have to deal with an ever-increasing number of empirical constants without any gain in accuracy or versatility of the models.
Finally, it should be noted that the division of all existing models of turbulence into semi-empirical models and models that employ transport equations for the characteristics of turbulence is really a matter of convention because models of the second type also rely on experiments to garner the functions and constants involved, and to determine the range of applicability. In this sense, all models are semi-empirical.

References

1 Prandtl, L. (1956) *Führer durch die Strömungslehre*, Braunschweig.
2 Batchelor, G.K. (1953) *The Theory of Homogeneous Turbulence*, Cambridge University Press, Cambridge.
3 Townsend, A.A. (1976) *The Structure of Turbulent Shear Flow*, Cambridge University Press, Cambridge.
4 Levich, V.G. (1962) *Physicochemical Hydrodynamics*, Prentice-Hall, Englewood Cliffs, N.J.
5 Abramovitch, G.N. (1960) *Theory of turbulent jets*, Physmatgis, Moscow.
6 Hinze, J.O. (1959) *Turbulence*, McGraw-Hill, New York.
7 Monin, A.S. and Yaglom, A.M. *Statistical Fluid Mechanics: Mechanics of Turbulence*, V.1 (1971), V.2 (1975), MIT Press, Cambridge, MA.
8 Loitsyanskii, L.G. (1970) *Mechanics of Fluid and Gas*, Nauka, Moscow.
9 Launder B.E. and Spalding D.B. (1972) *Mathematical Models of Turbulence*, Academic Press, London, New York.
10 Kuznetsov, Y.R. and Sabel'nikov, V.A. (1990) *Turbulence and Combustion*, Hemisphere, New York.
11 Landau, L.D. and Lifshiz E.M. (1987) *Fluid Mechanics*, Pergamon Press, Oxford.
12 Kolmogorov, A.N. (1941) Local structure of turbulence in incompressible fluid at very high Reynolds number. *Dokl. Akad. Nauk SSSR* **30**(4), 299–303 (in Russian).
13 Kolmogorov, A.N. (1941) Energy dissipation by local isotrope turbulence. *Dokl. Akad. Nauk USSR* **32**(1), 19–21 (in Russian).
14 Kolmogorov, A.N. (1941) To degeneracy of isotrope turbulence in incompressible viscous fluid. *Dokl. Akad. Nauk SSSR* **31**(6), 538–541 (in Russian).
15 Taylor, G.I. (1935) Statistical theory of turbulence. I–IV. *Proc. Roy. Soc. A.* **151**(874), 421–478.
16 Ginevski, A.S., Ioselevitch, V.A., Kolesnikov, A.V., Lapin, J.V., Pilipenko, V.N., and Sekundov, A.N. (1978) Meth-

ods of calculation of turbulent boundary layer. *Itogi Nauki i Techniki. VINITI Ser. Mechanics of Fluid and Gas* **11**, 155–304 (in Russian).

17 Lushik, V.G., Pavelev, A.A., and Jakubenko, A.E. (1988) Transport equations for characteristics of turbulence: Models and results of calculations. *Itogi Nauki i Techniki. VINITI. Ser. Mechanics of Fluid and Gas.* **22**, 3–61 (in Russian).

18 Lushik, V.G., Pavelev, A.A., and Jakubenko, A.E. (1994) Turbulent flows. Models and numerical investigations. *Fluid Dyn.* **4**, 4–27 (in Russian).

19 Kolmogorov, A.N. (1942) Equation of turbulent flow of incompressible fluid. *Izvestia Acad. Sci. USSR. Ser. Phys.* **6**(1–2), 56–58 (in Russian).

20 Rotta, J.C. (1951) Statistische Theorie nichthomogener Turbulenz. *Z. Phys.* **129**(5), 547–572, **131**(1), 51–77.

21 Glushko, G.S. (1965) Turbulent boundary layer on a plane plate in incompressible fluid. *Izvestia Acad. Nauk. SSSR, Ser. Fluid Dyn.* **4**, 13–23 (in Russian).

22 Glushko, G.S. (1970) Differential equation for scale of turbulence and calculation of turbulent boundary layer on a plane plate. *In Collected Articles: Turbulent Flows*, Moscow p. 37–44 (in Russian).

23 Bradshaw, P., Ferris, D.H., and Atwell, N.P. (1967) Calculation of boundary layer development using the turbulent energy equation. *J. Fluid Mech.* **28**(3), 593–616.

24 Kovasznay, L.S.G. (1967) Structure of the turbulent boundary layer. *Phys. Fluids.* **10**(9), 25–30.

25 Sekundov, A.N. (1971) Use of differential equations for turbulent viscosity in analysis of plane nonself-similar flows. *Izvestia Acad. Nauk SSSR. Ser. Fluid Dyn.* **5**, 114–127 (in Russian).

26 Davidov, B.I. (1961) To statistical dynamics of incompressible turbulent fluid. *Dokl. Akad. Nauk USSR* **136**(1), 47–50 (in Russian).

Appendix A
Foundations of Vectorial and Tensorial Analysis

The equations of continuum mechanics provide a mathematical formulation of physical laws independent of the choice of the system of coordinates that is invariant, where some physical parameters of the equations can change from one system of coordinates to the other. Thus, the physical laws and their mathematical form have a covariant shape.

The goal of vectorial and tensorial analysis consists of examining the characteristics of geometrical properties for the properties of physical quantities that are invariant under transition from one system of coordinates to another.

Physical parameters that characterize the state of the continuous medium are divided in scalars, vectors and tensors.

A parameter that only has one value at a given point and given time is called a scalar, for example, mass density, pressure, temperature and so on. Scalar values can change as a function of time in a given point and have different values at a given time though in different points. Therefore, scalars are functions $f(X^1, X^2, X^3, t)$ of the coordinates of the point and time t. From the definition of a scalar, it follows that a scalar does not depend on the system of coordinates.

A parameter that also has a direction in addition to a value at a given time and given point is called a vector, for example, force, velocity, electrical and magnetical field strength and so on. In a chosen system of coordinates, the vector coordinates are characterized via the cosines l^i of the angles between the vector and the coordinate axes and the absolute value (vector length) F. The vector is written as $\boldsymbol{F}(F^1, F^2, F^3)$, with the coordinates $F^i = Fl^i$ that uniquely determine the parameter $\boldsymbol{F}$ in the chosen system of coordinates $\{X^i\}$. The transition from one system of coordinates $\{X^i\}$ to the other, $\{\xi^i\}$, results in a change of l^i and thus of F^i. For the solution of practical problems, one uses different systems of coordinates (orthogonal and curvilinear ones) which are fixed in space or bound to objects that are moving in space and time. If the system of coordinates is bound to a continuum particle which is deformed during its movement, the system of coordinates is also deformed. Therefore, one needs to know how the coordinates of a vector change under transition from one system of coordinates to another.

In addition to scalars and vectors, there are physical objects called tensors. For example, consider the stress $\boldsymbol{T}$ experienced by the surface element S with unit normal vector $\boldsymbol{n}$ under the influence of an applied force. According to the definition,

Hydromechanics. Theory and Fundamentals. Emmanuil G. Sinaiski

ISBN: 978-3-527-41026-2

T is equal to the force F acting on the unit surface element. Therefore, according to this definition, one should take $T = F/S$. Since there is no procedure to divide by a vector, one has to write this relationship as the product

$$F = T \cdot S \,. \tag{A1}$$

The vectors F and S generally have different directions and thus one can write (A1) in component form as

$$F^i = \sum_{j=1}^{3} T^i_{\cdot j} S^j \,, \quad (i, j = 1, 2, 3) \,. \tag{A2}$$

Equations A1 and (A2) can be considered as the physical definition of a tensor of second rank. In the following, we will give the exact mathematical definition of a tensor.

From (A2), it follows that the stress tensor T is determined by nine components $T^i_{\cdot j}$ $(i, j = 1, 2, 3)$. Since the tensor T connects two vectors F and S, its components change during the transition from one system of coordinates to another.

It must be noted that the subdivision of parameters into scalars, vectors and tensors is a relative concept because scalars and vectors are themselves tensors of zeroeth and first rank.

A.1 Vectors

A space is called affine space if the same system of coordinates can be chosen in the whole space such that each object (vector, tensor) can at each point be expressed via the basis vectors of the system of coordinates. If one defines a scalar product for the space and also introduces a metric (metric tensor), the space becomes Euclidean. An orthogonal basis and straight coordinate axes define a Cartesian system of coordinates. By using a basis of linear independent non-orthogonal vectors that are directed along the tangents of the corresponding coordinate axes, one defines a curvilinear system of coordinates. A space for which a system of coordinates applicable to the whole space does not exist is called a Riemann space.

In the three dimensional Euclidean space and the Cartesian system of coordinates, the coordinates of the point $P(X^1, X^2, X^3)$ are defined via the corresponding lengths r^i of the projections of the line OP on the axes OX^1, OX^2, OX^3 such that each vector $\mathbf{r}$ with origin O can be expanded as a linear combination of unit vectors e_1, e_2, e_3, that is,

$$r = \sum_{i=1}^{3} r^i e_i \,. \tag{A3}$$

The quantities r^1, r^2, r^3 are called the components of the vector r. A vector is defined uniquely via its components (see Figure A.1). The vectors $\mathbf{e}_1, \mathbf{e}_2, \mathbf{e}_3$ constitute

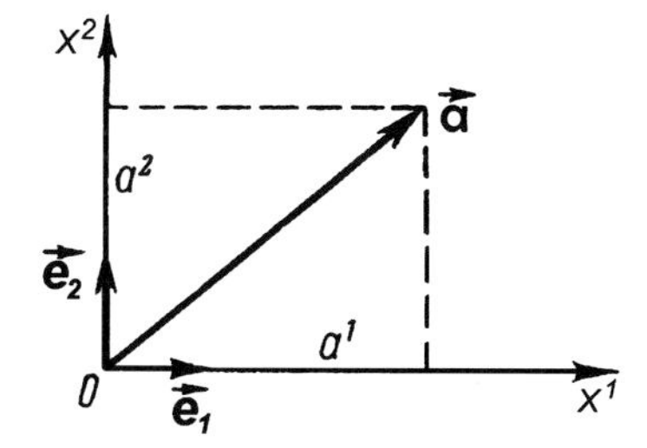

Figure A.1 Orthogonal Cartesian coordinates of the vector $\boldsymbol{a}$.

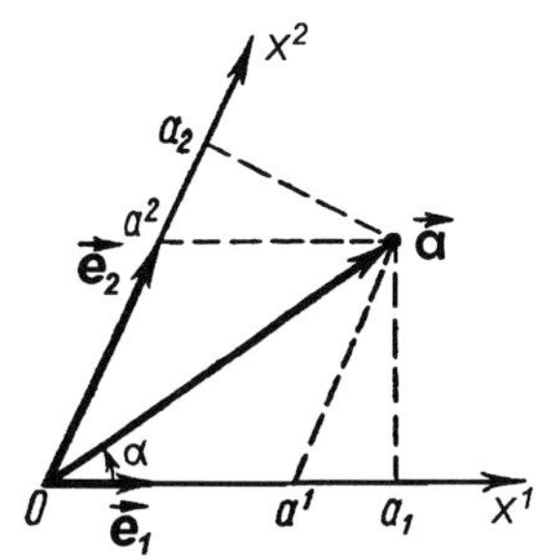

Figure A.2 Contravariant (a^1, a^2) and covariant (a_1, a_2) components of a vector $\boldsymbol{a}$.

a right-handed system if they have the same orientation as thumb, index finger and middle finger of the right hand, otherwise one refers to a left-handed system.

If the system of coordinates is not orthogonal, one can define the vectors $\boldsymbol{r}$ in two ways: via components r^i as per (A3) or via orthogonal projections $r_i = \boldsymbol{r} \cdot \boldsymbol{e}_i$ of the vector on the coordinate axes. The components r^i are called contravariant and the components r_i covariant vector coordinates (see Figure A.2).

It must be noted, that the vectors $\boldsymbol{e}_i$ do not have to be unit vectors since each vector can be decomposed into three vectors that are not in the same plane. To simplify the notation, (A3) is rewritten in the form $\boldsymbol{r} = \mathrm{r}^{\mathrm{i}}\boldsymbol{e}_{\mathrm{i}}$ with the normal convention that summation is assumed for each upper and lower indices. Such indices are called silent. Indices that are not repeated are called free indices. During algebraic transformations, silent indices can be renamed in an arbitrary manner, for example, $r^i \boldsymbol{e}_i = r^j \boldsymbol{e}_j$ and so on.

In practice, one frequently uses curvilinear systems of coordinates (cylindrical, spherical and so on). Therefore, we consider a curvilinear system of coordinates $\{\alpha^1, \alpha^2, \alpha^3\}$ (see Figure A.3) together with the orthogonal system of coordinates $\{X^1, X^2, X^3\}$ with unit vectors $\boldsymbol{e}_i$. Changing from the coordinates X^i to new coordinates α^j, the X^i are functions of the α^j:

$$X^i = X^i(\alpha^1, \alpha^2, \alpha^3) \,. \tag{A4}$$

We assume that these functions are differentiable. Then, the radius vector $\boldsymbol{r}$ that connects the origin of the orthogonal system of coordinates with the point P is a differentiable function $\boldsymbol{r}\left(\alpha^1, \alpha^2, \alpha^3\right)$. It can be easily shown that the vectors $\partial \boldsymbol{r}/\partial \alpha^i$

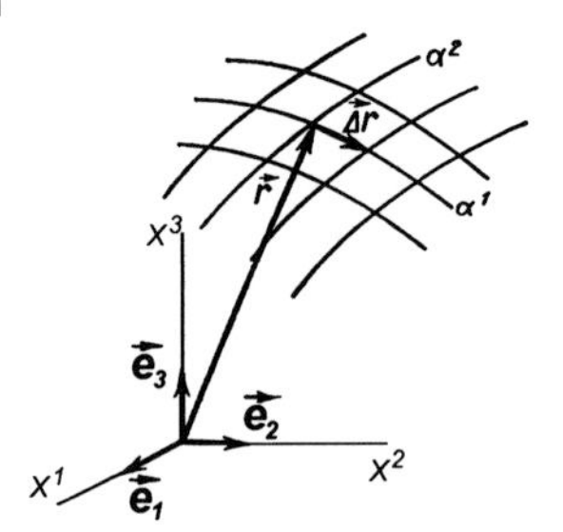

Figure A.3 Curvilinear system of coordinates.

are directed along the tangents on the coordinate lines α^i at the point P. Therefore, one can use the vectors

$$\boldsymbol{\varepsilon}_i = \frac{\partial \boldsymbol{r}}{\partial \alpha^i} \tag{A5}$$

as a local basis under the condition that they are not coplanar. This condition is fulfilled if the determinant does not vanish, that is,

$$D = \det \left| \partial \boldsymbol{r}/\partial \alpha^1, \partial \boldsymbol{r}/\partial \alpha^2, \partial \boldsymbol{r}/\partial \alpha^3 \right| \neq 0\,. \tag{A6}$$

The inequality (A6) means that the transformation (A4) can be inverted, that is, the new coordinates can be expressed in terms of the old coordinates

$$\alpha^i = \alpha^i \left(X^1, X^2, X^3 \right)\,, \tag{A7}$$

where the components of the Jacobian's $X^i_{.j} = \partial X^i/\partial \alpha^j$ and $Y^i_{.j} = \partial \alpha^i/\partial X^j$ of the transformations (A4) and (A7) are the inverse of each other. The basis sets $\boldsymbol{e}_i$ and $\boldsymbol{\varepsilon}_i$ can be expressed in terms of each other as follows:

$$\boldsymbol{\varepsilon}_j = X^i_{.j} \boldsymbol{e}_i\,, \quad \boldsymbol{e}_j = Y^i_{.j} \boldsymbol{\varepsilon}_i\,. \tag{A8}$$

Since the new system of coordinates is not necessarily orthogonal, each vector $\boldsymbol{r}$ can be expressed in the local basis through contravariant coordinates r^i or covariant r_j coordinates, with

$$r_j = \boldsymbol{r} \cdot \boldsymbol{\varepsilon}_j = r^i \boldsymbol{\varepsilon}_i \cdot \boldsymbol{\varepsilon}_j = r^i g_{ij}\,. \tag{A9}$$

The matrix with elements

$$g_{ij} = \boldsymbol{\varepsilon}_i \cdot \boldsymbol{\varepsilon}_j = \boldsymbol{\varepsilon}_i = \frac{\partial \boldsymbol{r}}{\partial \alpha^i} \cdot \frac{\partial \boldsymbol{r}}{\partial \alpha^j} \tag{A10}$$

is called fundamental matrix. It is symmetric and its determinant does not vanish, that is, $g = \left| g_{ij} \right| \neq 0$, and therefore the inverse matrix g^{ij} exists with

$$g_{ik} g^{kj} = \delta^{j\cdot}_{\cdot i}\,. \tag{A11}$$

Here, $\delta^{j\cdot}_{\cdot i}$ is the Kronecker symbol (0 for $i \neq j$; 1 for $i = j$).

From (A9)–(A11), one can obtain the relationship between contravariant and covariant coordinates of a vector $\boldsymbol{r}$, namely,

$$r_j = r^i g_{ij}\,, \quad r^i = r_j g^{ji}\,. \tag{A12}$$

It must be noted that we have $r_i = r^i$ in the Cartesian system of coordinates and

$$g_{ij} = g^{ij} = \delta^{j\cdot}_{\cdot i}\,.$$

The formulae (A12) show that one can raise or lower the indices of vector components via direct or inverse fundamental matrices. Sometimes this operation is called juggling of indices.

The scalar product can be expressed in terms of contravariant or covariant vector coordinates as follows:

$$\boldsymbol{a}\cdot\boldsymbol{b} = a^i b^j g_{ij} = a^i b_i = a_i b_j g^{ij} = a_i b^i\,. \tag{A13}$$

On the other hand, the scalar product is equal to

$$\boldsymbol{a}\cdot\boldsymbol{b} = |\boldsymbol{a}|\,|\boldsymbol{b}|\cos\vartheta\,.$$

By using (A13), one can obtain the vector length $\boldsymbol{a}$ and the angle ϑ between the vectors $\boldsymbol{a}$ and $\boldsymbol{b}$ as

$$|\boldsymbol{a}| = \left(a^i a^j g_{ij}\right)^{1/2}\,, \quad \cos\vartheta = \frac{a^i b^j g_{ij}}{|\boldsymbol{a}|\,|\boldsymbol{b}|}\,.$$

Together with the vectors $\boldsymbol{\varepsilon}_i$, we introduce the vectors $\boldsymbol{\varepsilon}^i$ via the following relationships

$$\boldsymbol{\varepsilon}^i = g^{ij}\boldsymbol{\varepsilon}_j\,. \tag{A14}$$

Since $\boldsymbol{\varepsilon}^i\cdot\boldsymbol{\varepsilon}_k = \delta^{i}_{\cdot k}, \boldsymbol{\varepsilon}^i\cdot\boldsymbol{\varepsilon}^k = g^{ik}$ and for $i \neq k\delta^{i}_{\cdot k} = 0$, the vectors $\boldsymbol{\varepsilon}^i$ and $\boldsymbol{\varepsilon}_k$ are orthogonal for $i \neq k$.

Therefore, they can be used as basis vectors of a curvilinear system of coordinates where the components of the vector $\boldsymbol{r}$ in this basis are covariant components, that is,

$$\boldsymbol{r} = r^i\boldsymbol{\varepsilon}_i = r_j\boldsymbol{\varepsilon}^j \tag{A15}$$

and $r^i = \boldsymbol{r}\cdot\boldsymbol{\varepsilon}^i, r_j = \boldsymbol{r}\cdot\boldsymbol{\varepsilon}_j$. The basis $\boldsymbol{\varepsilon}^i$ is called the reciprocal basis relative to the basis $\boldsymbol{\varepsilon}_j$.

The main difference between the two bases is that the basis $\boldsymbol{\varepsilon}_j$ is directly bound to the system of coordinates since the vectors $\boldsymbol{\varepsilon}_j$ are tangents on the coordinate lines while the basis $\boldsymbol{\varepsilon}^i$ has no connection with the system of coordinates. The latter one is also called a nonholonomic basis.

We now consider another operation with vectors, the vector product. The vector product $\boldsymbol{a}\times\boldsymbol{b}$, also sometimes denoted by $\boldsymbol{a}\wedge\boldsymbol{b}$, of the vectors $\boldsymbol{a}$ and $\boldsymbol{b}$ is defined to be a vector $\boldsymbol{c}$ of length $|\boldsymbol{a} \times \boldsymbol{b}| = |\boldsymbol{a}||\boldsymbol{b}| \sin(\boldsymbol{a}, \boldsymbol{b})$, which is orthogonal to the surface spanned by $\boldsymbol{a}$ and $\boldsymbol{b}$, and chosen such that $\boldsymbol{a}, \boldsymbol{b}, \boldsymbol{c}$ constitute a right-handed system. Using the decomposition into basis vectors $\boldsymbol{\varepsilon}^i$ or $\boldsymbol{\varepsilon}_i$, the vector $\boldsymbol{a} \times \boldsymbol{b}$ can be written as

$$\boldsymbol{a} \times \boldsymbol{b} = \sqrt{g}\varepsilon_{ijk} a^i b^j \boldsymbol{\varepsilon}^k = \frac{1}{\sqrt{g}} \varepsilon^{ijk} a_i b_j \boldsymbol{\varepsilon}_k . \tag{A16}$$

Here, $g = \det(g_{ij})$ and ε_{ijk}, ε^{ijk} are the permutation symbols

$$\varepsilon_{ijk} = \varepsilon^{ijk} = \begin{cases} 0 , & \text{if two or more of } i, j, k \text{ are equal} \\ 1 , & \text{if } i, j, k \text{ are different and constitute} \\ & \text{an even permutation} \\ -1 , & \text{if } i, j, k \text{ are different and constitute} \\ & \text{an even permutation} \end{cases} .$$

From (A16), it follows that $\boldsymbol{a} \times \boldsymbol{b} = -\mathbf{b} \times \boldsymbol{a}$. In addition, if the basis $\boldsymbol{\varepsilon}_i$ consists of unit vectors, we have $\boldsymbol{\varepsilon}_i \times \boldsymbol{\varepsilon}_j = \boldsymbol{\varepsilon}_k$ if $i \neq j \neq k$ and i, j, k constitute an even permutation.

By using the vector product, one can obtain the reciprocal basis set of $\boldsymbol{\varepsilon}^i$ relative $\boldsymbol{\varepsilon}_k$ and inversely the reciprocal basis of $\boldsymbol{\varepsilon}_i$ relative $\boldsymbol{\varepsilon}^k$, namely,

$$\begin{aligned} \boldsymbol{\varepsilon}^i &= \frac{\boldsymbol{\varepsilon}_j \times \boldsymbol{\varepsilon}_k}{E} = \frac{1}{\sqrt{g}} \varepsilon^{ijk} \boldsymbol{\varepsilon}_j \times \boldsymbol{\varepsilon}_k , \quad (i \neq j \neq k) , \\ \boldsymbol{\varepsilon}_i &= \frac{\boldsymbol{\varepsilon}^j \times \boldsymbol{\varepsilon}^k}{E_1} = \sqrt{g}\varepsilon_{ijk} \boldsymbol{\varepsilon}^j \times \boldsymbol{\varepsilon}^k , \quad (i \neq j \neq k) . \end{aligned} \tag{A17}$$

Here, E and E_1 are the scalar triple products of the basis vectors

$$E = \boldsymbol{\varepsilon}_1 \cdot \boldsymbol{\varepsilon}_2 \times \boldsymbol{\varepsilon}_3 = (\boldsymbol{\varepsilon}_1, \boldsymbol{\varepsilon}_2, \boldsymbol{\varepsilon}_3) ; \quad E_1 = \boldsymbol{\varepsilon}^1 \cdot \boldsymbol{\varepsilon}^2 \times \boldsymbol{\varepsilon}^3 = (\boldsymbol{\varepsilon}^1, \boldsymbol{\varepsilon}^2, \boldsymbol{\varepsilon}^3) ,$$

with

$$(\mathbf{X}, \mathbf{Y}, \mathbf{Z}) = \sqrt{g}\varepsilon_{ijk} X^i Y^j Z^k = \frac{1}{\sqrt{g}} \varepsilon^{ijk} X_i Y_j Z_k . \tag{A18}$$

For the basis vectors, we have

$$(\boldsymbol{\varepsilon}_1, \boldsymbol{\varepsilon}_2, \boldsymbol{\varepsilon}_3) = \sqrt{g} , \quad (\boldsymbol{\varepsilon}^1, \boldsymbol{\varepsilon}^2, \boldsymbol{\varepsilon}^3) = \frac{1}{\sqrt{g}} . \tag{A19}$$

The vectors $\boldsymbol{\varepsilon}^k$ and $\boldsymbol{\varepsilon}_i$ have the following properties

$$\begin{aligned} &\boldsymbol{\varepsilon}^i \times \boldsymbol{\varepsilon}_j = \delta^i_{\cdot j} , \quad E_1 = \boldsymbol{\varepsilon}^1 \cdot \boldsymbol{\varepsilon}^2 \times \boldsymbol{\varepsilon}^3 = E^{-1} , \\ &\boldsymbol{\varepsilon}^i \times \boldsymbol{\varepsilon}^j = \frac{1}{\sqrt{g}} \varepsilon^{ijk} \boldsymbol{\varepsilon}_k , \quad \boldsymbol{\varepsilon}_i \times \boldsymbol{\varepsilon}_j = \sqrt{g}\varepsilon_{ijk} \boldsymbol{\varepsilon}^k . \end{aligned} \tag{A20}$$

If the basis vectors $\boldsymbol{\varepsilon}_i$ are orthogonal, we have $\left|\boldsymbol{\varepsilon}^k\right| = |\boldsymbol{\varepsilon}_k|^{-1}$.

We now consider how the vector coordinates are transformed under the transition from the old curvilinear system of coordinates $\{\alpha^1, \alpha^2, \alpha^3\}$ to the new system of coordinates $\{\alpha^{1'}, \alpha^{2'}, \alpha^{3'}\}$, while the coordinates of both systems are related as follows:

$$\alpha^{i'} = \alpha^{i'}(\alpha^1, \alpha^2, \alpha^3) . \tag{A21}$$

We begin with the transformation of contravariant coordinates. If the Jacobian determinant of this transformation is non-vanishing, that is, if $|A^{i'}_{\cdot j}| = |\partial\alpha^{i'}/\partial\alpha_j| \neq 0$, the inverse transformation $\alpha^i = \alpha^i(\alpha^{1'}, \alpha^{2'}, \alpha^{3'})$ exists, whose Jacobian matrix $B^i_{\cdot j'}$ is the inverse of the matrix $A^{i'}_{\cdot j}$ with $A^{i'}_{.k} B^k_{.j'} = \delta^{i'}_{\cdot j'}$ and $B^i_{.k'} A^{k'}_{\cdot j} = \delta^i_{\cdot j}$. The vectors of new $\boldsymbol{\varepsilon}'_i$ and old $\boldsymbol{\varepsilon}_i$ local bases are expressed as follows in terms of each other

$$\boldsymbol{\varepsilon}'_i = \frac{\partial \boldsymbol{r}}{\partial \alpha^{i'}} = \frac{\partial \boldsymbol{r}}{\partial \alpha^k} B^k_{.i'} = \boldsymbol{\varepsilon}_k B^k_{.i'} , \quad \boldsymbol{\varepsilon}_i = \frac{\partial \boldsymbol{r}}{\partial \alpha^i} = \frac{\partial \boldsymbol{r}}{\partial \alpha^{k'}} A^{k'}_{.i} = \boldsymbol{\varepsilon}'_k A^{k'}_{.i} . \tag{A22}$$

Using the representations of the vector $\boldsymbol{r}$ in both systems of coordinates

$$\boldsymbol{r} = r^k \boldsymbol{\varepsilon}_k = r^{i'} \boldsymbol{\varepsilon}'_i = r^{i'} \boldsymbol{\varepsilon}_k B^k_{.i'} ,$$

one obtains the transformation law for the vector components

$$r^{k'} = A^{k'}_{\cdot i} r^i . \tag{A23}$$

Thus, under the transition from the old to the new system of coordinates, the contravariant vector components are transformed via the Jacobian matrix $A^{k'}_{\cdot i}$.

To obtain the transformation law for covariant vector components, we use the definition of covariant components $r_k = \boldsymbol{r} \cdot \boldsymbol{\varepsilon}_k$ and (A22). As a result, one obtains

$$r'_k = r_i B^i_{.k'} . \tag{A24}$$

This means that the covariant vector components r_k are transformed via the matrix $B^i_{\cdot k'}$ which is the transposed inverse of the matrix $A^{k'}_{\cdot i}$.

Since the basis vectors $\boldsymbol{\varepsilon}_i$ are transformed according to the law (A22), which is similar to (A24) for the transformation of covariant vector components, the basis $\boldsymbol{\varepsilon}_i$ is called the covariant basis of the system of coordinates.

To obtain the transformation law for the vectors of the reciprocal basis $\boldsymbol{\varepsilon}^i$, we use the decomposition of the vector $\boldsymbol{r}$ in terms of old and new bases and the relationships (A22). As a result, one obtains

$$vecr = r_i \boldsymbol{\varepsilon}^i = r_{k'} \boldsymbol{\varepsilon}^{k'} = r_i B^i_{.k'} \boldsymbol{\varepsilon}^{k'}$$

and

$$\boldsymbol{\varepsilon}^i = B^i_{.k'} \boldsymbol{\varepsilon}^{k'} , \quad \boldsymbol{\varepsilon}^{k'} = A^{k'}_{\cdot i} \boldsymbol{\varepsilon}^i . \tag{A25}$$

Thus, the vectors of the reciprocal basis $\boldsymbol{\varepsilon}^{k'}$ are transformed as contravariant vector components (see (A23)). Therefore, one also calls the reciprocal basis the contravariant basis. The transformation laws (A23) and (A24) can serve as formal definitions of covariant vector components.

Using (A21)–(A24), one can obtain the transformation laws of the components of the fundamental matrix and its inverse

$$g^{k'm'} = A^{k'}_{\cdot i} A^{m'}_{\cdot j} g^{ij} , \quad g_{k'm'} = B^{i}_{\cdot k'} B^{j}_{\cdot m'} g_{ij} . \tag{A26}$$

Until now, we have considered such parameters that change under transition from one system of coordinates to another. However, there are some parameters that do not change under transformation of the coordinates. Such parameters are called invariants. The simplest invariant quantity is a scalar. As far a vector is concerned, one can refer to it as a invariant physical object, for example, the speed at a given point does not depend on the system of coordinates. However, according to the laws (A23) and (A24), the components do. In the following, we will give some general methods that can be used to obtain the invariants. We also note that there already is one invariant for vectors in Euclidean space. This is the length of the vector.

We now consider the operation of differentiation for vectors. To this end, one has to consider a vector field instead of individual vectors. A continuous vector field is defined by giving, for each point $M(X^1, X^2, X^3)$, a vector $\boldsymbol{r}(M)$, which changes continuously from point to point. Then, the vector components $r^i(M)$ also depend on the point M. When moving from point M to the neighboring point M', the vector $\boldsymbol{r}$ changes by $\Delta \boldsymbol{r}$. This differential can be represented in terms of the basis vectors $\boldsymbol{\varepsilon}_k$.

We first consider the differentiation of a vector $\boldsymbol{r}$ in an orthogonal system of coordinates. Then, one can represent this vector in the form (A3) where the basis vectors are $\boldsymbol{e}_i = \text{const}$. One obtains

$$\frac{\partial \boldsymbol{r}}{\partial X^k} = \frac{\partial \left(r^i \boldsymbol{e}_i \right)}{\partial X^k} = \frac{\partial r^i}{\partial X^k} \boldsymbol{e}_i = \nabla_k r^i \boldsymbol{e}_i = \nabla \boldsymbol{r} . \tag{A27}$$

The quantities $\partial r^i / \partial X^k = \nabla^k_i r$ are called absolute derivatives. They are components of a tensor of second order. The absolute derivative of a scalar $r(M)$ is $\partial r / \partial X^i = \nabla_i r$ such that $\nabla_i r$ are the components of the vector gradient and that the differential of the scalar field $dr = \partial r^i / \partial X^i dX^i$ is a scalar and thus an invariant. According to the definition, the absolute derivative of a vector is the gradient of the vector field $\nabla \boldsymbol{r}$.

We now consider a curvilinear system of coordinates $\{\alpha^1, \alpha^2, \alpha^3\}$ with local basis $\boldsymbol{\varepsilon}_i$ in point M. In contrast to an orthogonal system of coordinates, the basis vectors $\boldsymbol{\varepsilon}_i$ for a curvilinear system of coordinates change during the transition to a neighboring point M'. We consider the partial derivatives of the basis vectors $\partial \boldsymbol{\varepsilon}_i / \partial \alpha^j$ and decompose them in terms of $\boldsymbol{\varepsilon}_k$:

$$\frac{\partial \boldsymbol{\varepsilon}_i}{\partial \alpha^j} = \Gamma^k_{ij} \boldsymbol{\varepsilon}_k . \tag{A28}$$

The coefficients Γ^k_{ij} (sometimes denoted by $\left\{ \begin{matrix} k \\ ij \end{matrix} \right\}$) are called Christoffel symbols of the second kind (connection coefficients). From (A5) and (A28), it follows that $\Gamma_{ij}^{k} = \Gamma_{ji}^{k}$. Therefore, the Christoffel symbols of the second kind for the Euclidean space are symmetric with respect to the lower indices:

$$\Gamma_{ij}^{k}\boldsymbol{\varepsilon}_k = \frac{\partial \boldsymbol{\varepsilon}_i}{\partial \alpha^j} = \frac{\partial^2 \boldsymbol{r}}{\partial \alpha^i \partial \alpha^j} = \frac{\partial^2 \boldsymbol{r}}{\partial \alpha^j \partial \alpha^i} = \Gamma_{ji}^{k}\boldsymbol{\varepsilon}_k \,. \tag{A29}$$

The Christoffel symbols of the second kind can be expressed in terms of the fundamental matrix g_{ij}. From (A10), it follows that

$$\begin{aligned} \frac{\partial g_{ij}}{\partial \alpha^k} &= \frac{\partial \boldsymbol{\varepsilon}_i}{\partial \alpha^k} \cdot \boldsymbol{\varepsilon}_j + \boldsymbol{\varepsilon}_i \cdot \frac{\partial \boldsymbol{\varepsilon}_j}{\partial \alpha^k} = \Gamma_{ik}^{m}\boldsymbol{\varepsilon}_m \cdot \boldsymbol{\varepsilon}_j + \Gamma_{jk}^{m}\boldsymbol{\varepsilon}_m \cdot \boldsymbol{\varepsilon}_i \\ &= \Gamma_{ik}^{m} g_{mj} + \Gamma_{jk}^{m} g_{mi} \end{aligned}$$

and

$$\frac{\partial g_{kj}}{\partial \alpha^i} = \Gamma_{ik}^{m} g_{mj} + \Gamma_{ji}^{m} g_{mk} \,, \qquad \frac{\partial g_{ik}}{\partial \alpha^j} = \Gamma_{ij}^{m} g_{mk} + \Gamma_{jk}^{m} g_{mi} \,.$$

By adding the last two equations, subtracting the equation on the line above them, using the symmetry of Γ_{ij}^{k} with respect to lower indices, multiplying both sides with $1/2g^{kl}$ and finally summing the result over k, one obtains

$$\Gamma_{ij}^{l} = \frac{1}{2} g^{kl} \left(\frac{\partial g_{kj}}{\partial \alpha^i} + \frac{\partial g_{ik}}{\partial \alpha^j} - \frac{\partial g_{ij}}{\partial \alpha^k} \right) . \tag{A30}$$

Therefore, in Euclidean space, the Christoffel symbols of the second kind are symmetric with respect to the lower indices and we also have $\Gamma_{ij}^{k} = 0$ in the Cartesian system of coordinates since the basis vectors are constant in this system of coordinates. In Riemannian space, we have $g_{ij} \neq \text{const}$ and $\Gamma_{ij}^{k} \neq 0$.

Since each vector can be expanded in terms of vectors of the covariant basis $\boldsymbol{\varepsilon}_i$ or the contravariant basis $\boldsymbol{\varepsilon}^i$, the derivatives of $\boldsymbol{r}$ in a curvilinear system of coordinates have a different form depending on the choice of the local basis. In the covariant basis, we have

$$\begin{aligned} \frac{\partial \boldsymbol{r}}{\partial \alpha^k} &= \frac{\partial (r^i \boldsymbol{\varepsilon}_i)}{\partial \alpha^k} = \frac{\partial r^i}{\partial \alpha^k}\boldsymbol{\varepsilon}_i + r^i \frac{\partial \boldsymbol{\varepsilon}_i}{\partial \alpha^k} \\ &= \frac{\partial r^i}{\partial \alpha^k}\boldsymbol{\varepsilon}_i + \Gamma_{ik}^{m} r^i \boldsymbol{\varepsilon}_m = \left(\frac{\partial r^m}{\partial \alpha^k} + \Gamma_{ik}^{m} r^i \right) \boldsymbol{\varepsilon}_m = \nabla_k r^m \boldsymbol{\varepsilon}_m \,. \end{aligned}$$

The expression $\nabla_k r^m$ is called the covariant derivative or absolute derivative of the covariant components of the vector $\boldsymbol{r}$. Sometimes, this is denoted by $r^m_{,k}$. Thus, the covariant derivative of the vector $\boldsymbol{r}$ in the covariant basis is given by

$$\frac{\partial \boldsymbol{r}}{\partial \alpha^k} = \left(\frac{\partial r^m}{\partial \alpha^k} + \Gamma_{ik}^{m} r^i \right) \boldsymbol{\varepsilon}_m = \nabla_k r^m \boldsymbol{\varepsilon}_m \,. \tag{A31}$$

In the contravariant basis $\boldsymbol{\varepsilon}^i$, we have $\boldsymbol{r} = r_i \boldsymbol{\varepsilon}^i$, $r_i = \boldsymbol{r} \cdot \boldsymbol{\varepsilon}_i$. Then, the covariant derivative of the covariant components is defined as follows:

$$\frac{\partial \boldsymbol{r}}{\partial \alpha^k} = \nabla_k r_m \boldsymbol{\varepsilon}^m \,. \tag{A32}$$

By calculating the scalar product with $\boldsymbol{\varepsilon}_n$ on both sides of (A31), one obtains

$$\frac{\partial \boldsymbol{r}}{\partial \alpha^k} \cdot \boldsymbol{\varepsilon}_n = \nabla_k r_n$$

and

$$\frac{\partial r_j}{\partial \alpha^k} = \frac{\partial \boldsymbol{r}}{\partial \alpha^k} \cdot \boldsymbol{\varepsilon}_j + \boldsymbol{r} \cdot \frac{\partial \boldsymbol{\varepsilon}_j}{\partial \alpha^k} = \nabla_k r_j \boldsymbol{\varepsilon}^m \boldsymbol{\varepsilon}_j + \boldsymbol{r} \cdot \Gamma^m_{jk} = \nabla_k r_j + \Gamma^m_{jk} r_m ,$$

from which it follows that

$$\frac{\partial \boldsymbol{r}}{\partial \alpha^k} = \nabla_k r_j \boldsymbol{\varepsilon}^j = \left(\frac{\partial r_j}{\partial \alpha^k} - \Gamma^m_{jk} r_m \right) \boldsymbol{\varepsilon}^j . \tag{A33}$$

Finally, there are a few useful relationships:

$$\nabla_i \varphi = \frac{\partial \varphi}{\partial \alpha^i} , \quad \nabla_i (v^k + w^k) = \nabla_i v^k + \nabla_i w^k , \quad \nabla_i g^{jk} = \nabla_i g_{jk} = 0 . \tag{A34}$$

The last equation means that the components of the fundamental matrix behave under covariant differentiation like a constant under normal differentiation. Thus, one can move g_{ij} in front of the derivative sign, for example,

$$\nabla_i \left(g_{jk} w^k \right) = g_{jk} \nabla_i w^k . \tag{A35}$$

In Cartesian coordinates, the covariant and normal derivatives coincide:

$$\nabla_i w^k = \frac{\partial w^k}{\partial X^i} . \tag{A36}$$

The derivatives of the vectors of the contravariant local basis are given by

$$\frac{\partial \boldsymbol{\varepsilon}^i}{\partial \alpha^j} = -\Gamma^i_{jk} \boldsymbol{\varepsilon}^k , \tag{A37}$$

$$\nabla_{j'} r^{i'} = A^{i'}_{\cdot i} B^j_{j'} \nabla_j r^i ; \quad \nabla_{j'} r_{i'} = B^i_{\cdot i'} B^j_{\cdot j'} \nabla_j r_i . \tag{A38}$$

The quantities $\Gamma_{ij,k} = g_{km} \Gamma^m_{ij}$ are called Christoffel symbols of the first kind. Then, it follows from (A30) that

$$\frac{\partial g_{ij}}{\partial \alpha^k} = \Gamma_{ki,j} + \Gamma_{kj,i} . \tag{A39}$$

Under a change of the system of coordinates, the Christoffel symbols of the first kind are transformed according to the law

$$\Gamma_{i'j',k'} = B^i_{\cdot i} B^j_{j'} B^k_{\cdot k'} \Gamma_{ij,k} + \frac{\partial^2 \alpha^l}{\partial \alpha^{i'} \partial \alpha^{j'}} B^k_{\cdot k'} g_{kl} . \tag{A40}$$

A.2 Tensors

We will now introduce new fundamental objects that are more complicated than vectors. These objects are called dyads and denoted by $\boldsymbol{ab}$. According to the definition, the components of a dyad constitute a matrix with elements $a^i b^j$. From the vectors of the covariant basis $\boldsymbol{\varepsilon}_i$, one can create the dyads $\boldsymbol{\varepsilon}_i\boldsymbol{\varepsilon}_j$ whose corresponding matrix elements are 1 on the crossing of the *i*-th line and *j*-th column and vanish in all other positions.

Dyadic products possess the following properties:

1. The distributive property is

$$\boldsymbol{\varepsilon}_i(a\boldsymbol{\varepsilon}_j + b\boldsymbol{\varepsilon}_k) = a\boldsymbol{\varepsilon}_i\boldsymbol{\varepsilon}_j + b\boldsymbol{\varepsilon}_i\boldsymbol{\varepsilon}_k ,$$

   where a and b are constants.
2. The factors in a dyadic product may not be exchanged. This means that the dyadic product is not commutative and we have $\boldsymbol{\varepsilon}_i\boldsymbol{\varepsilon}_j \neq \boldsymbol{\varepsilon}_j\boldsymbol{\varepsilon}_i$.
3. The dyad with exchanged factors is called a transposed dyad and is denoted by $(\boldsymbol{\varepsilon}_i\boldsymbol{\varepsilon}_j)^T = \boldsymbol{\varepsilon}_j\boldsymbol{\varepsilon}_i$.
4. The dyad is called symmetric if $\boldsymbol{\varepsilon}_i\boldsymbol{\varepsilon}_j = \boldsymbol{\varepsilon}_j\boldsymbol{\varepsilon}_i$.
5. For dyads, one can formally define scalar and vector products that depend on the positions of the vector and dyad in the product:

$$\boldsymbol{c} \cdot (\boldsymbol{\varepsilon}_i\boldsymbol{\varepsilon}_j) = (\boldsymbol{c} \cdot \boldsymbol{\varepsilon}_i)\boldsymbol{\varepsilon}_j \,; \quad (\boldsymbol{\varepsilon}_i\boldsymbol{\varepsilon}_j) \cdot \boldsymbol{c} = \boldsymbol{\varepsilon}_i(\boldsymbol{\varepsilon}_j \cdot \boldsymbol{c}) \,;$$
$$(\boldsymbol{\varepsilon}_i \boldsymbol{a}) \cdot (\boldsymbol{\varepsilon}_j) = (\boldsymbol{a} \cdot \boldsymbol{b}) \cdot (\boldsymbol{\varepsilon}_i\boldsymbol{\varepsilon}_j) \,.$$

   From the above, it follows that vectors that form a symmetric dyad are collinear.

The dyad defined by $\boldsymbol{E} = \boldsymbol{\varepsilon}^i\boldsymbol{\varepsilon}_i$ is a unit dyad.

From (A14), it follows that

$$\boldsymbol{E} = \boldsymbol{\varepsilon}^i\boldsymbol{\varepsilon}_i = g^{ij}\boldsymbol{\varepsilon}_j\boldsymbol{\varepsilon}_i = \boldsymbol{\varepsilon}_j\boldsymbol{\varepsilon}^j = g_{ij}\boldsymbol{\varepsilon}^i\boldsymbol{\varepsilon}^j \,.$$

By using the properties of dyads, one obtains

$$\boldsymbol{a} \cdot \boldsymbol{E} = \boldsymbol{a} \cdot \boldsymbol{\varepsilon}^i\boldsymbol{\varepsilon}_i = a^i\boldsymbol{\varepsilon}_i \,; \quad \boldsymbol{E} \cdot \boldsymbol{a} = \boldsymbol{\varepsilon}^i\boldsymbol{\varepsilon}_i \cdot \boldsymbol{a} = a_i\boldsymbol{\varepsilon}^i \,.$$

It is evident that the scalar product of a vector with the unit dyad produces the decomposition of the vector in basis vectors. For multiplication from the left, the decomposition takes place in terms of the basis $\boldsymbol{\varepsilon}_i$ and for multiplication from the right, in terms of the reciprocal basis.

Every dyad can be represented in terms of nine dyads of basis vectors as follows:

$$\boldsymbol{ab} = a^i\boldsymbol{\varepsilon}_i b^j\boldsymbol{\varepsilon}_j = a^i b^j\boldsymbol{\varepsilon}_i\boldsymbol{\varepsilon}_j \,.$$

Therefore, the dyads $\boldsymbol{\varepsilon}_i\boldsymbol{\varepsilon}_j$ are called dyadic basis.

We call the object

$$T = T^{ij}\boldsymbol{\varepsilon}_i\boldsymbol{\varepsilon}_j \tag{A41}$$

a tensor of second order and T^{ij} its contravariant components.

This tensor has nine components. From the properties of the dyad, it follows that the transposed tensor $\boldsymbol{T}^T = T^{ji}\boldsymbol{\varepsilon}_j\boldsymbol{\varepsilon}_i$ corresponds to a tensor with the transposed matrix $T^{ji} = (T^{ij})^T$.

Tensors are invariant objects, but their components depend on the system of coordinates. We now determine the transformation law of the tensor $\boldsymbol{T}$ during the transition from the $\{\alpha^1, \alpha^2, \alpha^3\}$-system of coordinates to the $\{\alpha^{1'}, \alpha^{2'}, \alpha^{3'}\}$-system of coordinates according to (A19). Since the tensor is an invariant object, we have

$$T^{i'j'}\boldsymbol{\varepsilon}_{i'}\boldsymbol{\varepsilon}_{j'} = T^{ij}\boldsymbol{\varepsilon}_i\boldsymbol{\varepsilon}_j = T^{ij}A^{i'}_{\cdot i}A^{j'}_{j}\boldsymbol{\varepsilon}_{i'}\boldsymbol{\varepsilon}_{j'}$$

and

$$T^{i'j'} = T^{ij}A^{i'}_{\cdot i}A^{j'}_{j}\ . \tag{A42}$$

Therefore, the components of the tensor $\boldsymbol{T} = T^{ij}\boldsymbol{\varepsilon}_i\boldsymbol{\varepsilon}_j$ are transformed according to the same law as contravariant vector components. Therefore, this tensor is called a contravariant tensor. The corresponding basis of dyads transform like covariant basis vectors:

$$\boldsymbol{\varepsilon}_{i'}\boldsymbol{\varepsilon}_{j'} = B^{i}_{\cdot i}B^{j}_{j'}\boldsymbol{\varepsilon}_i\boldsymbol{\varepsilon}_j\ . \tag{A43}$$

If we take the dyadic products of contravariant basis vectors $\boldsymbol{\varepsilon}^i\boldsymbol{\varepsilon}^j$ as a dyadic basis, one can introduce the tensor of second order

$$\boldsymbol{T} = T_{ij}\boldsymbol{\varepsilon}^i\boldsymbol{\varepsilon}^j\ , \tag{A44}$$

whose components are transformed according to the transformation laws of covariant vector components, that is,

$$T_{i'j'} = T_{ij}B^{i}_{\cdot i'}B^{j}_{\cdot j'}\ . \tag{A45}$$

This tensor is called covariant.

Its dyadic basis $\boldsymbol{\varepsilon}^i\boldsymbol{\varepsilon}^j$ is transformed according to the transformation law of contravariant basis vectors:

$$\boldsymbol{\varepsilon}^{i'}\boldsymbol{\varepsilon}^{j'} = A^{i'}_{\cdot i}A^{j'}_{\cdot j}\boldsymbol{\varepsilon}^i\boldsymbol{\varepsilon}^j\ . \tag{A46}$$

Since we have $\boldsymbol{\varepsilon}^i = g^{ij}\boldsymbol{\varepsilon}_j$ (see (A14)), a tensor of second order can be written in one of the following forms

$$\boldsymbol{T} = T_{ij}\boldsymbol{\varepsilon}^i\boldsymbol{\varepsilon}^j = T^{ij}\boldsymbol{\varepsilon}_i\boldsymbol{\varepsilon}_j = T_{i}^{\cdot j}\boldsymbol{\varepsilon}^i\boldsymbol{\varepsilon}_j = T^{i\cdot}_{\cdot j}\boldsymbol{\varepsilon}_i\boldsymbol{\varepsilon}^j\ ,$$

with $T_{ij} = T_{kl}g^{ik}g^{jl} = T_{k}^{\cdot j}g^{ik} = T^{i}_{\cdot j}g^{kj}$.

The tensor of second order $\boldsymbol{g} = g_{ij}\boldsymbol{\varepsilon}^i\boldsymbol{\varepsilon}^j$, where g_{ij} is the fundamental matrix (see (A10) is called the metric tensor.

As a basis, one can not only use the dyads $\boldsymbol{\varepsilon}^i\boldsymbol{\varepsilon}^j, \boldsymbol{\varepsilon}_i\boldsymbol{\varepsilon}_j, \boldsymbol{\varepsilon}^i\boldsymbol{\varepsilon}_j, \boldsymbol{\varepsilon}_i\boldsymbol{\varepsilon}^j$, but also the polidyads (triads $\boldsymbol{\varepsilon}_i\boldsymbol{\varepsilon}_j\boldsymbol{\varepsilon}_k, \boldsymbol{\varepsilon}^i\boldsymbol{\varepsilon}_j\boldsymbol{\varepsilon}_k, \ldots, \boldsymbol{\varepsilon}^i\boldsymbol{\varepsilon}^j\boldsymbol{\varepsilon}^k$ and so on). Thus, one can introduce tensors of a higher order into the discussion.

Using the transformation laws (A42) and (A45), one can give the exact definition of a tensor. An object $\boldsymbol{T}$ whose components are transformed under the transition from one system of coordinates to another one according to the transformation law

$$T^{i'_1 i'_2 \ldots i'_n}_{\ldots\ldots\ldots j'_1 j'_2 \ldots j'_m} = A^{i'_1}_{i_1} A^{i'_2}_{i_2} \ldots A^{i'_n}_{i_n} B^{j_1}_{j'_1} B^{j_2}_{j'_2} \ldots B^{j_m}_{j'_m} T^{i_1 i_2 \ldots i_n}_{\ldots\ldots\ldots j_1 j_2 \ldots j_m} \tag{A47}$$

is called a tensor of $(n+m)$-th order with m-fold covariant and n-fold contravariant components $T^{i_1 i_2 \ldots i_n}_{\ldots\ldots\ldots j_1 j_2 \ldots j_m}$.

The definition of a tensor is given for the transformation (A19). However, an object, that is, a tensor relative to one group of transformations, is not necessarily a tensor relative to another group of transformations. The transformation (A19) only satisfies the condition that it can be inverted. Such transformations are called general transformations. A special case of this transformation is the linear transformation

$$\alpha^{i'} = A^{i'}_{\cdot i}\alpha^i + A^{i'}, \quad \alpha^i = B^i_{\cdot i'}\alpha^{i'} + B^i . \tag{A48}$$

Such a transformation is called an affine transformation. An affine transformation in an orthogonal system of coordinates is effected via the matrices $B_{ii'}$ and $A_{i'i}$, whose elements are the cosines of the angles between the corresponding coordinates axis. We denote the transformation matrices not by $A^{i'}_{\cdot i}$ and $B^i_{\cdot i'}$ but by $B_{ii'}$ and $A_{i'i}$ because there is no difference between covariant and contravariant components in an orthogonal system of coordinates. For $B_{ii'} = A_{i'i}$, such a transformation is called an orthogonal transformation. During an orthogonal transformation, the components of the metric tensor are conserved because the scalar products between the basis vectors are unchanged.

If a transformation does not change the tensor components, one considers the tensor to be invariant with respect to this transformation group. For example, the unit tensor $\boldsymbol{I}$ with the components $\delta^i_{\cdot j}$ is invariant relative to the transformation group (A21).

Tensors that are invariant relative to the rotation group are called isotropic tensors. The isotropic tensors of second order are the unit tensor $\boldsymbol{I} = \delta^i_{\cdot j}\boldsymbol{\varepsilon}_i\boldsymbol{\varepsilon}^j$ and the metric tensor $\boldsymbol{g} = g_{ij}\boldsymbol{\varepsilon}^i\boldsymbol{\varepsilon}^j = g^{ij}\boldsymbol{\varepsilon}_i\boldsymbol{\varepsilon}_j$, the isotropic tensors of third order are the Levi–Civita tensors $\boldsymbol{\varepsilon}^{ijk}$ and $\boldsymbol{\varepsilon}_{ijk}$. The isotropic tensors of fourth order are the following combinations of metric tensors: $g^{ij}g^{mn}$; $g^{im}g^{jn} + g^{in}g^{jm}$; $g^{im}g^{jn} - g^{in}g^{jm}$, and therefore an arbitrary isotropic tensor of fourth order can be represented as a linear combination of these tensors. In particular, a contravariant tensor of fourth order that is symmetric with respect to the pairs i, j and m, n has the following form:

$$T^{ijmn} = \lambda g^{ij}g^{mn} + \mu(g^{im}g^{jn} + g^{in}g^{jm}) . \tag{A49}$$

In the following, we describe the operations that allow one to construct new tensors via algebraic operations from given tensors.

By multiplying the components $T^{i_1 i_2 \dots i_n}_{\dots\dots\dots j_1 j_2 \dots j_m}$ of a m-fold covariant and n-fold contravariant tensor $\boldsymbol{T}$ by a scalar a, the numbers $a\,T^{i_1 i_2 \dots i_n}_{\dots\dots\dots j_1 j_2 \dots j_m}$ are again the components of a m-fold covariant and n-fold contravariant tensors $\boldsymbol{S}$ with components

$$S^{i_1 i_2 \dots i_n}_{\dots\dots\dots j_1 j_2 \dots j_m} = a\,T^{i_1 i_2 \dots i_n}_{\dots\dots\dots j_1 j_2 \dots j_m} \,. \tag{A50}$$

By adding the components $P^{i_1 i_2 \dots i_n}_{\dots\dots\dots j_1 j_2 \dots j_m}$ and $Q^{i_1 i_2 \dots i_n}_{\dots\dots\dots j_1 j_2 \dots j_m}$ of two tensors $\boldsymbol{P}$ and $\boldsymbol{Q}$ with equal structure (equal number of lower and upper indices), their sum yields the components of a tensor $\boldsymbol{S}$ of the same structure with the components

$$S^{i_1 i_2 \dots i_n}_{\dots\dots\dots j_1 j_2 \dots j_m} = P^{i_1 i_2 \dots i_n}_{\dots\dots\dots j_1 j_2 \dots j_m} + Q^{i_1 i_2 \dots i_n}_{\dots\dots\dots j_1 j_2 \dots j_m} \,. \tag{A51}$$

Adding tensors with different structures to each other is not feasible.

By multiplying the components $P^{i_1 i_2 \dots i_n}_{\dots\dots\dots j_1 j_2 \dots j_m}$ and $Q^{i_1 i_2 \dots i_k}_{\dots\dots\dots j_1 j_2 \dots j_r}$, they constitute the components of a $(m + r)$-fold covariant and $(n + k)$-fold contravariant tensor $\boldsymbol{S}$ with components

$$S^{i_1 i_2 \dots i_{n+k}}_{\dots\dots\dots j_1 j_2 \dots j_{m+r}} = P^{i_1 i_2 \dots i_n}_{\dots\dots\dots j_1 j_2 \dots j_m} \; Q^{i_1 i_2 \dots i_k}_{\dots\dots\dots j_1 j_2 \dots j_r} \,. \tag{A52}$$

If we consider a tensor with one or more equal upper and lower indices, this means that summation over these indices takes place. As a result, one obtains a tensor whose order is reduced by the number of repeated indices, that is,

$$S^{iknp}_{\dots lk} = P^{inp}_{\dots l} \,. \tag{A53}$$

In agreement with this rule, we define the reduction of a tensor as an operation that consists of replacing a covariant or contravariant index by a corresponding contravariant or covariant index and summation over this index. One example is the reduction of a simply covariant and simply contravariant tensor of second degree $Q^{i}_{\cdot j}$ which yields the trace of the tensor $tr\boldsymbol{Q} = Q^{i}_{\cdot i}$.

The reduction of a tensor can not only be applied for individual tensors, but also for the multiplication of tensors that have equal number of lower and upper indices. Examples are the scalar product of two vectors $\boldsymbol{a} \cdot \boldsymbol{b} = a^i b_i$ and the full scalar product of two tensors of second degree $\boldsymbol{A} : \boldsymbol{B} = A^{ij} B_{ij}$. As a result, one obtains a scalar.

The permutation of indices for the tensor components consist of changing the indices of only upper or lower indices via changing their positions. The permutation of lower indices with upper indices and vice versa is not considered since this is not an invariant operation. As a result, the structure of the tensor does not change, though a new tensor of the same type is created, for example,

$$P^{\dots ij}_{pqr} = Q^{\dots ij}_{qrp} \,. \tag{A54}$$

The main reason for the permutation operation becomes evident when combined with addition and subtraction operations in order to obtain the operations denoted by symmetrization and anti-symmetrization.

One arbitrarily selects N indices from equivalent (either lower or upper) indices of a tensor. One applies all possible $N!$ permutations to these indices and forms the arithmetic mean of all permuted tensors. In the resulting tensor, all indices participating in the permutation are enclosed in round brackets:

$$Q^*_{i_1 \dots i_N} = Q_{(i_1 \dots i_N)} = \frac{1}{N!}\left(Q_{i_1 \dots i_N} + \dots + Q_{i_N \dots i_1}\right). \tag{A55}$$

If other indices are present between the indices that take part in the symmetrization, those indices are separated by a vertical bar, for example, $Q_{(i|mn|j}$. The most important cases for applications are $N = 2$ and $N = 3$

$$Q^*_{ij} = Q_{(ij)} = \frac{1}{2}\left(Q_{ij} + Q_{ji}\right),$$
$$Q^*_{ijk} = Q_{(ijk)} = \frac{1}{6}\left(Q_{ijk} + Q_{jki} + Q_{kij} + Q_{jik} + Q_{ikj} + Q_{kji}\right).$$

The tensor that is obtained as a result of the symmetrization is symmetric with respect to its indices, that is, $B_{ij} = B_{ji}$.

To anti-symmetrize a tensor, one proceeds as in the symmetrization except for applying a minus sign for odd permutations in the arithmetic mean. This procedure is symbolized by square brackets around the affected indices.

$$\tilde{Q}_{i_1 \dots i_N} = Q_{[i_1 \dots i_N]}. \tag{A56}$$

For the cases $N = 2$ and $N = 3$, we have

$$\tilde{Q}_{ij} = Q_{[ij]} = \frac{1}{2}\left(Q_{ij} - Q_{ji}\right),$$
$$\tilde{Q}_{ijk} = Q_{[ijk]} = \frac{1}{6}\left(Q_{ijk} + Q_{jki} + Q_{kij} - Q_{jik} - Q_{kji} - Q_{ikj}\right).$$

A tensor is called antisymmetric with respect to equivalent (for example) lower indices if its components stay unchanged for an even permutation of these indices while a factor -1 appears for odd permutations. For example, the components of the two-fold covariant antisymmetric tensor $\tilde{Q}_{ij} = Q_{[ij]}$ are a matrix with elements $\tilde{Q}_{ij} = -\tilde{Q}_{[ji]}$. The components of the threefold covariant antisymmetric tensor $\tilde{Q}_{ijk} = Q_{[ijk]}$ have the following properties:

$$\tilde{Q}_{ijk} = \tilde{Q}_{jki} = \tilde{Q}_{kij} = -\tilde{Q}_{jik} = -\tilde{Q}_{kji} = -\tilde{Q}_{ikj}.$$

The operation of antisymmetrization can be applied to all or just a few indices. In the first case, one obtains a tensor which is antisymmetric with respect to all indices, for example, $Q^{[i_1 \dots i_m]}$, that is called an m-vector.

A m-vector is called decomposable if there are m simply contravariant tensors (vectors) $q_1^{i_1}, q_2^{i_2}, \dots, q_m^{i_m}$ such that

$$Q^{[i_1 \dots i_m]} = q_1^{[i_1}, q_2^{i_2}, \dots, q_m^{i_m]} = \frac{1}{m!}\begin{vmatrix} q_1^{i_1} & \cdot & \cdot & q_1^{i_m} \\ q_2^{i_1} & \cdot & \cdot & q_2^{i_m} \\ \cdot & \cdot & \cdot & \cdot \\ q_m^{i_1} & \cdot & \cdot & q_m^{i_m} \end{vmatrix}.$$

If one refers to a tensor as antisymmetric, it means that this tensor is antisymmetric with respect to all indices.

The components of an antisymmetric tensor with at least two equal equivalent indices are zero, for example, $\tilde{Q}_{ii} = 0$. Therefore, all components of an antisymmetric tensor of $n + 1$-th degree in n-dimensional space are zero.

The properties of symmetry or antisymmetry are invariant with respect to the coordinate transformation (A19). Every tensor of second order having equivalent indices (Q_{ij} or Q^{ij}) can be written as the sum of a symmetric and an antisymmetric tensor:

$$Q = Q^* + \tilde{Q}\,, \tag{A57}$$

where $Q^*_{ij} = 0.5(Q_{ij} + Q_{ji})$, $\tilde{Q}_{ij} = 0.5(Q_{ij} - Q_{ji})$.

We now consider an antisymmetric tensor of second order Ω with components $\Omega_{ij} = -\Omega_{ji}$ in the Cartesian system of coordinates. For this tensor, we have $\Omega_{ii} = 0$ and for components with unequal indices, there are only three independent components $\omega_1 = \Omega_{23}, \omega_2 = \Omega_{31}$ and $\omega_3 = \Omega_{12}$. Therefore, each antisymmetric tensor of second order $\boldsymbol{\Omega} = \Omega_{ij}\boldsymbol{\varepsilon}^i\boldsymbol{\varepsilon}^j$ corresponds to a vector $\boldsymbol{\omega} = \omega_k\boldsymbol{\varepsilon}^k$. This vector is called an axial vector.

In a curvilinear system of coordinates, every antisymmetric tensor of second order $\boldsymbol{\Omega} = \Omega_{ij}\boldsymbol{\varepsilon}^i\boldsymbol{\varepsilon}^j$ corresponds to a vector $\boldsymbol{\omega} = \omega^k\boldsymbol{\varepsilon}_k$ with contravariant coordinates $\omega^k = \frac{1}{\sqrt{g}}\Omega_{ij}$, where k, i, j is obtained form 1, 2, 3 via cyclic permutation and g is the determinant of the metric tensor.

The components of an axial tensor are transformed according to the transformation law (A19) according to the formula $\omega^k = \omega^{k'} B^k_{.k'}\Delta/|\Delta|$, where $\Delta = |A^{i'}_{.j}|$.

If the transformation law (A19) conserves the orientation of basis vectors, we have $\Delta > 0$ and the transformation laws of the vector components ω^k and of the contravariant tensor components coincide.

However, if the transformation (A19) changes the orientation of the basis vectors via reflection, for example, if a right-handed system of coordinates changes to a left-handed system of coordinates, then $\Delta < 0$. Therefore, an axial vector is a non-invariant object and is therefore referred to as a pseudo vector.

An axial vector is connected to the vector product of two vectors $\boldsymbol{a}$ and $\boldsymbol{b}$.

Indeed, according to (A16), one obtains a vector $\boldsymbol{c}$ with coordinates

$$c^k = \frac{1}{\sqrt{g}}(a_i b_j - a_j b_i) = \frac{1}{\sqrt{g}}\Omega_{ij}\,.$$

This means that an axial vector is equal to the vector product of two polar vectors. The physical meaning of an axial vector is that of the instantaneous angular velocity of the rotation of a rigid body and half the curl of the velocity vector $\boldsymbol{\omega} = 0.5\nabla \times \boldsymbol{V}$, or in component form,

$$\boldsymbol{\omega} = \frac{0.5}{\sqrt{g}}\left(\nabla_j a_i - \nabla_i a_j\right) = \frac{0.5}{\sqrt{g}}\left(\frac{\partial a_i}{\partial X^j} - \frac{\partial a_j}{\partial X^i}\right)\,. \tag{A58}$$

Since the expression in brackets is an antisymmetric tensor of second order, the vector $\boldsymbol{\omega}$ is an axial vector.

We consider a transformation of the system of coordinates which transforms the basis vectors according to the law $\boldsymbol{\varepsilon}_{i'} = B^{i}_{.i'}\boldsymbol{\varepsilon}_i$, $\boldsymbol{\varepsilon}_i = A^{i'}_{.i}\boldsymbol{\varepsilon}_{i'}$ and the determinant of the transformation is larger than zero, that is, $|B^{i}_{.i'}| > 0$. The object $Q^{.j}_{i}$ is called a relative tensor density of weight p if it is transformed as follows

$$Q^{.j'}_{i'} = \left|B^{i}_{.i'}\right|^{p} B^{i}_{.i'}A^{j}_{.j'}Q^{.j}_{i} \,. \tag{A59}$$

For $p = 0$, this tensor is a true tensor or a tensor of zero weight. It is obvious that the product of two relative tensors of weights $-p$ and p yields a true one. For example, the metric tensor is transformed according to the law $g_{i'j'} = B^{i}_{.i'}B^{j}_{.j'}g_{ij}$. However, the determinant of the metric tensor is transformed according to

$$\left|g_{i'j'}\right| = \left|B^{i}_{.i'}\right|^{2}\left|g_{ij}\right| \quad \text{or} \quad \left|g'\right| = \left|B^{i}_{.i'}\right|^{2}|g| \,. \tag{A60}$$

Thus, the tensor g_{ij} is a true tensor, while g is a scalar of weight 2. Relative tensors transform unlike true tensors, which is why they are called pseudo tensors. We now consider the most useful pseudo tensors.

We define ε objects as quantities of the form $\varepsilon^{i_1\ldots i_m}$ and $\varepsilon_{i_1\ldots i_m}$, with $i_1, i_2, \ldots, i_m$ as a permutation of $1, 2, \ldots., m$, having the following properties:

$$\varepsilon_{i_1,i_2,\ldots,i_m} = \varepsilon^{i_1,i_2,\ldots,i_m} = \begin{cases} 0\,, & \text{if two or more of } i_1,\ldots,i_m \text{ are equal}\,,\\ 1\,, & \text{if } i_1,\ldots,i_m \text{ are different}\\ & \text{ and are an even permutation}\,,\\ -1\,, & \text{if } i_1,\ldots,i_m \text{ are different and are}\\ & \text{an odd permutation}\,. \end{cases} .$$

The ε-objects introduced above make it possible to simplify the calculation of determinants. The determinant $|\boldsymbol{a}|$ of a 3×3-matrix $a^{i}_{.j}$ can be written as follows:

$$|\boldsymbol{a}| = \varepsilon_{ijk}a^{i}_{.1}a^{j}_{.2}a^{k}_{.3} \,.$$

The properties of a determinant under exchange of columns are obtained from

$$|\boldsymbol{a}|\,\varepsilon_{mnl} = a^{i}_{.m}a^{j}_{.n}a^{k}_{.l} \quad \text{or} \quad |\boldsymbol{a}| = \frac{1}{6}\varepsilon^{mnl}\varepsilon_{ijk}a^{i}_{.m}a^{j}_{.n}a^{k}_{.l} \,. \tag{A61}$$

By $\delta^{i_1\ldots i_m}_{j_1\ldots j_m}$, we denote the determinant of the matrix with elements $\delta^{i_k}_{j_n}$, where $k, n = 1, \ldots, m$. Then, we have $\varepsilon^{i_1\ldots i_m}\varepsilon_{j_1\ldots j_m} = \delta^{i_1\ldots i_m}_{j_1\ldots j_m}$ and the previous relationship can be written in the form

$$|\boldsymbol{a}| = \frac{1}{6}\varepsilon^{mnl}\varepsilon_{ijk}a^{i}_{.m}a^{j}_{.n}a^{k}_{.l} = \frac{1}{6}\delta^{mnl}_{ijk}a^{i}_{.m}a^{j}_{.n}a^{k}_{.l} \,.$$

For the three dimensional case, we have

$$\begin{aligned} \varepsilon^{ijk}\varepsilon_{lmk} &= \delta^{ijk}_{lmk} = \delta^{ij}_{lm} = \delta^{i}_{l}\delta^{j}_{m} - \delta^{i}_{m}\delta^{j}_{l}\,,\\ \varepsilon^{ijk}\varepsilon_{ljk} &= \delta^{ijk}_{ljk} = \delta^{ij}_{lj} = 2\delta^{i}_{l}\,,\\ \varepsilon^{ijk}\varepsilon_{ijk} &= \delta^{ijk}_{ijk} = \delta^{ij}_{ij} = 2\delta^{i}_{l} = 6\,. \end{aligned} \tag{A62}$$

By rewriting (A59), we obtain

$$\varepsilon_{i'j'k'} = \frac{1}{|\boldsymbol{a}|}\varepsilon_{ijk} a^i_{i'} a^j_{j'} a^k_{k'} , \quad \text{or} \quad \varepsilon^{m'n'l'} = |\boldsymbol{a}|\, a^{m'}_m\, a^{n'}_n\, a^{l'}_l\, \varepsilon^{mnl} . \tag{A63}$$

One can consider these equations as the transformation laws for the components of the ε-objects during the transition to a new system of coordinates. Therefore, the ε-objects with components ε^{ijk} and ε_{ijk} are pseudo tensors of third order with weights of -1 and 1.

However, the objects

$$\bar{\varepsilon}_{ijk} = \sqrt{|g|}\varepsilon_{ijk} \quad \text{and} \quad \bar{\varepsilon}^{ijk} = \frac{1}{\sqrt{|g|}}\varepsilon^{ijk} \tag{A64}$$

are real tensors of third order. They are called Levi–Civita tensors.

Here, one writes $|g|$ instead of g because g changes its sign from positive to negative under transition from a right-handed to a left-handed system of coordinates. Therefore, $\sqrt{|g|}$ is a pseudo tensor of order zero and a scalar of weight 1 and the product $\sqrt{|g|}\varepsilon_{ijk}$ or the ratio $\varepsilon^{ijk}/\sqrt{|g|}$ is a true tensor.

We finally write down some useful relationships for Levi–Civita-tensors:

$$\begin{aligned} &\bar{\varepsilon}^{ij}\bar{\varepsilon}_{kj} = \delta^i_k , \quad \bar{\varepsilon}^{ij}\bar{\varepsilon}^{kl} g_{jl} = g^{ik} , \quad \bar{\varepsilon}_{ij} g^{jk}\bar{\varepsilon}_{kl} g^{lm} = -\delta^m_i , \\ &\bar{\varepsilon}^{ij}\bar{\varepsilon}^{kl} + \bar{\varepsilon}^{jl}\bar{\varepsilon}^{kj} = 2g^{ik}g^{jl} - g^{ij}g^{kl} - g^{il}g^{kj} . \end{aligned} \tag{A65}$$

We first consider the scalar product of a tensor $\boldsymbol{T} = T_{kl}\boldsymbol{\varepsilon}^k\boldsymbol{\varepsilon}^l$ with a vector $\boldsymbol{a} = a^i\boldsymbol{\varepsilon}_i$. From the properties of the dyadic product, it follows that this scalar product is not commutative. Therefore, one has to consider the multiplication of a dyad with a vector from the left or the right separately: In the first case, the vector $\boldsymbol{a}$ is multiplied with the first vector of the basis dyad, while the vector is multiplied with the second vector of the basis dyad in the second case:

$$\boldsymbol{a} \cdot \boldsymbol{T} = a^i\boldsymbol{\varepsilon}_i \cdot T_{kl}\boldsymbol{\varepsilon}^k\boldsymbol{\varepsilon}^l = a^i T_{kl}\delta^{\cdot k}_i\boldsymbol{\varepsilon}^l = a^i T_{il}\boldsymbol{\varepsilon}^l ; \tag{A66}$$

$$\boldsymbol{T} \cdot \boldsymbol{a} = T_{kl}\boldsymbol{\varepsilon}^k\boldsymbol{\varepsilon}^l \cdot a^i\boldsymbol{\varepsilon}_i = T_{kl}\delta^l_{\cdot i} a^i\boldsymbol{\varepsilon}^k = T_{ki}a^i\boldsymbol{\varepsilon}^k . \tag{A67}$$

As a result, in both cases, we obtain a vector, but with different coordinates $a^i T_{il}$ and $T_{ki}a^i$ respectively. If the tensor $\boldsymbol{T}$ is symmetric, we have $\boldsymbol{a} \cdot \boldsymbol{T} = \boldsymbol{T} \cdot \boldsymbol{a}$.

In a similar way, one obtains the scalar product of two tensors $\boldsymbol{T}$ and $\boldsymbol{P}$ of second order:

$$\boldsymbol{T} \cdot \boldsymbol{P} = T_{ij}\boldsymbol{\varepsilon}^i\boldsymbol{\varepsilon}^j \cdot P^{kl}\boldsymbol{\varepsilon}_k\boldsymbol{\varepsilon}_l = T_{ij}P^{kl}\delta^j_{\cdot k}\boldsymbol{\varepsilon}^i\boldsymbol{\varepsilon}_l = T_{ij}P^{jl}\boldsymbol{\varepsilon}^i\boldsymbol{\varepsilon}_l . \tag{A68}$$

This scalar product, however, does not agree with the usual notion that a scalar product should produce a scalar. Therefore, one introduces another scalar product for tensors of second order which is called a double scalar product or total scalar product. This operation is denoted by $\boldsymbol{T} : \boldsymbol{P}$ and is different from (A68) insofar as the corresponding basis vectors enter the scalar product in pairs, for example,

$\boldsymbol{\varepsilon}^i \cdot \boldsymbol{\varepsilon}_k$ and so on. As a result, one obtains a scalar:

$$\boldsymbol{T} : \boldsymbol{P} = T_{ij}\boldsymbol{\varepsilon}^i\boldsymbol{\varepsilon}^j : P^{kl}\boldsymbol{\varepsilon}_k\boldsymbol{\varepsilon}_l = T_{ij}P^{kl}\delta^j_{.k}\delta^j_{.l} = T_{kl}P^{kl} . \tag{A69}$$

It has been shown above that every antisymmetric tensor of second order in three dimensions corresponds to an axial vector, that is, the vector product of two vectors (see (A16)). On the other hand, every tensor can be represented as a decomposition (see Section 2.4) in terms of the dyadic basis $\boldsymbol{\varepsilon}^i\boldsymbol{\varepsilon}^j$ or $\boldsymbol{\varepsilon}_i\boldsymbol{\varepsilon}_j$.

To also represent antisymmetric tensors, the basis elements should have the anticommutative property, and therefore each pair of dyadic basis elements should satisfy equation $\boldsymbol{\varepsilon}_i\boldsymbol{\varepsilon}_j = -\boldsymbol{\varepsilon}_j\boldsymbol{\varepsilon}_i$. We denote dyadic basis elements that are similar to the vector product by $\boldsymbol{\varepsilon}^i \wedge \boldsymbol{\varepsilon}^j$ or $\boldsymbol{\varepsilon}_i \wedge \boldsymbol{\varepsilon}_j, \boldsymbol{\varepsilon}^i \wedge \boldsymbol{\varepsilon}_j, \boldsymbol{\varepsilon}_i \wedge \boldsymbol{\varepsilon}^j$. Therefore, if one writes, for example, $\boldsymbol{\varepsilon}^i \wedge \boldsymbol{\varepsilon}^j$, this means that $\boldsymbol{\varepsilon}^i \wedge \boldsymbol{\varepsilon}^j = -\boldsymbol{\varepsilon}^j \wedge \boldsymbol{\varepsilon}^i$.

The elements of the space of antisymmetric tensors with a dyadic basis are defined as $\boldsymbol{A} = A_{ij}\boldsymbol{\varepsilon}^i \wedge \boldsymbol{\varepsilon}^j$. Then, we have $\boldsymbol{A} = A_{ij}\boldsymbol{\varepsilon}^i \wedge \boldsymbol{\varepsilon}^j = -A_{ji}\boldsymbol{\varepsilon}^j \wedge \boldsymbol{\varepsilon}^i$. The objects $\boldsymbol{A}$ defined in this way have the general properties of tensors and, at the same time, of antisymmetric tensors. They are called external forms of second order. The corresponding basis dyads are called external basis dyads.

By generalizing this definition, one defines the external form of m-th order as follows:

$$\tilde{\boldsymbol{A}} = A_{i_1 i_2 \dots i_m}\boldsymbol{\varepsilon}^{i_1} \wedge \boldsymbol{\varepsilon}^{i_2} \wedge \dots \wedge \boldsymbol{\varepsilon}^{i_m} . \tag{A70}$$

As an example, we consider the external form of second order that is constructed from two vectors $\boldsymbol{X} = X_i\boldsymbol{\varepsilon}^i$ and $\boldsymbol{Y} = Y_j\boldsymbol{\varepsilon}^j$. Using the anticommutative property of external dyads, one obtains

$$\begin{aligned}
\tilde{\boldsymbol{A}} &= \boldsymbol{X} \wedge \boldsymbol{Y} = c_{ij}\boldsymbol{\varepsilon}^i \wedge \boldsymbol{\varepsilon}^j = X_i Y_j \boldsymbol{\varepsilon}^i \wedge \boldsymbol{\varepsilon}^j = c_{[ij]}\boldsymbol{\varepsilon}^i \wedge \boldsymbol{\varepsilon}^j = X_{[i} Y_{j]}\boldsymbol{\varepsilon}^i \wedge \boldsymbol{\varepsilon}^j \\
&= 2!\left[\frac{1}{2}(X_1Y_2 - Y_1X_2)\,\boldsymbol{\varepsilon}^1 \wedge \boldsymbol{\varepsilon}^2 + \frac{1}{2}(X_1Y_3 - Y_1X_3)\,\boldsymbol{\varepsilon}^1 \wedge \boldsymbol{\varepsilon}^3 \right.\\
&\quad \left. + \frac{1}{2}(X_2Y_3 - Y_2X_3)\,\boldsymbol{\varepsilon}^2 \wedge \boldsymbol{\varepsilon}^3\right] .
\end{aligned}$$

The tensor c_{ij} is antisymmetric, that is, $c_{ij} = -c_{ji}, c_{ii} = 0$, and it therefore corresponds to an antisymmetric matrix with vanishing diagonal elements

$$\begin{aligned}
&c_{23} = -c_{32} = X_1Y_2 - Y_1X_2 = \omega_1 , \quad c_{31} = -c_{13} = X_1Y_3 - Y_1X_3 = \omega_2 , \\
&c_{12} = -c_{21} = X_2Y_3 - Y_2X_3 = \omega_3 .
\end{aligned}$$

If the basis $\boldsymbol{\varepsilon}^j$ is orthonormal, the ω_i are the coordinates of the axial vector $\boldsymbol{\omega}$, meaning $\boldsymbol{\omega} = \boldsymbol{X} \times \boldsymbol{Y}$.

We now consider an external form of third order $\tilde{\boldsymbol{B}} = B_{ijk}\boldsymbol{\varepsilon}^i \wedge \boldsymbol{\varepsilon}^j \wedge \boldsymbol{\varepsilon}^k$. By expanding the sum and combining similar terms, one obtains

$$\begin{aligned}
\tilde{\boldsymbol{B}} &= (B_{123} + B_{312} + B_{231} - B_{213} - B_{321} - B_{132})\,\boldsymbol{\varepsilon}^1 \wedge \boldsymbol{\varepsilon}^2 \wedge \boldsymbol{\varepsilon}^3 \\
&= 3!\,B_{[123]}\,\boldsymbol{\varepsilon}^1 \wedge \boldsymbol{\varepsilon}^2 \wedge \boldsymbol{\varepsilon}^3 = B_{[ijk]}\boldsymbol{\varepsilon}^i \wedge \boldsymbol{\varepsilon}^j \wedge \boldsymbol{\varepsilon}^k ,
\end{aligned}$$

where $B_{[ijk]}$ is the coefficient of the external form which is alternating and therefore antisymmetric with respect to all indices.

If one considers the product of coordinates of the vectors $\boldsymbol{X}$, $\boldsymbol{Y}$ and $\boldsymbol{Z}$ components of the tensor as components of the tensor B_{ijk}, the alternating coefficient in an orthonormal right-handed coordinate system is equal to the mixed vector product of these vectors,

$$B_{[123]} = \frac{1}{3!} \begin{vmatrix} X_1 & X_2 & X_3 \\ Y_1 & Y_2 & Y_3 \\ Z_1 & Z_2 & Z_3 \end{vmatrix} = (\boldsymbol{X}, \boldsymbol{Y}, \boldsymbol{Z}) .$$

Every external form of m-th order can be expressed in a similar manner:

$$\tilde{\boldsymbol{B}} = m! B_{[12\ldots m]} \boldsymbol{\varepsilon}^1 \wedge \boldsymbol{\varepsilon}^2 \wedge \ldots \wedge \boldsymbol{\varepsilon}^m = B_{[i_1 i_2 \ldots i_m]} \boldsymbol{\varepsilon}^{i_1} \wedge \boldsymbol{\varepsilon}^{i_2} \wedge \ldots \wedge \boldsymbol{\varepsilon}^{i_m} . \tag{A71}$$

The external product of an arbitrary number of vectors is defined as the alternation of the coordinates of these vectors. Then, the external product is equal to the external form of these vectors, for example, for two vectors A and B , we have

$$\tilde{\mathrm{Q}} = \boldsymbol{A} \wedge \boldsymbol{B} = A_i B_i \boldsymbol{\varepsilon}^i \wedge \boldsymbol{\varepsilon}^j = A_{[i} B_{j]} \boldsymbol{\varepsilon}^i \wedge \boldsymbol{\varepsilon}^j .$$

In the general case of m vectors, we have

$$\begin{aligned} \tilde{\mathrm{Q}} &= \boldsymbol{a}^1 \wedge \boldsymbol{a}^2 \wedge \ldots \wedge \boldsymbol{a}^m = a^1_{i_1} a^2_{i_2} \ldots a^m_{i_m} \boldsymbol{\varepsilon}^{i_1} \wedge \boldsymbol{\varepsilon}^{i_2} \wedge \ldots \wedge \boldsymbol{\varepsilon}^{i_m} \\ &= a^1_{[i_1} a^2_{i_2} \ldots a^m_{i_m]} \boldsymbol{\varepsilon}^{i_1} \wedge \boldsymbol{\varepsilon}^{i_2} \wedge \ldots \wedge \boldsymbol{\varepsilon}^{i_m} , \end{aligned}$$

where the alternating coefficient is

$$a^1_{[i_1} a^2_{i_2} \ldots a^m_{i_m]} \boldsymbol{\varepsilon}^{i_1} \wedge \boldsymbol{\varepsilon}^{i_2} \wedge \ldots \wedge \boldsymbol{\varepsilon}^{i_m} = \frac{1}{m!} \begin{vmatrix} a^1_{i_1} & \cdot & a^1_{i_m} \\ \cdot & \cdot & \cdot \\ a^m_{i_1} & \cdot & a^m_{i_m} \end{vmatrix} .$$

According to the antisymmetric property, we have

$$\begin{aligned} &\boldsymbol{\varepsilon}^{i_1} \wedge \boldsymbol{\varepsilon}^{i_2} \wedge \ldots \wedge \boldsymbol{\varepsilon}^{i_m} = \frac{1}{m!} \varepsilon^{i_1 i_2 \ldots i_m} \varepsilon_{\alpha_1 \alpha_2 \ldots \alpha_m} \boldsymbol{\varepsilon}^{\alpha_1} \wedge \boldsymbol{\varepsilon}^{\alpha_2} \wedge \ldots \wedge \boldsymbol{\varepsilon}^{\alpha_m} \\ &= \delta^{i_1}_{[j_1 j_2 \ldots} \delta^{i_m}_{j_m]} \boldsymbol{\varepsilon}^{j_1} \wedge \boldsymbol{\varepsilon}^{j_2} \wedge \ldots \wedge \boldsymbol{\varepsilon}^{j_m} . \end{aligned}$$

We define the product of two external forms $\tilde{\mathrm{C}}$ and $\tilde{\boldsymbol{D}}$ of m-th and p-th order by the following external form

$$\tilde{\mathrm{Q}} = \tilde{\mathrm{C}} \wedge \tilde{\boldsymbol{D}} = C_{i_1 i_2 \ldots i_m} D_{i_{m+1} i_{m+2} \ldots i_{m+p}} \boldsymbol{\varepsilon}^{i_1} \wedge \boldsymbol{\varepsilon}^{i_2} \wedge \ldots \wedge \boldsymbol{\varepsilon}^{i_{m+p}} .$$

We now list some properties of external forms:

1. The external product of m vectors vanishes if at least two of them coincide.
2. For two external forms $\tilde{\boldsymbol{P}}$ and $\tilde{\mathrm{Q}}$ of p-th and q-th order, we have

$$\tilde{\boldsymbol{P}} \wedge \tilde{\mathrm{Q}} = (-1)^{pq} \tilde{\mathrm{Q}} \wedge \tilde{\boldsymbol{P}} .$$

3. An external form of order larger than the number of space dimensions vanishes.
4. When moving from the basis $\boldsymbol{\varepsilon}^{i}$ to the new basis $\boldsymbol{\varepsilon}^{i'}$, the basis polyads are, using the relationship $\boldsymbol{\varepsilon}^{i'} = A^{i'}_{.i}\boldsymbol{\varepsilon}^{i}$, transformed as follows:

$$\boldsymbol{\varepsilon}^{i'_1} \wedge \boldsymbol{\varepsilon}^{i'_2} \wedge \ldots \wedge \boldsymbol{\varepsilon}^{i'_n} = \frac{1}{n!}\mathrm{Det}\left(A^{i'_k}_{.i_k}\right)\boldsymbol{\varepsilon}^{i_1} \wedge \boldsymbol{\varepsilon}^{i_2} \wedge \ldots \wedge \boldsymbol{\varepsilon}^{i_m} .$$

Tensors correspond to different physical and geometric objects, and they are therefore invariant objects. In other words, their meaning does not depend on the system of coordinates.

However, their components change when moving from an old system of coordinates to a new system of coordinates, such that each tensor component in the new system of coordinates is expressed in terms of all old components according to the transformation law (A47). Characteristics that do not change when moving from one system of coordinates to the other one are of great interest. Such characteristics are called invariants. The simplest invariant is a scalar or tensor of zeroeth order. We now consider one after the other regarding the invariants of tensors of higher orders.

A tensor of first order is a vector. Its invariants are the length of the vector, the angle between two vectors, the scalar product of two vectors and others. Since those quantities are obtained via the scalar product between vectors $\boldsymbol{a}\cdot\boldsymbol{b} = a^{i}b^{j}g_{ij}$ which constitutes the contraction of tensor, the contraction of a tensor is the general method of obtaining the invariants of tensors of first order (vectors). The vectors only yield one independent invariant. All other invariants are functions of this invariant.

For tensors of second order, the tensor contraction is one of the methods to obtain an invariant. For example, the tensor trace $tr\boldsymbol{T} = T^{i}_{.i}$ is an invariant that is obtained by contracting the tensor T^{ij} with the metric tensor g_{ij}. The contractions $T^{i}_{.j}T^{i}_{.j}$ and $T^{i}_{.k}T^{k}_{.l}T^{l}_{.i}$ are also invariants. For a symmetric tensor of second order, these three invariants, the linear, quadratic and cubic invariants, are independent invariants. Other invariants are functions of these three invariants.

Permutation of the indices is another method to obtain invariants. We first consider the symmetrization that turns every tensor of second or higher order into a symmetric tensor $\boldsymbol{T}$ with components T_{ij}. Each symmetric tensor of second order in every point M corresponds to a central surface of second degree

$$F = \frac{1}{2}T_{ij}X^{i}Y^{j} = \text{const} ,$$

which is referred to as tensor surface of Cauchy surface. To bring this surface into the canonical form, one has to solve the characteristic equation $\left|\lambda\delta^{i}_{.j} - T^{i}_{.j}\right| = 0$, which yields the following cubic equation:

$$\lambda^{3} - I_1\lambda^{2} + I_2\lambda - I_3 = 0 , \tag{A72}$$

where $I_1 = T^{i}_{.i}$, $I_2 = \frac{1}{2}\left[\left(T^{i}_{.i}\right)^{2} - T^{i}_{.j}T^{j}_{.i}\right]$, $I_3 = \left|T^{i}_{.j}\right|$.

The quantity $|T^{i}_{.j}|$ is the determinant of the tensor matrix. Since the characteristic equation does not depend on the system of coordinates, the coefficients I_1, I_2 and I_3 are also invariants.

Now, we consider the method of obtaining invariants via alternation. Via alternation of a tensor of second or higher order, one obtains an antisymmetric tensor. We consider a tensor of third order $\boldsymbol{T}$ with components T_{ijk}. Then, we have $T_{123} = T_{231} = T_{312} = -T_{213} = -T_{321} = -T_{132}$. All other components with repeated indices vanish. Therefore, there is only one independent component, for example, T_{123}. We now consider an invariant that is produced by contracting the tensor T_{ijk} with the vectors X^i, Y^j and Z^k, meaning $I = T_{ijk}X^i, Y^j Z^k$. Since the components T_{ijk} are antisymmetric with respect to all indices, we have $(\boldsymbol{X}, \boldsymbol{Y}, \boldsymbol{Z})T_{123}$, where the positive sign corresponds to a right-handed and the negative sign corresponds to the left-hand system of coordinates. Since the mixed product is a determinant and therefore an invariant, T_{123} must be a relative invariant and not a true one since it changes sign under a transformation involving a mirror operation.

The alternation not only allows one to build relative invariants, but also absolute invariants. As an example, we consider a tensor of sixth order $T_{i_1 i_2 i_3 j_1 j_2 j_3}$. By alternating this tensor with respect to the first and second group of three indices, one obtains a tensor $T_{[i_1 i_2 i_3][j_1 j_2 j_3]}$ that is antisymmetric with respect to the first three as well as the second three indices.

If the tensor is a product of three equal tensors of second order $T_{i_1 i_2 i_3 j_1 j_2 j_3} = P_{i_1 j_1} P_{i_2 j_2} P_{i_3 j_3}$, one obtains, after alternation of the last three indices,

$$T_{i_1 i_2 i_3 [j_1 j_2 j_3]} = \frac{1}{6} \begin{vmatrix} P_{i_1 j_1} & P_{i_1 j_2} & P_{i_1 j_3} \\ P_{i_2 j_1} & P_{i_2 j_2} & P_{i_2 j_3} \\ P_{i_3 j_1} & P_{i_3 j_2} & P_{i_3 j_3} \end{vmatrix} .$$

The last tensor is antisymmetric not only with respect to $j_1 j_2 j_3$, but also with respect to $i_1 i_2 i_3$. Therefore,

$$T_{123[123]} = \frac{1}{6} \begin{vmatrix} P_{11} & P_{12} & P_{13} \\ P_{21} & P_{22} & P_{23} \\ P_{31} & P_{32} & P_{33} \end{vmatrix} = \frac{1}{6} |P_{ij}| \tag{A73}$$

is an absolute invariant. This means that the determinant of a tensor matrix of second order is an invariant.

In continuum mechanics, tensors of second order are most widely used. Every tensor of second order $\boldsymbol{T}$ can be written as sum of a symmetric tensor and an antisymmetric tensor. The symmetric tensor can in turn be divided into an isotropic tensor $\boldsymbol{P}$ and a symmetric traceless tensor $\boldsymbol{S}$. As a result, one obtains

$$\boldsymbol{T} = \boldsymbol{P} + \boldsymbol{S} + \boldsymbol{R} \tag{A74}$$

with the isotropic tensor $\boldsymbol{P} = p\boldsymbol{I}$, the scalar $p = 1/3 tr \boldsymbol{T}$, the unit tensor $\boldsymbol{I}$, the symmetric traceless tensor $\boldsymbol{S}$ with components $S^{i}_{.j} = (T^{i}_{.j} + T^{j}_{.i}) - 1/3 tr \boldsymbol{T} \delta^{i}_{.j}$ and the antisymmetric tensor $\boldsymbol{R}$ with components $R^{i}_{.j} = (S^{i}_{.j} - S^{j}_{.i})$. Equation A74 has the following component form

$$T^{i}_{.j} = \frac{1}{3} trT\delta^{i}_{.j} + S^{i}_{.j} + R^{i}_{.j} \quad \text{(A75)}$$

and the tensor form

$$T = pI + \frac{1}{2}(S + S^T) + \frac{1}{2}(S - S^T) \,. \quad \text{(A76)}$$

The goal of covariant differentiation is to derive a new tensor from a given tensor field via the process of differentiation. If $h(X)$ is a scalar tensor field, then

$$\nabla_i f = \frac{\partial f}{\partial X^i} \quad \text{(A77)}$$

is a simply covariant tensor field, the vector gradient.

The covariant derivative of first order of a vector is a tensor of second order

$$\nabla_i a^j = \frac{\partial a^j}{\partial X^i} + \Gamma^j_{ik} a^k \,, \quad \nabla_i a_j = \frac{\partial a_j}{\partial X^i} - \Gamma^k_{ij} a_k \,. \quad \text{(A78)}$$

The covariant derivative of a tensor of second order is a tensor whose order is one larger as the given tensor

$$T^{i}_{.j,k} = \nabla_k T^{i}_{.j} = \frac{\partial T^{i}_{.j}}{\alpha^k} + T^{l}_{.j}\Gamma^{i}_{lk} - T^{i}_{.l}\Gamma^{l}_{jk} \,. \quad \text{(A79)}$$

In a similar manner, the covariant derivative of an arbitrary tensor can be written. The covariant derivative of the metric tensor and the unit tensor vanish: $g_{ij,k} = g^{ij}_{..,k} = 0$ and $\delta^{i}_{.j,k} = 0$. Therefore, the metric tensor and the unit tensor behave like constants during differentiation in a orthogonal system of coordinates in Euclidean space. It must be noted that all Christoffel symbols vanish in a Cartesian system of coordinates.

In the following, we will use different products of vectors and tensors. One can summarize the discussion above by saying that there are three kinds of multiplication involving vectors.

The scalar product of two vectors $\boldsymbol{a}$ and $\boldsymbol{b}$ yields the scalar $\boldsymbol{a} \cdot \boldsymbol{b} = a_i b^i$, the vector product yields the vector $\boldsymbol{a} \times \boldsymbol{b} = 1/\sqrt{g}\varepsilon^{ijk} a_i b_j \boldsymbol{\varepsilon}_k$ and the tensor product (dyadic product) produces the tensor $\boldsymbol{ab} = a_i b_j \boldsymbol{\varepsilon}^i \boldsymbol{\varepsilon}^j$.

These three operations can also be applied to tensors. The scalar product of two tensors of order n and m yields a tensor of order $n + m - 2$, and the total scalar product gives a scalar. For example, for the scalar product of tensors of second order in Cartesian coordinates, one obtains the tensor $\boldsymbol{A} \cdot \boldsymbol{B} = A_{ij} B^{jk}$, while the total scalar product yields a scalar $\boldsymbol{A} : \boldsymbol{B} = A_{ij} B^{ij}$.

Sometimes, one introduces a vector product of two tensors, for example, the vector product of a vector and a tensor in Cartesian coordinates yields the tensor $\boldsymbol{a} \times \boldsymbol{A} = \varepsilon_{ijk} a^j A^{kl}, \boldsymbol{A} \times \boldsymbol{a} = \varepsilon_{ijk} a^k A^{mj}$. The tensor product of two tensors of orders n and m yields a tensor or order $n + m$, for example, $\boldsymbol{AB} = A_{ij} B^{km}$.

A.3
Curvilinear Systems of Coordinates and Physical Components

In mechanics and physics, one uses, together with Cartesian coordinates, curvilinear orthogonal systems of coordinates $\{\alpha^1, \alpha^2, \alpha^3\}$. In these systems of coordinates, three families of surfaces that are specified via the dependence of the radius vector on the coordinates $\boldsymbol{r}(\alpha^1, \alpha^2, \alpha^3)$ intersect under right angles. This vector equation is equivalent to three scalar equations $X^i = X^i(\alpha^1, \alpha^2, \alpha^3)$.

The vectors $\boldsymbol{\varepsilon}_i = \partial \boldsymbol{r} / \partial \alpha^i$ are tangents on the coordinate lines α_i and one can take them as the local basis in a given point of space. In practice, the local orthogonal bases are the most useful. In applications, one mostly uses cylindrical and spherical systems of coordinates.

In an orthogonal basis, the components of the metric tensor are given by $g_{ij} = g^{ij} = 0$ for $i \neq j$ and $g_{ii} = 1/g^{ii} = H_i^2$, where H_i are the Lamé parameters.

The parameters can obtained by determining the square of the distance between two points that are at an infinitesimal distance $ds^2 = g_{ii}\left(d\alpha^i\right)^2$ using the geometric method.

The determinant of the fundamental matrix is $g = (H_1 H_2 H_3)^2$.

From (A5), one obtains the coordinates of the orthogonal basis

$$\boldsymbol{\varepsilon}_i = \left(\frac{\partial X^1}{\partial \alpha^i}, \frac{\partial X^2}{\partial \alpha^i}, \frac{\partial X^3}{\partial \alpha^i}\right)$$

and the corresponding coordinates of the orthonormal basis $\boldsymbol{e}_i = \boldsymbol{\varepsilon}_i / H_i$. The Christoffel symbols of the second kind are obtained via the formulae (A29). The non-vanishing symbols are given by

$$\Gamma_{\beta\beta}^{\beta} = \frac{1}{2g_{\beta\beta}} \frac{\partial g_{\beta\beta}}{\partial \alpha^\beta} = \frac{1}{H_\beta} \frac{\partial H_\beta}{\partial \alpha^\beta} ;$$

$$\Gamma_{\beta\beta}^{\gamma} = -\frac{1}{2g_{\gamma\gamma}} \frac{\partial g_{\beta\beta}}{\partial \alpha^\gamma} = -\frac{H_\beta}{H_\beta^2} \frac{\partial H_\beta}{\partial \alpha^\gamma} , \quad (\gamma \neq \beta) ;$$

$$\Gamma_{\beta\gamma}^{\beta} = \frac{1}{2g_{\beta\beta}} \frac{\partial g_{\beta\beta}}{\partial \alpha^\gamma} = \frac{1}{H_\beta} \frac{\partial H_\beta}{\partial \alpha^\gamma} ; \quad \Gamma_{\beta\gamma}^{\gamma} = \frac{1}{2g_{\gamma\gamma}} \frac{\partial g_{\gamma\gamma}}{\partial \alpha^\beta} = \frac{1}{H_\gamma} \frac{\partial H_\gamma}{\partial \alpha^\beta} .$$

The non-vanishing Christoffel symbols of the first kind are given by

$$\Gamma_{\beta\beta,\beta} = \frac{1}{2} \frac{\partial g_{\beta\beta}}{\partial \alpha^\beta} = \frac{1}{H_\beta} \frac{\partial H_\beta}{\partial \alpha^\beta} ; \quad \Gamma_{\beta\gamma,\beta} = -\Gamma_{\beta\beta,\gamma} = \frac{1}{2} \frac{\partial g_{\beta\beta}}{\partial \alpha^\gamma} = H_\beta \frac{\partial H_\beta}{\partial \alpha^\gamma} .$$

We now consider these relationships in cylindrical and spherical coordinates.

In cylindrical coordinates $\alpha^1 = \rho, \alpha^2 = \varphi, \alpha^3 = X^3$, the position of the point P is described via the distance ρ of the projection of the point P on the surface (X^1, X^2) to the origin O, the inclination angle ϕ of the projection of the radius vector $\boldsymbol{OP}$ on the surface (X^1, X^2) relative to the coordinate axis OX^1, and the projection X^3 of the point P on the coordinate axis OX^3.

The transition from the Cartesian coordinates to cylindrical coordinates (A4) is realized by the following formulae:

$$X^1 = \alpha^1 \cos\alpha^2\,, \quad X^2 = \alpha^1 \sin\alpha^2\,, \quad X^3 = \alpha^3\,. \tag{A80}$$

From (A80), one obtains the inverse transformation via (A7)

$$\alpha^1 = \sqrt{(X_1)^2 + (X_2)^2}\,, \quad \alpha^2 = \arcsin(X_2/X_1)\,, \quad \alpha^3 = X_3\,; \tag{A81}$$

The local basis of coordinates

$$\boldsymbol{\varepsilon}_1(\cos\alpha^2, \sin\alpha^2, 0)\,, \quad \boldsymbol{\varepsilon}_2(-\alpha^1\sin\alpha^2, \alpha^1\cos\alpha^2, 0)\,, \quad \boldsymbol{\varepsilon}_3(0,0,1)\,; \tag{A82}$$

The squared distance between two points at an infinitesimal distance

$$ds^2 = \left(d\alpha^1\right)^2 + (\alpha^1)^2\left(d\alpha^2\right)^2 + \left(d\alpha^3\right)^2\,; \tag{A83}$$

The components of the metric tensor and the Lamé parameters

$$\begin{aligned} &g_{11} = g_{33} = 1\,, \quad g_{22} = (\alpha^1)^2\,, \quad H_1 = 1,\\ &H_2 = \alpha^1\,, \quad H_3 = 1\,, \quad g = (\alpha^1)^2\,; \end{aligned} \tag{A84}$$

The non vanishing Christoffel symbols of the second and first kind:

$$\Gamma^2_{21} = -\alpha^1\,, \quad \Gamma^1_{22} = \frac{1}{\alpha^1}\,, \quad \Gamma_{12,2} = -\Gamma_{22,1} = \alpha^1\,. \tag{A85}$$

In spherical coordinates $\alpha^1 = r$, $\alpha^2 = \varphi$, $\alpha^3 = \vartheta$, the position of a point P is described by its distance r from the origin of the coordinate system O, the angle of inclination ϕ between the projection of the radius vector OP on the (X^1, X^2)-surface and the coordinate axis OX^1, the angle of inclination ϑ between OP and the coordinate axis OX^3. The transition from the Cartesian to the spherical system of coordinates and vice versa is given by the formulae

$$X^1 = \alpha^1 \sin\alpha^2 \cos\alpha^3\,, \quad X^2 = \alpha^1 \sin\alpha^2 \sin\alpha^3\,, \quad X^3 = \alpha^1\cos\alpha^2\,,$$

$$\alpha^1 = \sqrt{(X_1)^2 + (X_2)^2 + (X_3)^2}\,, \quad \alpha^2 = \arctan\left(\frac{\sqrt{(X_1)^2 + (X_2)^2}}{X^3}\right),$$

$$\alpha^3 = \arctan\left(\frac{X^2}{X^1}\right).$$

We now write down the main equations for the spherical system of coordinates. The coordinates of the local basis vectors:

$$\begin{aligned} &\boldsymbol{\varepsilon}_1(\sin\alpha^2\cos\alpha^3, \sin\alpha^2\sin\alpha^3, \cos\alpha^2)\,,\\ &\boldsymbol{\varepsilon}_2(\alpha^1\cos\alpha^2\cos\alpha^3, \alpha^1\cos\alpha^2\sin\alpha^3, -\alpha^1\sin\alpha^2)\,,\\ &\boldsymbol{\varepsilon}_3(-\alpha^1\sin\alpha^2\sin\alpha^3, \alpha^1\sin\alpha^2\cos\alpha^3, 0), \end{aligned} \tag{A86}$$

the squared distance between two infinitesimally close points:

$$ds^2 = \left(d\alpha^1\right)^2 + \left(d\alpha^2\right)^2 + \left(\alpha^1 \sin\alpha^2 d\alpha^3\right)^2 , \tag{A87}$$

the components of the metric tensor and Lamé parameters:

$$\begin{aligned} &g_{11} = 1 , \quad g_{22} = (\alpha^1)^2 , \quad g_{33} = \left(\alpha^1 \sin\alpha^2\right)^2 , \quad g = (\alpha^1)^4 (\sin\alpha^2)^2 \\ &H_1 = 1 , \quad H_2 = \alpha^1 , \quad H_3 = \alpha^1 \sin\alpha^2 , \end{aligned} \tag{A88}$$

and the non-vanishing Christoffel symbols of the second and first kind:

$$\begin{aligned} &\Gamma^2_{22} = -\alpha^1 , \quad \Gamma^1_{33} = -\alpha^1 (\sin\alpha^2)^2 , \quad \Gamma^2_{33} = -\sin\alpha^2 \cos\alpha^2 , \\ &\Gamma^2_{21} = \frac{1}{\alpha^1} , \quad \Gamma^3_{31} = \frac{1}{\alpha^1} , \quad \Gamma^3_{32} = \cot\alpha^2 , \\ &\Gamma_{21,2} = -\Gamma_{22,1} = \alpha^1 , \quad \Gamma_{31,3} = -\Gamma_{33,3} = \alpha^1 (\sin\alpha^2)^2 , \\ &\Gamma_{32,3} = -\Gamma_{33,2} = (\alpha^1)^2 \sin\alpha^2 \cos\alpha^2 . \end{aligned} \tag{A89}$$

During the transition from Cartesian to spherical or cylindrical coordinates, one encounters the following problem. In Cartesian coordinates, all coordinates of point X^i, $i = 1, 2, 3$ have the same dimensions, while the coordinates in a curvilinear system of coordinates have different dimensions. For example, in cylindrical coordinates, $\alpha^1 = \rho$ and $\alpha^3 = z$ have dimensions of length, while $\alpha^2 = \varphi$ is a dimensionless quantities. In spherical coordinates, $\alpha^1 = r$ has the dimension of length, while $\alpha^2 = \varphi$ and $\alpha^3 = \vartheta$ are dimensionless quantities. This difference in dimensions leads to different dimensions of the basis vectors $\boldsymbol{\varepsilon}_i$ and therefore to different dimensions of the components of the physical objects of vectors and tensor type, for example, velocity, strain and strain velocity tensor.

The reason is that the local basis vectors are not normalized. Instead of the basis $\boldsymbol{\varepsilon}_i$, we now introduce the orthonormal basis $\boldsymbol{e}_i = \boldsymbol{\varepsilon}_i / |\boldsymbol{\varepsilon}_i|$.

Due to (A10) and (A11), one has

$$\begin{aligned} &|\boldsymbol{\varepsilon}_i| = \sqrt{g_{ii}} = 1/\sqrt{g^{ii}} , \quad \boldsymbol{e}_i = \boldsymbol{\varepsilon}_i/\sqrt{g_{ii}} = \boldsymbol{\varepsilon}_i \sqrt{g^{ii}} , \\ &\boldsymbol{e}^i = \boldsymbol{\varepsilon}^i / \left|\boldsymbol{\varepsilon}^i\right| = \boldsymbol{\varepsilon}^i/\sqrt{g^{ii}} = \boldsymbol{\varepsilon}^i \sqrt{g^{ii}} . \end{aligned}$$

Therefore, we have, due to (A14),

$$\boldsymbol{e}_i = \frac{\boldsymbol{\varepsilon}_i}{\sqrt{g_{ii}}} = \frac{g_{ik}\boldsymbol{e}^k}{\sqrt{g_{ii}}} = \frac{g_{ii}\boldsymbol{e}_i}{\sqrt{g_{ii}}} = \sqrt{g_{ii}}\boldsymbol{e}^i = \frac{\boldsymbol{e}^i}{\sqrt{g^{ii}}} = \boldsymbol{e}^i .$$

Therefore, the covariant and contravariant bases $\boldsymbol{e}_i$ and $\boldsymbol{e}^i$ coincide, and their components are dimensionless.

By decomposing a vector $\boldsymbol{a}$ with respect to the unit basis vectors $\boldsymbol{e}_i$ and $\boldsymbol{e}^i$, we obtain

$$\boldsymbol{a} = a^i |\boldsymbol{\varepsilon}_i| \frac{\boldsymbol{\varepsilon}_i}{|\boldsymbol{\varepsilon}_i|} = \tilde{a}^i \boldsymbol{e}_i , \quad \boldsymbol{a} = a_i \boldsymbol{\varepsilon}^i = a_i \left|\boldsymbol{\varepsilon}^i\right| \frac{\boldsymbol{\varepsilon}^i}{\left|\boldsymbol{\varepsilon}^i\right|} = \tilde{a}_i \boldsymbol{e}^i .$$

Since $\boldsymbol{e}_i = \boldsymbol{e}^i$, we have $\tilde{a}^i = \tilde{a}_i$. This means that the components of the vector $\boldsymbol{a}$ coincide in both orthonormal bases $\boldsymbol{e}_i$ and $\boldsymbol{e}^i$. The quantities $\tilde{a}^i = a^i |\boldsymbol{\varepsilon}_i| =$

$a^i\sqrt{g_{ii}} = a^i H_i$ and $\tilde{a}_i = a_i\left|\boldsymbol{\varepsilon}^i\right| = a_i\sqrt{g_{ii}} = a_i/H^i$ are called physical components of the vector $\boldsymbol{a}$.

In contrast to the corresponding covariant and contravariant vector components a_i and a^i, the physical components $\tilde{a}^i$ and $\tilde{a}_i$ are not transformed according to the vector laws (A21) and (A22) when moving from one system of coordinates to another. Thus, they can be considered as components of an object, that is, not a vector in the precise mathematical sense.

Similar considerations can be repeated for tensors using their representation as dyadic products. For example, the transition to the physical components of a tensor of second order $\boldsymbol{T}(T^{ij}_{.k})$ takes place according to the formula

$$\boldsymbol{T} = T^{ij}_{.k}\,\boldsymbol{\varepsilon}_i\boldsymbol{\varepsilon}_j\boldsymbol{\varepsilon}^k = T^{ij}_{.k}\left|\boldsymbol{\varepsilon}_i\right|\left|\boldsymbol{\varepsilon}_j\right|\left|\boldsymbol{\varepsilon}^k\right|\boldsymbol{e}_i\boldsymbol{e}_j\boldsymbol{e}^k = \tilde{T}^{ij}_{.k}\,\boldsymbol{e}_i\boldsymbol{e}_j\boldsymbol{e}^k\,,$$

where

$$\tilde{T}^{ij}_{.k} = T^{ij}_{.k}\left|\boldsymbol{\varepsilon}_i\right|\left|\boldsymbol{\varepsilon}_j\right|\left|\boldsymbol{\varepsilon}^k\right| = T^{ij}_{.k}\sqrt{\frac{g_{ii}g_{jj}}{g_{kk}}} = T^{ij}_{.k}\frac{H_i H_j}{H_k} \tag{A90}$$

are the physical tensor components.

The transition to physical components also changes the Christoffel symbols of the second kind

$$\tilde{\Gamma}^i_{jk} = \frac{H_i H_j}{H_k}\Gamma^i_{jk}\,. \tag{A91}$$

They are referred to as physical Christoffel symbols.

A.4
Calculation of Lengths, Surface Areas and Volumes

A curve in Euclidean space can be given in parametric form as

$$\boldsymbol{r} = \boldsymbol{r}(t) = X^i\boldsymbol{e}_i\,.$$

Then, an infinitesimal curve element is given by

$$dS = \sqrt{d\boldsymbol{r}\cdot d\boldsymbol{r}} = \sqrt{\frac{d\boldsymbol{r}}{dt}\cdot\frac{d\boldsymbol{r}}{dt}}\,dt = \sqrt{g_{ij}\frac{dX^i}{dt}\frac{dX^j}{dt}}\,dt\,, \tag{A92}$$

with the metric tensor $g_{ij} = \boldsymbol{e}_i\boldsymbol{e}_j$.

The length of the curve between the points $\boldsymbol{r}(t_0)$ and $\boldsymbol{r}(t_1)$ is

$$L = \int_{t_1}^{t_0}\sqrt{g_{ij}\frac{dX^i}{dt}\frac{dX^j}{dt}}\,dt\,. \tag{A93}$$

The line integral of a function along the curve C is defined by

$$\int_{t_1}^{t_0} F(X^1(t), X^2(t), X^3(t))\,dt\,. \tag{A94}$$

If the length of the curve S is taken as parameter t, and the curve is given by the parameter representation $\boldsymbol{r}(t)$, the curve integral (A94) is reduced to a definite integral via the relationship

$$\int_{t_1}^{t_0} F(\boldsymbol{r}) d S = \int_{t_1}^{t_0} F[\boldsymbol{r}(t)] \sqrt{g_{ij} \frac{d X^i}{dt} \frac{d X^j}{dt}} dt \,. \tag{A95}$$

A surface in Euclidean space can be represented in parameter form $\boldsymbol{r}(U^1, U^2)$ where $\boldsymbol{r}$ is the radius vector from the origin of the coordinate system to the point on the surface and where U^1, U^2 are the coordinates of the point on the surface in the curvilinear system of coordinates whose coordinate lines $U^1 = \text{const}$ and $U^2 = \text{const}$ lie on the surface and generally cover it in its entirety. A curve on the surface is defined in parameter form by $U^\alpha = U^\alpha(t)$. Through every point on the curve, one can draw two coordinate lines U^1 and U^2 and two tangential vectors $\boldsymbol{\varepsilon}_\alpha = \partial \boldsymbol{r}/\partial U^\alpha$ can be defined as local basis vectors. Then, the length of the infinitesimal element of the curve is given by $dS = a_{\alpha\beta} d U^\alpha d U^\beta$ $(\alpha, \beta = 1, 2)$, where $a_{\alpha,\beta}$ is the metric surface tensor. The length of the curve between the points $\boldsymbol{r}(t_0)$ and $\boldsymbol{r}(t_1)$ is given by

$$L = \int_{t_1}^{t_0} \sqrt{a_{\alpha\beta} \frac{d U^\alpha}{dt} \frac{d U^\beta}{dt}} dt \,. \tag{A96}$$

In Euclidean space, the scalar product of two vectors $\boldsymbol{a}$ and $\boldsymbol{b}$ is given by

$$\boldsymbol{a} \cdot \boldsymbol{b} = g_{ij} a^i b^j = g^{ij} a_i b_j = a_i b^j = a^i b_j \tag{A97}$$

with the metric tensor $\boldsymbol{g} = g_{ij} \boldsymbol{\varepsilon}^i \boldsymbol{\varepsilon}^j = g^{ij} \boldsymbol{\varepsilon}_i \boldsymbol{\varepsilon}_j$, the local basis vectors $\boldsymbol{\varepsilon}_i = \partial \boldsymbol{r}/\partial a_i$ and the radius vector of the point in space $\boldsymbol{r}$.

The vector product of these vectors is equal to

$$\boldsymbol{a} \times \boldsymbol{b} = \sqrt{g} \varepsilon_{ijk} a^i b^j \boldsymbol{\varepsilon}^k = \frac{1}{\sqrt{g}} \varepsilon^{ijk} a_i b_j \boldsymbol{\varepsilon}_k \tag{A98}$$

with the permutation symbols (ε-objects) ε_{ijk} and ε^{ijk}, and the determinant of the metric tensor g.

The area of the parallelogram spanned by the vectors $\boldsymbol{a}$ and $\boldsymbol{b}$ can be determined as the absolute value of the vector product

$$\sigma = |\boldsymbol{a} \times \boldsymbol{b}| = |\boldsymbol{a}|\,|\boldsymbol{b}| \sin \alpha \tag{A99}$$

or

$$\sigma = |\boldsymbol{a} \times \boldsymbol{b}| = \left| \sqrt{g} \varepsilon_{ijk} a^i b^j \boldsymbol{\varepsilon}^k \right| = \left| \frac{1}{\sqrt{g}} \varepsilon^{ijk} a_i b_j "_k \right| \,. \tag{A100}$$

If the vectors $\boldsymbol{a}$ and $\boldsymbol{b}$ are collinear with basis vectors $\boldsymbol{\varepsilon}_1$ and $\boldsymbol{\varepsilon}_2$, we have

$$\boldsymbol{a} = a^1 \boldsymbol{\varepsilon}_1 \,, \quad \boldsymbol{b} = b^2 \boldsymbol{\varepsilon}_2 \,, \quad |\boldsymbol{a}| = \sqrt{g_{11} a^1 a^1} \,, \quad |\boldsymbol{b}| = \sqrt{g_{22} b^2 b^2} \,,$$

$$\cos \alpha = \frac{\boldsymbol{a} \cdot \boldsymbol{b}}{|\boldsymbol{a}\,|\boldsymbol{b}||} = \frac{g_{12}}{\sqrt{g_{11} g_{22}}} \,, \quad \sin \alpha = \sqrt{\frac{g_{11} g_{22} - g_{12}}{g_{11} g_{22}}} = \frac{\sqrt{g}\sqrt{g^{33}}}{\sqrt{g_{11}}\sqrt{g_{22}}}$$

and from (A99), obtain

$$\sigma = \sqrt{g} a^1 b^2 \sqrt{g^{33}} \, . \tag{A101}$$

In the curvilinear system of coordinates (U^1, U^2) on the surface, the area of an infinitesimal surface element is given by

$$d\sigma = \left| \frac{\partial \boldsymbol{r}}{\partial U^1} \times \frac{\partial \boldsymbol{r}}{\partial U^2} \right| dU^1 dU^2 = \sqrt{a}\, dU^1 dU^2 \, , \tag{A102}$$

where $a = a(U^1, U^2)$ is the determinant of the metric tensor $a_{\alpha\beta}$ on the surface.

The area of a finite surface Σ is equal to

$$\Sigma = \int_\Sigma d\sigma = \int_\Sigma \left| \frac{\partial \boldsymbol{r}}{\partial U^1} \times \frac{\partial \boldsymbol{r}}{\partial U^2} \right| dU^1 dU^2 = \int_\Sigma \sqrt{a}\, dU^1 dU^2 \, . \tag{A103}$$

The surface integral of a function $F(\boldsymbol{r}) = F(U^1, U^2)$ is given by

$$\int_\Sigma F(\boldsymbol{r}) d\sigma = \int_\Sigma F(\boldsymbol{r}) \left| \frac{\partial \boldsymbol{r}}{\partial U^1} \times \frac{\partial \boldsymbol{r}}{\partial U^2} \right| dU^1 dU^2 = \int_\Sigma F(U^1, U^2) \sqrt{a}\, dU^1 dU^2 \, . \tag{A104}$$

The volume of the parallelepiped defined by the vectors $\boldsymbol{a}$, $\boldsymbol{b}$ and $\boldsymbol{c}$ is equal to the absolute value of the scalar triple product that is

$$V = |(\boldsymbol{a}, \boldsymbol{b}, \boldsymbol{c})| = |(\boldsymbol{a} \times \boldsymbol{b}) \cdot \boldsymbol{c}| = \left| \sqrt{g} \varepsilon_{ijk} a^i b^j c^k \right| = \left| \frac{1}{\sqrt{g}} \varepsilon^{ijk} a_i b_j c_k \right| \, . \tag{A105}$$

One defines the infinitesimal volume element as the volume of the parallelepiped which is defined by the infinitesimal vectors $d\boldsymbol{X}_1 = dX^1 \boldsymbol{\varepsilon}_1$, $d\boldsymbol{X}_2 = dX^2 \boldsymbol{\varepsilon}_2$, $d\boldsymbol{X}_3 = dX^3 \boldsymbol{\varepsilon}_3$ that are tangents to the coordinate lines.

During the transition to a new system of coordinates using the transformation law $X^{i'} = X^{i'}(X^1, X^2, X^3)$ with the Jacobian transformation matrix $A^{i'}_i = \partial X^{i'} / \partial X^i$ or via the inverse transformation law $X^i = X^i(X^{1'}, X^{2'}, X^{3'})$ with transformation matrix $B^j_{j'} = \partial X^j / \partial X^{j'}$, which is the inverse of the matrix $A^{i'}_i$, that is $A^{i'}_i B^k_{j'} = \delta^{i'}_{j'}$, the infinitesimal volume is transformed according to $dV' = |\mathrm{Det}(A^{i'}_i)| dV$ or $dV = |\mathrm{Det}(B^j_{j'})| dV'$, with

$$\mathrm{Det}\left(A^{i'}_i\right) = \varepsilon^{m'n'k'} \frac{\partial X_{m'}}{\partial X_1} \frac{\partial X_{n'}}{\partial X_2} \frac{\partial X_{k'}}{\partial X_3} \, , \quad \mathrm{Det}\left(B^j_{j'}\right) = \varepsilon_{mnk} \frac{\partial X_m}{\partial X_1} \frac{\partial X_n}{\partial X_2} \frac{\partial X_k}{\partial X_3} \, .$$

Since the metric tensor is transformed according to (A24), the determinants g and g' in the old and new coordinate systems are connected via the relationships $g' = (\mathrm{Det}(A^{i'}_i))^2 g$ and $g = (\mathrm{Det}(B^j_{j'}))^2 g'$.

Therefore, the volume element is equal to

$$dV = \left|d\boldsymbol{X}^1 \cdot (d\boldsymbol{X}^2 \times d\boldsymbol{X}^3)\right| = \left|\sqrt{g} dX_1 dX_2 dX_3\right| ,$$
$$dV' = \left|d\boldsymbol{X}^{1'} \cdot (d\boldsymbol{X}^{2'} \times d\boldsymbol{X}^{3'})\right| = \left|\sqrt{g'} dX_{1'} dX_{2'} dX_{3'}\right| .$$

A finite volume is given by

$$V = \int_V \sqrt{g} dX^1 dX^2 dX^3 . \tag{A106}$$

Therefore, the volume integral of a function $F(\boldsymbol{r})$ has the form

$$\int_V F(\boldsymbol{r}) dV = \int_V F(X^1, X^2, X^3) \sqrt{g} dX^1 dX^2 dX^3. \tag{A107}$$

It must be noted that we have $g = 1$ in a Cartesian system of coordinates.

A.5 Differential Operators and Integral Theorems

We now consider the differential vector operator ∇ whose components are covariant derivatives such that $\nabla = \boldsymbol{\varepsilon}^i \nabla_i$. Considering the vector operator as a regular vector, we apply the product operators mentioned above. By evaluating the scalar product of the vector operator ∇ with the vector $\boldsymbol{a}(\alpha^1, \alpha^2, \alpha^3)$, we obtain

$$\nabla \cdot \boldsymbol{a} = \nabla_i \boldsymbol{\varepsilon}^i \cdot a_j \boldsymbol{\varepsilon}^j = \nabla_i a_j g^{ij} = \nabla_i a^i \equiv a^i_{,i} . \tag{A108}$$

This operation is called divergence and is denoted by $\operatorname{div} \boldsymbol{a} = \nabla \cdot \boldsymbol{a}$. It must be noted that the divergence operation involves a contraction and one obtains an invariant (scalar). The divergence of a vector can also be written as

$$\nabla \cdot \boldsymbol{a} = \frac{1}{\sqrt{g}} \frac{\partial}{\partial \alpha^i} \left(\sqrt{g} a^i\right) . \tag{A109}$$

We now consider the scalar product of the vector operator ∇ with a tensor, for example, with a tensor of second order. Then, one obtains the vector

$$\nabla \cdot \boldsymbol{T} = \nabla_i \boldsymbol{\varepsilon}^i \cdot T^k_{.m} \boldsymbol{\varepsilon}_k \boldsymbol{\varepsilon}^m = \nabla_i T_{km} \delta^i_{.k} \boldsymbol{\varepsilon}^m = \nabla_i T^i_{.m} \boldsymbol{\varepsilon}^m . \tag{A110}$$

The vector product of the vector operator ∇ with a vector $\mathbf{a}$ is called curl and is defined by $\operatorname{curl} \boldsymbol{a} = \nabla \times \boldsymbol{a}$. The curl of a vector is given by a vector with the components

$$\nabla \times \boldsymbol{a} = \frac{1}{\sqrt{g}} \varepsilon^{ijk} \nabla_i a_j \boldsymbol{\varepsilon}_k = \frac{1}{\sqrt{g}} \varepsilon^{ijk} a_{j,i} \boldsymbol{\varepsilon}_k . \tag{A111}$$

The application of the operator $\nabla\times$ on a tensor of second order yields the tensor of second order

$$\nabla \times \boldsymbol{T} = \frac{1}{\sqrt{g}} \varepsilon^{ijk} \nabla_i T_{jl} \boldsymbol{\varepsilon}^l \boldsymbol{\varepsilon}_k \,. \tag{A112}$$

We now construct the tensor product of the vector operator ∇ and the scalar $\varphi(\alpha^1, \alpha^2, \alpha^3)$. As a result, one obtains the vector

$$\operatorname{grad} \varphi = \nabla \varphi = \nabla_i \varphi \boldsymbol{\varepsilon}_i \,, \tag{A113}$$

which is referred to as the gradient of a scalar.

In a similar manner, one can form the tensor product of the vector operator ∇ and a vector. As a result, one obtains the tensor

$$\nabla \boldsymbol{a} = \nabla_i a^j \boldsymbol{\varepsilon}^i \boldsymbol{\varepsilon}_j = \nabla_i a_j \boldsymbol{\varepsilon}^i \boldsymbol{\varepsilon}^j \,, \tag{A114}$$

which is called the gradient of a vector.

The transposition of the vector gradient gives

$$(\nabla \boldsymbol{a})^T = \nabla_j a_i \boldsymbol{\varepsilon}^i \boldsymbol{\varepsilon}^j \,. \tag{A115}$$

The symmetrization of the tensor $\nabla \boldsymbol{u}$ gives the tensor

$$\operatorname{def} \boldsymbol{u} = \frac{1}{2} \left[\nabla \boldsymbol{u} + (\nabla \boldsymbol{u})^T \right] = \frac{1}{2} \left(\nabla_i u_j + \nabla_j u_i \right) \boldsymbol{\varepsilon}^i \boldsymbol{\varepsilon}^j = \varepsilon_{ij} \boldsymbol{\varepsilon}^i \boldsymbol{\varepsilon}^j \,. \tag{A116}$$

If $\boldsymbol{u}$ is the displacement vector, then def$\boldsymbol{u}$ yields the strain tensor with components

$$\varepsilon_{ij} = \frac{1}{2} \left(\nabla_i u_j + \nabla_j u_i \right) \,. \tag{A117}$$

If $\boldsymbol{u}$ is the velocity vector, then def $\boldsymbol{u}$ yields the strain velocity tensor with components

$$e_{ij} = \frac{1}{2} \left(\nabla_i u_j + \nabla_j u_i \right) \,. \tag{A118}$$

The scalar product of the vector operator ∇ and of the vector operator grad yields a scalar which is referred to as the Laplace operator

$$\Delta \varphi = \nabla^2 \varphi = \nabla \cdot \nabla \varphi = \operatorname{div} (\operatorname{grad} \varphi) = g^{ij} \nabla_i \nabla_j \varphi = g^{ij} \left(\frac{\partial \varphi}{\partial \alpha^i} \right)_{,j} \,. \tag{A119}$$

Differential Operators in Curvilinear Coordinate Systems

In orthogonal curvilinear coordinates, the differential operators have the following form

$$\nabla \varphi = \frac{1}{H_1} \frac{\partial \varphi}{\partial \alpha^1} \boldsymbol{\varepsilon}_1 + \frac{1}{H_2} \frac{\partial \varphi}{\partial \alpha^2} \boldsymbol{\varepsilon}_2 + \frac{1}{H_3} \frac{\partial \varphi}{\partial \alpha^3} \boldsymbol{\varepsilon}_3 \,; \tag{A120}$$

$$\nabla \cdot \boldsymbol{a} = \frac{1}{H_1 H_2 H_3} \left[\frac{\partial}{\partial \alpha^1} \left(H_2 H_3 a^1 \right) + \frac{\partial}{\partial \alpha^2} \left(H_1 H_3 a^2 \right) + \frac{\partial}{\partial \alpha^3} \left(H_1 H_2 a^3 \right) \right] ; \tag{A121}$$

$$\nabla \times \boldsymbol{a} = \frac{1}{H_1 H_2 H_3} \begin{vmatrix} H_1 \boldsymbol{\varepsilon}_1 & H_2 \boldsymbol{\varepsilon}_2 & H_3 \boldsymbol{\varepsilon}_3 \\ \partial / \partial \alpha^1 & \partial / \partial \alpha^2 & \partial / \partial \alpha^3 \\ H_1 a^1 & H_2 a^2 & H_3 a^3 \end{vmatrix} ; \tag{A122}$$

$$\Delta \varphi = \frac{1}{H_1 H_2 H_3} \left[\frac{\partial}{\partial \alpha^1} \left(\frac{H_2 H_3}{H_1} \frac{\partial \varphi}{\partial \alpha^1} \right) + \frac{\partial}{\partial \alpha^2} \left(\frac{H_1 H_3}{H_2} \frac{\partial \varphi}{\partial \alpha^2} \right) \right.$$

$$\left. + \frac{\partial}{\partial \alpha^3} \left(\frac{H_1 H_2}{H_3} \frac{\partial \varphi}{\partial \alpha^3} \right) \right] . \tag{A123}$$

We now write down the expressions for the differential operators in cylindrical and spherical coordinates.

Cylindrical Coordinates:

$$\nabla \varphi = \frac{\partial \varphi}{\partial \alpha^1} \boldsymbol{\varepsilon}_1 + \frac{1}{\alpha^1} \frac{\partial \varphi}{\partial \alpha^2} \boldsymbol{\varepsilon}_2 + \frac{\partial \varphi}{\partial \alpha^3} \boldsymbol{\varepsilon}_3 ; \tag{A124}$$

$$\nabla \cdot \boldsymbol{a} = \frac{1}{\alpha^1} \left[\frac{\partial}{\partial \alpha^1} (\alpha^1 a^1) + \frac{\partial a^2}{\partial \alpha^2} + \alpha^1 \frac{\partial a^3}{\partial \alpha^3} \right] ; \tag{A125}$$

$$\nabla \times \boldsymbol{a} = \frac{1}{\alpha^1} \begin{vmatrix} \boldsymbol{\varepsilon}_1 & \alpha^1 \boldsymbol{\varepsilon}_2 & \boldsymbol{\varepsilon}_3 \\ \partial / \partial \alpha^1 & \partial / \partial \alpha^2 & \partial / \partial \alpha^3 \\ a^1 & \alpha^1 a^2 & a^3 \end{vmatrix} ; \tag{A126}$$

$$\Delta \varphi = \frac{1}{\alpha^1} \left[\frac{\partial}{\partial \alpha^1} \left(\alpha^1 \frac{\partial \varphi}{\partial \alpha^1} \right) + \frac{\partial}{\partial \alpha^2} \left(\frac{1}{\alpha^1} \frac{\partial \varphi}{\partial \alpha^2} \right) + \frac{\partial}{\partial \alpha^3} \left(\alpha^1 \frac{\partial \varphi}{\partial \alpha^3} \right) \right] . \tag{A127}$$

Spherical Coordinates:

$$\nabla \varphi = \frac{\partial \varphi}{\partial \alpha^1} \boldsymbol{\varepsilon}_1 + \frac{1}{\alpha^1} \frac{\partial \varphi}{\partial \alpha^2} \boldsymbol{\varepsilon}_2 + \frac{1}{\alpha^1 \sin \alpha^2} \frac{\partial \varphi}{\partial \alpha^3} \boldsymbol{\varepsilon}_3 ; \tag{A128}$$

$$\nabla \cdot \boldsymbol{a} = \frac{1}{(\alpha^1)^2 \sin \alpha^2} \left\{ \frac{\partial}{\partial \alpha^1} \left[(\alpha^1)^2 \sin \alpha^2 a^1 \right] \right.$$

$$\left. + \frac{\partial}{\partial \alpha^2} (\alpha^1 \sin \alpha^2 a^2) + \frac{\partial}{\partial \alpha^3} (\alpha^1 a^3) \right\} ; \tag{A129}$$

$$\nabla \times \boldsymbol{a} = \frac{1}{(\alpha^1)^2 \sin \alpha^2} \begin{vmatrix} \boldsymbol{\varepsilon}_1 & \alpha^1 \boldsymbol{\varepsilon}_2 & \alpha^1 \sin \alpha^2 \boldsymbol{\varepsilon}_3 \\ \partial / \partial \alpha^1 & \partial / \partial \alpha^2 & \partial / \partial \alpha^3 \\ a^1 & \alpha^1 a^2 & \alpha^1 \sin \alpha^2 a^3 \end{vmatrix} ; \tag{A130}$$

$$\Delta \varphi = \frac{1}{(\alpha^1)^2 \sin \alpha^2} \left\{ \frac{\partial}{\partial \alpha^1} \left[(\alpha^1)^2 \sin \alpha^2 \frac{\partial \varphi}{\partial \alpha^1} \right] \right.$$

$$\left. + \frac{\partial}{\partial \alpha^2} \left(\sin \alpha^2 \frac{\partial \varphi}{\partial \alpha^2} \right) + \frac{\partial}{\partial \alpha^3} \left(\frac{1}{\sin \alpha^2} \frac{\partial \varphi}{\partial \alpha^3} \right) \right\} . \tag{A131}$$

In applications, one often uses the following relationships:

$$\frac{\partial \varphi}{\partial n} = \nabla \varphi \cdot \boldsymbol{n} \quad \text{Derivative in the direction of } \boldsymbol{n} \tag{A132}$$

$$\nabla (\varphi \psi) = \varphi \nabla \psi + \psi \nabla \varphi ; \tag{A133}$$

$$\nabla (\varphi \boldsymbol{a}) = \nabla \varphi \cdot \boldsymbol{a} + \varphi \nabla \cdot \boldsymbol{a} ; \tag{A134}$$

$$\nabla \cdot (\boldsymbol{a} \times \boldsymbol{b}) = (\nabla \times \boldsymbol{a}) \cdot \boldsymbol{b} - \boldsymbol{a} \cdot (\nabla \times \boldsymbol{b}) \; ; \tag{A135}$$

$$\Delta\varphi(r) = \frac{d^2\varphi}{dr^2} + \frac{2}{r}\frac{d\varphi}{dr} \; ; \tag{A136}$$

$$\nabla \cdot (\nabla \times \boldsymbol{a}) = 0 \; ; \tag{A137}$$

$$\nabla \times (\nabla\varphi) = 0 \; ; \tag{A138}$$

$$\nabla \times (\varphi \boldsymbol{a}) = \nabla\varphi \times \boldsymbol{a} + \varphi\nabla \times \boldsymbol{a} \; . \tag{A139}$$

We consider a finite, simply connected volume V which is bounded by a closed regular surface Σ with external normal $\boldsymbol{n} = n_i \boldsymbol{\varepsilon}^i$ $(n_i n_j g^{ij} = 1)$ and with orientated surface element $d\boldsymbol{S} = \boldsymbol{n} dS$. For continuous vector functions $\boldsymbol{a}(\alpha^1, \alpha^2, \alpha^3)$ that have continuous derivatives in $V + \Sigma$, we have the following integral theorems:

- Gauss theorem in in vector form:

$$\int_V \nabla \cdot \boldsymbol{a} dV = \int_\Sigma \boldsymbol{n} \cdot \boldsymbol{a} dS = \int_\Sigma \boldsymbol{a} \cdot d\boldsymbol{S} \tag{A140}$$

or in component form

$$\int_V a^i_{,i} dV = \int_V g^{ij} a_{i,j} dV = \int_\Sigma a^i n_i dS = \int_\Sigma a_i n^i dS = \int_\Sigma g^{ij} a_i n_j dS \; . \tag{A141}$$

- Stokes theorem:

$$\int_V \nabla \times \boldsymbol{a} dV = \int_\Sigma \boldsymbol{n} \times \boldsymbol{a} dS = \int_\Sigma \boldsymbol{a} \times d\boldsymbol{S} \; , \tag{A142}$$

$$\int_\Sigma (\nabla \times \boldsymbol{a}) \cdot \boldsymbol{n} dS = \int_C \boldsymbol{a} \cdot d\boldsymbol{r} \; . \tag{A143}$$

In the last integral, a regular closed curve C is inside the volume V, the surface Σ is bounded by the curve and the vector $d\boldsymbol{r} = d\alpha^i \boldsymbol{\varepsilon}_i$ is the tangent on the curve.

Equations A140–(A143) are valid not only for vectors, but also for tensors. They are particularly valid for tensors of second order:

$$\int_V \nabla \cdot \boldsymbol{T} dV = \int_\Sigma \boldsymbol{n} \cdot \boldsymbol{T} dS \quad \text{or} \quad \int_V T^{ij}_{..,i} dV = \int_\Sigma T^{ij} n_i d\Sigma \; , \tag{A144}$$

$$\int_\Sigma (\nabla \times \boldsymbol{T}) \cdot \boldsymbol{n} dS = \int_c \boldsymbol{T} \cdot d\boldsymbol{r} \text{ or } \int_\Sigma \frac{1}{\sqrt{g}} \varepsilon^{ijk} T_{jm,i} n_k dS = \int_C T_{im} d\alpha^i \; . \tag{A145}$$

In applications, vector fields with the properties $\boldsymbol{a} = \nabla\varphi, \nabla \cdot \boldsymbol{a} = 0$ or $\nabla \times \boldsymbol{a} = 0$ are used most frequently. The first one is called a potential field, the second one a solenoidal field and the last one an irrotational field. If a vector field $\boldsymbol{a}(\alpha^1, \alpha^2, \alpha^3)$ is a potential field, we have $\boldsymbol{a} = \nabla\varphi$ and according to (A138), $\nabla \times \boldsymbol{a} = 0$.

Therefore, a potential vector field is at the same time irrotational. The converse assertion is also true: if a vector field $\boldsymbol{a}$ is irrotational, it is a potential field up to an additive constant, and thus $\boldsymbol{a} = \nabla\varphi$.

From the theorem (A140) and the relation (A132), it follows that

$$\int_V \Delta\varphi dV = \int_\Sigma \nabla\varphi \cdot \boldsymbol{n} dS = \int_\Sigma \frac{\partial\varphi}{\partial n} dS \, . \tag{A146}$$

For a solenoidal vector field $\boldsymbol{a}(\alpha^1, \alpha^2, \alpha^3)$, we have $\nabla\cdot\boldsymbol{a} = 0$. If it is also a potential field, the potential ϕ satisfies the Laplace equation $\Delta\varphi = 0$. From Gauss theorem (A140), it follows that the flux of a solenoidal vector field through a closed surface Σ vanishes:

$$\int_\Sigma \boldsymbol{n} \cdot \boldsymbol{a} dS = 0 \, . \tag{A147}$$

According to (A137), a solenoidal vector field $\boldsymbol{a}$ can be written in the form $\boldsymbol{a} = \nabla \times \boldsymbol{b}$.

For an irrotational vector field, we have $\nabla \times \boldsymbol{a} = 0$ and it follows from the Stokes theorem (A143) that the circulation of vector $\boldsymbol{a}$ over the closed contour C vanishes:

$$\int_C \boldsymbol{a} \cdot d\boldsymbol{r} = 0 \, . \tag{A148}$$

Appendix B
Some Differential Geometry

B.1
Curves on a Plane

We define a simple curve segment as a set of points on the plane with coordinates (X, Y) in a Cartesian system of coordinates $\{X, Y\}$ which satisfy the equation

$$Y = f(X) \text{ for } X_1 \leqslant X \leqslant X_2 \,, \tag{B1}$$

where $f(X)$ is a unique, continuous and sufficiently differentiable function. The geometrical meaning of this definition is that a simple curve segment is obtained from the straight line (X_1, X_2) on the X-axis as the result of a continuous deformation according to (B1).

A curve on the plane is defined as the set of points whose coordinates satisfy the equation

$$F(X, Y) \,. \tag{B2}$$

A point M on the curve (B2) is called an ordinary point if the curve has in a neighborhood of this point, for example, a rectangle, the form of a straight line. A point, for which the curve does not have the form of a straight line, irrespective of how small the neighborhood is chosen, is called singular point. Examples of singular points of a curve are points where the curve intersects itself and those points where the derivatives $Y'(X)$ or $X'(Y)$ do not exist.

If, for a given point on the curve $M(X_0, Y_0)$, the derivatives F'_x and F'_y do not both vanish at the same time, the point M is an ordinary point according to the theorem on implicit functions and (B2) can be solved uniquely in the form (B1). Singular points can only occur where we have $F'_x = 0$ and $F'_y = 0$ simultaneously. It must be noted that this condition is sufficient, but not necessary.

The construction of a curve in the vicinity of a regular point can be investigated by the expansion of the function $Y = f(X)$ in a Taylor series:

$$Y = Y_0 + Y' \frac{\Delta X}{1!} + Y'' \frac{(\Delta X)^2}{2!} + \ldots + Y_0^{(n)} \frac{(\Delta x)^n}{n!} + R_n \,, \tag{B3}$$

where $\Delta X = X - X_0$, $R_n = (Y^{(n+1)})(\xi)(\Delta X)^{(n+1)}/(n+1)!$ is the remainder term and we have $X_0 - \Delta X < \xi < X_0 + \Delta X$.

Hydromechanics. Theory and Fundamentals. Emmanuil G. Sinaiski

ISBN: 978-3-527-41026-2

By restricting the series (B3) to two terms, one obtains

$$Y \approx Y_0 + Y' \frac{\Delta X}{1!} . \tag{B4}$$

The geometrical meaning of (B4) is that it provides the equation of the tangent on the curve $Y = f(X)$ in the point M, and the behavior of the curve in the vicinity of an ordinary point is approximated by the line (tangent) (B4) up to a quantity of first order. One can, for example, judge the increase of the decrease of the function via the inclination angle between the tangent and the abscissa axis.

By restricting the series (B3) to three terms, one obtains

$$Y \approx Y_0 + Y' \frac{\Delta X}{1!} + Y'' \frac{(\Delta X)^2}{2!} . \tag{B5}$$

The third term in (B5) describes the difference between the ordinate of the curve and the ordinate of the tangent. Therefore, the sign of the second derivative determines whether the convexity of the curve is directed upwards or downwards. A point, for which $Y'' = 0$, is called the turning point. Then, the replacement of the curve equation via the equation of tangent (B5) is accurate up to quantities of third order. If all derivatives of second up to n-th order vanish at X_0, the curve can be approximated by the line (tangent) accurate to $n + 1$-th order.

B.2 Vectorial Definition of Curves

In addition to the curve equations in the form of (B1) and (B2), there is another method to define curves via a vector valued function. A vector function $\boldsymbol{a}(t)$ is defined via assigning to each value of t from the interval (t_1, t_2) a unique vector $\boldsymbol{a}(t)$. The vector function with an initial point at the origin of coordinates O and the end point $M(X, Y, Z)$ is denoted by $\boldsymbol{r}(t)$ and called the radius vector, where t is a parameter. During the change of t from t_1 to t_2, the vector describes a curve from the point $M_1(t_1)$ to the point $M_2(t_2)$, and the coordinates of each point $X = X(t)$, $Y = Y(t)$ and $Z = Z(t)$ are functions of t, and $\boldsymbol{r}$ can be considered as parametric form of the curve.

In vector analysis, the definition of the derivative of $\boldsymbol{r}(t)$ is given. If the derivative $\boldsymbol{r}'(t)$ exists at $t = t_0$, then the tangent on the curve exists in the point $M_0(t_0)$ and is directed along the vector $\boldsymbol{r}'(t_0)$.

The derivative of a vector function has the following properties:

$$
\begin{aligned}
&1. \quad (\boldsymbol{a} + \mathbf{b})' = \boldsymbol{a}' + \boldsymbol{b}' ; \\
&2. \quad \left[m(t)\boldsymbol{a}\right]' = m'(t)\boldsymbol{a} + m(t)\boldsymbol{a}' ; \\
&3. \quad (\boldsymbol{a} \cdot \boldsymbol{b})' = \boldsymbol{a}'\boldsymbol{b} + \boldsymbol{a}\boldsymbol{b}' ; \\
&4. \quad (\boldsymbol{a} \times \boldsymbol{b})' = \left(\boldsymbol{a}' \times \boldsymbol{b}\right) + \left(\boldsymbol{a} \times \boldsymbol{b}'\right) .
\end{aligned}
\tag{B6}
$$

The definition of a curve in terms of coordinates can be obtained from the parametric definition via the expansion of $\boldsymbol{r}(t)$ in the basis of the Cartesian system of coordinates,

$$\boldsymbol{r}(t) = X(t)i + Y(t)j + Z(t)k \,. \tag{B7}$$

The definitions of a curve segment and of regular as well as singular points for a plane curve can be generalized to a curve in space. We define a simple curve segment in space as the set of points (X, Y, Z) in the Cartesian system of coordinates $\{X, Y, Z\}$ that satisfy the equation

$$Y = f(X) \,, \quad Z = g(X) \quad \text{for} \quad X_1 \leqslant X \leqslant X_2 \,, \tag{B8}$$

where $f(X)$ and $g(X)$ are unique, continuous and sufficiently differentiable functions. If $X = X(t)$, then we have $Y = Y(t)$ and $Z = Z(t)$. The point $M(X_0, Y_0, Z_0)$ is an ordinary point under the condition that the derivatives $X'(t_0)$, $Y'(t_0)$ and $Z'(t_0)$ do not vanish at the same time. Otherwise, the point is a singular point. Since

$$\boldsymbol{r}'(t) = X'(t)\boldsymbol{i} + Y'(t)\boldsymbol{j} + Z'(t)\boldsymbol{k} \,,$$

this means that $\boldsymbol{r}'(t) \neq 0$.

The representation of the vector function in the form (B7) allows one to expand it into a Taylor series in the vicinity of the parameter value t_0:

$$\boldsymbol{r}(t) = \boldsymbol{r}(t_0) + \boldsymbol{r}'(t_0)\frac{\Delta t}{1!} + \boldsymbol{r}''(t_0)\frac{(\Delta t)^2}{2!} + \ldots + \boldsymbol{r}^{(n)}(t_0)\frac{(\Delta t)^n}{n!} + \boldsymbol{r}_n \,, \tag{B9}$$

where $\Delta t = t - t_0$, $\boldsymbol{r}_n = \boldsymbol{Q}_n(\Delta t)^{(n+1)}/(n+1)!$ is the remainder term with $|\boldsymbol{Q}_n| < C_n = \text{const}$.

Using the expansion (B9), one can investigate the behavior of the spatial curve in the vicinity of the point M_0. Keeping the first three terms in the expansion (B9), one obtains

$$\boldsymbol{r}(t) = \boldsymbol{r}(t_0) + \boldsymbol{r}'(t_0)\frac{\Delta t}{1!} + \boldsymbol{Q}_1\frac{(\Delta t)^2}{2!} \,. \tag{B10}$$

On the left-hand side, we have the displacement vector that connects the points $M(t)$ and $M_0(t_0)$. This vector has two components. The first one is the vector-differential $d\boldsymbol{r} = \boldsymbol{r}'(t_0)dt$ which is directed along the tangent on the curve in the point M_0, and the second one is the vector $\boldsymbol{Q}_1(\Delta t)^2/2$ whose absolute value is of second order.

We now show a few important properties that play a decisive role in the following:

1. We assume, that the vector function $\boldsymbol{m}(t)$ has constant absolute value, that is, $|\boldsymbol{m}(t)| = \text{const}$. Then, we have $\boldsymbol{m}^2(t) = \text{const}$ and $\boldsymbol{m}(t) \cdot \boldsymbol{m}'(t) = 0$. Therefore, $\boldsymbol{m}(t)$ and $\boldsymbol{m}'(t)$ are orthogonal. This especially applies for a unit vector function.

2. The rotational velocity of the vector function $\boldsymbol{m}(t)$ is defined as the velocity of change of the angle $d\phi/dt$ of the vector $\boldsymbol{m}(t)$ in the point $M(t)$. Then, we have $d\varphi/dt = |\boldsymbol{m}'(t)|$. If the parameter t changes by dt, the unit vector $\boldsymbol{m}$ turns by the angle $\Delta\varphi = |d\boldsymbol{m}|$ accurate to infinitesimal quantities of higher order.
3. We consider a unit vector $\boldsymbol{e}$ on the plane with the initial point in the coordinate origin O making an angle ϕ to the X-axis. Then, we have

$$\boldsymbol{e}(\varphi) = \cos\varphi\boldsymbol{i} + \sin\varphi\boldsymbol{j}\ , \quad \boldsymbol{e}'(\varphi) = -\sin\varphi\boldsymbol{i} + \cos\varphi\boldsymbol{j} = \boldsymbol{e}(\varphi + \pi/2)\ .$$

If we consider a circle with radius R and the center in the point O, the points on the circle satisfy $\boldsymbol{r} = R\,\boldsymbol{e}(\phi)$, $\phi = S/R$, where S is the arc length of the circle segment with angle ϕ. Then, one obtains $dS/d\phi = 1/R$, $\dot{\boldsymbol{r}} = d\boldsymbol{r}/dS = \boldsymbol{e}'(\varphi)$, $\ddot{\boldsymbol{r}} = d^2\boldsymbol{r}/dS^2 = -\boldsymbol{e}(\varphi)/R$.

From this, it follows that the first derivative $\dot{\boldsymbol{r}}$ with respect to the angle of the unit vector that is rotating in the plane is given by the vector $\boldsymbol{e}'(\varphi)$, which is turned by an angle $\pi/2$ relative to $\boldsymbol{e}$ and that the second derivative $d^2\boldsymbol{r}/dS^2$ is directed toward the center of the circle and has the absolute value of $1/R$.

We now determine the arc length between the points that correspond to the parameter values t_0 and t. To do this, we divide the interval (t_0, t) into n subintervals and construct the integral sum over the line segments

$$\sum_i |\boldsymbol{r}'(t)|(t_{i+1} - t_i)\ .$$

In the limit of $n \to \infty$, we obtain the arc length as

$$S = \int_{t_0}^{t} |\boldsymbol{r}'(t)|dt\ . \tag{B11}$$

The definition of a curve in the form $\boldsymbol{r}(t)$ does not restrict the choice of parameter. In problems from mechanics, one usually takes the time t as parameter. However, when investigating geometrical properties, it is convenient to take the arc length S as parameter. Thus, one has to define an initial point and a certain direction on the curve so that the curve must be orientated. Then, it follows from (B11), that $S = S(t)$ and $\boldsymbol{r}(t) = \boldsymbol{r}(S)$, where S grows monotonically from its initial value $S_0 = S(t_0)$, and one takes $S_0 = 0$ for convenience. It is obvious that

$$\left|\frac{d\boldsymbol{r}}{dS}\right| = 1\ . \tag{B12}$$

Using (B7) and (B12), one obtains

$$(dS)^2 = (dX)^2 + (dY)^2 + (dZ)^2 \tag{B13}$$

and

$$S = \int_{t_0}^{t} \left[X'(t)^2 + Y'(t)^2 + Z'(t)^2\right]^{1/2} dt\ . \tag{B14}$$

In the following, we will denote the derivative of $\boldsymbol{r}$ with respect to S by $\dot{\boldsymbol{r}}$. Since $|\dot{\boldsymbol{r}}| = 1$ and the vector $\dot{\boldsymbol{r}}$ is a tangent on the curve, it is the unit tangent vector in the point M.

We now consider the problem of the proximity order of two curves that have a common point M_0. The curves are given by the equations $\boldsymbol{r}_1 = \boldsymbol{r}_1(S)$ and $\boldsymbol{r}_2 = \boldsymbol{r}_2(S)$. We assume that $S = 0$ at the common point of the curves. We now move from the point M_0 along both curves by the same small distance to the points M_1 and M_2. The deviation of one curve from the other is characterized by the vector $\boldsymbol{m}_1\boldsymbol{m}_2$, which is obtained by a Taylor expansion of $\boldsymbol{r}_1$ and $\boldsymbol{r}_2$ as

$$\boldsymbol{m}_1\boldsymbol{m}_2 = (\dot{\boldsymbol{r}}_2 - \dot{\boldsymbol{r}}_1)\frac{S}{1!} + (\ddot{\boldsymbol{r}}_2 - \ddot{\boldsymbol{r}}_1)\frac{S^2}{2!} + \ldots + (\boldsymbol{r}_2^{(n)} - \boldsymbol{r}_1^{(n)})\frac{S^n}{n!} + \ldots \tag{B15}$$

The vectors $\dot{\boldsymbol{r}}_1$ and $\dot{\boldsymbol{r}}_2$ are directed along the corresponding curves. Therefore, if they are not collinear, the curves intersect at the point M_0 and the deviation is of first order in S. If the vectors $\dot{\boldsymbol{r}}_1$ and $\dot{\boldsymbol{r}}_2$ are collinear, that is, the curves have a common tangent in the point M_0 and $\dot{\boldsymbol{r}}_2 = \dot{\boldsymbol{r}}_1$, the expansion (B15) starts with the second order term and the deviation also is of second order in S. In this case, one says that the curves have a contact of first order. If we have $\dot{\boldsymbol{r}}_1 = \dot{\boldsymbol{r}}_2, \ldots, \boldsymbol{r}_1^{(n)} = \boldsymbol{r}_2^{(n)}$, the deviation is at least of $n + 1$-th order and the curves have a contact of n-th order. Using the representation (B7), one can write corresponding equations in coordinate form.

If the curves are represented in the form (8.11), that is, if the first curve corresponds to the equations $Y = f_1(X), Z = g_1(X)$ and the second curve to the equations $Y = f_2(X), Z = g_2(X)$, one obtains, by the expansion of the functions $f_i(X)$ and $g_i(X)$ in the vicinity of the point M_0 in a Taylor series, the following necessary and sufficient conditions for the curve contact of n-th order:

$$f_1^{(i)}(X_0) = f_2^{(i)}(X_0)\,, \quad g_1^{(i)}(X_0) = g_2^{(i)}(X_0)\,, \quad (i = 0, 1, \ldots, n)\,. \tag{B16}$$

These contact conditions of two curves allow one to solve the following two tasks: to find a certain curve among a family of curves that has contact to a) a given order or b) the maximal possible order with a given curve.

B.3
Curvature of a Curve in the Plane

The most important characteristic of the behavior of a curve in the vicinity of a given point is the curvature. The elementary curve, that is, the straight line, does not have any curvature. The other elementary curve, the circle, does have a constant curvature. The curvature of other curves can change from point to point.

In the previous section, it was shown that a curve can be approximated in the vicinity of an ordinary point via a tangent, which then has contact of first order. Therefore, we approximate the curve $X = X(t), Y = Y(t)$ in the vicinity of an ordinary point M_0 by a circle $(X - a)^2 + (Y - b)^2 = R^2$. To determine the three unknowns a, b and R, three conditions corresponding to (B16) for $i = 0, 1, 2$ have to be satisfied that correspond to a contact of second order. As a result one obtains

$$a = X_0 + Y_0' \frac{X_0'^2 + Y_0'^2}{X_0'' Y_0' - Y_0'' X_0'} , \quad b = Y_0 + X_0' \frac{X_0'^2 + Y_0'^2}{Y_0'' X_0' - X_0'' Y_0'} ,$$
$$R = \frac{\left(X_0'^2 + Y_0'^2\right)^{3/2}}{\left|X_0'' Y_0' - Y_0'' X_0'\right|} . \tag{B17}$$

The circle with the parameters (B18) is called an osculating circle. The point with coordinates (a, b) is called the center of curvature and R is called the radius of curvature of the curve in a point with $\boldsymbol{r}' \neq 0$ and $\boldsymbol{r}'' \neq 0$. If M_0 is an ordinary point, we have $X_0'' Y_0' - Y_0'' X_0' \neq 0$. In a point where $X_0'' Y_0' - Y_0'' X_0' = 0$, the radius of curvature is equal to infinity and such a point is a singular point and is also called a rectification point. In this point, the approximating line (tangent) has a contact of second order with the curve. If the curve equation is given in the form $Y = Y(X)$, the formulae () read

$$a = X_0 - Y_0' \frac{1 + Y_0'^2}{Y_0''} , \quad b = Y_0 + \frac{1 + Y_0'^2}{Y_0''} , \quad R = \frac{\left(1 + Y_0'^2\right)^{3/2}}{\left|Y_0''\right|} . \tag{B18}$$

The reciprocal of the radius of curvature $K = 1/R$ in the point M is called curvature. It characterizes the degree of curvature. Indeed, if one moves from the point M along the curve by a small arc length ΔS, the increment of the angle between the tangent and the X-axis is equal to $\Delta\alpha$. At $\tan\alpha = Y'(t)/X'(t)$, from (B12) and B.3, in the limit of $\Delta S \to 0$, we obtain

$$K = \frac{d\alpha}{dS} . \tag{B19}$$

The curvature of every straight line is equal to zero and that of a circle is a constant. The necessary and sufficient conditions for a point to be a rectification point is $K = 0$.

The notion of the osculating circle makes it possible to introduce for every ordinary point of the curve a local orthogonal system of coordinates with its origin in the point M and the axes in the directions of the tangent, and the normal on the curve in the point M. The equation of the curve is $\boldsymbol{r} = \boldsymbol{r}(S)$ and the unit vectors in the directions of the tangent and the normal are denoted by $\boldsymbol{t}$ and $\boldsymbol{n}$. Since, according to property one in Section B.2, the vector $\dot{\boldsymbol{t}} = \ddot{\boldsymbol{r}}$ is orthogonal to $\boldsymbol{t}$, we agree that the normal vector $\boldsymbol{n}$ is directed along $\dot{\boldsymbol{t}}$ in such a way that $\boldsymbol{n}$ points to the center of curvature C. During movement of M along the curve, the parameter t is changing, and therefore we have $\boldsymbol{t} = \boldsymbol{t}(S)$ and $\boldsymbol{n} = \boldsymbol{n}(S)$. According to property two in Section B.2, we have $|d\alpha/dS| = |\dot{\boldsymbol{t}}|$, and from (B19), we obtain that $|\dot{\boldsymbol{t}}| = K$ and

$$\dot{\boldsymbol{t}} = K\boldsymbol{n} . \tag{B20}$$

The vector $\dot{\boldsymbol{n}}$ is orthogonal to $\boldsymbol{n}$ and therefore collinear with $\boldsymbol{t}$. Because of $\boldsymbol{t}\cdot\boldsymbol{n} = 0$, we have

$$\dot{\boldsymbol{n}} = -K\boldsymbol{t} \,. \tag{B21}$$

Equations B20 and (B21) are called Frenets formula for plane curves. The geometrical meaning of Frenets formula is that they represent the rotation of the vectors $\boldsymbol{t}$ and $\boldsymbol{n}$ as a solid body from the point M to an infinitesimally close point. If one provides the equation of a curve and carries out mathematical calculations with it, one uses a certain system of coordinates that can actually be chosen arbitrarily and whose choice depends on the details of the solution for the concrete task. The solution of physical problems requires the independence of the physical parameter from the choice of the system of coordinates. When investigating the geometrical properties of curves and planes, it is also necessary that these properties are independent of the choice of the system of coordinates.

With respect to a plane curve, this means that one needs an equation for the curve that does not depend on the system of coordinates. Such an equation is called a natural curve equation and represents the dependence of the curvature from the arc length $K = K(S)$.

This assertion is based on the following theorem: If two curves are only different as far as their position on the plane is concerned, then they have the same natural curve equation and vice versa, or if two curves have the same natural equations, they only differ with respect to their position in the plane and these curves can coincide via a suitable rotation and translation, that is, by an orthogonal transformation of the system of coordinates.

B.4 Curves in Space

We now generalize the results obtained in the previous section for curves in space.

Those curves are defined by the vectorial representation

$$\boldsymbol{r} = \boldsymbol{r}(t) = X(t)\boldsymbol{i} + Y(t)\boldsymbol{j} + Z(t)\boldsymbol{k} \tag{B22}$$

or the parametric representation

$$X = X(t), Y = Y(t), Z = Z(t) \,. \tag{B23}$$

The derivative $\boldsymbol{r}'(t) = X'(t)\boldsymbol{i} + Y'(t)\boldsymbol{j} + Z'(t)\boldsymbol{k}$ is a vector which is directed along the tangent to the curve in the point $M(X(t), Y(t), Z(t))$. The equation of the tangent has the following form

$$\frac{X - X(t)}{X'(t)} = \frac{Y - Y(t)}{Y'(t)} = \frac{Z - Z(t)}{Z'(t)} \,, \tag{B24}$$

where X, Y and Z are the coordinates of the point on the tangent.

While the plane curve in two dimensions only has one normal in the point M, the spatial curve has an infinite number of normals that are lying in the normal plane. The equation for this plane is

$$X'(t)\left[X - X(t)\right] + Y'(t)\left[Y - Y(t)\right] + Z'(t)\left[Z - Z(t)\right] = 0\,, \qquad \text{(B25)}$$

where X, Y and Z are the coordinates within the normal plane.

In ordinary points, we have $\boldsymbol{r}' \neq 0$. In the following, we assume that $\boldsymbol{r}' \neq 0$ and $\boldsymbol{r}'' \neq 0$, that is, we only consider ordinary points with the exception of turning points. We also assume that the vectors $\boldsymbol{r}'$ and $\boldsymbol{r}''$ are not collinear.

We now discuss the problem of contact between a curve and a plane. We consider an arbitrary point $M(t)$ on the curve and construct a plane through this point. This plane is uniquely determined by the coordinates of the point M and the unit normal $\boldsymbol{m}$, which must not be confused with the normal to the tangential vectors mentioned above.

By increasing the parameter t by a small increment Δt, the point $M(t)$ is moved to the neighboring point $M' = M(t + \Delta t)$. The deviation of the curve from the plane is measured by the length of the vector PM' that is orthogonal to the plane. The proximity of the curve to the plane PE in the vicinity of the point M can be judged by the expansion order of the length PM'. One says that the curve in the point M has contact of n-th order with the plane if PM' is a quantity of $n + 1$-th order with respect to Δt. Since we have $\Delta S \sim \Delta t$ according to (B12), the contact order with respect to Δt is identical to the contact order with respect to S, that is, one can always move from the parameters t to the arc length.

Because the distance PM is equal to the projection of MM' on the direction $\boldsymbol{m}$, and using (B9), we obtain

$$PM' = \boldsymbol{m} \cdot \boldsymbol{MM'} = \boldsymbol{m} \cdot \boldsymbol{r}'(t)\frac{\Delta t}{1!} + \boldsymbol{m} \cdot \boldsymbol{r}''(t)\frac{(\Delta t)^2}{2!} + \cdots \qquad \text{(B26)}$$

We now consider the different cases of the position of the plane relative to the curve:

1. For $\boldsymbol{m} \cdot \boldsymbol{r}'(t) \neq 0$, the expansion (B26) starts with the first term. The deviation of the plane from the curve is of first order in Δt and the contact is of 0th order. The tangent does not lay in the plane. It intersects the plane.
2. For $\boldsymbol{m} \cdot \boldsymbol{r}'(t) = 0$, the expansion (B26) starts with the second term. The deviation of the plane from the curve is of second order under the condition that $\boldsymbol{m} \cdot \boldsymbol{r}''(t) \neq 0$. Since $\boldsymbol{r}'(t)$ is parallel to the tangent $\boldsymbol{t}$ and orthogonal to the normal $\boldsymbol{m}$, the tangent lies in the plane. The contact between the curve and the plane is of first order. It is obvious that there are an infinite number of tangential planes and corresponding normals.
3. We have $\boldsymbol{m} \cdot \boldsymbol{r}'(t) = 0$ and $\boldsymbol{m} \cdot \boldsymbol{r}''(t) = 0$. That means that from the family of planes containing the tangent of type two, that plane is chosen, for which $\boldsymbol{m} \cdot \boldsymbol{r}''(t) = 0$. Such a plane must contain the point M and the two vectors $\boldsymbol{r}'$ and $\boldsymbol{r}''$ that start in M and which are not collinear. This plane is called the osculating plane. The expansion (B26) starts with the third term and the deviation PM' of the plane from the curve is of third order.

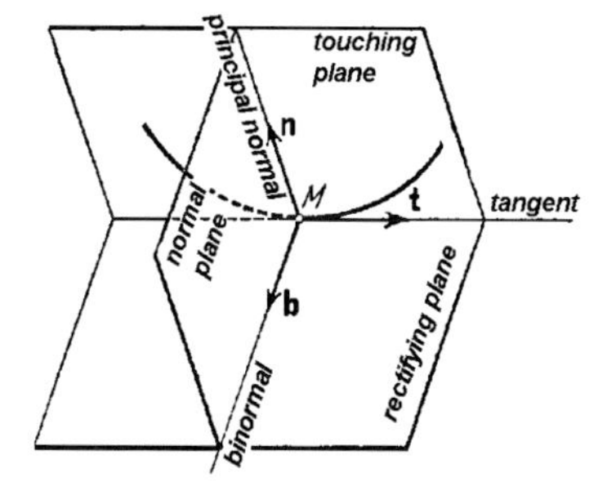

Figure B.1 Local orthogonal system of coordinates at the point M.

Therefore, for every point M on a spatial curve, for which $\boldsymbol{r}'' \neq 0$ and the vectors $\boldsymbol{r}'$ and $\boldsymbol{r}''$ are not collinear, one can construct an osculating plane such that the deviation between the plane and the curve is of third order in the distance Δt. Then, the vectors $\boldsymbol{r}'$ and $\boldsymbol{r}''$ that lie in this plane and the normal vector $\boldsymbol{m}$ on this plane satisfy the following conditions

$$\boldsymbol{m} \cdot \boldsymbol{r}'(t) = 0\,, \quad \boldsymbol{m} \cdot \boldsymbol{r}''(t) = 0\,. \tag{B27}$$

One can connect a local orthogonal system of coordinates to the point M. As parameter, we take t as the arc length S. Then, as shown in Section B.2, $\boldsymbol{t} = \dot{\boldsymbol{r}}$ is the unit tangential vector on the curve. One can take as a local basis vectors the tangent vector $\boldsymbol{t}$ and two normal vectors: the main normal vector $\boldsymbol{n}$ that has the direction of $\ddot{\boldsymbol{r}}$ and the binormal vector $\boldsymbol{b} = \boldsymbol{t} \times \boldsymbol{n}$. The curve will have contact of second order with the plane spanned by $\boldsymbol{t}$ and $\boldsymbol{n}$ which is also called the osculating plane, while $\boldsymbol{n}$ and $\boldsymbol{b}$ span the normal plane, and $\boldsymbol{t}$ and $\boldsymbol{b}$ lie in the rectification plane (see Figure B.1).

We assume that the three vectors $\boldsymbol{t}$, $\boldsymbol{n}$ and $\boldsymbol{b}$ constitute a local right-handed basis. The coordinate planes are the osculating plane with the vectors $\boldsymbol{t}$ and $\boldsymbol{n}$, the normal plane with the vectors $\boldsymbol{n}$ and $\boldsymbol{b}$, and the rectification plane with the vectors $\boldsymbol{t}$ and $\boldsymbol{b}$. The vectors $\boldsymbol{t}$, $\boldsymbol{n}$ and $\boldsymbol{b}$ are called the accompanying trihedron or the Frenet trihedron. Using the methods of analytical geometry, it is easy to obtain equations for the coordinate planes and coordinate axes with respect to a fixed coordinate system which is not bound to the curve as well as relationships between the local basis vectors $\boldsymbol{t}, \boldsymbol{n}$, and $\boldsymbol{b}$. The latter ones are called Frenets formula (see Section B.5). For example, the equation of the ocular plane is written as the orthogonality of the vectors $\boldsymbol{R} - \boldsymbol{r}$ and $\boldsymbol{r}' \times \boldsymbol{r}''$:

$$(\boldsymbol{R} - \boldsymbol{r}) \cdot (\boldsymbol{r}' \times \boldsymbol{r}'') = 0 \quad \text{or} \quad (\boldsymbol{R} - \boldsymbol{r}, \boldsymbol{r}', \boldsymbol{r}'') = 0\,. \tag{B28}$$

In Section B.3, the osculating circle was defined for a plane curve, that is, the circle that has contact of second order with the circle. One can also find such a circle for the spatial curve at the point M. It lies in the osculating plane. Because of property three (see Section B.2), the radius of this circle is equal to $R = 1/|\ddot{\boldsymbol{r}}(S)|$, where the vector $\ddot{\boldsymbol{r}}(S)$ is directed along the main normal $\boldsymbol{n}$ toward the center of the circle C, meaning $\boldsymbol{m}C = R\boldsymbol{n}$ (see Figure B.2). Therefore, one can consider

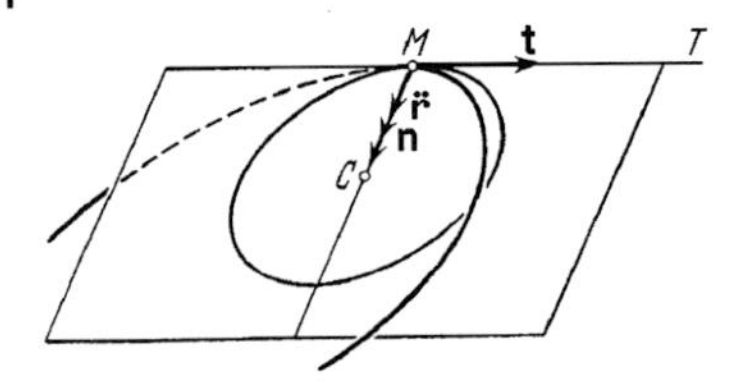

Figure B.2 Osculating circle of a spatial curve.

the spatial curve within an infinitesimal neighborhood of the point M as a circle accurate to infinitesimal quantities of third order that deviates from the tangent in the positive direction of the main normal.

B.5
Curvature of Spatial Curves

We assume that the spatial curve is given in parametric form $\boldsymbol{r} = \boldsymbol{r}(S)$ where S is the arc length. Then, one can introduce three orthogonal unit vectors $\boldsymbol{t}$, $\boldsymbol{n}$ and $\boldsymbol{b}$ for every point which are directed along the tangent, the normal and the binormal, where $\boldsymbol{t} = \dot{\boldsymbol{r}}$, $\boldsymbol{n}$ is collinear with the vector $\ddot{\boldsymbol{r}}$ and $\boldsymbol{b} = \boldsymbol{t} \times \boldsymbol{n}$.

According to (B19), we define the curvature K of a spatial curve as the angular velocity of the tangent vector $\boldsymbol{t}$ with respect to the arc length S. According to properties two and three of Section B.2, we have

$$K = \left|\frac{d\alpha}{dS}\right| = |\dot{\boldsymbol{t}}| = |\ddot{\boldsymbol{r}}| = \frac{1}{R}\,, \tag{B29}$$

where R is the radius of the osculating circle which is referred to as the radius of curvature of a spatial curve. For $K = 0$, we have $\ddot{\boldsymbol{r}} = 0$, $R = \infty$ and the osculating circle degenerates to a straight line. Therefore, every point M where the curvature vanishes is a rectification point. If $K = 0$ for all points on the curve, the curve is a straight line.

The relationships between the vectors $\boldsymbol{t}$,$\boldsymbol{n}$ and $\boldsymbol{b}$ are given by the Frenet formula. The first of the Frenet formula follows from the equations $\dot{\boldsymbol{t}} = \ddot{\boldsymbol{r}} = \boldsymbol{n}/R = K\boldsymbol{n}$ and was already obtained above as (B20):

$$\dot{\boldsymbol{t}} = K\boldsymbol{n}\,. \tag{B30}$$

The vector $\boldsymbol{n}$ is called the vector of curvature for the curve. For the binormal vector, we have

$$\boldsymbol{b} = \boldsymbol{t} \times \boldsymbol{n}\,, \quad \dot{\boldsymbol{b}} = \dot{\boldsymbol{t}} \times \boldsymbol{n} + \boldsymbol{t} \times \dot{\boldsymbol{n}} = K(\boldsymbol{n} \times \boldsymbol{n}) + \boldsymbol{t} \times \dot{\boldsymbol{n}} = \boldsymbol{t} \times \dot{\boldsymbol{n}}\,.$$

According to property one of Appendix B.2, we have $\dot{\boldsymbol{n}} \perp \boldsymbol{n}$ and $\boldsymbol{t} \perp \boldsymbol{n}$. Therefore, $\dot{\boldsymbol{b}}$ is collinear to $\boldsymbol{n}$ and one can assume

$$\dot{\boldsymbol{b}} = -\kappa\boldsymbol{n}\,, \tag{B31}$$

where κ is the called the torsion of the spatial curve. This formula is referred to as the third Frenet formula.

To obtain the second Frenet formula, we use the formulae

$$\dot{\boldsymbol{n}} \perp \boldsymbol{n}\,; \quad \boldsymbol{n} = \boldsymbol{b} \times \boldsymbol{t}\,; \quad \boldsymbol{n} \times \boldsymbol{t} = -\boldsymbol{b}\,; \quad \boldsymbol{b} \times \boldsymbol{n} = -\boldsymbol{t}\,;$$

$$\dot{\boldsymbol{n}} = \dot{\boldsymbol{b}} \times \boldsymbol{t} + \boldsymbol{b} \times \dot{\boldsymbol{t}} = (-\kappa \boldsymbol{n} \times \boldsymbol{t}) + (\boldsymbol{b} \times K\boldsymbol{n}) = \kappa \boldsymbol{b} - K\boldsymbol{t}\,.$$

Thus, the second Frenet formula is

$$\dot{\boldsymbol{n}} = \kappa \boldsymbol{b} - K\boldsymbol{t}\,. \tag{B32}$$

Therefore, the Frenet formulae provide a representation of the first derivatives of the basis vectors $\dot{\boldsymbol{t}}$, $\dot{\boldsymbol{n}}$ and $\dot{\boldsymbol{b}}$ in terms of the basis vectors $\boldsymbol{t}$, $\boldsymbol{n}$ and $\boldsymbol{b}$.

In the Frenet formulae, there are two parameters K and κ. The geometrical meaning of the curvature K is the absolute value of the velocity of rotation for the tangent $\boldsymbol{t}$ relative to the arc length S along the curve. The geometrical meaning of the torsion κ is the absolute value of the velocity of rotation of the binormal $\boldsymbol{b}$ at the point M or of the osculating plane relative with respect to the arc length S along the curve. The equation $K = 0$ is a necessary and sufficient condition for the curve to be a straight line, while the equation $\kappa = 0$ is a necessary and sufficient condition for the curve to be a plane curve that lies in its osculating plane.

During the movement of a point M along the curve in the direction of increasing S, the relationships (B30)–(B32) are maintained, and the Frenet trihedron moves like a solid body with the angular velocity vector $\boldsymbol{\Omega} = \kappa \boldsymbol{t} + K\boldsymbol{b}$ in the rectification plane.

The first term $\kappa \boldsymbol{t}$ represents the rotation from $\boldsymbol{n}$ to $\boldsymbol{b}$, from $\kappa > 0$ and from $\boldsymbol{b}$ to $\boldsymbol{n}$ from $\kappa < 0$ around the axis $\kappa \boldsymbol{t}$ with angular velocity $|\kappa|$. The second term $K\boldsymbol{b}$ represents the rotation of $\boldsymbol{t}$ to $\boldsymbol{n}$ along the axis $\boldsymbol{b}$ with angular velocity K.

Using (B29)–(B32), one can obtain the main geometrical parameters: the local basis vectors $\boldsymbol{t}, \boldsymbol{n}, \boldsymbol{b}$, the curvature K and the torsion κ. If the curve is given in the parametric form $\boldsymbol{r}(S)$, we have

$$\boldsymbol{t} = \dot{\boldsymbol{r}}\,, \quad \boldsymbol{n} = \frac{\ddot{\boldsymbol{r}}}{|\ddot{\boldsymbol{r}}|}\,, \quad \boldsymbol{b} = \frac{\dot{\boldsymbol{r}} \times \ddot{\boldsymbol{r}}}{|\ddot{\boldsymbol{r}}|}\,, \tag{B33}$$

$$K = |\ddot{\boldsymbol{r}}|\,, \quad \kappa = \frac{(\dot{\boldsymbol{r}}, \ddot{\boldsymbol{r}}, \dddot{\boldsymbol{r}})}{|\ddot{\boldsymbol{r}}|^2}\,. \tag{B34}$$

These formulae can be expressed in coordinate form via (B7). If instead of S another parameter is used, for example, t, which is related to S via $t = t(S)$, the above mentioned parameters can be determined by the chain rule $dr/dS = (dr/dt)(dt/dS)$.

To obtain the geometrical parameter of the spatial curve in the case of a parametric definition of the curve, it is necessary to choose a certain system of coordinates. Like plane curves, the spatial curve can be defined by the natural equations $K = K(S)$ and $\kappa = \kappa(S)$, whose form is independent of the system of coordinates.

The role of natural equations is described by the following theorem: In order for two curves to be identical expect for their positions in space, it is necessary and sufficient that their natural equations agree, including the arc lengths measured from the initial points.

Given two arbitrary continuous functions $\phi(S)$ and $\psi(S)$, which have the same interval of definition ($S_1 < S < S_2$), it is possible to build a spatial curve with the natural equations $K = \phi(S)$ and $\kappa = \psi(S)$ where the curve is determined exactly up to its position in space, independent of the system to coordinates.

B.6
Surfaces in Space

A surface in three-dimensional space can be defined in coordinate form or in parametric form. We first consider the definition of a surface in coordinate form. We assume that a surface is given in the form

$$F(X, Y, Z) = 0 . \tag{B35}$$

We assume that F_x, F_y and F_z do not vanish at the same time for any point of the surface. We now assume that $F_z \neq 0$. Then, the surface equation can be written in explicit form as

$$Z = f(X, Y) . \tag{B36}$$

In the vicinity of an ordinary point, the surface equation can be written in the form (B36), where f is a unique function that has well defined derivatives of at least first order. A piece of the surface that includes the vicinity of an ordinary point is called an ordinary surface piece. Those points where F_x, F_y and F_z vanish at the same time are called singular points.

We now consider a curve in parametric form. The condition that the point $M[X(t), Y(t), Z(t)]$ lies on the surface is

$$F[X(t), Y(t), Z(t)] = 0 . \tag{B37}$$

B differentiating this equation, one obtains $F_x X'(t) + F_y Y'(t) + F_z Z'(t) = 0$. Since $\nabla F = F_x \boldsymbol{i} + F_y \boldsymbol{j} + F_z \boldsymbol{k}$, we rewrite the last equation as

$$\nabla F \cdot \boldsymbol{r}'(t) = 0 . \tag{B38}$$

The vector $\boldsymbol{r}'(t)$ is the tangent on the curve, and the curve is chosen arbitrarily on the surface. Therefore, the vector ∇F is orthogonal to the tangent. That means that all tangents to the curves through a given point lie in the same plane. This plane is called the tangential plane because the vector ∇F is the normal vector of the surface in the point M. The equations of the tangential surface and the normal are given by

$$F_x[X - X(t)] + F_y[Y - Y(t)] + F_z[Z - Z(t)] = 0 \tag{B39}$$

and

$$\frac{X - X(t)}{F_x} = \frac{Y - Y(t)}{F_y} = \frac{Z - Z(t)}{F_z} . \tag{B40}$$

The parametric method for defining a surface is the most common method for the investigation of the surfaces and of curves on these surfaces. In this case, one needs two parameters instead of the one parameter t needed (see (B22)) when defining a curve in space.

We assume that the vector function $\boldsymbol{r}$ is a function of two scalar arguments U and V with the intervals of definition $U_1 < U < U_2$ and $V_1 < V < V_2$:

$$\boldsymbol{r} = \boldsymbol{r}(U, V) = X(U, V)\boldsymbol{i} + Y(U, V)\boldsymbol{j} + Z(U, V)\boldsymbol{k} . \tag{B41}$$

The initial point of the radius vector is at the origin of the Cartesian system of coordinates and the endpoint of the radius vector describes, by change of parameters, a geometrical set of points that corresponds to a surface in space. We consider the partial derivatives

$$\boldsymbol{r}_u = X_u(U, V)\boldsymbol{i} + Y_u(U, V)\boldsymbol{j} + Z_u(U, V)\boldsymbol{k} ,$$

$$\boldsymbol{r}_v = X_v(U, V)\boldsymbol{i} + Y_v(U, V)\boldsymbol{j} + Z_v(U, V)\boldsymbol{k} \tag{B42}$$

and assume that the vectors $\boldsymbol{r}_u$ and $\boldsymbol{r}_v$ in the point M are not collinear.

That means that the coordinates of these vectors are not proportional to each other, and at least one of the determinants of second order of the matrix

$$D = \begin{pmatrix} X_u & Y_u & Z_u \\ X_v & Y_v & Z_v \end{pmatrix}$$

does not vanish.

According to the theorem of implicit functions, the system of equations $X = X(U, V)$ and $Y = U(U, V)$ can be solved for U and V, that is, $U = U(X, Y)$, $V = V(X, Y)$. As a result, one obtains that $Z = Z[U(X, Y), V(X, Y)] = f(X, Y)$.

Therefore, it follows from the non-collinearity of the vectors $\boldsymbol{r}_u$ and $\boldsymbol{r}_v$ at the point M, that M is an ordinary point in a vicinity where we have a one-to-one relationship between the points of the surface and the parameters U and V. Therefore, one can consider the parameters U and V as coordinates of the surface points. Therefore, they are called curvilinear coordinates where the curves $U =$ const and $V =$ const, which in general are not straight lines, correspond to the coordinate grid on the surface.

The choice of the curvilinear system of coordinates on a surface is arbitrary. The transition from a curvilinear system of coordinates on the surface to other coordinates U', V' is achieved by a one-to-one transformation $U' = U'(U, V)$, $V' = V'(U, V)$ under the condition that the transformation matrix (Jacobian)

$$D' = \begin{pmatrix} U'_u & V'_u \\ U'_v & V'_v \end{pmatrix}$$

is not degenerate, that is, the determinant $|D'|$ does not vanish.

As examples of curvilinear coordinates on a plane serve polar coordinates $U = \rho$ and $V = \phi$, or curvilinear coordinates on a sphere $U = \phi$ and $V = \vartheta$, which are called geographical longitude and latitude. In the first case, the coordinate grid

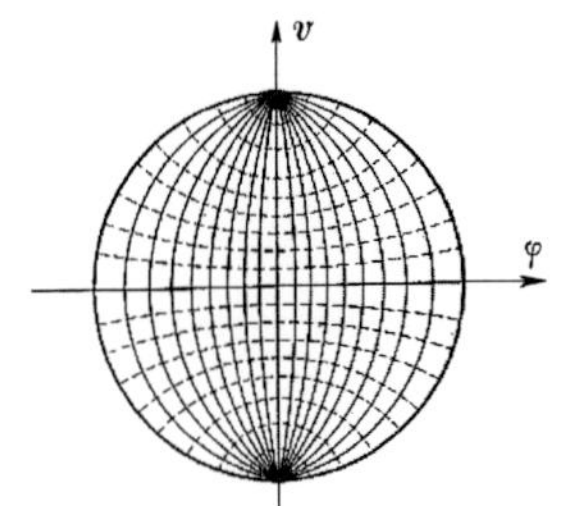

Figure B.3 Curvilinear coordinate grid on a sphere.

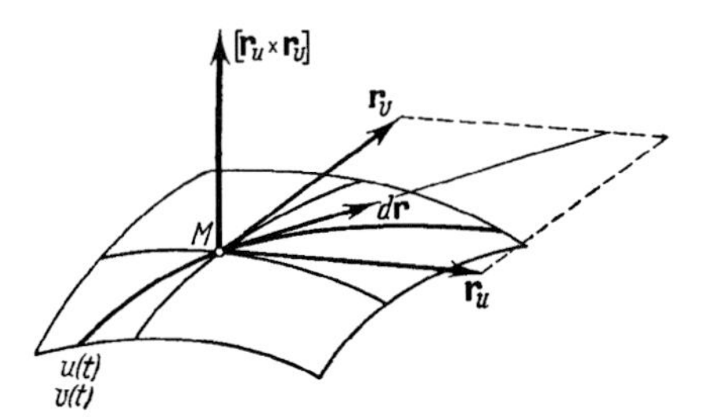

Figure B.4 Local basis vectors at point M on a surface.

ϕ = const and ρ = const represents the lines and the circle originating from the origin of the Cartesian system coordinates, while in the second case, the curves ϕ = const and ϑ = const are meridians and circles of latitude on the sphere (see Figure B.3).

The curves on a surface can be defined in parametric form as follows:

$$r = r[U(t), V(t)] . \tag{B43}$$

The coordinate lines on the surface constitute a grid of two families of lines: V-lines along which $dU = 0$, meaning that U = const and U-lines, along which $dV = 0$, which means that V = const. Two lines of the coordinate grid pass through each point of the surface.

We now consider the tangents on the curves (B43). In every ordinary surface point M through which an arbitrary curve passes, the corresponding differential vector $d\boldsymbol{r}$ is directed along the tangent on this curve. Since

$$d\boldsymbol{r} = \boldsymbol{r}_u dU + \boldsymbol{r}_v dV , \tag{B44}$$

the vectors of $d\boldsymbol{r}$ of the corresponding lines lie on the same tangential plane (see Figure B.4).

The vectors $\boldsymbol{r}_u$ (the tangential vector on the coordinate line U), and $\boldsymbol{v}_v$ (the tangential vector on the coordinate line V) that emanate from a surface point constitute the tangential plane in the point M. Since the vectors $\boldsymbol{r}_u$ and $\boldsymbol{r}_v$ are uniquely defined by the choice of the point M and the choice of the curvilinear system of coordinates, the lines passing through the same point M only differ in the value of dV/dU. Therefore, the tangent on a specific curve is defined by the vectors $\boldsymbol{r}_u$, $\boldsymbol{r}_v$ and the parameter dV/dU.

The normal vector $\boldsymbol{n}$ in the point M is a vector that is orthogonal to the tangential plane in such a manner that $\boldsymbol{r}_u$, $\boldsymbol{r}_v$ and $\boldsymbol{n}$ constitute an orthogonal system. Therefore, we have $\boldsymbol{n} = \boldsymbol{r}_u \times \boldsymbol{r}_v$. The equation of the tangential plane has the form of complanarity conditions of the vectors $\boldsymbol{R} - \boldsymbol{r}$, $\boldsymbol{r}_u$, $\boldsymbol{r}_v$, that is,

$$\left[(\boldsymbol{r} - \boldsymbol{r}), \boldsymbol{r}_u, \boldsymbol{r}_v\right] = 0 \,, \tag{B45}$$

where $\boldsymbol{r}$ is the radius vectors of the current point in the tangential plane. The equation of the normal can be written as the collinear condition for the vectors $\boldsymbol{R} - \boldsymbol{r}$ and $\boldsymbol{r}_u \times \boldsymbol{r}_v$.

B.7 Fundamental Forms of the Surface

We consider a point $M(U, V)$ of the surface curve $\boldsymbol{r}[U(t), V(t)]$. During the movement along the curve, the main part of the increment of the radius vector is equal to $d\boldsymbol{r}$, where $d\boldsymbol{r}$ is the tangential vector on the curve in the point M and is defined by (B44). According to (B12), we have $dS = |d\boldsymbol{r}|$ where S is the arc length, and according to (B44), one obtains

$$dS^2 = d\boldsymbol{r}^2 = (\boldsymbol{r}_u dU + \boldsymbol{r}_v dV)^2 = \boldsymbol{r}_u^2 dU^2 + 2\boldsymbol{r}_u \cdot \boldsymbol{r}_v dU dV + \boldsymbol{r}_v^2 dV^2 \,.$$

With the definitions

$$a_{uu}(U, V) = \boldsymbol{r}_u^2 = \boldsymbol{r}_u \cdot \boldsymbol{r}_u \,,$$
$$a_{uv}(U, V) = \boldsymbol{r}_u \cdot \boldsymbol{r}_v \,,$$
$$a_{vv}(U, V) = \boldsymbol{r}_v^2 = \boldsymbol{r}_v \cdot \boldsymbol{r}_v \,,$$

the expression for dS^2 is rewritten as follows:

$$dS^2 = a_{uu} dU^2 + 2a_{uv} dU dV + a_{vv} dV^2 \,. \tag{B46}$$

The expression on the right-hand side of (B46) represents the quadratic form with respect to the differentials dU and dV, and is referred to as the first fundamental form of the plane. The first fundamental form plays a crucial rule in surface geometry because it serves to determine the arc length of infinitesimal curves on surfaces, the angle between two intersecting curves, the arc length of curves and finite surface areas. Since $dS^2 > 0$, the first fundamental form is positive definite. Therefore, we have $a_{uu} > 0$ and the determinant $a = a_{uu}a_{vv} - a_{uv}^2$ is positive.

We now determine the angle α between two coordinate lines that intersect in the surface point $M(U, V)$. Along the coordinate line U, we have $dV = 0$ and $(d\boldsymbol{r})_u = \boldsymbol{r}_u dU$, and along the coordinate line V, we have $dU = 0$ and $(d\boldsymbol{r})_v = \boldsymbol{r}_v dV$.

Each of these vectors is the tangent on the respective coordinate line. Therefore, α is equal to the angle between the corresponding tangents. As a result, one obtains

$$\cos\alpha = \frac{\boldsymbol{r}_u \cdot \boldsymbol{r}_v}{|\boldsymbol{r}_u|\,|\boldsymbol{r}_v|} = \frac{a_{uv}}{\sqrt{a_{uu}a_{vv}}} \,. \tag{B47}$$

The arc length from point $M_1(t_1)$ to point $M_2(t_2)$ is given by

$$L = \int_{t_1}^{t_2} dS = \int_{t_1}^{t_2} \sqrt{(a_{uu} U'(t)^2 + 2a_{uv} U'(t)^2 V'(t)^2 + a_{vv} V'(t)^2) dt} \,. \quad \text{(B48)}$$

The surface area of the surface element $d\sigma$ is equal to

$$d\sigma = |\boldsymbol{r}_u dU \times \boldsymbol{r}_v dV| = |\boldsymbol{r}_u| |\boldsymbol{r}_v| \sin \alpha \, dU dV = \sqrt{a_{uu} a_{vv} - a_{uv}^2} \, dU dV \,.$$

The surface area σ of the surface Σ is defined as the limit of the corresponding sum of surface elements. Thus, one obtains

$$\sigma = \int_{\Sigma} d\sigma = \int_{\Sigma} \sqrt{a_{uu} a_{vv} - a_{uv}^2} \, dU dV \,. \quad \text{(B49)}$$

The expressions (B48) and (B49) for the arc length and the surface area only depend on the first fundamental form. During the transition to another system of coordinates U', V', the first fundamental form is transformed in such a way that the form of dS^2 and σ remains invariant, that is,

$$dS^2 = a'_{u'u'} dU'^2 + 2a_{u'v'} dU' dV' + a_{v'v'} dV'^2 \,,$$

$$\sigma = \int_{\Sigma'} \sqrt{a'_{u'u} a'_{v'v'} - a'^2_{u'v'}} \, dU' dV' \,. \quad \text{(B50)}$$

If only the first fundamental form is given in a curvilinear system of coordinates, that is, a_{uu}, a_{uv} and a_{vv} are known, many geometrical properties of the surface can be determined without knowledge of the surface equations and thus of the surface form. The surface properties that can be defined with the help of the first fundamental form alone constitute the internal geometry of the surface.

We now determine the second fundamental form of the surface. For this purpose, we consider a curve on a surface in the vicinity of a point $M(U, V)$ (see Figure B.5). We use as parameter, the arc length S such that the coordinates of the curve are $U(S)$, $V(S)$ and the curve equation is $\boldsymbol{r} = \boldsymbol{r}[U(S), V(S)]$. For a displacement to the infinitesimally close point M' along the curve, the increment of the arc length is ΔS and the increment of the radius vector $\Delta \boldsymbol{r}$ is equal to

$$MM' = \Delta \boldsymbol{r} = d\boldsymbol{r} + \frac{1}{2} \ddot{\boldsymbol{r}} (\Delta S)^2 + \ldots \,; \quad d\boldsymbol{r} = \dot{\boldsymbol{r}} dS \,. \quad \text{(B51)}$$

Previously, it was sufficient to only know the first part of the expansion (B51) in order to determine the first fundamental form. Then, the vector $\boldsymbol{MM'} \approx d\boldsymbol{r}$ coincides in its direction with the tangent on the plane up to an infinitesimal quantity of first order.

Now, taking into account the second term of the expansion (B51), $\boldsymbol{MM'}$ deviates from the tangential plane and the deviation is equal, accurate to an infinitesimal

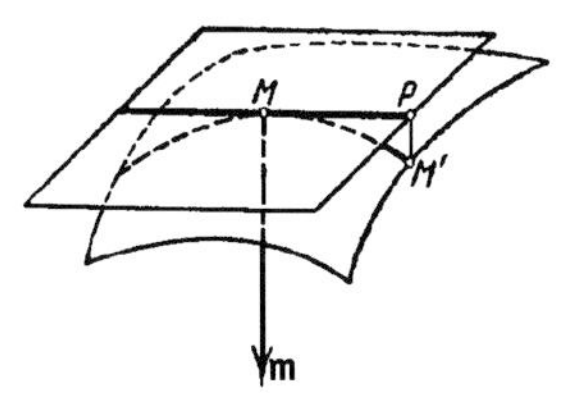

Figure B.5 Tangential plane and normal to the surface in the point M.

quantity of second order, to the projection of the vector $\ddot{\boldsymbol{r}}(\Delta S)^2/2$ on to the normal in the point M.

Define $\boldsymbol{m}$ as the unit normal vector such that $\boldsymbol{PM'} = l\boldsymbol{m}$. Then, we have $l = \ddot{\boldsymbol{r}} \cdot \boldsymbol{m}(\Delta S)^2/2$ and

$$l = \frac{(\Delta S)^2}{2}\left(\boldsymbol{r}_{uu} \cdot \boldsymbol{m}\dot{U}^2 + 2\boldsymbol{r}_{uv} \cdot \boldsymbol{m}\dot{U}\dot{V} + \boldsymbol{r}_{vv} \cdot \boldsymbol{m}\dot{V}^2\right) . \tag{B52}$$

The vector $\boldsymbol{m}$ is given by

$$\boldsymbol{m} = \frac{\boldsymbol{r}_u \times \boldsymbol{r}_v}{|\boldsymbol{r}_u \times \boldsymbol{r}_v|} = \frac{\boldsymbol{r}_u \times \boldsymbol{r}_v}{\sqrt{a_{uu}a_{vv} - a_{uv}^2}} . \tag{B53}$$

From the above equations, one obtains

$$l = \frac{1}{2}\left(b_{uu}dU^2 + 2b_{uv}dUdV + b_{vv}dV^2\right) \tag{B54}$$

with

$$b_{uu} = \frac{(\boldsymbol{r}_{uu}, \boldsymbol{r}_u, \boldsymbol{r}_v)}{\sqrt{a}} , \quad b_{uv} = \frac{(\boldsymbol{r}_{uv}, \boldsymbol{r}_u, \boldsymbol{r}_v)}{\sqrt{a}} ,$$
$$b_{vv} = \frac{(\boldsymbol{r}_{vv}, \boldsymbol{r}_u, \boldsymbol{r}_v)}{\sqrt{a}} , \quad a = a_{uu}a_{vv} - a_{uv}^2 .$$

The quadratic form on the right-hand side of (B54) is called the second fundamental form of the surface. It characterizes the main part of the deviation of the curve from the tangential plane in an infinitesimal vicinity of the point M that lies in the tangential plane. The second fundamental form can also be written as

$$b_{uu}dU^2 + 2b_{uv}dUdV + b_{vv}dV^2 = -d\boldsymbol{r} \cdot d\boldsymbol{m} . \tag{B55}$$

There also exists the third fundamental form that characterizes the change of the normal vector $\boldsymbol{m}$ to the surface at the point M during the movement along a curve that passes through the point M. We consider a curve on the surface that is given in the form $\boldsymbol{r} = \boldsymbol{r}[U(S), V(S)]$.

Then, the normal vector $\boldsymbol{m}$ can be considered as a function $\boldsymbol{m}(S)$. According to property two of Section B.2, the unit normal vector turns, accurate to an infinitesimal quantity, by the angle $\Delta\varphi = |d\boldsymbol{m}|$ during the transition along the curve from the point M to the infinitesimally close point M'.

We now consider two tangential planes in the points M and M'. The normal vector to the first plane is $\boldsymbol{m}$ and the normal vector to the second plane is $\boldsymbol{m} + d\boldsymbol{m}$.

The angle between those tangential planes is equal to the angle between their normal vectors and is given by $d\varphi = |d\boldsymbol{m}|$. Therefore, $|d\boldsymbol{m}|$ characterizes the angle of rotation of the tangential plane during the displacement of the curve along the plane to a infinitesimally close point. Since $d\boldsymbol{m} = \boldsymbol{m}_u dU + \boldsymbol{m}_v dV$, we have

$$d\varphi^2 = d\boldsymbol{m}^2 = c_{uu} dU^2 + 2c_{uv} dU dV + c_{vv} dV^2 \,. \tag{B56}$$

This quadratic form is called the third fundamental form of the surface. The fundamental forms (B46), (B54) and (B56) are connected by the following relationship:

$$d\boldsymbol{m}^2 - 2H d\boldsymbol{m} \cdot d\boldsymbol{r} - K d\boldsymbol{r}^2 = 0 \,. \tag{B57}$$

Here, H is the average curvature and K is the total curvature of the surface in the point M.

We now consider the special case when the surface is given in the form

$$Z = f(X, Y) \,. \tag{B58}$$

We now consider X and Y as the curvilinear coordinates U and V on the plane. Then, the equation of the surface can be written in the form

$$\boldsymbol{r} = X\boldsymbol{i} + Y\boldsymbol{j} + f(X, Y)\boldsymbol{k} \,. \tag{B59}$$

Then, the following relationships apply:

$$\boldsymbol{r}_x = \boldsymbol{i} + f_x \boldsymbol{k} \,, \quad \boldsymbol{r}_y = \boldsymbol{j} + f_y \boldsymbol{k} \,, \quad \boldsymbol{r}_{xx} = f_{xx} \boldsymbol{k} \,,$$
$$\boldsymbol{r}_{xy} = f_{xy} \boldsymbol{k} \,, \quad \boldsymbol{r}_{yy} = f_{yy} \boldsymbol{k} \,.$$

The coefficients of the first fundamental form are given by

$$a_{xx} = \boldsymbol{r}_x^2 = 1 + f_x^2 \,, \quad a_{xy} = \boldsymbol{r}_x \cdot \boldsymbol{r}_y = f_x f_y \,, \quad a_{yy} = \boldsymbol{r}_y^2 = 1 + f_y^2$$

and the form itself is

$$dS^2 = \left(1 + f_x^2\right) dX^2 + 2 f_x f_y dX dY + \left(1 + f_y^2\right) dY^2 \,. \tag{B60}$$

The determinant of the first fundamental form is $a = a_{uu} a_{vv} - a_{uv}^2$. The unit normal vector on the surface is equal to

$$\boldsymbol{m} = \frac{\boldsymbol{k} - f_x \boldsymbol{i} - f_y \boldsymbol{j}}{\sqrt{1 + f_x^2 + f_y^2}} \,.$$

The coefficients of the second fundamental form are given by

$$b_{xx} = \boldsymbol{r}_{xx} \cdot \boldsymbol{m} = \frac{f_{xx}}{\sqrt{1 + f_x^2 + f_y^2}} \,, \quad b_{xy} = \boldsymbol{r}_{xy} \cdot \boldsymbol{m} = \frac{f_{xy}}{\sqrt{1 + f_x^2 + f_y^2}} \,,$$
$$b_{yy} = \boldsymbol{r}_{yy} \cdot \boldsymbol{m} = \frac{f_{yy}}{\sqrt{1 + f_x^2 + f_y^2}}$$

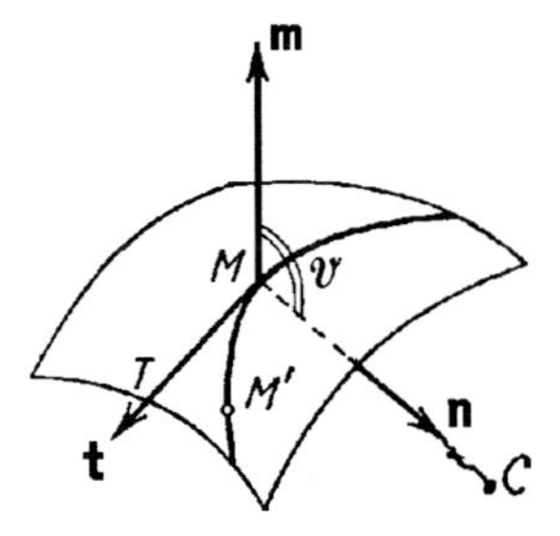

Figure B.6 The main normal $\boldsymbol{n}$ to a curve in space.

and the second fundamental form and its determinant b are equal to

$$l = \frac{1}{2\sqrt{1 + f_x^2 + f_y^2}} \left(f_{xx} d X^2 + 2 f_{xy} d X d Y + f_{yy} d Y^2 \right) , \tag{B61}$$

$$b = \frac{f_{xx} f_{yy} - f_{xy}^2}{1 + f_x^2 + f_y^2} .$$

The total curvature is

$$K = \frac{f_{xx} f_{yy} - f_{xy}^2}{\left(1 + f_x^2 + f_y^2\right)^2} . \tag{B62}$$

B.8
Curvature of a Curve on the Surface

In Section B.5, the curvature K of a spatial curve was defined and in Section B.4, it was shown how for each point of every spatial curve, the osculating circle can be found that lies in the osculating plane, having its center in curvature center C and a radius of curvature R such that $K = 1/R$.

The radius of this circle is given by $R = 1/|\ddot{\boldsymbol{r}}(S)|$, where the vector $\ddot{\boldsymbol{r}}(S)$ is directed to the center of curvature along the main normal $\boldsymbol{n}$ such that $\boldsymbol{MC} = R\boldsymbol{n}$ (see Figure B.6).

We assume that the the curves are given in the parametric form (B43).

In the point M, the unit tangential vector $\boldsymbol{t} = \dot{\boldsymbol{r}}$, the normal vector $\boldsymbol{m}$ to the plane and the main normal vector $\boldsymbol{n}$ exist. By denoting the angle between $\boldsymbol{m}$ and $\boldsymbol{n}$ as ϑ, we obtain from the first Frenet formula (B30) that $\dot{\boldsymbol{r}} \cdot \boldsymbol{m} = K \cos\vartheta$. Therefore, according to (B55) and (B52), one obtains

$$K \cos\vartheta = \frac{b_{uu} d U^2 + 2 b_{uv} d U d V + b_{vv} d V^2}{a_{uu} d U^2 + 2 a_{uv} d U d V + a_{vv} d V^2} . \tag{B63}$$

On the right-hand side of (B63), the ratio of second and first fundamental forms appears, whose coefficients at the point M depend on U and V.

If a curve passing through M is given, then the tangent and therefore $d U/d V$, the osculating plane and ϑ are known. Thus, all curves that pass through M and have the same tangent and osculating plane in the point M have the same curvature K.

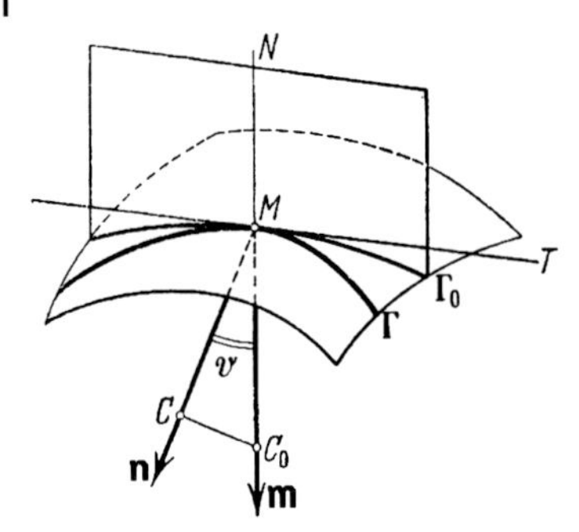

Figure B.7 Intersection of the normal plane with a surface.

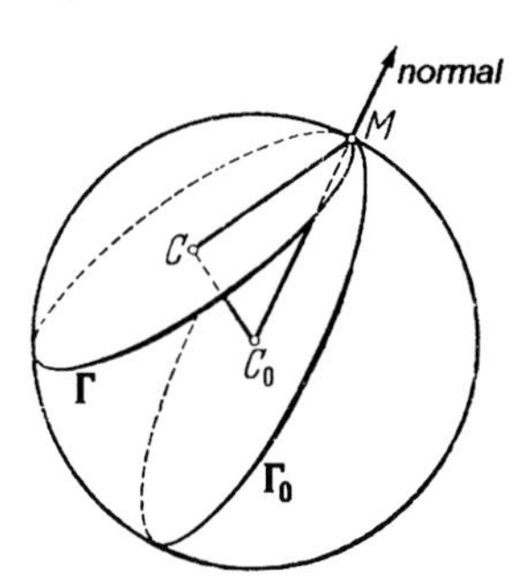

Figure B.8 The normal section Γ_0 of a sphere.

We consider different curves Γ (see Figure B.7) that pass through a fixed point M on the surface and have a common tangent M with the unit tangential vector $\boldsymbol{t}$, but different osculating planes. One of these curves is Γ_0, which is created via the intersection of the normal plane TMN with the surface under consideration. This curve is called the normal section. The angle ϑ in (B63) is the angle between the normal plane and the osculating plane for the curve Γ. Since the curve Γ_0 is a plane curve, the osculating plane coincides with the normal plane and the curvature center $m C_0$ of the normal section Γ_0 lies in the direction of the vector $\boldsymbol{m}$. Since we have $\vartheta = 0$ for the normal section, it follows from (B63) that its curvature is

$$K_0 = \frac{b_{uu} d U^2 + 2 b_{uv} d U d V + b_{vv} d V^2}{a_{uu} d U^2 + 2 a_{uv} d U d V + a_{vv} d V^2} \,. \tag{B64}$$

For any other curve Γ, we have $\vartheta \neq 0$, and, comparing (B64) and (B63), one obtains

$$K_0 = K \cos\vartheta \text{ or } R = R_0 \cos\vartheta \,. \tag{B65}$$

In (B65), Meusnier's theorem is expressed. It must be noted that we have $C C_0 \perp M C$ because of $R = M C$ and $R_0 = M C_0$.

As example, we consider a point M on a sphere. The normal to its surface is directed inwards to the center (see Figure B.8). The normal section Γ_0 is the major circle of the sphere whose center coincides with the curvature center of the curve Γ_0.

We now take as curve Γ the intersection of the sphere with another plane. This is a circle with curvature center C in the center of this circle. The angle ϑ is the angle between those two intersections. It is obvious that $R = R_0 \cos\vartheta$, which corresponds to (B65). Thus, the curvature K of arbitrary surface curves depends, according to (B37), on the curvature of the normal section K_0, which in turn is defined by (B64). If $K_0 = 0$, the tangential direction of the normal section is called an asymptotic direction.

We now assume that $K_0 \neq 0$. Then, for every point of the surface M two directions exist that are characterized by two orthogonal unit tangential vectors $\boldsymbol{t}_1$ and $\boldsymbol{t}_2$ to corresponding normal sections. These directions are called main directions and their curvatures K_1 and K_2 are called main curvatures at the point M. Without loss of generality, we may assume that $K_2 \geq K_1$.

The curvature $\tilde{K}$ for a normal section M with unit tangential vector $\boldsymbol{t}$ that makes an angle ϕ with $\boldsymbol{t}_1$ is expressed in terms of the main curvatures by Euler's formula

$$\tilde{K} = K_1 \cos^2\varphi + K_2 \sin^2\varphi \,. \tag{B66}$$

From (B66), it follows that $\tilde{K}$ increases monotonically with ϕ from its minimal value K_1 for $\phi = 0$ to its maximal value K_2 for $\phi = \pi/2$.

If one moves from a given point M on the surface along a curve in a main direction, $d\boldsymbol{m}$ becomes, according to Rodrigues formulae, proportional to $d\boldsymbol{r}$:

$$d\boldsymbol{m} = -K_1 d\boldsymbol{r} \,, \quad d\boldsymbol{m} = -K_2 d\boldsymbol{r} \,. \tag{B67}$$

These equations make it possible to obtain the main directions and main curvatures for a surface, for which the first and second fundamental forms are given.

To find the main directions from the condition

$$d\boldsymbol{m} = K d\boldsymbol{r} \quad \text{or} \quad \boldsymbol{m}_u dU + \boldsymbol{m}_v dV = -K(\boldsymbol{r}_u dU + \boldsymbol{r}_v dV) \,,$$

we first multiply this equation by $\boldsymbol{r}_u$ and then with $\boldsymbol{r}_v$, make use of the quadratic forms (B46) and (B53), obtaining the two equations

$$b_{uu} dU + b_{uv} dV = K(a_{uu} dU + a_{uv} dV) \,,$$

$$b_{uv} dU + b_{vv} dV = K(a_{uv} dU + a_{vv} dV) \,. \tag{B68}$$

The system of equations (B68) is a homogeneous system of equations:

$$(b_{uu} - K a_{uu}) dU + (b_{uv} - K a_{uv}) dV = 0 \,,$$
$$(b_{uv} - K a_{uv}) dU + (b_{vv} - K a_{vv}) dV = 0 \,.$$

In order for non-trivial solutions to exist, the determinant of this system of equations must vanish, that is,

$$\left(a_{uu} a_{vv} - a_{uv}^2\right) K^2 + (2 b_{uv} a_{uv} - a_{uu} b_{vv} - a_{vv} b_{uu}) K + \left(b_{uu} b_{vv} - b_{uv}^2\right) = 0 \,.$$

This means that the roots of this equation satisfy the following conditions:

$$K_1 K_2 = \frac{b_{uu} b_{vv} - b_{uv}^2}{a_{uu} a_{vv} - a_{uv}^2} \,, \quad K_1 + K_2 = \frac{a_{uu} b_{vv} + a_{vv} b_{uu} - 2 a_{uv} b_{uv}}{a_{uu} a_{vv} - a_{uv}^2} \,. \tag{B69}$$

The product of the main curvatures is called total or Gauss curvature K, and half their sum is called the mean curvature H. They are equal to

$$K = K_1 K_2 \,, \quad H = \frac{K_1 + K_2}{2} \,. \tag{B70}$$

By eliminating from (B68) the Gauss curvature K, one obtains

$$(b_{uu} a_{uv} - b_{uv} a_{uu})\, d U^2 + (b_{uu} a_{vv} - b_{vv} a_{uu})\, d U d V + (b_{uv} a_{vv} - b_{vv} a_{uv})\, d V^2 = 0 \,. \tag{B71}$$

From this equation, one can obtain the values of $d U/d V$ for the main directions. There are three possibilities:

1. If the coefficients of the first and second fundamental forms are proportional to each other, that is,

$$\frac{b_{uu}}{a_{uu}} = \frac{b_{uu}}{a_{uu}} = \frac{b_{vv}}{a_{vv}} = K_0 \,, \tag{B72}$$

   all coefficients in (B71) vanish and the equation is satisfied for arbitrary $d U$ and $d V$, that is, every tangential direction is the main direction and all normal sections in the point M have the same curvature. Such a point is called a spherical point.
2. If at least one of the coefficients for $d U^2$ or $d V^2$ does not vanish, there will be two different solutions. This means that there will be two main directions in the point M:

$$\frac{d U}{d V} = f_1(U, V) \,, \quad \frac{d U}{d V} = f_2(U, V) \tag{B73}$$

   with different main curvatures.
3. If the coefficients of $d U^2$ and $d V^2$ in (B71) vanish, but the coefficient of $d U d V$ is unequal to zero, which is only possible for $a_{uv} = b_{uv} = 0$, the equation yields $d U d V = 0$. This gives two main directions: along the one, we have $d U = 0$ and along the other, we have $d V = 0$. Therefore, the main directions coincide with the coordinate lines on the surface. The condition $a_{uv} = \boldsymbol{r}_u \cdot \boldsymbol{r}_v = 0$ means that the tangential vectors $\boldsymbol{r}_u$ and $\boldsymbol{r}_v$ of the corresponding coordinate lines are orthogonal, and the condition $b_{uv} = 0$ means that the directions of the coordinate lines are adjunct.

From the sign of the total curvature, one can obtain the behavior of a surface in the vicinity of a point M. There are three possibilities:

1. $K > 0$, that means both main curvatures K_1 and K_2 have the same sign. According to (B64), K is the ratio of two fundamental forms. Since the first fundamental form in the denominator is positive definite, the second fundamental form in the numerator also has to be positive, that is, $b_{uu} b_{vv} - b_{uv}^2 > 0$. A point

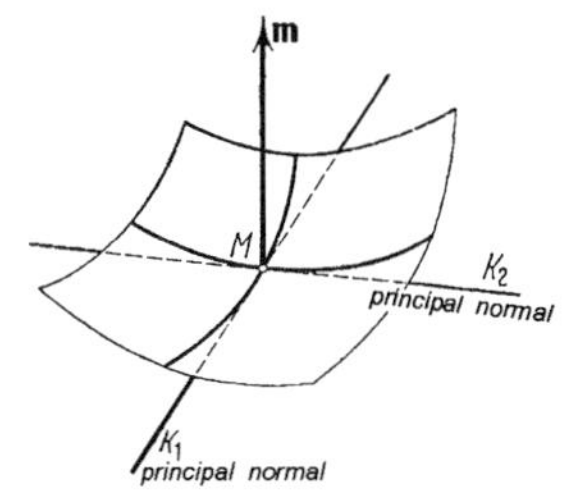

Figure B.9 Elliptical point.

in which $K > 0$ is called the elliptical point (see Figure B.9). In this point, both main sections are bent to the same side: for $K_1 > 0$ and $K_2 > 0$ in the direction of the normal vector $\boldsymbol{m}$, and for $K_1 < 0$ and $K_2 < 0$, in the direction of $-\boldsymbol{m}$. Since K_1 is the minimum and K_2 is the maximum value of the curvature, there is no curve through the point M with $K = 0$, meaning there is no asymptotic direction in an elliptical point.

2. $K < 0$ and $b_{uu}b_{vv} - b_{uv}^2 < 0$. Such a point is called hyperbolic point. In this point, the main curvatures K_1 and K_2 have different signs. Therefore, one main direction is bent in the direction $\boldsymbol{m}$, while the other one is bent in the direction $-\boldsymbol{m}$. As a result, the surface in the vicinity of the point M has the form of a saddle. Since $K_1 < K < K_2$, there is at least one curve with $K = 0$. Since the normal sections whose tangents at the point M are symmetric relative to the main direction have the same curvatures, there are two asymptotic directions.
3. $K = 0$ and $b_{uu}b_{vv} - b_{uv}^2 = 0$. Such a point is called parabolic point. One of the main directions has zero curvature. Therefore, M is a turning point and this direction coincides with an asymptotic direction. If there are parabolic points on a surface, they constitute a line that separates elliptic points from parabolic points.

The curvature line on a surface is called a line, for which the tangent in every point is along one of the main directions. For a curve to be a curvature line, it is necessary and sufficient that condition (B71) is satisfied. Then, the differential equation (B72) yields in every point of the surface two families of orthogonal curvature lines.

In case three ($dUdV = 0$), the curvature lines coincide with the coordinate lines on the surface. For example, the curvilinear system of coordinates with geographical latitude and longitude on a sphere has this property. It is obvious that not every system of coordinates has this property.

B.9 Internal Geometry of a Surface

It has been noted above that a curve on a plane (Section B.3) or on in space (Section B.5) can be defined by a natural equation, and that the form of the curve does not depend on the system of coordinates, and the curve and its geometrical properties are defined exactly up to the position on the plane or in space. To formulate such a task for a curve on a surface, we introduce the notions of bending (continuous deformation) of the surface and of an internal geometry.

We consider two surfaces Σ and Σ_1 in space. The bending of surface Σ into another surface Σ_1 is defined as the one-on-one mapping from Σ to Σ_1, during which the length of the curve L on Σ is the same as the length of the corresponding curve L_1 on Σ_1. The term bending corresponds to the idea of a surface as a bendable, non-stretchable and incompressible film that can be deformed in space in such a way that the curves that lay on the surface do not change their lengths. Every sufficiently smooth surface can be bent, but a bending mapping does not exist for every pair of surfaces. Thus, a plane can be bent to obtain a cone or a cylinder, but not a sphere.

The internal geometry of surfaces is a branch of geometry in which the geometrical properties of surfaces are examined only from the knowledge of the first fundamental form (B46). For the determination of the length of a curve, it is only necessary to know the first fundamental form, but to obtain the curvature of a curve, one also needs to know the second fundamental form. Therefore, it follows from the definition of the bending of a surface, that the first fundamental form does not change during bending, that is, the internal geometry of the surface remains unchanged during bending. Therefore, one can say that the internal surface geometry investigates such properties of the surface that are invariant under bending. This leads to the question as to what is needed in addition to the first fundamental form for a surface to behave as a solid body or for two surfaces to coincide except for the position in space.

For this purpose, we now move to tensor relations. We have to assume for the moment that the surface under consideration lies in the three-dimensional Euclidean space in which a uniform Cartesian system of coordinates $X^i, (i = 1, 2, 3)$, an orthonormal basis $\boldsymbol{e}_i$ and a metric tensor $g_{ij} = \boldsymbol{e}_i \cdot \boldsymbol{e}_j$ can be introduced for the whole space. We consider a surface on which a curvilinear system of coordinates U^α $(\alpha = 1, 2)$ is chosen. The surface has the parametric form

$$X^i = X^i(U^1, U^2) \tag{B74}$$

and it can be considered to be a two-dimensional surface in the three-dimensional space. The coordinate grid on the surface is chosen in an arbitrary manner such that through every point on the surface, two coordinate curves $U^1 = \text{const}$ and $U^2 = \text{const}$ pass. The radius vector in the point M is equal to $\boldsymbol{r}(U^1, U^2)$.

In the point M, there are two non-collinear vectors $\boldsymbol{\varepsilon}_1 = \partial \boldsymbol{r}/\partial U^1$ and $\boldsymbol{\varepsilon}_2 = \partial \boldsymbol{r}/\partial U^2$ that are directed along the tangents on the corresponding coordinate lines $U^1(U^2 = \text{const})$ and $U^2(U^1 = \text{const})$, and lie in the tangential plane. These vectors and the unit normal vector $\boldsymbol{\varepsilon}_n = \boldsymbol{m}$ (see (B53) can be taken as a local basis such that every vector in the point M of the surface that does not lie in the surface can be written as

$$\boldsymbol{a} = a^i \boldsymbol{\varepsilon}_i\,, \quad (i = 1, 2, 3). \tag{B75}$$

Every vector in the point M that does lie in the surface can be written as

$$\boldsymbol{a} = a^{\alpha} \boldsymbol{\varepsilon}_{\alpha} , \quad (\alpha = 1, 2) . \tag{B76}$$

In the following, we will assume Greek indices for components of a surface vector and other objects connected to the surface.

We now introduce on the surface new curvilinear coordinates $U^{\alpha'}$ $(\alpha' = 1, 2)$ that are connected to the coordinates U^{α} via the relationship $U^{\alpha'} = U^{\alpha'}(U^1, U^2)$ with a non-degenerate Jacobian $A^{\alpha'}_{\alpha} = \partial U^{\alpha}/\partial U^{\alpha'}$, that means $A = |A^{\alpha'}_{\alpha}| \neq 0$. Then, the inverse matrix $B^{\alpha}_{\alpha'}$ such that $A^{\alpha}_{\gamma} B^{\gamma}_{\beta} = \delta^{\alpha}_{\beta}$, and the inverse transformation $U^{\alpha} = U^{\alpha}(U^{1'}, U^{2'})$ exist. The differentials dU^{α} are transformed as contravariant surface vectors, $dU^{\alpha'} = A^{\alpha'}_{\alpha} dU^{\alpha}$, the partial derivatives of scalar function as covariant surface vectors, $\partial f/\partial U^{\alpha} = B^{\alpha}_{\alpha'} \partial f/\partial U^{\alpha'}$ and the surface tensors as

$$A^{\lambda'}_{\mu'\nu'} = A^{\lambda'}_{.\alpha} B^{\beta}_{.\mu'} B^{\gamma}_{.\nu'} A^{\alpha}_{\beta\gamma} . \tag{B77}$$

The surface (B74) is characterized by the first and second fundamental forms of the surface, that is,

$$dS^2 = a_{\alpha\beta} U^{\alpha} U^{\beta} , \quad l = \frac{1}{2} b_{\alpha\beta} U^{\alpha} U^{\beta} , \tag{B78}$$

where $a_{\alpha\beta} = \boldsymbol{\varepsilon}_{\alpha} \cdot \boldsymbol{\varepsilon}_{\beta}$ is the metric tensor of the surface and $b_{\alpha\beta} = \partial \boldsymbol{\varepsilon}_{\alpha}/\partial U^{\beta} \cdot \boldsymbol{m}$ is a covariant surface tensor of second order.

The Euclidean three-dimensional space in which the surface is situated is described by the metric tensor $g_{ij} = \boldsymbol{e}_i \cdot \boldsymbol{e}_j$, and in an orthonormal basis $\boldsymbol{e}_k$, we have $g_{ij} = \delta_{ij}$. If the surface is given in coordinate form $X^i = X^i(U^j)$, we have $a_{\alpha\beta} = \sum_{i=1}^{3} \partial X^i/\partial U^{\alpha} \partial X^i/\partial U^{\beta}$ and the metric tensors for the space and for the surface are connected by the following relationship

$$a_{\alpha\beta} = g_{ij} \frac{\partial X^i}{\partial U^{\alpha}} \frac{\partial X^i}{\partial U^{\beta}} . \tag{B79}$$

Since the determinant $a = \left| a_{\alpha\beta} \right|$ does not vanish, the inverse matrix $a^{\alpha\beta}$ to $a_{\alpha\beta}$ exists, whose elements are the components of the contravariant metric tensor $a^{\alpha\beta}$ $(a^{\alpha\gamma} a_{\gamma\beta} = \delta^{\alpha}_{\beta})$.

The elements are explicitly given by

$$a^{11} = \frac{a_{22}}{a} , \quad a^{12} = a^{21} = -\frac{a_{12}}{a} , \quad a^{22} = \frac{a_{11}}{a} , \quad a = a_{11} a_{22} - (a_{12})^2 .$$

With the help of the metric tensor, one can raise and lower tensor indices, for example, $A^{\alpha} = a^{\alpha\beta} A_{\beta}$ and introduce the Levy–Civita tensors for the surface $\sqrt{a}\varepsilon_{\alpha\beta}$ and $\varepsilon^{\alpha\beta}/\sqrt{a}$, where $\varepsilon_{\alpha\beta}$ and $\varepsilon^{\alpha\beta}$ are the permutation symbols. By using the results of Appendix A.1, one can obtain all mathematical operations with surface vectors and tensors by replacing g_{ij} by a_{ij}.

However, there is a crucial difference between the geometry of the surface (Riemannian geometry) and the geometry of the surrounding space (Euclidean geometry). The geometry of the Euclidean space is completely characterized by the metric tensor g_{ij}, which from the viewpoint of differential geometry, has the meaning of the first fundamental form. To describe the geometry of the Riemannian space completely, it is not sufficient to only know the metric tensor of the surface or the first fundamental form $a_{\alpha,\beta}$. One also needs to know the tensor $b_{\alpha\beta}$ or the second fundamental form of the surface.

It was shown previously that the fundamental forms define the important geometrical characteristic curvature. Therefore, the Riemannian space is, unlike the Euclidean space, also characterized by the curvature. Since the curvature is a characteristic of the surface of the Riemannian space and has no connection with the external Euclidean space, the curvature must be an invariant object of the surface, that is, a tensor. Therefore, the curvature in the Riemannian space is a tensor which is called curvature tensor.

To define the curvature of a curve on a surface, one has to use, in addition to the first fundamental form $a_{\alpha,\beta}$, the second fundamental form $b_{\alpha,\beta}$ which depends on $\partial \boldsymbol{\varepsilon}_\alpha / \partial U^\beta$. We now decompose $\partial \boldsymbol{\varepsilon}_\alpha / \partial U^\beta$ in terms of the local basis vector in the point M:

$$\frac{\partial \boldsymbol{\varepsilon}_\alpha}{\partial U^\beta} = \Gamma^\gamma_{\alpha\beta} \boldsymbol{\varepsilon}_\gamma + \Lambda_{\alpha\beta} \boldsymbol{m} \,, \quad (\alpha, \beta, \gamma = 1, 2) \,. \tag{B80}$$

The coefficients $\Gamma^\gamma_{\alpha\beta}$ are the Christoffel symbols of the second kind. They are symmetric with respect to the lower indices, $\Gamma^\gamma_{\alpha\beta} = \Gamma^\gamma_{\beta\alpha}$, and are related to the Christoffel symbols of the first kind $\Gamma_{\alpha\beta,\gamma}$ by

$$\Gamma^\delta_{\alpha\beta} = a^{\gamma\delta} \Gamma_{\alpha\beta,\gamma} \,. \tag{B81}$$

Since $\Gamma_{\alpha\beta,\gamma} = \Gamma_{\beta\alpha,\gamma}$, there are only six different coefficients of $\Gamma_{\alpha\beta,\gamma}$. The Christoffel symbols of the first kind can be expressed by the coefficients of the first fundamental form as

$$\Gamma_{\alpha\beta,\gamma} = \frac{1}{2} \left(\frac{\partial a_{\gamma\alpha}}{\partial U^\beta} + \frac{\partial a_{\beta\gamma}}{\partial U^\alpha} - \frac{\partial a_{\alpha\beta}}{\partial U^\gamma} \right) \,. \tag{B82}$$

To obtain the coefficients $\Lambda_{\alpha\beta}$, we evaluate the scalar product with $\boldsymbol{m}$ of both sides of (B80) and take into account that $\boldsymbol{\varepsilon}_\gamma \cdot \boldsymbol{m} = 0$. Then, one obtains $\Lambda_{\alpha\beta} = b_{\alpha\beta}$. The formulae (B80) are called the first group of derivation formulae or Gauss formulae. They represent the change of the vectors $\boldsymbol{\varepsilon}_\gamma$ along the coordinate lines.

The second group of derivation formulae gives the change of the normal vectors $\boldsymbol{m}$ along the coordinate lines. We expand $\partial \boldsymbol{m} / \partial U^\alpha$ in terms of the local basis vectors in the point M,

$$\frac{\partial \boldsymbol{m}}{\partial U^\alpha} = -d^\beta_\alpha \boldsymbol{\varepsilon}_\beta, \tag{B83}$$

and evaluate the scalar product with $\boldsymbol{\varepsilon}_\delta$ on both sides. One obtains $b_{\alpha\beta} = d^\delta_\alpha a_{\delta\beta}$ or

$$d^\delta_\beta = b_{\alpha\beta} a^{\alpha\delta} \,.$$

Thus, (B80) and (B83) express the first partial derivatives of the local basis vectors $\boldsymbol{\varepsilon}_1$, $\boldsymbol{\varepsilon}_2$ and $\boldsymbol{m}$ with respect to U^1 and U^2 in terms of those same basis vectors.

Without the knowledge of the derivatives of second and higher order, one can investigate the complete behavior of the surface in the vicinity of an arbitrary point since this is completely defined by the values of the coefficients of the first and second fundamental forms.

Now, the question posed at the beginning of this section can be answered. Two surfaces are only different with respect to their position in space if and only if they have identical first and second fundamental forms. If the first fundamental forms coincide and the second fundamental forms only differ by their sign, it means that the first surface can be obtained by translation, rotation and a mirror operation from the second surface.

Now, we discuss the restrictions that apply to the coefficients of the fundamental forms. Only one condition applies to the coefficients of the first fundamental form, namely, that the first fundamental forms must be positive definite:

$$a = a_{11}a_{22} - (a_{12})^2 > 0\,, \quad \text{and} \quad a_{11} > 0\,. \tag{B84}$$

The coefficients of the second fundamental form must satisfy the Gauss formulae

$$\begin{aligned} b = b_{11}b_{22} - (b_{12})^2 &= \frac{\partial^2 a_{12}}{\partial U^1 \partial U^2} - \frac{1}{2}\frac{\partial^2 a_{11}}{\partial U^2 \partial U^2} - \frac{1}{2}\frac{\partial^2 a_{22}}{\partial U^1 \partial U^1} \\ &+ \Gamma^{\gamma}_{12}\Gamma^{\delta}_{12}a_{\gamma\delta} - \Gamma^{\alpha}_{11}\Gamma^{\beta}_{22}a_{\alpha\beta} \end{aligned} \tag{B85}$$

and the Peterson–Kodazzi formulae

$$\frac{\partial b_{\alpha 1}}{\partial U^2} - \frac{\partial b_{\alpha 2}}{\partial U^1} = -\Gamma^1_{\alpha 1}b_{12} - \Gamma^2_{\alpha 1}b_{22} + \Gamma^1_{\alpha 2}b_{11} + \Gamma^2_{\alpha 2}b_{21}\,. \tag{B86}$$

From these formulae, if follows that the coefficients of the second fundamental form depend on the coefficients of the first fundamental form.

From (B69) for the total curvature K and from (B84) for the determinant of the first fundamental form, it follows that K is defined via the first fundamental form alone, although the main curvatures K_1 and K_2 each depend on the coefficients of both fundamental forms. Therefore, the total curvature K is conserved during the bending of a surface, although the main curvatures K_1 and K_2 both change. Therefore, a surface that can be bent into a plane must have a vanishing total curvature ($K = 0$) and it must therefore be a rollable surface (cone, cylinder).

We now determine the curvature tensor. According to (A32), the covariant derivative of the covariant components A_γ of vector $\boldsymbol{a}$ is equal to

$$\nabla_\alpha A_\gamma = A_{\gamma,\alpha} = \frac{\partial A_\gamma}{\partial U^\alpha} - \Gamma^{\beta}_{\gamma\alpha}A_\beta\,, \quad \nabla_\beta A_\gamma = A_{\gamma,\beta} = \frac{\partial A_\gamma}{\partial U^\beta} - \Gamma^{\delta}_{\gamma\beta}A_\delta\,.$$

The mixed derivatives of second order are given by

$$A_{\gamma,\alpha\beta} = \frac{\partial A_{\gamma,\alpha}}{\partial U^\beta} - \Gamma^\delta_{\gamma\beta} A_{\delta,\alpha} - \Gamma^\delta_{\alpha\beta} A_{\gamma,\delta} ,$$

$$A_{\gamma,\beta\alpha} = \frac{\partial A_{\gamma\beta}}{\partial U^\alpha} - \Gamma^\delta_{\gamma\alpha} A_{\delta,\beta} - \Gamma^\delta_{\beta\alpha} A_{\gamma,\beta} .$$

In Euclidean space, we have $A_{\gamma,\alpha\beta} = A_{\gamma,\beta\alpha}$, but in Riemannian space, the difference $A_{\gamma,\alpha\beta} - A_{\gamma,\beta\alpha}$ is generally unequal to zero such that

$$A_{\gamma,\alpha\beta} - A_{\gamma,\beta\alpha} = R^{\delta}_{\gamma\alpha\beta} A_\delta . \tag{B87}$$

The tensor of fourth order $R^{\delta}_{\gamma\alpha\beta}$ is called the Riemann tensor of the first kind and is defined by

$$R^{\delta}_{\gamma\alpha\beta} = \frac{\partial}{\partial U^\alpha} \Gamma^\delta_{\beta\gamma} - \frac{\partial}{\partial U^\beta} \Gamma^\delta_{\gamma\alpha} + \Gamma^\varepsilon_{\gamma\beta} \Gamma^\delta_{\varepsilon\alpha} - \Gamma^\varepsilon_{\gamma\alpha} \Gamma^\delta_{\varepsilon\beta} . \tag{B88}$$

It is antisymmetric with respect to the indices α and β, that is, $R^{\delta}_{\gamma\alpha\beta} = -R^{\delta}_{\gamma\beta\alpha}$ is defined only in terms of the first fundamental form $a_{\alpha\beta}$ and therefore belongs to the internal geometry. It is obvious that $R^{\delta}_{\gamma\alpha\beta} = 0$ in Euclidean space since all Christoffel symbols vanish in that space.

The Riemann tensor of the second kind is defined as the tensor $R_{\lambda\gamma\alpha\beta} = a_{\lambda\delta} R^{\delta}_{\gamma\alpha\beta}$. It is antisymmetric with respect to the index pairs λ, α and γ, β. In the two-dimensional space, only four of the 16 components are unequal to zero, and only one of those is independent because $R_{1212} = -R_{2112} = -R_{1221} = R_{2121} = Ka$, where K is the total Gauss curvature. The scalar $1/4\varepsilon^{\lambda\alpha}\varepsilon^{\beta\gamma} R_{\lambda\gamma\alpha\beta} = K/a$ is equal to the total curvature divided by the determinant of the first quadratic fundamental form.

B.10 Surface Vectors

A surface vector is defined to be a vector $\boldsymbol{a}$ whose initial point M lies on the surface and which lies on the tangential surface (see Figure B.10). Two coordinate lines U^1 and U^2 pass through the point M. If these lines are given in parametric form, $\boldsymbol{r} = \boldsymbol{r}(t)$, $\boldsymbol{a}$ can be expanded in terms of the vectors $\boldsymbol{\varepsilon}_1 = \partial \mathbf{r}/\partial U^1$ and $\boldsymbol{\varepsilon}_2 = \partial \mathbf{r}/\partial U^2$ as

$$\boldsymbol{a} = a^\alpha \boldsymbol{\varepsilon}_\alpha , \quad (\alpha = 1, 2) . \tag{B89}$$

We assume that the vector $\boldsymbol{a}$ is a tangent on the surface curve C which is defined by the equations $U^1 = U^1(t)$ and $U^2 = U^2(t)$. Then, it is collinear to the vector $d\boldsymbol{r}/dt = \boldsymbol{\varepsilon}_\alpha dU^\alpha/dt$. During the bending of the surface Σ into the surface Σ^*, which conserves the coordinate grid, the curve C becomes the curve C^*, and the tangential vectors $\boldsymbol{\varepsilon}_\alpha$ on the coordinate lines of Σ become the tangential vectors $\boldsymbol{\varepsilon}^*_\alpha$ on the coordinate lines of Σ^*.

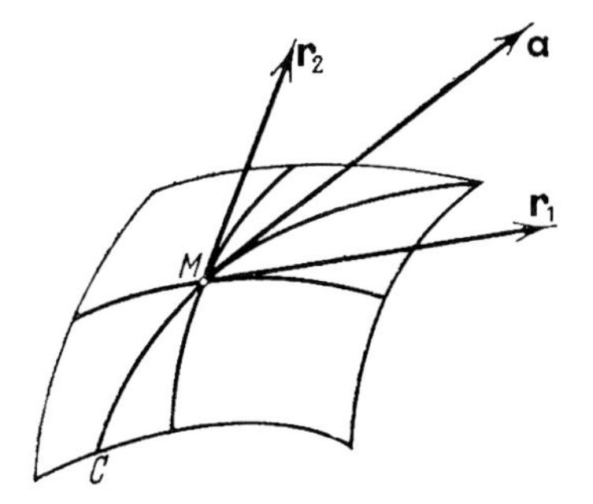

Figure B.10 Surface vector.

Since the first fundamental form is conserved under bending the length of a surface vector is also conserved, that is, $|\boldsymbol{a}| = |\boldsymbol{a}^*|$. Therefore, during the bending of the surface which keeps the values of U^1 and U^2 unchanged for every point, the coordinates and the lengths of a surface vector are unchanged.

A vector in three-dimensional space is understood to be a vector that can be attached to every point. Under parallel displacement, such a vector does not change, and its coordinates and length remain the same. Surface vectors can not be displaced in a parallel manner as in Euclidean space.

Indeed, if one moved a vector $\boldsymbol{a}$ in a parallel manner as a free vector to the point M' along the curve C, it would not lie on the tangential plane in the point M' and thus it would not be a surface vector. From the above, it follows that the notion of a free vector is not useful for surface vectors. However, one can introduce the notion of parallel displacement for a surface vector while all properties of the surface vectors are conserved. We now show how this can be done.

We consider a vector field along a surface curve C, which is given by the equation $U^\alpha(t)$. That means every point $M(t)$ of this curve corresponds to a certain surface vector $\boldsymbol{a}(t)$ which is a tangent to the surface, but not necessarily to the curve (see Figure B.11). During the transition along the curve from the point $M(t)$ to the infinitesimally close point $M'(t + dt)$, the vector $\boldsymbol{a}$ becomes

$$\boldsymbol{a}(t + dt) = \boldsymbol{a}(t) + \mathrm{a}'(t)dt + \ldots$$

The vector $\boldsymbol{a}$ in the point M' experiences the increment $d\boldsymbol{a} = \boldsymbol{a}'(t)dt$, which is generally not tangential to the surface. Therefore, the vector $\boldsymbol{a} + d\boldsymbol{a}$ generally does not belong to the surface vector field. To resolve this contradiction, one has to introduce the notion of the differential $D\boldsymbol{a}$ of a vector such that $D\boldsymbol{a}$ is a tangential vector on the surface.

We decompose the vector $d\boldsymbol{a}$ in the point M into the normal and tangential components

$$(\boldsymbol{a} + d\boldsymbol{a})_m = [(\boldsymbol{a} + d\boldsymbol{a}) \cdot \boldsymbol{m}]\boldsymbol{m} = (d\boldsymbol{a} \cdot \boldsymbol{m})\boldsymbol{m} ,$$
$$(\boldsymbol{a} + d\boldsymbol{a})_t = \boldsymbol{a} + d\boldsymbol{a} - (d\boldsymbol{a} \cdot \boldsymbol{m})\boldsymbol{m} .$$

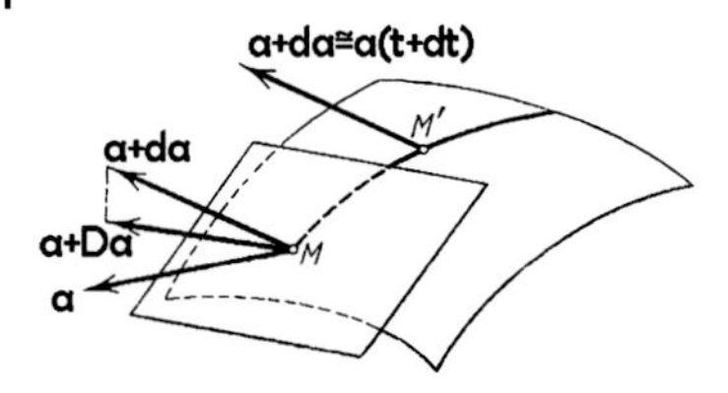

Figure B.11 Definition of the parallel displacement for a surface vector.

Since the vector $(\boldsymbol{a} + d\boldsymbol{a})_t$ is a surface vector, one can assume that $(\boldsymbol{a} + d\boldsymbol{a})_t = \boldsymbol{a} + D\boldsymbol{a}$. The vector $D\boldsymbol{a} = d\boldsymbol{a} - (d\boldsymbol{a} \cdot \boldsymbol{m})\boldsymbol{m}$ is a surface vector and is called the absolute differential.

If we have $D\boldsymbol{a} = 0$, we say that the vector $\boldsymbol{a}$ undergoes parallel displacement, and the equation $(\boldsymbol{a} + d\boldsymbol{a})_t = \boldsymbol{a}$ applies. The condition $D\boldsymbol{a} = 0$ means that $d\boldsymbol{a} = (d\boldsymbol{a} \cdot \boldsymbol{m})\boldsymbol{m}$. Thus, the vector $\boldsymbol{a}$ only has a normal component.

From (B89) and (B80), one obtains that

$$d\boldsymbol{a} = da^\alpha \boldsymbol{\varepsilon}_\alpha + a^\alpha \frac{\partial \boldsymbol{\varepsilon}_\alpha}{\partial U^\beta} = da^\alpha \boldsymbol{\varepsilon}_\alpha + a^\alpha \left(\Gamma^\gamma_{\alpha\beta} \boldsymbol{\varepsilon}_\gamma + b_{\alpha\beta} \boldsymbol{m} \right) dU^\beta .$$

Since the absolute differential $D\boldsymbol{a}$ is a surface vector, we have

$$D\boldsymbol{a} = (d\boldsymbol{a})_t = \left(da^\gamma + a^\alpha \Gamma^\gamma_{\alpha\beta} dU^\beta \right) \boldsymbol{\varepsilon}_\gamma . \tag{B90}$$

The absolute derivative with respect to the curve parameter t is equal to

$$\frac{D\boldsymbol{a}}{dt} = \left(\frac{da^\gamma}{dt} + a^\alpha \Gamma^\gamma_{\alpha\beta} \frac{dU^\beta}{dt} \right) \boldsymbol{\varepsilon}_\gamma . \tag{B91}$$

If a surface vector undergoes parallel displacement along a curve, we have $D\boldsymbol{a} = 0$ and

$$da^\gamma = -a^\alpha \Gamma^\gamma_{\alpha\beta} dU^\beta . \tag{B92}$$

We now consider the parallel displacement of a surface vector along the surface curve $U^\alpha(t)$ of finite length. In the initial point $M_0[U^1(0), U^2(0)]$, the vector $\boldsymbol{a}_0$ is given. During the parallel displacement of the vector $\boldsymbol{a}$ from the point $M(t)$ to the point $M'(t+dt)$, the vector $\boldsymbol{a}$ becomes the vector $\boldsymbol{a}+d\boldsymbol{a}$ where the vector $d\boldsymbol{a}$ is parallel to the normal vector in the point M, and its components are defined by (B92). The corresponding vector components $a^\alpha(t)$ are obtained from the following system of differential equations:

$$\frac{da^\alpha}{dt} = -a^\alpha \Gamma^\gamma_{\alpha\beta} \frac{dU^\beta}{dt} , \quad a^\alpha(0) = a^\alpha_0 . \tag{B93}$$

We now compare the properties of the parallel displacement of a vector in the three-dimensional Euclidean space and in the Riemannian space, namely, on a surface. In Euclidean space, one can introduce a uniform system of coordinates for the whole space which is Cartesian and for which Γ^k_{ij} vanishes everywhere. Therefore, the vector components a^i are constant when undergoing parallel displacement. In Riemannian space, we have $\Gamma^k_{ij} \neq 0$ and the vector components a^i change during parallel displacement. Another peculiarity is that the result of the parallel displacement does not only depend on the initial position of the vector and the initial and final position, but also on the displacement path. During parallel displacement of a vector along a closed contour around a surface, the vector turns by the angle

$$\Delta\varphi = \int_{\Sigma} K d\sigma \tag{B94}$$

with the surface Σ inside the contour and the total curvature K. For a rollable surface (cone, cylinder), we have $K = 0$ and $\Delta\phi = 0$. The common properties of the free vectors and the surface vectors are the conservation of the vector length and of the angle between the vectors during parallel displacement. From (B90)–(B93), it follows that the operations of absolute differentiation and parallel displacement belong to the internal geometry of the surface. Therefore, during the bending, a surface Σ is transformed into a surface Σ^*, a surface curve C into a surface curve C^*, the absolute differential $D\boldsymbol{a}$ of vector field $\mathbf{a(t)}$ along C into the absolute differential $D^*\boldsymbol{a}^*$ of a field $\boldsymbol{a}^*(t)$ along C^*, and the property of parallel displacement is conserved.

This means, that the operations of absolute differentiation and parallel displacement are invariant relative to the bending transformation (continuous deformation).

B.11
Geodetic Lines on a Surface

We consider a curve C on a surface Σ. During the bending of the surface, C is also bent and its curvature changes. It is possible to decompose the curvature into two components, one of which changes during bending while the second one remains unchanged, that is, is invariant. Since the first quadratic form is invariant during bending, the second component, which is called geodetic curvature, is part of the internal geometry. We now construct in the point M a unit tangential vector $\boldsymbol{t} = \dot{\boldsymbol{r}}$ on the surface curve C, then tangential line MT, unit normal vector $\boldsymbol{m}$ and the unit vector $\mathbf{u} = \boldsymbol{m} \times \boldsymbol{t}$ (see Figure B.12).

The vector $\boldsymbol{u}$ lies in the tangential plane, the main normal $\boldsymbol{n}$ lies in the osculating plane and the curvature vector $\dot{\boldsymbol{t}} = K\boldsymbol{n}$ (see (B30)) is directed toward the center of curvature of the curve. We now decompose the curvature vector in components along the vectors $\boldsymbol{m}$ and $\boldsymbol{u}$. The first component is called the normal curvature vector and the second one, $(K\boldsymbol{n})_g$, the vector of geodetic curvature (see Figure B.12). The length of the normal curvature vector is called normal curvature and the length of the vector of geodetic curvature is called geodetic curvature. The normal curvature vector coincides with the curvature vector of the normal section with the same

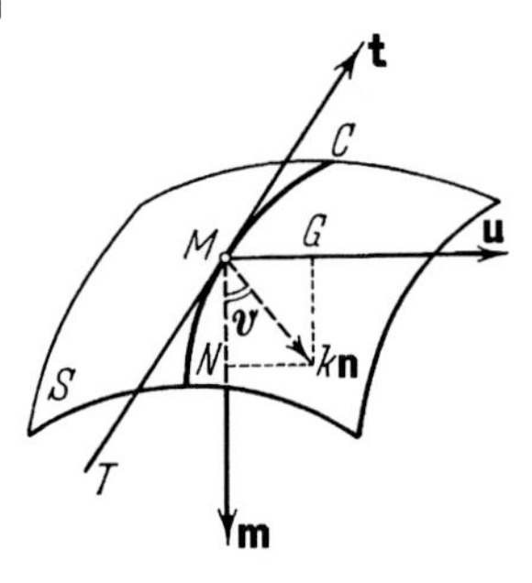

Figure B.12 Geodetic line on a surface.

tangent MT. Therefore, the normal curvature is equal to the curvature K_0 of the normal section of the curve C in the point M (see (B64)). The normal curvature changes during the bending.

The vector of the geodetic curvature coincides with the orthogonal projection of the curvature vector $K\boldsymbol{n}$ on the tangential vector of the surface in the point M. According to the first Frenet formula (B30), we have $K\boldsymbol{n} = \dot{\boldsymbol{t}} = \ddot{\boldsymbol{r}}$. Since the curve is part of the surface, we have $\boldsymbol{r} = \boldsymbol{r}(U^1, U^2)$, $U^\alpha = U^\alpha(S)$ and

$$\boldsymbol{t} = \dot{\boldsymbol{r}} = \frac{dU^\alpha}{dS}\boldsymbol{\varepsilon}^\alpha , \quad \ddot{\boldsymbol{r}} = \frac{\partial \boldsymbol{\varepsilon}^\alpha}{\partial U^\beta}\frac{dU^\alpha}{dS}\frac{dU^\beta}{dS} + \boldsymbol{\varepsilon}^\gamma \frac{d^2 U^\gamma}{dS^2} , \tag{B95}$$

where S is the arc length of the curve.

By substituting the first derivation formula (B80) for $\partial \boldsymbol{\varepsilon}^\alpha / \partial U^\beta$ into (B95) and projecting onto the tangential plane, one obtains the vector of the geodetic curvature

$$(K\boldsymbol{n})_g = \ddot{\boldsymbol{r}} = \left(\frac{d^2 U^\gamma}{dS^2} + \Gamma^\gamma_{\alpha\beta} \frac{dU^\alpha}{dS}\frac{dU^\beta}{dS} \right) \boldsymbol{\varepsilon}^\gamma . \tag{B96}$$

Taking into account the relationships $|\boldsymbol{t}| = 1, (K\boldsymbol{n})_g \perp \boldsymbol{t}$ and $K_g = |(K\boldsymbol{n})_g| = |\boldsymbol{t} \times (K\boldsymbol{n})_g|$, and evaluating the vector product of the first equation of (B95) with (B96), one obtains

$$K_g = \sqrt{a} \left| \frac{dU^1}{dS} \left(\frac{d^2 U^2}{dS^2} + \Gamma^2_{\alpha\beta} \frac{dU^\alpha}{dS}\frac{dU^\beta}{dS} \right) \right. \\ \left. - \frac{dU^2}{dS} \left(\frac{d^2 U^1}{dS^2} + \Gamma^1_{\alpha\beta} \frac{dU^\alpha}{dS}\frac{dU^\beta}{dS} \right) \right| . \tag{B97}$$

If the values of the curvilinear coordinates in every point of the surface stay unchanged during bending, then the right-hand side of (B97) does not change, that is, the geodetic curvature is an invariant of the bending.

We now consider the geodetic on a surface. A geodetic on the surface is a curve, for which the geodetic curvature vanishes in every point, that is $K_g = 0$ and the vector of the geodetic curvature $(K\boldsymbol{n})_g$ vanishes. This is possible if, and only if the curvature vector vanishes, ($K\boldsymbol{n} = 0, K = 0$, the curve is a straight line) or the curvature vector $K\boldsymbol{n}$ is directed along the normal to the surface. The geodetics for a sphere are the major circles since their normals pass through the center of the sphere and coincide with the normals of the sphere. The geodetics for a plane are straight lines. The differential equation of the geodetic $U^\alpha = U^\alpha(S)$ is obtained from (B96) and the condition $(K\boldsymbol{n})_g = 0$ as

$$\frac{d^2U^\gamma}{dS^2} = -\Gamma^\gamma_{\alpha\beta}\frac{dU^\alpha}{dS}\frac{dU^\beta}{dS}\,, \quad (\gamma = 1, 2) \tag{B98}$$

or

$$\frac{d^2U^\gamma}{dS^2} = -a^{\delta\gamma}\Gamma_{\alpha\beta,\delta}\frac{dU^\alpha}{dS}\frac{dU^\beta}{dS}\,. \tag{B99}$$

These equations are two ordinary differential equations of second order. Since the Christoffel symbols $\Gamma^\gamma_{\alpha\beta}$ depend on the coefficients of the first fundamental form $a_{\alpha\beta}$, one has to add to (B98) the equation $dS^2 = a_{\alpha\beta}dU^\alpha dU^\beta$, which makes the solution more difficult.

Therefore, one normally does not use the equation for the functions $U^\gamma(S)$, but for $U^2 = f(U^1)$. Then, (B97) for the geodetic curvature becomes

$$K_g = \sqrt{a}\frac{\left|f'' + \Gamma^2_{11} + (2\Gamma^2_{12} - \Gamma^1_{11})f' + (\Gamma^2_{22} - 2\Gamma^1_{12})f'^2 - \Gamma^1_{22}f'^3\right|}{\left(a_{11} + 2a_{12}f' + a_{22}f'^2\right)^{3/2}}\,. \tag{B100}$$

From the condition $K_g = 0$, one obtains the equation for the geodetic as

$$f'' = -\Gamma^2_{11} - \left(2\Gamma^2_{12} - \Gamma^1_{11}\right)f' + \left(-\Gamma^2_{22} + 2\Gamma^1_{12}\right)f'^2 + \Gamma^1_{22}f'^3\,. \tag{B101}$$

If at any point the values $(U^1)_0$, $(U^2)_0 = f[(U^1)_0]$ and $(dU^2/dU^1)_0$ are given, one obtains a solution that depends on the parameters of the points on the surface $(U^1)_0$, $(U^2)_0$ and on the parameter of the direction $(dU^2/dU^1)_0$. This means that through a given point on the surface in a given direction, only one geodetic $U^2 = f(U^1)$ passes. Therefore, one can define a two parameter family of geodetics on a surface. Given a certain direction $(dU^2/dU^1)_0$, one obtains a one parameter family of geodetics which is similar to a family of straight lines on the plane.

Since the geodetic curvature is invariant under bending, it will vanish after bending if it vanished before bending. This means that a family of geodetics on a surface will be transformed into a family of geodetics on a surface under a bending transformation. The geodetics are often used as a family of curvilinear coordinate lines on the surface. As the other family, one takes lines that are orthogonal to the geodetic curves. These curvilinear coordinates are called semigeodetic coordinates. We consider a semigeodetic system of coordinates U^2, U^1 such that the coordinate lines U^1 are the geodetics and the coordinate lines U^2 are orthogonal to the coordinate lines U^1.

The family of geodetics is described via $U^2 = f(U^1)$, where the function f satisfies the differential equation (B101). Along the geodetic, we have $U^2 = \text{const}$, and thus $f(U^1) = \text{const}$, $f'(U^1) = f''(U^1) = 0$. Then, it follows from (B100) that for a point on the geodetic $\Gamma^2_{11} = 0$. From the orthogonality of the coordinate lines U_1 and U_2, one obtains that the corresponding tangents are also orthogonal, that is, $a_{12} = a_{21} = \boldsymbol{\varepsilon}_1 \cdot \boldsymbol{\varepsilon}_2 = 0$. From (B81) and (B82), it follows that $\Gamma_{11,2} = 0$ and $\partial a_{11}/\partial U^2 = 0$. Thus, we have $a_{11} = a_{11}(U^1)$ and the first fundamental form takes on the shape

$$dS^2 = a_{11}(U^1)(dU^1)^2 + a_{22}(U^1, U^2)(dU^2)^2 \,.$$

In semigeodetic coordinates, the expression for the main characteristics of the surface are simplified. The Christoffel symbols of the first and the second kind are

$$\Gamma_{11,1} = 0\,, \quad \Gamma_{12,1} = 0\,, \quad \Gamma_{22,1} = -\frac{1}{2}\frac{\partial a_{22}}{\partial U^1}\,,$$
$$\Gamma_{11,2} = 0\,, \quad \Gamma_{12,2} = \frac{1}{2}\frac{\partial a_{22}}{\partial U^1}\,, \quad \Gamma_{22,2} = \frac{1}{2}\frac{\partial a_{22}}{\partial U^1}\,,$$
$$\Gamma^1_{11} = 0\,, \quad \Gamma^1_{12} = 0\,, \quad \Gamma^1_{22} = -\frac{1}{2}\frac{\partial a_{22}}{\partial U^1}\,,$$
$$\Gamma^2_{11} = 0\,, \quad \Gamma^2_{12} = \frac{1}{2a_{22}}\frac{\partial a_{22}}{\partial U^1}\,, \quad \Gamma^2_{22} = \frac{1}{2a_{22}}\frac{\partial a_{22}}{\partial U^2}\,.$$

The total curvature K has the form

$$K = -\frac{1}{\sqrt{a_{22}}}\frac{\partial^2 \sqrt{a_{22}}}{\partial (U^1)^2}\,.$$

The differential equation of the geodetics is also simplified to

$$f'' = -\frac{1}{a_{22}}\frac{\partial a_{22}}{\partial U^1} f' - \frac{1}{2a_{22}}\frac{\partial a_{22}}{\partial U^2} f'^2 - \frac{1}{2}\frac{\partial a_{22}}{\partial U^1} f'^3\,.$$

The geodetics possess extraordinary properties. The first of these follows from the definition of the geodetic: $(K\boldsymbol{n})_g = 0$. Since the geodetic curvature is equal to the projection of the curvature vector $K\boldsymbol{n} = d\boldsymbol{t}/dS$ on the tangential plane and since the projection of the vector $\boldsymbol{t}$ on the same plane is equal to the absolute differential $D\boldsymbol{t}$, the tangential vector $\boldsymbol{t}$ undergoes parallel displacement along the geodetic. The converse is also true: if the tangential vector undergoes parallel displacement along a curve, then this curve is a geodetic.

The second property is that among all curves that connect to given points on a surface, the geodetic has the minimum length. To prove this, we consider a curve $U^\alpha(t)$ on the surface that connects two points at $t = a$ and $t = b$. The arc length L of the curve between these two points is given by

$$L = \int_a^b \left(a_{\alpha\beta}\dot{U}^\alpha \dot{U}^\beta\right)^{1/2} dt\,, \quad \dot{U}^\alpha = \frac{dU^\alpha}{dt}\,, \quad \dot{U}^\beta = \frac{dU^\beta}{dt}\,. \tag{B102}$$

We introduce the function $\Phi^2(U, \dot{U}) = a_{\alpha\beta}\dot{U}^\alpha \dot{U}^\beta$ and consider other curves with the equations $U^\alpha = U^\alpha(t) + \varepsilon V^\alpha(t)$, where $V^\alpha(a) = V^\alpha(b) = 0$. The arc length of the curve between the points $t = a$ and $t = b$ is given by

$$L = \int_a^b \Phi\left[U^\alpha(t) + \varepsilon V^\alpha(t), \dot{U}^\alpha(t) + \varepsilon \dot{V}^\alpha(t)\right] dt \,.$$

We assume that the curve $U^\alpha(t)$ has minimum length L among the curves defined above. The condition for the minimum has the following form

$$\left(\frac{dL}{d\varepsilon}\right)_{\varepsilon=0} = \int_a^b \left(\frac{\partial \Phi}{\partial U^\alpha} V^\alpha + \frac{\partial \Phi}{\partial \dot{U}^\alpha} \dot{V}^\alpha\right) dt = 0 \,.$$

After applying integration by parts and taking into account the boundary conditions $V^\alpha(a) = V^\alpha(b) = 0$, one obtains the following differential equation for Φ:

$$\frac{d}{dt}\left(\frac{\partial \Phi}{\partial \dot{U}^\alpha}\right) - \frac{\partial \Phi}{\partial U^\alpha} = 0 \,.$$

From the following properties of the function Φ:

$$2\Phi \frac{\partial \Phi}{\partial U^\alpha} = \frac{\partial a_{\beta\gamma}}{\partial U^\alpha} \dot{U}^\beta \dot{U}^\gamma$$

and

$$\Phi \frac{\partial \Phi}{\partial \dot{U}^\alpha} = a_{\alpha\beta} \dot{U}^\beta ,$$

we obtain that

$$\frac{1}{\Phi}\left(a_{\alpha\beta} \ddot{U}^\beta + \frac{\partial a_{\alpha\beta}}{\partial U^\gamma} \dot{U}^\beta \dot{U}^\gamma\right) - \frac{a_{\alpha\beta} \dot{U}^\beta}{\Phi^2} \frac{d\Phi}{dt} - \frac{1}{2\Phi} \frac{\partial a_{\beta\gamma}}{\partial U^\alpha} \dot{U}^\beta d\dot{U}^\gamma = 0 \,.$$

This equation is valid for arbitrary parameters t. By choosing the arc length S as parameter, we then have, according to (B79),

$$\Phi^2(U, \dot{U}) = \sqrt{a_{\alpha\beta} \dot{U}^\alpha \dot{U}^\beta} = 1 \,.$$

As a result, one obtains the following equation for $U^\alpha(t)$:

$$a_{\alpha\beta} \ddot{U}^\beta + \left(\frac{\partial a_{\alpha\beta}}{\partial U^\gamma} - \frac{1}{2} \frac{\partial a_{\beta\gamma}}{\partial U^\alpha}\right) \dot{U}^\beta \dot{U}^\gamma = 0 \,.$$

With the help of (B81), one can show that this equation coincides with the equation for the geodetic (B98)

$$\frac{d^2 U^\alpha}{dS^2} = -\Gamma^\alpha_{\beta\gamma} \frac{dU^\beta}{dS} \frac{dU^\gamma}{dS} \,. \tag{B103}$$

B.12
Vector Fields on the Surface

If a surface vector $\boldsymbol{a}$ is given in every point of the surface with the coordinates $a^{\alpha} = \quad a^{\alpha}(U^1, U^2)$, then one says that a vector field is given on the surface. In the case when a scalar field $\varphi(U^1, U^2)$ is given on the surface, the partial derivatives $\varphi_{,\alpha}(U^1, U^2) = \nabla_{\alpha}\varphi = \partial\varphi/\partial U^{\alpha}$ are the components of the covariant vector gradient that corresponds to the vector field $\varphi_{,\alpha}(U^1, U^2)$. Using the contravariant metric tensor $a^{\alpha,\beta}$, one can obtain the field of the contravariant vector gradient $\varphi^{,\beta} = a^{\alpha\beta}\varphi_{,\alpha}$.

We consider a curve $U(S)$ on the surface. The potential ϕ along this curve is a function of S. Then, we have

$$\frac{d\varphi}{dS} = \frac{\partial\varphi}{\partial U^{\alpha}}\frac{dU^{\alpha}}{dS} = \varphi_{,\alpha}\frac{dU^{\alpha}}{dS} \,.$$

We denote by $\boldsymbol{a} = a^{\beta}\boldsymbol{\varepsilon}_{\beta}$ a surface vector whose projection on the tangential vectors $\boldsymbol{\varepsilon}_{\alpha}$ to the coordinate lines U^1 and U^2 are $\varphi_{,\alpha} = \boldsymbol{a}\cdot\boldsymbol{\varepsilon}_{\alpha}$.

Since $\boldsymbol{\varepsilon}_{\alpha}\cdot\boldsymbol{\varepsilon}_{\beta} = a_{\alpha\beta}$, one obtains $a_{\alpha\beta}a^{\beta} = \varphi_{,\alpha}$, $a^{\alpha} = a^{\alpha\beta}\varphi_{,\beta}$ and

$$\frac{d\varphi}{dS} = \varphi_{,\alpha}\frac{dU^{\alpha}}{dS} = \boldsymbol{a}\cdot\boldsymbol{\varepsilon}_{\alpha}\frac{dU^{\alpha}}{dS} = \boldsymbol{a}\cdot\boldsymbol{t} \,, \tag{B104}$$

where $\boldsymbol{t}$ is the unit tangential vector for the curve C. From (B104), it follows that the projection of the vector $\boldsymbol{a}$ on the tangent of the curve C is equal to the derivative of the potential φ along the curve. If one takes as C the equipotential line of the potential $\varphi(U^1, U^2) = 0$, we have $d\varphi/dS = 0$ and $\boldsymbol{a}\cdot\boldsymbol{t} = 0$ along C. This means that $\boldsymbol{a}\perp\boldsymbol{t}$, that is, the vector $\mathbf{a}$ is in every point orthogonal to the equipotential line.

If we assume C to be a curve, that is, a tangential to the vector $\boldsymbol{a}$ in a given point, and denote the arc length in the direction of $\boldsymbol{a}$ by S, we have $\boldsymbol{t}\cdot\mathbf{a} = |\boldsymbol{a}|$ and $d\varphi/dS = |\boldsymbol{a}|$. Therefore, $|\boldsymbol{a}|$ is the velocity of potential change in the direction orthogonal to the equipotential line since $d\varphi/dS > 0$, φ grows in the direction S, that is, in the direction $\boldsymbol{a}$. Therefore, in every point, a vector $\boldsymbol{a}(U^1, U^2)$ exists which is tangential to the surface along the normal to the equipotential line in the direction of growth of φ and whose absolute value is $d\varphi/dS$ (see Figure B.13).

The vector field $\boldsymbol{a}(U^1, U^2)$ defined in this way is called the gradient field of the scalar field $\varphi(U^1, U^2)$ and is designated by

$$\boldsymbol{a} = \nabla\varphi = \varphi_{,\alpha}{}^{"\alpha} \,. \tag{B105}$$

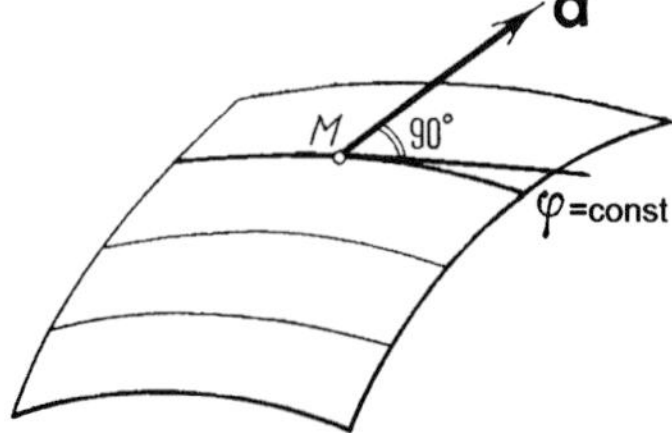

Figure B.13 Equipotential lines.

Since the vector $\boldsymbol{a}$ only depends on the first fundamental form, the scalar field $\varphi(U^1, U^2)$ and the corresponding vector field $\boldsymbol{a}(U^1, U^2)$ remain unchanged during the bending of the surface.

We now consider two scalar fields $\varphi(U^1, U^2)$ and $\psi(U^1, U^2)$ on the surface. Two of these scalar fields correspond gradient fields $\boldsymbol{a}(U^1, U^2)$ and $\boldsymbol{b}(U^1, U^2)$, which can be decomposed into the tangential vectors of the local basis: $\boldsymbol{a} = a^\alpha \boldsymbol{\varepsilon}_\alpha$ and $\boldsymbol{b} = b^\beta \boldsymbol{\varepsilon}_\beta$.

By evaluating the scalar product of these vectors, we obtain

$$\boldsymbol{a} \cdot \boldsymbol{b} = a^\alpha b^\beta \boldsymbol{\varepsilon}_\alpha \cdot \boldsymbol{\varepsilon}_\beta = a_{\alpha\beta} a^\alpha b^\beta = a^{\alpha\beta} \varphi_{,\alpha} \psi_{,\beta} = \nabla(\varphi, \psi) .$$

This relationship is called the mixed differential parameter of Beltrami.

For $\varphi = \psi$, one obtains

$$\nabla(\varphi, \varphi) = a^{\alpha\beta} \varphi_{,\alpha} \varphi_{,\beta} , \tag{B106}$$

which is referred to as the first differential parameter of Beltrami.

According to (A109), the divergence of the surface vector is given by

$$\nabla \cdot \boldsymbol{a} = a^\alpha_{.\alpha} = \frac{1}{\sqrt{a}} \frac{\partial}{\partial a^\alpha} \left(\sqrt{a} a^\alpha\right) . \tag{B107}$$

If one transforms the contravariant vector a^α into the covariant vector a_α according to the formula $a^\beta = a_\alpha a^{\alpha\beta}$ and assumes that a_α is a gradient vector, it follows from (B107) that

$$\nabla \cdot \boldsymbol{a} = \nabla \cdot (\nabla \varphi) = \nabla^2 \varphi = \Delta \varphi = a^{\alpha\beta} \varphi_{,\alpha\beta} = \frac{1}{\sqrt{a}} \frac{\partial}{\partial a^\beta} \left(\sqrt{a} a^{\alpha\beta} \varphi_{,\alpha}\right) . \tag{B108}$$

This expression provides the Laplace operator on the surface.

To obtain the curl of the surface vector $\nabla \times \boldsymbol{a}$, one has to use (A111). Since the covariant vector a_α, $(\alpha = 1, 2)$ only has two components, the vector $\nabla \times \boldsymbol{a}$ only has one non-vanishing component

$$\varepsilon^{\alpha\beta} a_{\beta,\alpha} = \frac{1}{\sqrt{a}} \left(\frac{\partial a_2}{\partial \alpha^1} - \frac{\partial a_1}{\partial \alpha^2} \right) . \tag{B109}$$

In analogy with (A141) and (A142), one can formulate the corresponding integral theorems for surface vectors. Gauss theorem and Stokes theorem take on the form

$$\int_\Sigma a^\alpha_{.\alpha} d\sigma = \int_C a^\alpha n_\alpha dS , \quad \int_\Sigma \varepsilon^{\alpha\beta} a_{\beta,\alpha} d\sigma = \int_C a_\beta t^\beta dS , \tag{B110}$$

where $t^\beta = t_\beta a^{\beta\alpha}$ and t_β are the components of the unit tangential vectors for the curve C, and $n_\alpha = \varepsilon_{\alpha\beta} t^\beta$ are the components of the unit normal vector fur the curve on the surface Σ.

From (B110), it follows that

$$\int_\Sigma a^{\alpha\beta} \varphi_{,\alpha\beta} d\sigma = \int_C \varphi_{,\alpha} n_\alpha dS . \tag{B111}$$

B.13
Hybrid Tensors

During the investigation of surfaces and their geometrical objects, two systems of coordinates are used. The first system of coordinates (X^1, X^2, X^3) is a system that is fixed in the three- dimensional space that surrounds the surface and can be Cartesian or curvilinear. The second system of coordinates (U^1, U^2) is bound to the surface and can be introduced in every point of the surface. Therefore, it is called the local system of coordinates and can be considered as a movable system of coordinates.

The local coordinate system consists of three basis vectors, two of which, namely, $\boldsymbol{t}_1$ and $\boldsymbol{t}_2$, lie in the tangential plane and the third is the normal to the surface $\boldsymbol{m}$. The set of vectors in space contains three groups of vectors: the surface vectors that lie on the tangential plane to the surface, free space vectors and hybrid tensors that are fixed to the surface in one point, but do not lie in the tangential plane. These vectors have non-vanishing normal components.

The same applies to tensors. There are also three groups of tensors. The surface tensors, that can be expressed via the dyads $\boldsymbol{t}_1\mathbf{t}_2$, space tensors, that are written in terms of the dyads $\boldsymbol{e}_i\mathbf{e}_j$ $(i, j = 1, 2, 3)$ of the basis vectors of a fixed system of coordinates, and the hybrid tensors, that are expressed in terms of the dyads $\boldsymbol{e}_i\mathbf{e}_j$ and $\boldsymbol{t}_1\mathbf{t}_2$. The tensors are invariant objects whose components are transformed according to the transformation laws (B77) for the surface tensors and (A47) for the space tensors. The components of the hybrid tensors are transformed according to (B77) or (A47), depending on which group they belong to.

The equation of a surface in three-dimensional space can be written in parametric form as

$$X^i = X^i(U^1, U^2)\,, \quad (i = 1, 2, 3)\,, \tag{B112}$$

where U^1, U^2 are parameters that can be considered as curvilinear coordinates on the surface.

We assume that the functions $X^i(U^1, U^2)$ are continuous and sufficiently differentiable. One can consider (B112) as three algebraic equations for two unknowns U^1 and U^2. By eliminating the parameters U^1 and U^2 from (B112), one obtains the usual equation for a surface in space

$$X^3 = X^3(X^1, X^2)\,. \tag{B113}$$

It must be noted that (B112) cannot be considered as a transformation law since it cannot be inverted. As previously, we denote the space object by Latin indices and the surface objects by Greek indices.

We now consider the expressions $X^i_\alpha = \partial X^i/\partial U^\alpha$. During the transformation of the space system of coordinates X^i to another system of coordinates $X^{i'}$ for a fixed index α, these expressions are transformed according to the space tensor law of contravariant components as $X^{i'}_\alpha = \partial X^{i'}/\partial X^i X^i_\alpha$ and during the transformation of the surface system of coordinates for a fixed index i according to the surface tensor law for covariant components as $X^i_{\alpha'} = \partial U^\alpha / \partial U^{\alpha'} X^i_\alpha$. Therefore, X^i_α are components of a hybrid tensor that connect the space and surface systems of coordinates.

The three-dimensional Euclidean space is characterized by the metric tensor g_{ij}, while the surface is characterized by the metric surface tensor $a_{\alpha,\beta}$. The square of the differential arc length can be written both in space coordinates as well as in plane coordinates:

$$d S^2 = a_{\alpha\beta} d U^\alpha d U^\beta = g_{ij} d X^i d X^j = g_{ij} X^i_\alpha X^j_\beta d U^\alpha d U^\beta \,.$$

This yields the following relationship between the space tensor and the metric surface tensor:

$$a_{\alpha\beta} = g_{ij} X^i_\alpha X^j_\beta \,. \tag{B114}$$

In Section B.6, it was shown that the space radius vector $d\boldsymbol{r}$ with coordinates $d X^i = X^i_\alpha d U^\alpha$ is a tangential vector on the surface $\boldsymbol{r}(U^1, U^2)$. Therefore, the space vector $a^i = X^i_\alpha a^\alpha$ that corresponds to the surface vectors a^α is also a tangential vector on the surface. Taking the two surface vectors $\boldsymbol{\varepsilon}_1 = \partial \boldsymbol{r}/\partial U^1$ and $\boldsymbol{\varepsilon}_2 = \partial \boldsymbol{r}/\partial U^2$, which are tangential to the surface coordinate lines U^1 and U^2, the normal vector $\boldsymbol{m}$ to the surface is equal to

$$m_i = \frac{1}{2} \varepsilon^{\alpha\beta} \varepsilon_{ijk} \frac{\partial X^j}{\partial U^\alpha} \frac{\partial X^k}{\partial U^\beta} \,. \tag{B115}$$

This vector is a space vector. The covariant derivative of a hybrid tensor T^i_α is given by

$$\nabla_\beta T^i_\alpha = T^i_{\alpha,\beta} = \frac{\partial T^i_\alpha}{\partial U^\beta} - \Gamma^\gamma_{\alpha\beta} T^i_\gamma + \Gamma^i_{jk} T^j_\alpha X^k_\beta \,. \tag{B116}$$

Appendix C
Foundations of Probability Theory

C.1
Events and Set of Events

Many processes in nature and technology are subject to chance. This means that it is not always possible to predict what result the process will have. However, it turns out that one can man make predictions about such processes if one has observed a sufficiently large number of them under constant conditions. The theory of probability provides mathematical models for random phenomena in reality. A random experiment or a trial is a process which has different possible results, such that one can not predict which of these results will occur. An experiment is also characterized by the property that it can be repeated, at least in principle, for an arbitrary number of times. Of special significance, are the possible mutually exclusive results of an experiment. The mutually exclusive results of an experiment are called its elementary events. The set of elementary events will be denoted by E.

In addition to elementary events, one is often also interested in events of a more complex nature, for example, a particle is located in a small (elementary) volume around a point $\boldsymbol{X}$ or N particles are within a small volume around a point $\boldsymbol{X}$ or N_1 particles of one type and N_2 particles of a second type are within a small volume around a point $\boldsymbol{X}$. These cases represent concrete realizations of the events.

A mathematical model for the general notion of an event is provided by the following definition. A certain experiment is given and E is the set of its elementary events. Every subset $A \subseteq E$ is then called an event. An event A occurs exactly if one of the elementary events which make up U occurs. Also, the improper subset of E, namely, E itself and the empty set $\emptyset$ are interpreted as events in line with the general definition. Since E consists of all elementary events, and for every experiment, exactly one elementary event occurs, E always occurs. Therefore, E is called the certain event. The empty set $\emptyset$ does not contain an elementary event and therefore never occurs. It is also called the impossible event.

Since the events are considered as subsets of the set E, the set operators can be applied both to sets as well as to the events:

Hydromechanics. Theory and Fundamentals. Emmanuil G. Sinaiski

ISBN: 978-3-527-41026-2

1. If $A_1, A_2, \ldots, A_n$ are events, then the union $A_1 \cup A_2 \cup \ldots \cup A_n$ is again an event since it is a subset of E. This event is called the sum of events $A_1, A_2, \ldots, A_n$ and occurs exactly if at least one of the events occurs.
2. In the same manner, $A_1 \cap A_2 \cap \ldots \cap A_n$ is also an event. This event is called the product of the events $A_1, A_2, \ldots, A_n$ and occurs exactly if all A_i occur at the same time. Often, it is also denoted by $A_1 A_2 \ldots A_n$.
3. Two events A_1 and A_2 are called exclusive if $A_1 \cap A_2 = \emptyset$, that is, if A_1 and A_2 cannot occur simultaneously. If A is an event, then the complement $\bar{A} = E \backslash A$ is also a subset of E, that is, an event that occurs exactly if A does not occur. $\bar{A}$ is called the complementary event to A or, in short, the complement of A such that A and $\bar{A}$ are always exclusive events ($A \cap \bar{A} = \emptyset$).

Therefore, every event is the element of a certain space. In the following, we will use the vector spaces that are given by the vectors $\boldsymbol{X}(X_1, X_2, \ldots, X_n)$, where X_i are real numbers.

For example, we consider the set $A(\Delta V, \boldsymbol{X})$ of events that consist of a particle being inside of a certain volume element ΔV with its center at the point $\boldsymbol{X}$. It turns out that one can decide whether a particle is in the vicinity of a point $\boldsymbol{X}$, but it is impossible to decide whether the particle is exactly at the point $\boldsymbol{X}$. By denoting the event by $\omega(\boldsymbol{x})$, that consists of a particle being inside of the volume element δV with its center at $\boldsymbol{X}$, one can ascribe a certain probability to the condition $\omega(\boldsymbol{X}) \subseteq A(\Delta V, \boldsymbol{X})$ which has the meaning of the probability, that is, the particle is within the volume element ΔV with its center in the point $\boldsymbol{X}$.

C.2 Probability

The probability $P(A)$ with respect to the set of events is introduced as a function of an event A that satisfies the following axioms:

1. $P(A) \geq 0$ for all A;
2. $P(E) = 1$;
3. If A_i is a finite or countable sequence of non-intersecting or disjunct sets, that is, $A_i \cap A_j = \emptyset$ unless $i = j$, then we have

$$P\left(\bigcup_{i=1}^{n} A_i\right) = \sum_{i=1}^{n} P(A_i) .$$

From these axioms, the relationships

4. $P(\bar{A}) = 1 - P(A)$;
5. $P(\emptyset) = 0$; follow.

Together with the probability of an event, one introduces the concept of relative frequency of a random event. To understand the difference between probability and relative frequency, we consider an arbitrarily chosen event $\omega \subseteq A$. By repeating the experiment N times, the event ω occurs M times. Then, the relative frequency is given by $\nu = M/N$. If one increases the number of experiments N, the relative frequency approaches the probability $P(A)$ such that for $N \gg 1$, the relative frequency can be considered to be equivalent to the probability without large error.

C.3
Common and Conditional Probability, Independent Events

Axiom three is applicable to the case of a finite or countably infinite number of sets. However, one often deals with an infinite number of sets, for example, with the set of particle position in space during the movement of the particle under the influence of external forces. We denote the position of a particle in space by $\boldsymbol{X}$. The probability of finding the particle exactly at point $\boldsymbol{X}$, namely, in a set with one element, vanishes, as noted above, but the probability of the particle being in a small vicinity ΔV of the point $\boldsymbol{X}$ does not vanish. The volume ΔV constitutes the union of an infinite number of sets of the type $\boldsymbol{X}$. If one tries to use Axiom three with the case under consideration, one obtains an undetermined quantity of the type $0 \cdot \infty$ since $P(\boldsymbol{X}) = 0$ and $N = \infty$. Therefore, axiom three cannot be used and the probability of presence of the particle in the vicinity ΔV is not eqal to the sum of probabilities of occurrence for all sets of the type $\boldsymbol{X}$ inside of ΔV.

Axiom three is applicable to finite and mutually exclusive events, that is, events that belong to disjunct sets. We now consider the case of intersecting sets and of events that can be part of two or more sets at the same time. Such events are called common events. We consider two sets A and B whose intersection $A \cap B$ is not empty. The event $\omega \subseteq A \cap B$ if $\omega \subseteq A$ and $\omega \subseteq B$. Then, the probability of the common event is written as

$$P(A \cap B) = P(\omega \subseteq A \text{ and } \omega \subseteq B) . \tag{C1}$$

Two situations will serve as an example of common events. The first event is defined by having particles of the first kind and N_2 particles of the second kind inside of the volume element ΔV around a point $\boldsymbol{X}$ N_1. The probability of this event is the common probability of the two events. The second example consists of finding at time t_1 within the volume element ΔV around a point $\boldsymbol{X}$ N_1 particles of the first kind and N_2 particles of the second kind and at time t_2 N_1 particles of the first kind and n_2 particles of the second kind. The probability of this event is the common probability of both events at times t_1 and t_2.

Sometimes, one has to consider events that occur under certain conditions. For example, one is interested in the event that consists of the the particle being inside the volume element ΔV around a point $\boldsymbol{X}$ at time t_1 under the condition that it was inside of the volume element ΔV_0 around the point $\boldsymbol{X}$ at time t_0. In reality, one has to consider the set of all events C that consist of the particle being inside of the volume element ΔV at time t. However, the particle can arrive in this volume from arbitrary initial positions with different probabilities.

Therefore, one is not interested in all events, but in a part of the events in set C. The probability of this event is called conditional probability. It is defined as the probability of an event ω, with $\omega \subseteq A$ under the condition $\omega \subseteq B$ and is given by

$$P(A|B) = P(A \cap B)/P(B) . \tag{C2}$$

The theory of stochastic processes is founded on the concept of the probability of common events. In this context, we show an important property of common events. We consider some sets B_i that subdivide E into subsets such that

$$B_i \cap B_j = \emptyset \quad \text{and} \quad \bigcup_i B_i = E .$$

One can show that we have

$$\bigcup_i (A \cap B_i) = A \cap \left(\bigcup_i B_i \right) = A \cap E = A .$$

By axiom three, one obtains

$$\sum_i P\,(A \cap B_i) = P \left[\bigcup_i \left(A \bigcup_i B_i \right) \right] = P(A) .$$

Then, it follows from (C2) that

$$\sum_i P\,(A|B_i)\; P(B_i) = P(A) . \tag{C3}$$

Therefore, the summation of all mutually exclusive possibilities (sets B_i) removes these from the consideration.

In the following, the concept of independent events will play an important role. Two random events A and B are independent of each other if the occurrence of one of them does not influence the probability of the other one, that is,

$$P\,(A|B) = P(A) . \tag{C4}$$

Equation C2 then becomes

$$P\,(A \cap B) = P(A) P(B) . \tag{C5}$$

C.4
Random Variables

The concept of a random variable plays an important role in the theory of stochastic processes. The random variable $F(\boldsymbol{X})$ is defined as a function in the space of random events $\boldsymbol{X}$. In the following, we will denote the event by $\boldsymbol{X}$. Examples of random variables are the position, the momentum and the orientation of a particle in space during its movement under the influence of a random external forces (Brownian motion, turbulent flows). The concept of a random variable simplifies the operations with functions of random variables, for example, the calculation of the distribution of random variables, of mean values and other statistical characteristics of the distribution of random variables. The introduction of random variables allows us to consider stochastic differential equations in order to investigate the change of random variables in space and time, as it is usually done for deterministic systems using differential equations.

C.5
Distribution of Probability Density and Mean Values

The probability of occurrence of a certain event lies between zero and one. If the events are mutually exclusive, then the sum of all probabilities is one. Therefore, at least one of the events occurs.

For the application in statistical mechanics, continuous random variables, namely, variables that have a continuous set of values, are of paramount interest. Therefore, the probability to obtain an arbitrary value inside of the continuum of possible values is equal to zero, but the total sum of probabilities is one. Therefore, one does consider the probability that a variable has a fixed value, though, the probability that the value of a variable is within an infinitesimal interval/volume close to a fixed value. The probability of this event is an infinitesimal quantity of the same order as the length of the interval or the measure of the set $d\boldsymbol{X}$. Therefore, one can introduce the probability $p(x)d\boldsymbol{X}$ that the random variable lines in the interval $\boldsymbol{X}, \boldsymbol{X} + d\boldsymbol{X}$, namely,

$$P\left[\boldsymbol{X} \in (\boldsymbol{X}, \boldsymbol{X} + d\boldsymbol{X})\right] = P(\boldsymbol{X})d\boldsymbol{X} . \tag{C6}$$

The function $p(\boldsymbol{X})$ is called a distribution of probability or probability density in short. Axiom three for the sum of probabilities can be written for the case of continuous random variables in the following form:

$$\int\limits_X P(\boldsymbol{X})\, d\boldsymbol{X} = 1 , \tag{C7}$$

where X is the general n-dimensional space of the random variable $\boldsymbol{X}$.

The probability density is also needed to define the statistical characteristics of the distribution of the random variable $\boldsymbol{X}$. The most important characteristic is the mean value of the random function $\langle f \rangle$

$$\langle f \rangle = \int\limits_X f(\boldsymbol{X})\, p(\boldsymbol{X}) d\boldsymbol{X} . \tag{C8}$$

If the vector $\boldsymbol{X}$ in a vector in a n-dimensional space, (C7) and (C8) can be represented in coordinate form as follows

$$\int_{-\infty}^{\infty}\int_{-\infty}^{\infty}\dots\int_{-\infty}^{\infty} p\,(X_1, X_2, \dots, X_n)\, dX_1 dX_2 \dots dX_n = 1\,,$$

$$\langle f \rangle = \int_{-\infty}^{\infty}\int_{-\infty}^{\infty}\dots\int_{-\infty}^{\infty} f\,(X_1, X_2, \dots, X_n)\, p\,(X_1, X_2, \dots, X_n)\, dX_1 dX_2 \dots dX_n\,.$$

The other important statistical characteristic is the dispersion σ^2 that characterizes the mean square deviation of the random function from its mean value which is given by

$$\sigma^2 = \int_X (f - \langle f \rangle)^2 p(\boldsymbol{X}) d\boldsymbol{X}\,. \tag{C9}$$

The probability density frequently has the form of function with a sharp maximum in a point $\boldsymbol{X} = \boldsymbol{X}_0$. In the limit, it is equal to ∞ for $\boldsymbol{X} = \boldsymbol{X}_0$ and vanishes for $\boldsymbol{X} \neq \boldsymbol{X}_0$. This situation occurs through the idealization of a process, for example, if a continuously distributed mass in a small volume element around a point $\boldsymbol{X}_0$ is replaced by a mass that is concentrated at the point $\boldsymbol{X}_0$. The the mass density is unequal to zero only at the point $\boldsymbol{X}_0$ and the integral of the form (C8) corresponds to the total mass in the region X. In order for the integral (C8) of such a function to exist, one has to extend the usual concept of a function which can be achieved via the introduction of generalized functions.

C.6
Generalized Functions

The simplest and most commonly used generalized function is the Dirac's delta function $\delta(\boldsymbol{X} - \boldsymbol{X}_0)$, which can be defined as the lines of the following sequence:

$$\delta(\boldsymbol{X} - \boldsymbol{X}_0) = \lim_{m\to\infty} \left(\frac{m}{\sqrt{\pi}}\right)^n \exp\left[-m^2 (\boldsymbol{X} - \boldsymbol{X}_0)^2\right]\,. \tag{C10}$$

The vector $\boldsymbol{X}$ has the components $X_1, X_2, \dots, X_n$, and therefore, one can write (C10) for every one-dimensional sequence that corresponds to $X_i - X_i^0$. Then, the equation

$$\delta(\boldsymbol{X} - \boldsymbol{X}_0) = \delta\left(X_1 - X_1^0\right) \delta\left(X_2 - X_2^0\right) \dots \delta\left(X_n - X_n^0\right) \tag{C11}$$

must be satisfied.

Since the limit of the right-hand side of (C10) is 0 for $\boldsymbol{X} \neq \boldsymbol{X}_0$ and equal to ∞ for $\boldsymbol{X} = \boldsymbol{X}_0$, a generalized function is different from a normal function that takes on a certain value at a certain point. However, for a generalized function, the scalar product between the generalized function $\delta(\boldsymbol{X}-\boldsymbol{X}_0)$ and an ordinary function $\phi(\boldsymbol{X}$ makes sense:

$$\begin{aligned}\left[\delta(\boldsymbol{X}-\boldsymbol{X}_0),\varphi(\boldsymbol{X})\right] &= \int_X \delta(\boldsymbol{X}-\boldsymbol{X}_0),\varphi(\boldsymbol{X})d\boldsymbol{X} \\ &= \lim_{m\to\infty}\int_X \left(\frac{m}{\sqrt{\pi}}\right)^n \exp\left[-m^2\left(\boldsymbol{X}-\boldsymbol{X}_0\right)^2\right]\varphi(\boldsymbol{X})d\boldsymbol{X} \\ &= \varphi(\boldsymbol{X}_0)\,.\end{aligned}$$

Therefore, we have

$$\int_X \delta(\boldsymbol{X}-\boldsymbol{X}_0)\varphi(\boldsymbol{X})d\boldsymbol{X} = \varphi(\boldsymbol{X}_0)\,. \tag{C12}$$

If $\phi(\boldsymbol{X}) = 1$, one obtains

$$\int_X \delta(\boldsymbol{X}-\boldsymbol{X}_0)d\boldsymbol{X} = 1\,. \tag{C13}$$

In the case $\phi(\boldsymbol{X}) = \boldsymbol{X}$, it follows from (C12) that

$$\int_X \delta(\boldsymbol{X}-\boldsymbol{X}_0)\boldsymbol{X}d\boldsymbol{X} = \boldsymbol{X}_0\,. \tag{C14}$$

Therefore, $\delta(\boldsymbol{X}-\boldsymbol{X}_0)$ is a probability density for which the random variable takes on the mean value $\boldsymbol{X}_0$. In the one-dimensional case, the following formal relationships apply:

$$(X-X_0)\delta(X-X_0) = 0\,, \quad X\delta(X-X_0) = X_0\delta(X-X_0)\,. \tag{C15}$$

Using $\varphi(\boldsymbol{X}) = (\boldsymbol{X}-\boldsymbol{X}_0)^2$, one obtains

$$\int_X \delta(\boldsymbol{X}-\boldsymbol{X}_0)(\boldsymbol{X}-\boldsymbol{X}_0)^2 d\boldsymbol{X} = (\boldsymbol{X}_0-\boldsymbol{X}_0)^2 = 0\,. \tag{C16}$$

Since the left-hand side of (C16) can be considered as the dispersion of the probability density $\delta(\boldsymbol{X}-\boldsymbol{X}_0)$, $\delta(\boldsymbol{X}-\boldsymbol{X}_0)$ has a vanishing dispersion. Therefore, a probability density in the form of a delta function describes a situation when $\boldsymbol{X} = \boldsymbol{X}_0$ applies with certainty.

Another one-dimensional sequence that converges to a delta function at the limit is

$$\delta(X-X_0) = \lim_{\varepsilon\to 0}\frac{\varepsilon}{\pi(X-X_0)^2+\varepsilon^2}\,. \tag{C17}$$

Another important property of the delta function is that it can be differentiated as many times as desired. The effect of the derivative of the delta function can be obtained by integration by parts of the integral $\int_X \varphi(X) d/dX \delta(X - X_0) dX$ and application of (C12)

$$\int\limits_X \delta'(X - X_0)\varphi(X) dX = -\int\limits_X \delta(X - X_0)\varphi'(X) dX = -\varphi'(X_0) . \tag{C18}$$

The derivatives of the one-dimensional delta function are given by

$$\delta^{(r)}(X) = (-1)^r r! \frac{\delta(X)}{X^r} , \quad (X \neq 0) . \tag{C19}$$

In the n-dimensional case, one introduces the concept of the vector gradient of the delta function which is defined as the generalized vector function

$$\frac{\partial \delta(\mathbf{X} - \mathbf{X}_0)}{\partial \mathbf{X}} = \left(\frac{\partial \delta}{\partial X_1}, \frac{\partial \delta}{\partial X_2}, \ldots, \frac{\partial \delta}{\partial X_n} \right) . \tag{C20}$$

C.7
Methods of Averaging

Many physical processes show the peculiarity of chaotic fluctuations. Therefore, the dependencies of the physical parameters on time and the space coordinates have a complicated and involved character. In addition, it turns out that repeating the experiments with identical initial and boundary conditions yields different values of the physical parameters. Therefore, when investigating random physical processes, one has to replace the unordered (chaotic) characteristics by the smoothed and regular average values of these characteristics using the method of averaging.

In practice, one uses time- and space averaging according to the formulae

$$\langle f(\mathbf{X}, t) \rangle_t = \frac{1}{T} \int\limits_{-T/2}^{T/2} \omega\,(\mathbf{X}, \tau)\, f\,(\mathbf{X}, t + \tau)\, d\tau \tag{C21}$$

and

$$\langle f(\mathbf{X}, t) \rangle_V = \frac{1}{V} \int\limits_V \omega\,(\mathbf{x}, t)\, f\,(\mathbf{X} + \mathbf{x}, t)\, d\mathbf{x} , \tag{C22}$$

with the weight function $\omega\,(\mathbf{x}, t)$.

The averaging can also be of a mixed nature in space and time:

$$\langle f(\mathbf{X}, t) \rangle_{V,T} = \frac{1}{VT} \int\limits_V \int\limits_{-T/2}^{T/2} \omega\,(\mathbf{x}, \tau)\, f\,(\mathbf{X} + \mathbf{x}, t + \tau)\, d\tau d\mathbf{x} . \tag{C23}$$

In some applied problems, the autocorrelation function is introduced and defined by

$$G(\tau) = \lim_{T \to \infty} \frac{1}{T} \int\limits_0^T f(t) f(t + \tau) dt , \tag{C24}$$

which is equal to the mean value of the product of the same fluctuating function taken at times t and $t + \tau$ at the same space point $\boldsymbol{X}$ over the time interval $(0, T)$.

The disadvantage of these three kinds of averaging is that it can only be used for one realization of a process. By repeating one and the same processes for identical boundary and initial conditions, one obtains many realizations of the process and its parameters. In this case, one speaks of the statistic ensemble of similar processes under the same boundary and initial conditions. By averaging the values of a parameter $u(\boldsymbol{X})$ that was obtained for all the experiments conducted, one obtains a mean value $\langle u(\boldsymbol{X}, t)\rangle$ which is called the average of probability theory over the ensemble. Sometimes, the averaging over the ensemble is sufficiently stable, that is, the results of the experiments only show a small deviation from the corresponding mean values for a sufficiently large number of experiments.

We assume that the random quantity u is characterized by the probability density $p(u)$. If one is interested in the value of $\boldsymbol{u}$ at one and the same point M, then $p(u)d(u)$ is the probability that the value u is part of the interval $u, u + du$. Therefore, the average value of the ensemble is equal to

$$\langle u(\boldsymbol{X}, t)\rangle = \int_{-\infty}^{\infty} u p(u) du \,. \tag{C25}$$

In a similar way, the mean value of an arbitrary function F is defined with respect to the ensemble as

$$\left\langle F\left[u(\boldsymbol{X}, t)\right]\right\rangle = \int_{-\infty}^{\infty} F(u) p(u) du \,. \tag{C26}$$

We now assume that the values of u are now measured at different times in different points, that is, the values $u_1, u_2, \ldots, u_n$ are given in N space time points $M_1(X_1, t_1), M_2(X_2, t_2), \ldots, M_N(X_N, t_N)$. By introducing an N-dimensional probability density, $p(u_1, u_2, \ldots, u_N)du_1, du_2, \ldots, du_N$ is the probability that the u_i lie in the intervals $(u_i, u_i + du_i)$. Thus, the mean value of an arbitrary function F is given by

$$\langle F\rangle = \int_{-\infty}^{\infty}\int_{-\infty}^{\infty} \cdots \int_{-\infty}^{\infty} F(u_1, u_2, \ldots, u_N) p(u_1, u_2, \ldots, u_N) du_1 du_2 \ldots du_N \,. \tag{C27}$$

By introducing the N-dimensional vector $\boldsymbol{u}$ with the components $u_1, u_2, \ldots, u_N$, one can write the expression (C27) in a more compact manner as

$$\langle F\rangle = \int_{-\infty}^{\infty} F(\boldsymbol{u}) p(\boldsymbol{u}) d\boldsymbol{u} \,. \tag{C28}$$

Multidimensional probability distributions are especially important for the investigation of a system of N particles in the random field of external forces. By denoting the coordinates of the i-th particle by u_i, $p(\boldsymbol{u})$ takes on the meaning of a multi-particle distribution of probability density. Of special interest for applications, are the one and two particle distributions. The latter is also sometimes referred to as pair distribution.

Multidimensional distributions of the probability density have to satisfy the following conditions:

1. $p(\boldsymbol{u}) \geqslant 0$;
2. $\int\limits_{-\infty}^{\infty} p(\boldsymbol{u})d\boldsymbol{u} = 1$;
3. $p(u_1, u_2, \ldots, u_N) = p(u_{i_1}, u_{i_2}, \ldots, u_{i_N})$, where $i_1, i_2, \ldots, i_N$ are permutations of the numbers $1, 2, \ldots, N$ in arbitrary order.
4. $p(u_1, u_2, \ldots, u_n) = \int\limits_{-\infty}^{\infty}\int\limits_{-\infty}^{\infty} \cdots \int\limits_{-\infty}^{\infty} p(u_1, u_2, \ldots, u_n, u_{n+1}, \ldots, u_N)$ $du_{N+1}du_{N+2} \ldots du_N, (n < N)$;
5. for independent random quantities $u_1, u_2, \ldots, u_N$, we have

$$p(u_1, u_2, \ldots, u_N) = p(u_1)p(u_2) \ldots p(u_N)\,. \tag{C29}$$

Properties three and four are referred to as properties of symmetry and coordination.

We finally discuss the relationship between the different methods of averaging. In practice, one uses either the time or the space averaging since the averaging over the ensemble requires a large number of experiments. In statistical mechanics, the average over the totality of all possible states is often replaced by time or space averaging. This is done based on the hypothesis that the mean values will converge against the mean values of the ensemble in the limit of increasing the averaging interval or volume to infinity. This assumption is called the ergodic hypothesis or, if it can be proven, the ergodic theorem.

C.8
Characteristic Function

Sometimes, one uses, instead of the probability density $p(u_1, u_2, \ldots, u_N)$, its Fourier transform

$$\varphi(\rho_1, \rho_2, \ldots, \rho_N) =$$

$$\int\limits_{-\infty}^{\infty}\int\limits_{-\infty}^{\infty} \cdots \int\limits_{-\infty}^{\infty} \exp\left\{i \sum_{k=1}^{N} \rho_k u_k\right\} p(u_1, u_2, \ldots, u_N)du_1 du_2, \ldots du_N$$

or in vector notation

$$\varphi(\boldsymbol{\rho}) = \int\limits_{-\infty}^{\infty} e^{i\boldsymbol{\rho}\cdot\boldsymbol{u}} p(\boldsymbol{u})d\boldsymbol{u}\,. \tag{C30}$$

Here, $\boldsymbol{\rho}$ is the N-dimensional vector with components $\rho_1, \rho_2, \ldots, \rho_N$. The function ϕ is called the characteristic function or the generating function for the moments. According to (C27), it can be written in the following form:

$$\varphi(\boldsymbol{\rho}) = \left\langle e^{i\boldsymbol{\rho}\cdot\boldsymbol{u}} \right\rangle . \tag{C31}$$

If the characteristic function is known, one can obtain the probability density by an inverse Fourier transform as

$$p(\boldsymbol{u}) = \frac{1}{(2\pi)^N} \int_{-\infty}^{\infty} e^{-i\boldsymbol{\rho}\cdot\boldsymbol{u}} \varphi(\boldsymbol{\rho}) d\boldsymbol{\rho} . \tag{C32}$$

This means that the knowledge of the characteristic function is equivalent to knowing the probability density.

The properties of the characteristic function can be easily obtained from the properties of the probability density. From property two, that is, the normalization of the probability density, one obtains

$$\varphi(\boldsymbol{0}) = 1 . \tag{C33}$$

For independent random variables, we have, according to (C29),

$$\varphi(\boldsymbol{\rho}) = \varphi(\rho_1)\varphi(\rho_2)\ldots\varphi(\rho_N) . \tag{C34}$$

From properties three and four, one obtains the conditions of symmetry and coordination for the characteristic function as

$$\varphi(\rho_1, \rho_2, \ldots, \rho_N) = \varphi(\rho_{i_1}, \rho_{i_2}, \ldots, \rho_{i_N}) , \tag{C35}$$

$$\varphi(\rho_1, \rho_2, \ldots, \rho_n) = \varphi(\rho_1, \rho_2, \ldots, \rho_n, \underbrace{0, 0, \ldots 0}_{N-n}) , \quad (n < N) , \tag{C36}$$

where $i_1, i_2, \ldots, i_N$ are an arbitrary permutation of $1, 2, \ldots, N$.

Property (C36) easily allows one to obtain the characteristic function for a smaller dimension from the one for N dimensions and finally the probability density via inverse Fourier transform. Therefore, one can immediately give the probability density in terms of a single function that is called the characteristic functional for the values of the random variable in all possible points. For a one-dimensional random function $u(\boldsymbol{X})$, which is given on the finite interval $a \leqslant C \leqslant b$, the characteristic functional is equal to

$$\Phi\left[\rho(X)\right] = \left\langle \exp\left\{ i \int_a^b \rho(X)u(X)dX \right\} \right\rangle , \tag{C37}$$

where ρ is a certain function that has to be chosen such that the integral in (C37) exists.

If one assumes

$$\rho(X) = \sum_{i=1}^{N} \rho_i \delta(X - X_i) ,$$

after integration, one obtains

$$\Phi\left[\rho(X)\right] = \left\langle \exp\left\{ i \sum_{k=1}^{N} \rho_k X_k \right\} \right\rangle = \varphi(\rho_1, \rho_2, \ldots, \rho_N) . \tag{C38}$$

Therefore, the characteristic function is transformed into the characteristic function of the multidimensional probability density for the random quantities $u(X_1), u(X_1), \ldots, u(X_N)$.

C.9
Moments and Cumulants of Random Quantities

To solve concrete problems, the multidimensional distributions of the probability density have to be given. This leads to the problem of defining these distributions with sufficient accuracy and using them further. Therefore, during the solution of applied problems, one is restricted to the consideration of simpler statistical parameters that describe only part of the statistical properties of the processes.

The most important parameters among these are the moments. We consider a system of N random quantities $u_1, u_2, \ldots, u_N$ with a N-dimensional probability density $p(u_1, u_2, \ldots, u_N)$. The following expressions

$$\begin{aligned} B_{k_1 k_2 \ldots k_N} &= \left\langle u_1^{k_1} u_2^{k_2} \ldots u_N^{k_N} \right\rangle \\ &= \int_{-\infty}^{\infty} \int_{-\infty}^{\infty} \cdots \int_{-\infty}^{\infty} u_1^{k_1} u_2^{k_2} \ldots u_N^{k_N} p(u_1, u_2, \ldots, u_N) du_1 du_2 \ldots du_N \end{aligned} \tag{C39}$$

are called the moments of the random quantities $u_1, u_2, \ldots, u_N$. Here, $k_1 k_2 \ldots k_N$ are non-negative integers whose sum $K = k_1 + k_2 + \cdots + k_N$ is called the degree of the moment. A special case involves the mean values of the quantities $u_1, u_2, \ldots, u_M$ which are given by the moments of first degree.

In addition to the ordinary moments (C39), some of their combinations are also used. Quite often, the central moments are used, that is, the moments of the differences between the random quantities and their mean values defined by

$$b_{k_1 k_2 \ldots k_N} = \left\langle (u_1 - \langle u_1 \rangle)^{k_1} (u_2 - \langle u_2 \rangle)^{k_2}, \ldots, (u_N - \langle u_N \rangle)^{k_N} \right\rangle . \tag{C40}$$

For $N = 1$ and $k_1 = 2$, we have $b_2 = \sigma^2$ (see (C9)). If u_1 is the velocity of the turbulent flow in the point X_i, the differences $u_i - \langle u_i \rangle$ describe velocity fluctuations in those points. Therefore, the central moments characterize the statistical properties of random velocity fluctuations. If the quantities describe the random properties of particles during their movement under the influence of random external forces, the central moments characterize the statistical properties of the particle movement in the field of random external forces.

By expanding the brackets in (C40) and using the definition (C39), one obtains the relationship between the central and the ordinary moments. For $N = 1$, we have

$$b_1 = 0\,, \quad b_2 = B_2 - B_1^2\,, \quad b_3 = B_3 - 3B_1B_2 + 2B_1^3\,,$$
$$b_4 = B_4 - 4B_1B_3 + 6B_1^2B_2 - 2B_2^4 \quad \text{and so on.} \tag{C41}$$

For $\langle u \rangle = 0$, the central and the ordinary moments coincide. The combinations of the central moments defined by

$$s = \frac{b_3}{b_2^{3/2}}\,; \quad \delta = \frac{b_4}{b_2^2} - 3 \tag{C42}$$

are statistical characteristics of the random quantities and are called asymmetry and excess respectively.

One can express the moments of the random quantities $u_1, u_2, \ldots, u_N$ by the corresponding characteristic function $\varphi(\rho_1, \rho_2, \ldots, \rho_N)$. By comparing (C39) and (C30), one obtains

$$B_{k_1k_2\ldots k_N} = (-i)^k \left.\frac{\partial^K \varphi(\rho_1, \rho_2, \ldots, \rho_N)}{\partial\rho_1^{k_1}\partial\rho_2^{k_2}\ldots\partial\rho_N^{k_N}}\right|_{\rho_1=\rho_2=\ldots=\rho_N=0} . \tag{C43}$$

This means that the moments are coefficients of the expansion of the characteristic function in the Taylor series

$$\varphi(\rho_1, \rho_2, \ldots, \rho_N) = \sum_{k_1k_2\ldots k_N} i^k \frac{B_{k_1k_2\ldots k_N}}{k_1!k_2!\ldots k_N!}\rho^{k_1}\rho^{k_2}\ldots\rho^{k_N} . \tag{C44}$$

If the moments are known, one can obtain, via (C46), the characteristic function and then form the probability density by (C32). This means that the moments define the distribution of the probability density uniquely.

Another combination of the central moments are the cumulants (semi-invariants) $S_{k_1k_2\ldots k_N}$.

We introduce the logarithm of the characteristic function which is referred to as the generating function of the cumulants:

$$\psi(\rho_1, \rho_2, \ldots, \rho_N) = \ln\left[\varphi(\rho_1, \rho_2, \ldots, \rho_N)\right] . \tag{C45}$$

The cumulants are the coefficients of the expansion of the generating function ψ into a Taylor series just as the moments are obtained as the coefficients of the

Taylor series (C44),

$$\psi(\rho_1, \rho_2, \ldots, \rho_N) = \sum_{k_1 k_2 \ldots k_N} i^k \frac{S_{k_1 k_2 \ldots k_N}}{k_1! k_2! \ldots k_N!} \rho^{k_1} \rho^{k_2} \ldots \rho^{k_N} , \tag{C46}$$

$$S_{k_1 k_2 \ldots k_N} = (-i)^k \left. \frac{\partial^K \psi(\rho_1, \rho_2, \ldots, \rho_N)}{\partial \rho_1^{k_1} \partial \rho_2^{k_2} \ldots \partial \rho_N^{k_N}} \right|_{\rho_1 = \rho_2 = \ldots = \rho_N = 0} . \tag{C47}$$

Since $\varphi(0, 0, \ldots, 0) = 1$, the cumulants can be expressed in terms of ordinary moments B_i and central moments b_i as

$$\begin{aligned} &S_1 = b_1 , \quad S_2 = B_2 - B_1^2 = b_2 , \quad S_3 = B_3 - 3 B_1 B_2 + 2 B_1^3 = b_3 , \\ &S_4 = B_4 - 4 B_1 B_3 - 3 B_2^2 + 12 B_1^2 B_2 - 6 B_1^4 = b_4 - 3 b_2^2 , \\ &S_5 = b_5 - 10 b_2 b_3 \quad \text{and so on,} \end{aligned} \tag{C48}$$

and vice versa, the moments via the cumulants as

$$B_1 = S_1 , \quad B_2 = S_2 + S_1^2 , \quad B_3 = S_3 + 3 S_1 S_2 + S_1^3 \quad \text{and so on.} \tag{C49}$$

In many cases, the cumulants are very convenient characteristics of the probability density. For a one-dimensional distribution $p(u)$, we have the following relationships for moments and cumulants:

$$B_n = \int_{-\infty}^{\infty} p(u) u^n du = \left(\frac{1}{i} \frac{d}{d\rho} \right)^n \varphi(\rho)|_{\rho=0} , \tag{C50}$$

$$S_n = \left(\frac{1}{i} \frac{d}{d\rho} \right)^n \psi(\rho)|_{\rho=0} . \tag{C51}$$

Between the moments and cumulants, a recurrence relation exists that has the following form for a one-dimensional distribution:

$$B_0 = 1 , \quad B_n = \sum_{k=1}^{n} \frac{(n-1)!}{(k-1)!(n-k)!} S_k B_{n-k} , \quad (n = 1, 2, \ldots) . \tag{C52}$$

The formulae (C52) follow from (C39).

C.10 Correlation Functions

In the statistical mechanics of disperse media and in the theory of turbulence, one deals with random fields that are described by random functions $u(M)$ of a point M in space time. In agreement with (C39), one refers to the following expressions

$$B_{uu\ldots u}(M_1, M_2, \ldots, M_k) = \langle u(M_1) u(M_2) \ldots u(M_k) \rangle \tag{C53}$$

as moments of the degree k.

In general, some of the points M_i can coincide. The number of different points is called the type of the moment. Sometimes, they are also referred to as multi-point moments.

The mean value of product of some random functions of different fields that are statistically related is called a mixed moment. For example, we consider the velocity field of a turbulent flow $\boldsymbol{u}(u_1, u_2, u_3)$. The components of a vector can be considered as separate, but connected random fields. The components can be taken at the same point or in different points. Therefore, moments of different types and degrees are considered. In statistical mechanics, the two point moments of the second degree defined by

$$B_{ij}(M_1, M_2) = \langle u_i(M_1), u_j(M_2) \rangle \tag{C54}$$

have special significance.

These moments are called correlation functions or short correlations. The components B_{ij} constitute a tensor of second order that can be written in matrix form as

$$\boldsymbol{B}(M_1, M_2) = \langle \boldsymbol{u}(M_1), \boldsymbol{u}^{\mathrm{T}}(M_2) \rangle ,$$

where T stands for the operation of transposing a matrix.

If $M_1, M_2, \ldots M_n$ are points in space time, the corresponding moments and correlations are called space time moments and space time correlations. One often deals with correlations of random quantities at the same time in different points or with correlations at different times in the same point. The first are called space correlations and the latter time correlations.

We now mention some properties of the correlation functions. The correlation function $B_{uu}(M_1, M_2) = \langle u(M_1) u(M_2) \rangle$ is symmetric with respect to the points M_1 and M_2, that is,

$$B_{uu}(M_1, M_2) = B_{uu}(M_2, M_1) . \tag{C55}$$

The quadratic form with the coefficients $B_{uu}(M_i, M_j)$ is positive definite, meaning

$$\sum_{i=1}^{n} \sum_{j=1}^{n} B_{uu}(M_i, M_j) c_i c_j \geqslant 0 , \tag{C56}$$

for real quantities c_i, non-negative integers n for an arbitrary choice of the points $M_1, M_2, \ldots, M_n$.

For $n = 2$, it follows from (C56) that

$$|B_{uu}(M_1, M_2)| \leqslant |B_{uu}(M_1, M_1)|^{1/2} \, |B_{uu}(M_2, M_2)|^{1/2} . \tag{C57}$$

In addition to two point correlations of identical random functions $u(M_1)$ and $u(M_2)$, we consider the two point correlations of different random functions $u(M_1)$

and $v(M_2)$. The mixed two point correlation $B_{uv}(M_1, M_2)$ is called the mutual correlation function. It has similar properties, for example,

$$B_{uv}(M_1, M_2) = B_{vu}(M_2, M_1) . \tag{C58}$$

Moments of higher degree also exist in this case.

In a similar way, one can define the central moments of second degree as

$$\begin{aligned} b_{uu} &= \langle [u(M_1) - \langle u(M_1) \rangle] [u(M_2) - \langle u(M_2) \rangle] \rangle \\ &= B_{uu}(M_1, M_2) - \langle u(M_1) \rangle \langle u(M_2) \rangle , \end{aligned} \tag{C59}$$

$$\begin{aligned} b_{uv} &= \langle [u(M_1) - \langle u(M_1) \rangle] [v(M_2) - \langle v(M_2) \rangle] \rangle \\ &= B_{uv}(M_1, M_2) - \langle u(M_1) \rangle \langle v(M_2) \rangle . \end{aligned} \tag{C60}$$

The dispersion of the distribution is equal to

$$\sigma_u^2(M) = b_{uu}(M, M) , \quad \sigma_v^2(M) = b_{vv}(M, M) . \tag{C61}$$

These correlation functions characterize statistical relations of the deviations of the random functions from the mean values in different points. Therefore, they are sometimes called the correlation functions of the fluctuations.

Another statistical parameter is the correlation coefficient defined by

$$\lambda_{uu} = \frac{b_{uu}(M_1, M_2)}{\sigma_u(M_1)\sigma_u(M_2)} \quad \text{or} \quad \lambda_{uv} = \frac{b_{uv}(M_1, M_2)}{\sigma_u(M_1)\sigma_v(M_2)} . \tag{C62}$$

According to the Schwartz inequality, we have $|\lambda_{uu}| < 1$ and $|\lambda_{uv}| < 1$. If the correlation coefficient vanishes, this means that there is no correlation between the fluctuations in different space points.

An important property that follows from the intuitive physical perspective is the decay of the statistical connection between the random properties in different points of the space time with increasing distance between the points. Therefore, the correlation function has to vanish for an infinite distance of the points from each other. The distance of these points is either the geometrical distance, that is, $|\boldsymbol{X}_2 - \boldsymbol{X}_1| \to \infty$ for $t_2 = t_1$, or the distance of the points in time, that is, $|t_2 - t_1| \to \infty$ for $\boldsymbol{X}_2 = \boldsymbol{X}_1$, or both of them combined.

C.11
Poisson, Bernoulli and Gaussian Distributions

We consider the three most frequently used distributions in statistical physics using the example of the random walk of a particle in the theory of Brownian motion.

The problem of the random walk is defined as follows. We consider a particle that moves along a straight line via steps of length one, where each step takes place with an equal probability of 0.5 forward or backward. We put the coordinate axis along the straight line with the origin at the initial point of the particle movement.

Then, the possible values of the particle coordinates after N steps are $-N, -N+1, \ldots, 0, 1, \ldots, N-1, N$. The probability $P(M, N)$ that after N steps the particle will be at the point with coordinate m is given by the Bernoulli distribution

$$P(M, N) = C^N_{(N+M)/2} \left(\frac{1}{2}\right)^N , \tag{C63}$$

where $C^N_{(N+M)/2} = N!/[(\frac{1}{2}(N+M))!(\frac{1}{2}(N-M))!]$ is the binomial coefficient. The mean value of the displacement and the quadratic mean of the displacement are equal to $\langle M \rangle = 0$, $\sqrt{|M^2|} = \sqrt{N}$. In the limit of $N \gg 1$ and $M \ll N$, the asymptotic formula

$$P(M, N) = \left(\frac{2}{\pi N}\right)^{1/2} \exp\left(-\frac{M^2}{2N}\right) \tag{C64}$$

applies.

We consider a small volume v which is part of a much larger volume V. We consider N particles that are distributed randomly inside of the volume V. The probability that n particles are inside of v is given by the Bernoulli distribution

$$P_N(n) = \frac{N!}{n!(N-n)!} \left(\frac{v}{V}\right)^n \left(1 - \frac{v}{V}\right)^{N-n} . \tag{C65}$$

The mean value of the particle number n is equal to

$$\langle n \rangle = N \left(\frac{v}{V}\right) = \nu .$$

In the limit of $N \to \infty$ and $V \to \infty$ under the condition that ν stays finite, the distribution (C65) becomes the Poisson distribution

$$P(n) = \frac{\nu^n e^{-\nu}}{n!} . \tag{C66}$$

In the case of $\nu \gg 1$ and $n \sim \nu$, the Poisson distribution approaches the distribution

$$P(n) = \left(\frac{1}{2\pi\nu}\right)^{1/2} \exp\left(-\frac{(n-\nu)^2}{2\nu}\right). \tag{C67}$$

The distributions (C64) and (C67) are special cases of the Gaussian or normal distribution. This distribution has, for the general N-dimensional case, the following form:

$$p(u_1, u_2, \ldots, u_N) = C \exp\left\{-\frac{1}{2} \sum_{j=1}^{N} \sum_{k=1}^{N} g_{ik}(u_j - a_j)(u_k - a_k)\right\} .$$

With arbitrary real numbers a_j, the components g_{ik} of the positive definite matrix $\|g_{ik}\|$, the normalization constant $C = g^{1/2}/(2\pi)^{N/2}$ and the determinant g of the matrix $\|g_{ik}\|$. The constants a_j and g_{ik} can be expressed in terms of the moments of first and second degree (see (C39)) as follows:

$$\langle u_j \rangle = a_j , \quad b_{ij} = \langle (u_j - \langle u_j \rangle)(u_k - \langle u_k \rangle) \rangle = \frac{G_{jk}}{g} , \tag{C68}$$

$$B_{ij} = \langle u_j u_k \rangle = \frac{G_{jk}}{g} + a_j a_k , \tag{C69}$$

where $G_{jk} = \partial g \partial g_{jk}$ is the algebraic complement to the element g_{jk} of the matrix $\|g_{ik}\|$. This means that the matrices $\|g_{ik}\|$ and $\|b_{ij}\|$ are the inverse of each other.

A Gaussian distribution can be written in matrix form as

$$p(\boldsymbol{u}) = \frac{1}{(2\pi)^{N/2}} \exp\left(-\frac{1}{2}(\boldsymbol{u} - \langle \boldsymbol{u} \rangle)^{\mathrm{T}} \boldsymbol{b}^{-1} (\boldsymbol{u} - \langle \boldsymbol{u} \rangle) \right) , \tag{C70}$$

where T stands for transposition and $\boldsymbol{b} = |b_{ij}|$.

It follows from (C68)–(C70), that the first two moments of the probability density define the probability density uniquely and therefore all statistical characteristics are also defined in terms of these moments. A field of a random quantity whose probability density has a Gaussian (normal) density is called a Gaussian field.

Therefore, the complete statistical description of a random Gaussian field $\boldsymbol{u}(M) = (u_1(M), u_2(M), \ldots, u_N(M))$ is achieved by specifying the mean values and correlation functions of the random quantities. Starting from the condition that all central moments of uneven degree vanish, all other moments can be expressed in terms of the moments of second degree as follows:

$$\begin{aligned} b_{k_1 k_2 \ldots k_N} &= \left\langle (u_1 - \langle u_1 \rangle)^{k_1} (u_2 - \langle u_2 \rangle)^{k_2}, \ldots, (u_N - \langle u_N \rangle)^{k_N} \right\rangle \\ &= \sum b_{i_1 i_2} b_{i_3 i_4} \ldots b_{i_{2k-1} i_{2K}} , \end{aligned} \tag{C71}$$

where $k_1 + k_2 + \ldots + k_N = 2K$ and the lower pairs of indices have to be chosen from $1, 2, \cdots, 2K$ in such a way that the first index is larger then the second one. For example, we have

$$\begin{aligned} b_{1111} &= \langle (u_1 - \langle u_1 \rangle)(u_2 - \langle u_2 \rangle), (u_3 - \langle u_3 \rangle), (u_4 - \langle u_4 \rangle) \rangle \\ &= b_{12} b_{34} + b_{13} b_{24} + b_{14} b_{23} . \end{aligned}$$

During the examination of random quantities that are subject to a normal distribution, it is convenient to use the characteristic function because of its simple form (see (C30)) which is given by

$$\begin{aligned} &\varphi(\rho_1, \rho_2, \ldots, \rho_N) \\ &= \int_{-\infty}^{\infty} \int_{-\infty}^{\infty} \cdots \int_{-\infty}^{\infty} \exp\left\{ i \sum_{k=1}^{N} \rho_k u_k \right\} p(u_1, u_2, \ldots, u_N) du_1 du_2 \ldots du_N \\ &= \exp\left\{ i a_k \rho_k - \frac{1}{2} \sum_{j=1}^{N} \sum_{k=1}^{N} b_{jk} \rho_j \rho_k \right\} . \end{aligned} \tag{C72}$$

From (C72), the cumulants can be found. It turns out that the cumulants of the first and second degree of a Gaussian distribution are given by a_j and b_{jk}, while the cumulants of higher degrees vanish and each linear combination of random quantities with normal distribution also has a normal distribution. The Gaussian distribution is of great importance in practice because of the two following reasons. Firstly, the behavior of many random quantities is sufficiently well approximated by a Gaussian distribution, and secondly, a quantity which is the sum of a large number of independent components with arbitrary distributions which is often encountered in statistical mechanics tends to a Gaussian distribution due to the central limit theorem.

We now consider the one-dimensional Gaussian distribution

$$p(u) = \frac{1}{\sqrt{2\pi}\sigma} \exp\left(-\frac{u^2}{2\sigma^2}\right), \quad \sigma^2 = \langle u^2 \rangle . \tag{C73}$$

Using (C30) and (C45), one obtains

$$\varphi(\rho) = \exp\left(-\frac{\rho^2\sigma^2}{2}\right), \quad \psi(\rho) = -\frac{\rho^2\sigma^2}{2} . \tag{C74}$$

Then, from (C50) and (C51), one obtains that

$$B_1 = S_1 = 0 , \quad B_2 = S_2 = \sigma^2 , \quad S_{n>2} = 0 . \tag{C75}$$

The recurrence relation (C52) has the following form:

$$B_n = (n-1)\sigma^2 B_{n-2} , \tag{C76}$$

from which one obtains

$$B_{2n+1} = 0 , \quad B_{2n} = (2n-1)!!\sigma^{2n} . \tag{C77}$$

Consider a random quantity C with Gaussian distribution and a certain function $f(X)$ that satisfies the condition $f(X)\exp(-X^2/2\sigma^2) \to 0$ for $X \to \pm\infty$. Then, we have

$$\begin{aligned} \langle X f(X) \rangle &= \frac{1}{\sqrt{2\pi}\sigma} \int_{-\infty}^{\infty} X f(X) \exp\left(-\frac{X^2}{2\sigma^2}\right) dX \\ &= \frac{\sigma}{\sqrt{2\pi}} \int_{-\infty}^{\infty} \frac{d f(X)}{dX} \exp\left(-\frac{X^2}{2\sigma^2}\right) dX = \sigma^2 \left\langle \frac{d f(X)}{dX} \right\rangle . \end{aligned} \tag{C78}$$

A similar expression results for the Gaussian random vector $\boldsymbol{X} = (X_1, X_2, \ldots, X_n)$ with a multidimensional distribution of the form (C73):

$$\langle X_i f(\boldsymbol{X}) \rangle = B_{ij} \left\langle \frac{d f(\boldsymbol{X})}{dX_j} \right\rangle , \tag{C79}$$

where $B_{ij} = \langle X_i X_j \rangle$ are the components of the correlation matrix.

C.12
Stationary Random Functions and Homogeneous Random Fields

In Section C.7, the concept of the ergodic hypothesis was introduced according to which the time or space averages become identical to the ensemble averages in the limit of an infinite time or space interval. These necessary conditions for the ergodic hypothesis lead to special classes of the random functions $u(\boldsymbol{X})$ which are of large importance in the theory of Brownian motion and turbulent flows.

We first consider the time averaging and assume for simplicity that the random function only depends on time. The averaging over time is indicated by $\langle u(t)\rangle_t$. Then, we have, according to (C21),

$$\langle u(t)\rangle_T = \frac{1}{T}\int\limits_{-T/2}^{T/2} u\,(t+\tau)\,d\tau\ . \tag{C80}$$

In the limit of $T \to \infty$, $\langle u(t)\rangle_T$ approaches the mean value of probability theory $\langle u(t)\rangle$.

It turns out that the necessary condition for the above is

$$\langle u(t)\rangle_T = U = \text{const}\ . \tag{C81}$$

To obtain this condition, one has to consider the difference of time averages of a random quantity at different times t and t_1:

$$\begin{aligned}\langle u(t)\rangle_T - \langle u(t_1)\rangle_T &= \frac{1}{T}\left\{\int\limits_{-T/2}^{T/2} u\,(t+\tau)\,d\tau - \int\limits_{-T/2}^{T/2} u\,(t_1+\tau)\,d\tau\right\}\\ &= \frac{1}{T}\left\{\int\limits_{-T/2+t}^{T/2+t_1} u\,(s)\,ds - \int\limits_{-T/2+t_1}^{T/2+t_1} u\,(s)\,ds\right\}\ .\end{aligned} \tag{C82}$$

For $T \to \infty$, the right-hand side of (C82) approaches zero, from which the condition (C81) follows.

Averaging the product $u(t)u(t_1) = u(t)u(t+s)$ where $s = t_1 - t$ and taking the limit of $T \to \infty$, one concludes that the correlation function $(B_{uu}(t,t_1))_T$ is equal to the mean value over the ensemble $B_{uu}(t,t_1) = \langle u(t)u(t_1)\rangle$ if and only if for each t_1 and $t_2 > t_1$, the following condition is satisfied

$$B_{uu}(t_2,t_1) = B_{uu}(t_2 - t_1)\ . \tag{C83}$$

For the moment of degree N, the condition (C83) becomes

$$B_{uu\ldots u}(t_1,t_2,\ldots,t_N) = B_{uu\ldots u}(t_2-t_1,\ldots,t_N-t_1)\ . \tag{C84}$$

To obtain the average of probability theory for the random quantities $u(t_1)$, $u(t_2)$, $\ldots$, $u(t_N)$ via time averaging, only those probability functions $u(t)$ can be used for which the N-dimensional probability density $p_{t_1 t_2 \ldots t_N}(u_1, u_2, \ldots, u_N)$ for arbitrary N and $t_1, t_2, \ldots, t_N$ does not depend on N parameters, but on the $N-1$ parameters $t_2 - t_1, t_3 - t_1, \ldots, t_N - t_1$, that is, it satisfies the condition

$$p_{t_1 t_2 \ldots t_N}(u_1, u_2, \ldots, u_N) = p_{t_2 - t_1, \ldots, t_N - t_1}(u_1, u_2, \ldots, u_N) . \tag{C85}$$

It must be noted that from (C85), we obtain (C81) and (C83), and for a Gaussian random function, (C81) and (C83) lead to (C84) and (C85).

Condition (C85) defines a class of functions of time for which the probability density does not change if all t_i are moved by an arbitrary time interval. These functions are called stationary random functions or stationary random processes. Examples are the characteristics (velocity, pressure, temperature and so on) of a stationary turbulent flow. The velocity vector $\boldsymbol{u}$ with components $u_1(t), u_2(t), \ldots, u_N(t)$ in different points, whose probability density does not change under a simultaneous shift of all times $t_1, t_2, \ldots, t_N$ by the same quantity τ, is a multidimensional stationary random process.

The mixed moments of the functions $u_i(t)$ only depend on the differences of the corresponding time values. For example, all correlation functions $B_{jk}(t_1, t_2) = \langle u_j(t_1) u_k(t_2) \rangle$ only depend on $\tau = t_1 - t_2$. We now consider the space averaging of random quantities. To this end, one has to consider the random functions $u(\boldsymbol{X})$ of the space coordinates $\boldsymbol{X}(X_1, X_2, X_3)$. According to (C22), the space averages are defined as follows

$$\langle u(\boldsymbol{X}) \rangle_{ABC} = \frac{1}{ABC} \int\limits_{-A/2}^{A/2} \int\limits_{-B/2}^{B/2} \int\limits_{-C/2}^{C/2} u(X_1 + \xi_1, X_2 + \xi_2, X_3 + \xi_3) d\xi_1 d\xi_2 d\xi_3 . \tag{C86}$$

Just as for the problem of the time averaging, one obtains that if $A \to \infty$, $B \to \infty$ and $C \to \infty$ or even if only one of these limiting cases occurs, $\langle u(\boldsymbol{X}) \rangle_{ABC}$ must approach $\langle u(\boldsymbol{X}) \rangle$ for the ergodic hypothesis to be valid.

The corresponding necessary conditions are similar to the conditions (C81), (C83)–(C85), if one replaces t by $\boldsymbol{X}$ and reads

$$\langle u(\boldsymbol{X}) \rangle = U = \text{const} , \tag{C87}$$

$$B_{uu}(\boldsymbol{X}_1, \boldsymbol{X}_2) = B_{uu}(\boldsymbol{X}_2 - \boldsymbol{X}_1) , \tag{C88}$$

$$p_{x_1 x_2 \ldots x_N}(u_1, u_2, \ldots, u_N) = p_{x_2 - x_1, \ldots, x_N - x_1}(u_1, u_2, \ldots, u_N) , \tag{C89}$$

where $B_{uu}(\boldsymbol{X}_1, \boldsymbol{X}_2) = \langle u(\boldsymbol{X}_1) u(\boldsymbol{X}_2) \rangle$.

A random field $u(\boldsymbol{X})$, that has the properties (C87)–(C89), is referred to as a statistically homogeneous field. As an example, we consider the averaged characteristics of a turbulent flow that are independent of the space coordinates. It is obvious that this homogeneity does not apply to the whole space. For example, for a turbulent flow of a fluid in a tube, the averaged hydrodynamic parameters are nearly homogeneous in the center of the tube, but inhomogeneous close to the wall.

The conditions of stationarity and homogeneity are generally not sufficient for the time and space averaging to converge toward the average of probability theory. The necessary and sufficient conditions are given by the ergodic theorem. The convergence, on average, exists if and only if the following conditions for the correlation functions of the fluctuations $b_{uu}(\tau)$ are satisfied:

$$\lim_{T\to\infty} \frac{1}{T} \int_0^T b_{uu}(\tau) d\tau = 0 \, . \tag{C90}$$

The characteristic averaging time T can be estimated by the corresponding correlation time T_1 as

$$T_1 = \frac{1}{b_{uu}(0)} \int_0^\infty b_{uu}(\tau) d\tau \, . \tag{C91}$$

If $U = \langle u(t) \rangle$ is the mean value of probability theory and $\langle u(t) \rangle_T$ is the time average, then for large enough T, the following asymptotic formula for the mean square deviation from the mean value of probability theory applies:

$$\left\langle \left| \langle u(t) \rangle_T - U \right|^2 \right\rangle \approx 2 \frac{T_1}{T} b_{uu}(0) \, . \tag{C92}$$

If the desired accuracy of replacing U by $\langle u(t) \rangle_T$ is given, one obtains from (C92) the characteristic averaging time. A similar estimate can be made in the case of space averaging:

$$\left\langle \left| \langle u \rangle_V - U \right|^2 \right\rangle \approx 2 \frac{V_1}{V} b_{uu}(0) \, , \tag{C93}$$

where $\langle u \rangle_V$ is the average over the volume V and V_1, the correlation volume, defined by

$$V_1 = \frac{1}{b_{uu}(0)} \int_{-\infty}^{\infty} \int_{-\infty}^{\infty} \int_{-\infty}^{\infty} b_{uu}(r) dr_1 dr_2 dr_3 \, . \tag{C94}$$

C.13
Isotropic Random Fields

A scalar random field $u(\boldsymbol{X})$ is called an isotropic field if all corresponding finite probability densities $p_{x_1 x_2 \dots x_N}(u_1, u_2, \dots, u_N)$ are unchanged by rotation of the system of points $\boldsymbol{X}_1, \boldsymbol{X}_2, \dots, \boldsymbol{X}_n$ around any axis through the origin, and also unchanged by reflection of the system with respect to any plane that passes through the origin.

For applications, homogeneous and isotropic fields are of paramount interest. Therefore, we define isotropic fields as fields for which the probability density

$p_{x_1 x_2 \ldots x_N}(u_1, u_2, \ldots, u_N)$ is not changed during arbitrary parallel displacement, rotation and reflection of the system of points $\boldsymbol{X}_1, \boldsymbol{X}_2, \ldots, \boldsymbol{X}_n$.

Condition (C87) means that the mean value of a homogeneous random function $\langle u(\boldsymbol{X})\rangle$ has to be constant. This constant is often assumed to be zero. For this, one replaces $u(\boldsymbol{X})$ by $u'(\boldsymbol{X}) = u(\boldsymbol{X}) - \langle u(\boldsymbol{X})\rangle$. The correlation function $B(\boldsymbol{X},\boldsymbol{X}') = \langle u(\boldsymbol{X})u(\boldsymbol{X}')\rangle$ of an isotropic field has, for two point pairs, $(\boldsymbol{X}, \boldsymbol{X}')$ and $(\boldsymbol{X}_1, \boldsymbol{X}_1')$, that can be made to coincide via rotation and displacement, that is, the same values.

If the distance r between the points $\boldsymbol{X}$ and $\boldsymbol{X}'$ is equal to the distance between $\boldsymbol{X}_1$ and $\boldsymbol{X}_1'$, then we have $B(\boldsymbol{X}, \boldsymbol{X}') = B(\boldsymbol{X}_1, \boldsymbol{X}_1')$, and thus the correlation function $B(\boldsymbol{X}, \boldsymbol{X}')$ only depends on $r = |\boldsymbol{X}' - \boldsymbol{X}| = |\boldsymbol{r}|$, such that

$$\langle u(\boldsymbol{X})u(\boldsymbol{X}')\rangle = B(r) . \tag{C95}$$

The application of the method of harmonic analysis to random processes and random fields, that is, the representation of a random function in the form of Fourier series (for a function on a finite interval) or in the form of a Fourier integral (for a function on a infinite interval), has turned out be very fruitful.

For stationary random functions and homogeneous random fields, which by their definition can not decay at infinity, and for which a representation as a Fourier series of Fourier integral is possible, the spectral decomposition has a clear physical meaning: it represents the superposition of harmonic oscillations (for stationary random processes) and plane waves (for homogeneous random fields). The integral representation of the correlation function has the following form:

$$B(\boldsymbol{r}) = \int e^{i\boldsymbol{k}\cdot\boldsymbol{r}} F(\boldsymbol{k}) d\boldsymbol{k} , \tag{C96}$$

$$F(\boldsymbol{k}) = \frac{1}{8\pi^3} \int e^{-i\boldsymbol{k}\cdot\boldsymbol{r}} B(\boldsymbol{r}) d\boldsymbol{r} , \tag{C97}$$

where $F(\boldsymbol{r}$ is the spectrum of the homogeneous field and $\boldsymbol{k}$ is the wave vector.

Since an isotropic field satisfies the condition (C95), the spectrum does not depend on $\boldsymbol{k}$, but on $|\boldsymbol{k}|$. Using spherical coordinates

$$x = r \sin\vartheta \cos\varphi , \quad y = r \sin\vartheta \sin\varphi , \quad z = r\cos\vartheta ,$$

one obtains

$$\begin{aligned} F(\boldsymbol{k}) &= \frac{1}{8\pi^3} \int_{-\infty}^{\infty} e^{-i\boldsymbol{k}\cdot\boldsymbol{r}} B(\boldsymbol{r}) d\boldsymbol{r} = \frac{1}{8\pi^3} \int_0^{\infty} \int_{-\pi}^{\pi} \int_0^{\pi} e^{-ikr\cos\vartheta} B(r) r^2 \sin\vartheta \, d\vartheta \, d\varphi \, dr \\ &= \frac{1}{2\pi^2} \int_0^{\infty} \frac{\sin kr}{kr} B(r) r^2 dr = F(k) . \end{aligned} \tag{C98}$$

In a similar manner, one obtains

$$B(r) = 4\pi \int_0^{\infty} \frac{\sin kr}{kr} F(k) k^2 dk . \tag{C99}$$

We introduce, instead of $F(\boldsymbol{k})$, the function

$$E(\boldsymbol{k}) = \frac{1}{8\pi^3} \int\limits_{|\boldsymbol{k}|=k} F(\boldsymbol{k}) d S(k) , \tag{C100}$$

where $S(k)$ is a surface element of the sphere $|\boldsymbol{k}| = k$.

Taking $\boldsymbol{r} = 0$ in the formula (C96) and by taking into account (C95), one obtains:

$$B(0) = \langle (u(\boldsymbol{X}))^2 \rangle = \int\limits_0^\infty E(k) d k . \tag{C101}$$

If u is the velocity, for example, the velocity of turbulent flow, then $B(0)$ has the meaning of the total energy of the random field $u\boldsymbol{X}$. Therefore, $E(k)dk$ is the energy of plane waves with wave numbers in the interval $[k, k + dk]$.

The above can be generalized to the case of an isotropic multidimensional random field $u(\boldsymbol{X}) = (u_1(\boldsymbol{X}), u_2(\boldsymbol{X}), \ldots, u_n(\boldsymbol{X}))$. Such a field is characterized by the following correlation matrix:

$$\| B_{ij} \| = \langle u_i(\boldsymbol{X}) u_j(\boldsymbol{X} + \boldsymbol{r}) \rangle , \tag{C102}$$

whose components depend on $|\boldsymbol{r}| = r$ and can be written as follows:

$$B_{ij}(r) = 4\pi \int\limits_0^\infty \frac{\sin kr}{kr} F_{ij}(k) k^2 dk , \tag{C103}$$

$$F_{ij}(k) = \frac{1}{2\pi^2} \int\limits_0^\infty \frac{\sin kr}{kr} B_{ij}(r) r^2 dr . \tag{C104}$$

The above definition of an isotropic random field can be applied to a scalar random function, for example, the pressure $p(\boldsymbol{X})$, the temperature $T(\boldsymbol{X})$, the one-dimensional velocity $u(\boldsymbol{X})$ or to a random vector field, for example, the three-dimensional velocity $\boldsymbol{u}(\boldsymbol{X})$ and hydrodynamic fields in the form of both scalar and vector fields.

From the definition of a random vector field and the theory of invariants, it follows that the correlation tensor $B_{ij}(\boldsymbol{r})$ can be written as a linear combination of the constant invariant tensor δ_{ij} and of the tensor $\boldsymbol{r}_i \boldsymbol{r}_j$, with coefficients that depend on the single invariant r as

$$B_{ij}(\boldsymbol{r}) = A_1(r) \boldsymbol{r}_i \boldsymbol{r}_j + A_2(r) \delta_{ij} . \tag{C105}$$

C.14
Stochastic Processes, Markovian Processes and the Integral Equation of Chapman–Kolmogorov

The random behavior of a system in time is also called a stochastic process. This means that one considers a quantity which changes as a function of time in a random manner. Examples of stochastic processes are the Brownian motion of particles under the influence of external forces or the movement of a fluid in a turbulent flow. One of the random quantities is the space position $\boldsymbol{X}$ of a particle at different times. We denote the positions of the particle at times $t_1, t_2, \ldots$ by $\boldsymbol{X}_1, \boldsymbol{X}_2, \ldots$ We assume that a common probability density $p(\boldsymbol{X}_1, t_1; \boldsymbol{X}_2, t_2; \boldsymbol{X}_3, t_3; \ldots)$ exists. Then, the quantity

$$p(\boldsymbol{X}_1, t_1; \boldsymbol{X}_2, t_2; \boldsymbol{X}_3, t_3; \ldots) d\boldsymbol{X}_1 d\boldsymbol{X}_2 d\boldsymbol{X}_3 \ldots$$

is the probability for the particle being in the interval $(\boldsymbol{X}_1 + d\boldsymbol{X}_1)$ at time t_1, in the interval $(\boldsymbol{X}_2 + d\boldsymbol{X}_2)$ at time t_2 and so on. If the particle is moving under the influence of random forces, for example, the collisions with surrounding molecules as for the Brownian motion, it changes the direction of its movement millions of times per second. We consider two consecutive positions of the particle $\boldsymbol{X}_i$ and $\boldsymbol{X}_{i+1}$ at times t_i and t_{i+1} such that $\Delta t_i = t_{i+1} - t_i$ is very small compared to the characteristic time scale of the process, but sufficiently large compared to the time interval between consecutive collisions of the particle with the particles of the surrounding molecules. Then, one can assume that the position $\boldsymbol{X}_{i+1}$ of the particle at time t_{i+1} depends on the position $\boldsymbol{X}$ of the particle at time t_i and does not depend on the position of the particle at previous times $t < t_i$, that is, that the particle forgets its past very quickly during a chaotic small scale random walk. Such a random process is called a Markovian process.

Therefore, one can say that a Markovian process is a stochastic process founded on the principle of independence of the future from the past for a given present. Here, the past refers to all events that have been observed up to the fixed present time t. This means that we are dealing with random variables with statistically independent changes. A Markovian process can also be described by using the concept of conditional probabilities. For this, we consider an ordered sequence of times defined by $t_1 \geqslant t_2 \geqslant t_3 \geqslant \ldots \geqslant \tau_1 \geqslant \tau_2 \geqslant \ldots$. We divide the times into previous $\tau_1, \tau_2, \ldots$ and future times $\ldots, t_3, t_2, t_1$ and denote by $\boldsymbol{Y}_1, \boldsymbol{Y}_2, \ldots$ the values of the random quantities at past times $\tau_1 \tau_2 \ldots$ and by $\ldots, \boldsymbol{X}_3, \boldsymbol{X}_2, \boldsymbol{X}_1$ the values of the random quantities a future times $\ldots, t_3, t_2, t_1$. The probability density of the events $\boldsymbol{X}_1, \boldsymbol{X}_2, \ldots$ under the condition that the events $\boldsymbol{Y}_1, \boldsymbol{Y}_2, \ldots$ have occurred is (the conditional probability density) then written as $p(\boldsymbol{X}_1, t_1; \boldsymbol{X}_2, t_2; \boldsymbol{X}_3, t_3; \ldots | \boldsymbol{Y}_1, \tau_1; \boldsymbol{Y}_2, \tau_2; \ldots)|$. In agreement with the Markovian principle, we assume that the conditional probability is determined only by the state of the system at the most recent time. Then, we have

$$\begin{aligned} & p(\boldsymbol{X}_1, t_1; \boldsymbol{X}_2, t_2; \boldsymbol{X}_3, t_3; \ldots | \boldsymbol{Y}_1, \tau_1; \boldsymbol{Y}_2, \tau_2; \ldots) \\ & = p(\boldsymbol{X}_1, t_1; \boldsymbol{X}_2, t_2; \boldsymbol{X}_3, t_3; \ldots | \boldsymbol{Y}_1, \tau_1) \,. \end{aligned} \tag{C106}$$

This equation means that every conditional probability for a sequence of events can be expressed in terms of the conditional probability of a simple event of the type $p(\boldsymbol{X}_1, t_1 | \mathbf{Y}_1, \tau_1)$. Indeed, using the definition (C2) of the conditional probability, one obtains

$$p(\boldsymbol{X}_1, t_1; \boldsymbol{X}_2, t_2 | \boldsymbol{Y}_1, \tau_1) = p(\boldsymbol{X}_1, t_1 | \boldsymbol{X}_2, t_2; \boldsymbol{Y}_1, \tau_1) p(\boldsymbol{X}_2, t_2 | \boldsymbol{Y}_1, \tau_1) \,.$$

Then, one applies it again to the first factor on the right-hand side, which yields, with the help of the postulate (C106),

$$p(\boldsymbol{X}_1, t_1; \boldsymbol{X}_2, t_2 | \boldsymbol{Y}_1, \tau_1) = p(\boldsymbol{X}_1, t_1 | \boldsymbol{X}_2, t_2) p(\boldsymbol{X}_2, t_2 | \boldsymbol{Y}_1, \tau_1) . \tag{C107}$$

By extending this procedure, one obtains for N consecutive events

$$\begin{aligned} p(\boldsymbol{X}_1, t_1; \boldsymbol{X}_2, t_2; \ldots; \boldsymbol{X}_N, t_N) &= p(\boldsymbol{X}_1, t_1 | \boldsymbol{X}_2, t_2) p(\boldsymbol{X}_2, t_2 | \boldsymbol{X}_3, t_3) \\ &\quad \times \ldots \times p(\boldsymbol{X}_{N-1}, t_{N-1} | \boldsymbol{X}_N, t_N) . \end{aligned} \tag{C108}$$

As expected, the Markovian principle results in the independence of conditional events. Using property four from Appendix C.7 for two consecutive events and the expression (C2) in Appendix C.3 for conditional probability, one obtains

$$p(\boldsymbol{X}_1, t_1) = \int p(\boldsymbol{X}_1, t_1; \boldsymbol{X}_2, t_2) d\boldsymbol{X}_2 = \int p(\boldsymbol{X}_1, t_1 | \boldsymbol{X}_2, t_2) d\boldsymbol{X}_2 . \tag{C109}$$

Similar relationships for conditional probabilities can be written as

$$\begin{aligned} p(\boldsymbol{X}_1, t_1; \boldsymbol{X}_3, t_3) &= p(\boldsymbol{X}_1, t_1; \boldsymbol{X}_2, t_2 | \boldsymbol{X}_3, t_3) d\boldsymbol{X}_2 \\ &= p(\boldsymbol{X}_1, t_1 | \boldsymbol{X}_2, t_2; \boldsymbol{X}_3, t_3) p(\boldsymbol{X}_2, t_2 | \boldsymbol{X}_3, t_3) d\boldsymbol{X}_2 . \end{aligned}$$

Since we have $t_1 \geqslant t_2 \geqslant t_3$, the Markovian principle allows one to drop the dependence on $\boldsymbol{X}_3$ in the first factor of the integrand to obtain

$$p(\boldsymbol{X}_1, t_1 | \boldsymbol{X}_3, t_3) = \int p(\boldsymbol{X}_1, t_1 | \boldsymbol{X}_2, t_2) p(\boldsymbol{X}_2, t_2 | \boldsymbol{X}_3, t_3) d\boldsymbol{X}_2 . \tag{C110}$$

This integral equation is called the Chapman–Kolmogorov equation. It forms the basis of stochastic processes.

When considering Markovian random processes, it is very important to know whether the random quantity has a continuous or discrete range and whether the trajectory $\boldsymbol{X}(t)$ is a continuous function of t. As an example, we consider the behavior of a molecule in a diluted gas that is characterized by the velocity $\boldsymbol{V}(t)$ and the position $\boldsymbol{X}(t)$. It is obvious that the range of velocity change is continuous, although the function $\boldsymbol{V}(t)$ can be discontinuous if the collision of the molecules is modeled as the elastic collision of solid spheres. However, the position of the molecule is a continuous function even in this case. In reality, the molecules do not interact like solid spheres. Between the molecules exists a molecular interaction which leads to a continuous deflection of the molecule trajectory during the collision. The characteristic collision time is much smaller than the time intervals that make up a Markovian chain.

Thus, it can be said that the Markovian process can be considered as the approximation of the real process on a large time scale where the concept of continuity of the random function does not make sense. Independent of the way that the collision of molecules is modeled, it results in a jump of $\boldsymbol{V}$ on the large time scale. In the same way, trajectories will not necessarily be continuous on this time scale.

Another example is a chemical reaction during which molecules of a certain kind can vanish or can be created. Then, the random quantity will be the concentration of molecules that changes discontinuously on a large time scale.

In connection with the above, the following Lindenberg condition for the continuity of the random quantity $X(t)$ appears intuitive. If for any $\epsilon > 0$, the equation

$$\lim_{\Delta t \to 0} \frac{1}{\Delta t} \int_{|X-Z|>\varepsilon} p(X, t + \Delta t | Z, t) dX = 0 \tag{C111}$$

holds uniformly in Z, t and Δt, then realization of $X(t)$ is a continuous function of t with probability one.

This means that the probability that the difference between X and Z is larger than ϵ tends to zero for $\Delta t \to 0$ faster than Δt. One can show that the Gaussian probability density

$$p(X, t + \Delta t | Z, t) = \frac{1}{(4\pi D \Delta t)^{1/2}} \exp\left\{-\frac{(X - Z)^2}{4 D \Delta t}\right\} , \tag{C112}$$

which is the Einstein solution of the problem of Brownian motion, satisfies condition (C111), while the probability density

$$p(X, t + \Delta t | Z, t) = \frac{\Delta t}{\pi \left[(X - Z)^2 + (\Delta t)^2\right]} , \tag{C113}$$

which corresponds to the Cauchy process, does not satisfy the condition (C111).

Both distributions above approach $\delta(X - Z)$ (see (C10) and (C17)), and satisfy (C110). Therefore, solutions of the integral equation of Chapman–Kolmogorov can be both continuous as well as discontinuous distributions of the probability density.

C.15 Differential Equations of Chapman–Kolmogorov, Kolmogorov–Feller, Fokker–Planck and Liouville

For the solution of concrete problems, one does not use the integral equations of Chapman–Kolmogorov, but the corresponding differential equations that can be determined from (C110) under some additional assumptions. It must be noted that the usual assumption of a continuous random process leads to the Fokker–Planck equation. As already shown above, discontinuous solutions of the Chapman–Kolmogorov integral equation also exist. Therefore, the desired differential equation must not only have continuous, but also discontinuous solutions. Therefore, we assume that the following conditions are satisfied:

1. $$\lim_{\Delta t \to 0} \frac{1}{\Delta t} p(X, t + \Delta t | Z, t) = W(X|Z, t) \tag{C114}$$

uniformly in $\boldsymbol{X}, \boldsymbol{Z}, \boldsymbol{t}$ for $|\boldsymbol{X} - \boldsymbol{Z}| \geqslant \varepsilon$ and that the limit is independent of ϵ.

2. $$\lim_{\Delta t \to 0} \frac{1}{\Delta t} \int_{|X-Z<\varepsilon|} (X_i - Z_i)\, p(\boldsymbol{X}, t + \Delta t | \boldsymbol{Z}, t) d\boldsymbol{X} = A_i(\boldsymbol{Z}, t) + O(\varepsilon) \; ; \tag{C115}$$

3. $$\lim_{\Delta t \to 0} \frac{1}{\Delta t} \int_{|X-Z<\varepsilon|} (X_i - Z_i)\left(X_j - Z_j\right) p(\boldsymbol{X}, t + \Delta t | \boldsymbol{Z}, t) d\boldsymbol{X} = D_{ij}(\boldsymbol{Z}, t) + O(\varepsilon) \, . \tag{C116}$$

In conditions two and three, we assume that the integrals converge uniformly with respect to $\boldsymbol{Z}, \epsilon$ and t. Condition one is responsible for the continuity of the process. For $W(\boldsymbol{X}|\boldsymbol{Z}, t) = 0$, the process has continuous and for $W(\boldsymbol{X}|\boldsymbol{Z}, t) \neq 0$, discontinuous trajectories.

We consider the mean value of a twice differentiable random function $f(\boldsymbol{X})$ with respect to a probability density that satisfies the conditions (C114)–(C116):

$$\langle f(\boldsymbol{X}) \rangle = \int f(\boldsymbol{X}) p(\boldsymbol{X}, t | \boldsymbol{Y}, t') d\boldsymbol{X} \, .$$

This function has the time derivative

$$\frac{\partial \langle f(\boldsymbol{X}) \rangle}{\partial t} = \lim_{\Delta t \to 0} \frac{1}{\Delta t} \int \left[p(\boldsymbol{X}, t + \Delta t | \boldsymbol{Y}, t') - p(\boldsymbol{X}, t | \boldsymbol{Y}, t')\right] f(\boldsymbol{X}) d\boldsymbol{X} \, .$$

We assume $\boldsymbol{X}_1 = \boldsymbol{X}$, $t_1 = t + \Delta t$, $\boldsymbol{X}_2 = \boldsymbol{Z}$, $t_2 = t$, $\boldsymbol{X}_3 = \boldsymbol{Y}$, $t_3 = t'$ in (C110).

This means that we insert, between the position $\boldsymbol{Y}$ at times t' and $\boldsymbol{X}$ at time $t + \Delta t$, another position $\boldsymbol{Z}$ at the intermediate time $t(t' < t < t + \Delta t)$. Then, the above equation becomes

$$\begin{aligned} &\frac{\partial}{\partial t}\left\{\int f(\boldsymbol{X}) p(\boldsymbol{X}, t | \boldsymbol{Y}, t') d\boldsymbol{X}\right\} \\ &= \lim_{\Delta t \to 0} \frac{1}{\Delta t}\left\{\int d\boldsymbol{X} \int f(\boldsymbol{X}) p(\boldsymbol{X}, t + \Delta t | \boldsymbol{Z}, t) p(\boldsymbol{Z}, t | \boldsymbol{Y}, t') d\boldsymbol{Z} \right. \\ &\quad \left. - \int f(\boldsymbol{Z}) p(\boldsymbol{Z}, t | \boldsymbol{Y}, t') d\boldsymbol{Z}\right\} . \end{aligned} \tag{C117}$$

In the last term on the right-hand side, the integration variable $\boldsymbol{X}$ has been replaced by $\boldsymbol{Z}$. One now divides the integration range of $\boldsymbol{X}$ into the two regions $|\boldsymbol{X} - \boldsymbol{Z}| \geqslant \varepsilon$ und $|\boldsymbol{X} - \boldsymbol{Z}| < \varepsilon$. In the integral over the region $|\boldsymbol{X} - \boldsymbol{Z}| < \varepsilon$, one expands the function $f(\boldsymbol{X})$ into the Taylor series

$$\begin{aligned} f(\boldsymbol{X}) &= f(\boldsymbol{Z}) + \sum_i \frac{\partial f(\boldsymbol{Z})}{d Z_i}(X_i - Z_i) \\ &\quad + \sum_i \sum_j \frac{1}{2} \frac{\partial^2 f(Z)}{\partial Z_i \partial Z_j}(X_i - Z_i)(X_j - Z_j) + |\boldsymbol{X} - \boldsymbol{Z}|^2 R(\boldsymbol{X}, \boldsymbol{Z}) \, . \end{aligned} \tag{C118}$$

The last term on the right-hand side of (C118) is the residual term that satisfies the condition $R(\boldsymbol{X}, \boldsymbol{Z}) \to 0$ for $|\boldsymbol{X} - \boldsymbol{Z}| \to 0$.

By substituting (C118) on the right-hand side of (C117) and grouping the terms, one obtains

$$\begin{aligned}
&\frac{\partial}{\partial t}\left\{\int f(\boldsymbol{X}) p(\boldsymbol{X}, t|\boldsymbol{Y}, t')d\boldsymbol{X}\right\} \\
&= \lim_{\Delta t \to 0} \frac{1}{\Delta t}\left\{\int\limits_{|X-Z|<\varepsilon} \left[\sum_i \frac{\partial f(\boldsymbol{Z})}{d Z_i}(X_i - Z_i)\right.\right. \\
&\left. + \sum_i \sum_j \frac{1}{2}\frac{\partial^2 f(\boldsymbol{Z})}{\partial Z_i \partial Z_j}(X_i - Z_i)(X_j - Z_j)\right] \\
&\times p(\boldsymbol{X}, t + \Delta t|\boldsymbol{Z}, t) p(\boldsymbol{Z}, t|\boldsymbol{Y}, t')d\boldsymbol{X} d\boldsymbol{Z} \\
&+ \int\limits_{|X-Z|<\varepsilon} |\boldsymbol{X} - \boldsymbol{Z}|^2 R(\boldsymbol{X}, \boldsymbol{Z}) p(\boldsymbol{X}, t + \Delta t|\boldsymbol{Z}, t) p(\boldsymbol{Z}, t|\boldsymbol{Y}, t')d\boldsymbol{X} d\boldsymbol{Z} \\
&+ \int\limits_{|X-Z|\geqslant\varepsilon} f(\boldsymbol{X}) p(\boldsymbol{X}, t + \Delta t|\boldsymbol{Z}, t) p(\boldsymbol{Z}, t|\boldsymbol{Y}, t')d\boldsymbol{X} d\boldsymbol{Z} \\
&+ \int\limits_{|X-Z|<\varepsilon} f(\boldsymbol{Z}) p(\boldsymbol{X}, t + \Delta t|\boldsymbol{Z}, t) p(\boldsymbol{Z}, t|\boldsymbol{Y}, t')d\boldsymbol{X} d\boldsymbol{Z} \\
&\left. - \int f(\boldsymbol{Z}) p(\boldsymbol{Z}, t|\boldsymbol{Y}, t')d\boldsymbol{X} d\boldsymbol{Z}\right\} .
\end{aligned} \tag{C119}$$

We now consider, one by one, the integral terms on the right-hand side of (C119). Since $p(\boldsymbol{X}, t + \Delta t|\boldsymbol{Z}, t)$ is the probability density, we note that

$$\int p(\boldsymbol{X}, t + \Delta t|\boldsymbol{Z}, t)d\boldsymbol{X} = 1 .$$

By using the above equation and the assumption of uniform convergence, one can exchange the order of integration and, taking the limit of $\delta t \to 0$, transform the last term into the following form

$$\begin{aligned}
\int f(\boldsymbol{Z}) p(\boldsymbol{Z}, t|\boldsymbol{Y}, t')d\boldsymbol{X} &= \int f(\boldsymbol{Z}) p(\boldsymbol{Z}, t|\boldsymbol{Y}, t')d\boldsymbol{Z} \int p(\boldsymbol{X}, t + \Delta t|\boldsymbol{Z}, t)d\boldsymbol{X} \\
&= \iint f(\boldsymbol{Z}) p(\boldsymbol{X}, t + \Delta t|\boldsymbol{Z}, t) p(\boldsymbol{Z}, t|\boldsymbol{Y}, t')d\boldsymbol{X} d\boldsymbol{Z} .
\end{aligned}$$

To transform the first term, one uses conditions 2 and 3 and obtains

$$\lim_{\Delta t \to 0} \frac{1}{\Delta t} \left\{ \int\limits_{|X-Z<\varepsilon|} \left[\sum_i \frac{\partial f(Z)}{d Z_i} (X_i - Z_i) \right.\right.$$

$$\left. + \sum_i \sum_j \frac{1}{2} \frac{\partial^2 f(Z)}{\partial Z_i \partial Z_j} (X_i - Z_i)(X_j - Z_j) \right]$$

$$\left. \times\ p(\boldsymbol{X}, t + \Delta t | \boldsymbol{Z}, t) p(\boldsymbol{Z}, t | \boldsymbol{Y}, t') d\boldsymbol{X} d\boldsymbol{Z} \right\}$$

$$= \int \left[\sum_i A_i(\boldsymbol{Z}, t) \frac{\partial f}{\partial Z_i} + \sum_i \sum_j \frac{1}{2} D_{ij}(\boldsymbol{Z}, t) \frac{\partial^2 f}{\partial Z_i \partial Z_j} \right]$$

$$\times\ p(\boldsymbol{Z}, t | \boldsymbol{Y}, t') d\boldsymbol{Z} + O(\varepsilon)\,.$$

The second term tends to zero because of the condition $R(\boldsymbol{X}, \boldsymbol{Z}) \to 0$ for $\varepsilon \to 0$ since we have $|\boldsymbol{X} - \boldsymbol{Z}| \to 0$.

If one also splits the integration domain in the last integral into two regions $|\boldsymbol{X} - \boldsymbol{Z}| \geqslant \varepsilon$ and $|\boldsymbol{X} - \boldsymbol{Z}| < \varepsilon$, the last three terms can be, according to condition one, transformed to the following form:

$$\lim_{\Delta t \to 0} \frac{1}{\Delta t} \left\{ \int\limits_{|X-Z| \geqslant \varepsilon} f(\boldsymbol{X}) p(\boldsymbol{X}, t + \Delta t | \boldsymbol{Z}, t) p(\boldsymbol{Z}, t | \boldsymbol{Y}, t') d\boldsymbol{X} d\boldsymbol{Z} \right.$$

$$+ \int\limits_{|X-Z| < \varepsilon} f(\boldsymbol{Z}) p(\boldsymbol{X}, t + \Delta t | \boldsymbol{Z}, t) p(\boldsymbol{Z}, t | \boldsymbol{Y}, t') d\boldsymbol{X} d\boldsymbol{Z}$$

$$- \int\limits_{|X-Z| \geqslant \varepsilon} f(\boldsymbol{Z}) p(\boldsymbol{X}, t + \Delta t | \boldsymbol{Z}, t) p(\boldsymbol{Z}, t | \boldsymbol{Y}, t') d\boldsymbol{X} d\boldsymbol{Z}$$

$$\left. - \int\limits_{|X-Z| < \varepsilon} f(\boldsymbol{Z}) p(\boldsymbol{X}, t + \Delta t | \boldsymbol{Z}, t) p(\boldsymbol{Z}, t | \boldsymbol{Y}, t') d\boldsymbol{X} d\boldsymbol{Z} \right\}$$

$$= \int\limits_{|X-Z| \geqslant \varepsilon} \left[f(\boldsymbol{X})\, W(\boldsymbol{X} | \boldsymbol{Z}, t) p(\boldsymbol{Z}, t | \boldsymbol{Y}, t') \right.$$

$$\left. - f(\boldsymbol{Z})\, W(\boldsymbol{X} | \boldsymbol{Z}, t) p(\boldsymbol{Z}, t | \boldsymbol{Y}, t') \right] d\boldsymbol{X} d\boldsymbol{Z} = \ldots$$

Nothing changes if one exchanges $\boldsymbol{X}$ and $\boldsymbol{Z}$ in the first term. Therefore, we have

$$\ldots = \int\limits_{|X-Z| \geqslant \varepsilon} f(\boldsymbol{Z}) \left[W(\boldsymbol{Z} | \boldsymbol{X}, t) p(\boldsymbol{X}, t | \boldsymbol{Y}, t') \right.$$

$$\left. - W(\boldsymbol{X} | \boldsymbol{Z}, t) p(\boldsymbol{Z}, t | \boldsymbol{Y}, t') \right] d\boldsymbol{X} d\boldsymbol{Z}\,.$$

In the limit of $\varepsilon \to 0$, one finally obtains

$$\frac{\partial}{\partial t}\left\{\int f(\boldsymbol{X})p(\boldsymbol{X},t|\boldsymbol{Y},t')d\boldsymbol{X}\right\}$$

$$= \int \left[\sum_i A_i(\boldsymbol{Z},t)\frac{\partial f}{\partial Z_i} + \sum_i\sum_j \frac{1}{2}D_{ij}(\boldsymbol{Z},t)\frac{\partial^2 f}{\partial Z_i \partial Z_j}\right] p(\boldsymbol{Z},t|\boldsymbol{Y},t')d\boldsymbol{Z}$$

$$+ f(\boldsymbol{Z})d\boldsymbol{Z}\left[W(\boldsymbol{Z}|\boldsymbol{X},t)p(\boldsymbol{X},t|\boldsymbol{Y},t') - W(\boldsymbol{X}|\boldsymbol{Z},t)p(\boldsymbol{Z},t|\boldsymbol{Y},t')\right]d\boldsymbol{X}\,. \tag{C120}$$

In order for a solution of (C120) to exist, it is necessary that the integral

$$\int W(\boldsymbol{Z}|\boldsymbol{X},t)p(\boldsymbol{X},t|\boldsymbol{Y},t')d\boldsymbol{X}$$

exists. Condition (C114) defines the function $W(\boldsymbol{Z}|\boldsymbol{X},t)$ only if $\boldsymbol{X} \neq \boldsymbol{Z}$.

By integrating the first term on the right-hand side of (C120) by parts, one obtains

$$\int f(\boldsymbol{Z})\frac{\partial p(\boldsymbol{Z},t|\boldsymbol{Y},t')}{\partial t}d\boldsymbol{Z}$$

$$= \int f(\boldsymbol{Z})\left\{-\sum_i \frac{\partial}{\partial Z_i}\left[A_i(\boldsymbol{Z},t)p(\boldsymbol{Z},t|\boldsymbol{Y},t')\right]\right.$$

$$+\sum_i\sum_j \frac{1}{2}\frac{\partial^2}{\partial Z_i \partial Z_j}\left[D_{ij}(\boldsymbol{Z},t)p(\boldsymbol{Z},t|\boldsymbol{Y},t')\right]$$

$$\left.+\int\left[W(\boldsymbol{Z}|\boldsymbol{X},t)p(\boldsymbol{X},t|\boldsymbol{Y},t') - W(\boldsymbol{X}|\boldsymbol{Z},t)p(\boldsymbol{Z},t|\boldsymbol{Y},t')\right]d\boldsymbol{X}\right\}d\boldsymbol{Z} + \ldots,$$

where the dots denote the surface integrals over the boundary enclosing the region under consideration.

Under the integrals, we have the function $f(Z)$ that can be chosen arbitrarily. Therefore, one can assume that it vanishes on the boundary of the region. As a result, one obtains the following differential equation:

$$\frac{\partial p(\boldsymbol{Z},t|\boldsymbol{Y},t')}{\partial t} = -\sum_i \frac{\partial}{\partial Z_i}\left[A_i(\boldsymbol{Z},t)p(\boldsymbol{Z},t|\boldsymbol{Y},t')\right]$$

$$+\sum_i\sum_j \frac{1}{2}\frac{\partial^2}{\partial Z_i \partial Z_j}\left[D_{ij}(\boldsymbol{Z},t)p(\boldsymbol{Z},t|\boldsymbol{Y},t')\right]$$

$$+\int\left[W(\boldsymbol{Z}|\boldsymbol{X},t)p(\boldsymbol{X},t|\boldsymbol{Y},t')\right.$$

$$\left.- W(\boldsymbol{X}|\boldsymbol{Z},t)p(\boldsymbol{Z},t|\boldsymbol{Y},t')\right]d\boldsymbol{X}\,. \tag{C121}$$

This equation is called the differential equation of Chapman–Kolmogorov.
We now consider a few special cases of this equation.

Discontinuous Processes and the Kolmogorov–Feller Equation

The Kolmogorov–Feller equation follows from the Chapman–Kolmogorov equation under the condition that $A_i = 0$ and $D_{ij} = 0$, and reads

$$\frac{\partial p(\boldsymbol{Z}, t|\boldsymbol{Y}, t')}{\partial t} = \int \left[W(\boldsymbol{Z}|\boldsymbol{X}, t) p(\boldsymbol{X}, t|\boldsymbol{Y}, t') - W(\boldsymbol{X}|\boldsymbol{Z}, t) p(\boldsymbol{Z}, t|\boldsymbol{Y}, t') \right] d\boldsymbol{X} . \tag{C122}$$

Using as initial condition $p(\boldsymbol{Z}, t|\boldsymbol{Y}, t') = \delta(\boldsymbol{Y} - \boldsymbol{Z})$ for $t = t'$, the approximate solution for a small time Δt is

$$p(\boldsymbol{Z}, t|\boldsymbol{Y}, t') \approx \delta(\boldsymbol{Y} - \boldsymbol{Z}) \left[1 - \int W(\boldsymbol{X}|\boldsymbol{Z}, t) \Delta t \, d\boldsymbol{X} \right] + W(\boldsymbol{Z}|\boldsymbol{Y}, t) \Delta t .$$

This means that for each Δt, the particle will be in its initial position $\boldsymbol{Y}$ with probability $1 - \int W(\boldsymbol{X}|\boldsymbol{Z}, t) \Delta t \, d\boldsymbol{X}$ and the distribution of particles leaving $\boldsymbol{Y}$ is given by the function $W(\boldsymbol{Z}|\boldsymbol{Y}, t)$. Therefore, the trajectory $\boldsymbol{X}(t)$ consists of straight lines $\boldsymbol{X} =$ const that are alternating with jumps whose distribution is given by $W(\boldsymbol{X}|\boldsymbol{Z}, t)$. Therefore, the process is discontinuous and the trajectories have discontinuities in a discrete set of points.

Diffusion Processes and the Fokker–Planck Equation

If the process is continuous, we have $W(\boldsymbol{X}|\boldsymbol{Z}, t) = 0$ and the Chapman–Kolmogorov equation becomes the Fokker–Planck equation

$$\begin{aligned} \frac{\partial p(\boldsymbol{Z}, t|\boldsymbol{Y}, t')}{\partial t} &= -\sum_i \frac{\partial}{\partial Z_i} \left[A_i(\boldsymbol{Z}, t) p(\boldsymbol{Z}, t|\boldsymbol{Y}, t') \right] \\ &\quad + \sum_i \sum_j \frac{1}{2} \frac{\partial^2}{\partial Z_i \partial Z_j} \left[D_{ij}(\boldsymbol{Z}, t) p(\boldsymbol{Z}, t|\boldsymbol{Y}, t') \right] . \end{aligned} \tag{C123}$$

The corresponding process is called a diffusion process. The vector $\boldsymbol{A}(\boldsymbol{X})$ is called the drift vector. It is similar to the velocity vector in the convective term of the transport equation. The matrix $\boldsymbol{D}(\boldsymbol{Z}, t) = \| D_{ij}(\boldsymbol{Z}, t) \|$ is called the dispersion matrix. According to the definition (C116), it is non-negative.

We consider the development of a process with the initial condition

$$p(\boldsymbol{Z}, t|\boldsymbol{Y}, t') = \delta(\boldsymbol{Y} - \boldsymbol{Z})$$

at $t = t'$. We assume that $\Delta t \ll 1$ and both A_i as well as D_{ij} change very little as compared to p. Then, (C123) is reduced to

$$\frac{\partial p(\boldsymbol{Z}, t|\boldsymbol{Y}, t')}{\partial t} = -\sum_i A_i(\boldsymbol{Z}, t)\frac{\partial}{\partial Z_i}\left[p(\boldsymbol{Z}, t|\boldsymbol{Y}, t')\right] + \sum_i \sum_j \frac{1}{2} D_{ij}(\boldsymbol{Z}, t)\frac{\partial^2 f}{\partial Z_i \partial Z_j}\left[p(\boldsymbol{Z}, t|\boldsymbol{Y}, t')\right] , \tag{C124}$$

where $t - t' = \Delta t \ll 1$ and A_i as well as D_{ij} are on this small time interval independent of the initial position $\boldsymbol{Y}$ and the time t.

Then, the solution of (C124) is given by

$$p(\boldsymbol{Z}, t + \Delta t|\boldsymbol{Y}, t) = \frac{1}{(2\pi)^{1/2}\,|\boldsymbol{D}(\boldsymbol{Y}, t)|^{1/2}\,(\Delta t)^{1/2}} \times \exp\left\{-\frac{1}{2}\frac{\left[\boldsymbol{Z} - \boldsymbol{Y} - \boldsymbol{A}(\boldsymbol{Y}, t)\Delta t\right]^{\mathrm{T}}\left[\boldsymbol{D}(\boldsymbol{Y}, t)\right]^{-1}\left[\boldsymbol{Z} - \boldsymbol{Y} - \boldsymbol{A}(\boldsymbol{Y}, t)\Delta t\right]}{\Delta t}\right\} . \tag{C125}$$

Here, $|\boldsymbol{D}(\boldsymbol{Y}, t)|$ is the determinant of the matrix $\boldsymbol{D}(\boldsymbol{Y}, t)$.

One sees that the initial diffusion process takes place according to a Gaussian distribution function with regular drift velocity $\boldsymbol{A}(\boldsymbol{Y}, t)$ (see (C70)) on which fluctuations with the correlation matrix $\boldsymbol{D}(\boldsymbol{Y}, t)\Delta t$ are superimposed. The trajectories of this system have the form

$$\boldsymbol{Z}(t + \Delta t) = \boldsymbol{Y}(t) + \boldsymbol{A}(\boldsymbol{Y}(t), t)\Delta t + \eta(t)\,(\Delta t)^{1/2} , \tag{C126}$$

where $\eta(t)$ is a random vector quantity whose mean value and correlation matrix are

$$\langle \eta(t)\rangle = 0 , \quad \left\langle \eta(t)\eta^{\mathrm{T}}(t)\right\rangle = \boldsymbol{D}(\boldsymbol{Y}, t) . \tag{C127}$$

This means that the trajectories are continuous everywhere since $\boldsymbol{Z}(t+\Delta t) \to \boldsymbol{Z}(t)$ for $\Delta t \to 0$, but because of the last term, which is proportional to $(\Delta t)^{1/2}$, they are not differentiable.

Since $\boldsymbol{Z}(t + \Delta t) - \boldsymbol{Y}(t) \sim \Delta \boldsymbol{Z}(t)$ is the random increment of the position, one obtains from (C126) the stochastic differential equation

$$\frac{\Delta \boldsymbol{Z}}{\Delta t} = \boldsymbol{A}(\boldsymbol{Y}(t), t) + \eta(t)\,(\Delta t)^{-1/2} . \tag{C128}$$

This equation is the basis for the description of the particle movement under the influence of an external random force.

In the three-dimensional case, one can write the Gaussian distribution in a simple form. We introduce a Cartesian system of coordinates such that the directions of the coordinate axes coincide with the main axes of the diffusion tensor $\boldsymbol{D}$. We denote the main values of the diffusion tensor by D_{ii}. Then, the distribution (C125) has the following form:

$$p(Z, t + \Delta t | Y, t) = \frac{1}{(2\pi)^{3/2} \left[D_{11}(t) D_{22}(t) D_{33}(t) \right]^{1/2}} \times \exp \left\{ -\frac{(Z_1 - Y_1)^2}{2D_{11}(t)} - \frac{(Z_2 - Y_2)^2}{2D_{22}(t)} - \frac{(Z_3 - Y_3)^2}{2D_{33}(t)} \right\} . \tag{C129}$$

We now introduce the probability current

$$J_i(Z, t) = A_i(Z, t) p(Z, t | Y, t') - \frac{1}{2} \sum_j \frac{\partial f}{\partial Z_j} \left[D_{ij}(Z, t) p(Z, t | Y, t') \right] .$$

Then, the Fokker–Planck equation can be written in the usual form of a conservation equation:

$$\frac{\partial p(Z, t | Y, t')}{\partial t} + \sum_i \frac{\partial J_i(Z, t)}{\partial Z_i} = 0 . \tag{C130}$$

The concept of a probability current allows the formulation of boundary conditions for the Fokker–Planck equation. Using the example of the random movement of a particle in a finite region R with a surface boundary Σ or in an infinite domain, we consider the possible boundary conditions.

a) Absorbing boundary
It is assumed that as soon as a particle reaches the boundary, it vanishes, that is, it leaves the system, for example, adheres to the surface or reacts with the boundary surface. Therefore, the probability to find the particle at the boundary vanishes and the boundary condition for the probability density is

$$p(Z, t | Y, t') = 0 \quad \text{for} \quad Z \in \Sigma . \tag{C131}$$

b) Reflecting boundary
If the particle can not leave the region R, the probability current in the direction normal to the boundary surface Σ must vanish. Therefore, the boundary condition is

$$\boldsymbol{n} \cdot J_i(Z, t) = 0 \quad \text{for} \quad Z \in \Sigma , \tag{C132}$$

where $\boldsymbol{n}$ is the normal vector to the surface Σ.

c) Surface of discontinuity
We consider the boundary Σ between two media with different properties. At this boundary, the coefficients A_i and D_{ij} experience a jump, but the particles have to be able to cross unhindered through Σ. Therefore, the probability

and normal component of the probability current have to be continuous at the boundary:

$$\begin{aligned} p(\mathbf{Z},t|\mathbf{Y},t')\,|_{\Sigma+} &= p(\mathbf{Z},t|\mathbf{Y},t')\,|_{\Sigma-}\,, \\ \boldsymbol{n}\cdot \boldsymbol{J}_i(\mathbf{Z},t)\,|_{\Sigma+} &= \boldsymbol{n}\cdot \boldsymbol{J}_i(\mathbf{Z},t)\,|_{\Sigma-}\,. \end{aligned} \tag{C133}$$

d) Conditions at infinity
   If the process takes place in an infinite domain, then, depending on the problem, one of the following conditions has to be satisfied:

$$\begin{aligned} &p(\mathbf{Z},t) = 0 \text{ or } 1 \text{ for } |\mathbf{Z}| \to \infty\,; \\ \text{or } &\frac{\partial p(\mathbf{Z},t)}{\partial \mathbf{Z}} \to 0 \text{ for } |\mathbf{Z}| \to \infty\,. \end{aligned} \tag{C134}$$

In one dimension, the Fokker–Planck equation with scalar coefficients A and D has the following form:

$$\begin{aligned} \frac{\partial p(Z,t|Y,t')}{\partial t} &= -\frac{\partial}{\partial Z}\left[A(Z,t)p(Z,t|Y,t')\right] \\ &\quad + \frac{1}{2}\frac{\partial^2}{\partial Z^2}\left[D(Z,t)p(Z,t|Y,t')\right]. \end{aligned}$$

This equation can be rewritten as

$$\begin{aligned} \frac{\partial p(Z,t|Y,t')}{\partial t} &= \frac{\partial}{\partial Z}\left[\frac{1}{2}D(Z,t)\frac{\partial}{\partial Z}p(Z,t|Y,t')\right] \\ &\quad - \frac{\partial}{\partial Z}\left(\left[A(Z,t) - \frac{1}{2}\frac{\partial D}{\partial Z}\right]p(Z,t|Y,t')\right)\,. \end{aligned} \tag{C135}$$

We now compare this equation with the molecular diffusion equation

$$\frac{\partial C}{\partial t} = \frac{\partial (\nu C)}{\partial Z} + \frac{\partial}{\partial Z}\left(D_m \frac{\partial C}{\partial Z}\right), \tag{C136}$$

which describes the change of the concentration C of a substance in the solution where ν is the velocity of the substance under the influence of an external force and D_m is the coefficient of molecular diffusion.

One sees that

$$D = 2D_m\,, \quad A = \nu + \frac{1}{2}\frac{\partial D}{\partial Z}\,. \tag{C137}$$

Therefore, $D/2$ has the meaning of the coefficient of molecular diffusion and A the average velocity of particle displacement. The latter one consist of two components: the velocity under the influence of the external force and the inhomogeneity of the medium. It must be noted that (C136) does not deal with the probability but with the concentration C of the particles. However, if one uses C/N instead of C, where N is number density, one can consider C/N as p.

The one-dimensional Fokker–Planck equation with $A = 0$ and $D = 1$

$$\frac{\partial p(\mathbf{Z}, t|\mathbf{Y}, t')}{\partial t} = \frac{1}{2}\frac{\partial^2}{\partial Z^2}\left[p(\mathbf{Z}, t|\mathbf{Y}, t')\right] \tag{C138}$$

corresponds to the Wiener process.

With the initial condition $p(\mathbf{Z}, t|\mathbf{Y}, t') = \delta(\mathbf{Y} - \mathbf{Z})$ for $t = t'$, the solution is equal to

$$p(Z, t|Y, t') = \frac{1}{(2\pi)^{1/2}} \exp\left\{-\frac{(Z - Y)^2}{2(t - t')}\right\}. \tag{C139}$$

In the multidimensional case, one obtains

$$\frac{\partial p(\mathbf{Z}, t|\mathbf{Y}, t')}{\partial t} = \frac{1}{2}\sum_i \frac{\partial^2}{\partial Z_i^2}\left[p(\mathbf{Z}, t|\mathbf{Y}, t')\right] \tag{C140}$$

and

$$p(\mathbf{Z}, t|\mathbf{Y}, t') = \frac{1}{(2\pi)^{n/2}} exp\left\{-\frac{(\mathbf{Z} - \mathbf{Y})^2}{2(t - t')}\right\}. \tag{C141}$$

Sometimes, the Wiener process is referred to as Brownian motion.

Deterministic Processes and the Liouville Equation

For $W(\mathbf{X}|\mathbf{Y}, t) = 0$ and $D_{ij} = 0$, the Chapman–Kolmogorov equation (C121) becomes the Liouville equation

$$\frac{\partial p(\mathbf{Z}, t|\mathbf{Y}, t')}{\partial t} = -\sum_i \frac{\partial}{\partial Z_i}\left[A_i(\mathbf{Z}, t) p(\mathbf{Z}, t|\mathbf{Y}, t')\right]. \tag{C142}$$

This equation describes a deterministic movement, for example, the movement of a particle whose trajectory can be uniquely determined from the initial conditions. We consider the trajectory $\mathbf{X}(t)$ which is the solution of the characteristic equation

$$\frac{d\mathbf{X}(t)}{dt} = A(\mathbf{X}, t)$$

with the initial condition $\mathbf{X}(\mathbf{Y}, t') = \mathbf{Y}$.

Here, $\mathbf{A}$ is a vector with components A_i. We now show that the solution of (C142) with the initial condition $p(\mathbf{Z}, t|\mathbf{X}, t') = \delta(\mathbf{Z} - \mathbf{X})$ is the function $p(\mathbf{Z}, t|\mathbf{Y}, t') = \delta[\mathbf{Z} - \mathbf{X}(\mathbf{Y}, t)]$.

By substituting this function into (C142), one obtains

$$\begin{aligned}\frac{\partial\delta[Z - X(Y,t)]}{\partial t} &= -\sum_i \left\{\frac{\partial\delta\left[Z - X(Y,t)\right]}{\partial Z_i}\frac{dX_i(Y,t)}{dt}\right\} \\ &= -\sum_i \left\{\frac{\partial\delta[Z - X(Y,t)]}{\partial Z_i} A_i[X(Y,t),t]\right\} \\ &= -\sum_i \frac{\partial}{\partial Z_i}\{A_i[X(Y,t),t]\delta[Z - X(Y,t)]\} \\ &= -\sum_i \frac{\partial}{\partial Z_i}\{A_i(Z,t)\delta[Z - X(Y,t)]\} \ .\end{aligned}$$

This example corresponds to the movement of a single particle. For a number of particles, the Liouville equation has another form.

The system of material points is given. These points move in accordance with Newton's second law:

$$\ddot{X} = \frac{d^2 X}{dt^2} = F_i \quad \text{or} \quad \dot{V}_i = \frac{dV_i}{dt} = F_i \ ; \quad \dot{X}_i = \frac{dX_i}{dt} = V_i \ , \tag{C143}$$

where X_i, V_i, F_i are the radius vector, the velocity and the force that act on the i-th particle by other particles and through an external force. The initial positions and the velocities of the particles are given by

$$X_i(0) = X_i^0 \ , \quad V_i(0) = V_i^0 \ . \tag{C144}$$

To determine the trajectories of all N particles, one has to integrate the system of equations (C144). If the number of particles is very large, that is, $N \gg 1$, this is practically impossible and one must therefore apply statistical methods.

We introduce the density of the common probability that at a fixed time particle one is in the interval $(X_1 + dX_1, V_1 + dV_1)$ of the generalized coordinates, that is, (X, V), particle two is in the interval $(X_2 + dX_2, V_2 + dV_2), \ldots$ and particle N is in the interval $(X_N + dX_N, V_N + dV_n)$.

If the trajectories (i.e., the functions $X_i(t)$ and V_i) of the particles are known, that probability density vanishes if at least for one of the particles we have $X_i \neq X_i(t)$ or $V_i \neq V_i(t)$. This means that probability is reduced to certainty and the probability density must have the form of a product of delta functions (see (C11)):

$$p(X,V,t) = \prod_{i=1}^{N} \delta\left[X_i - X_i(t)\right]\delta\left[V_i - V_i(t)\right] \ . \tag{C145}$$

We now show that the distribution (C145) satisfies the Liouville equation. Using the property (C19) of the delta function, one obtains

$$\frac{\partial p(\boldsymbol{X}, \boldsymbol{V}, t)}{\partial t} = -\sum_{j=1}^{N} \left\{ \prod_{k=1,k\neq j}^{N} \delta\left[\boldsymbol{X}_k - \boldsymbol{X}_k(t)\right] \delta\left[\boldsymbol{V}_k - \dot{\boldsymbol{X}}_k(t)\right] \right\} \dot{\boldsymbol{X}}_j$$
$$\times \frac{\partial \delta\left[\boldsymbol{X}_j - \boldsymbol{X}_j(t)\right]}{\partial \boldsymbol{X}_j} \delta\left[\boldsymbol{V}_j - \dot{\boldsymbol{X}}_j(t)\right]$$
$$- \sum_{j=1}^{N} \left\{ \prod_{k=1,k\neq j}^{N} \delta\left[\boldsymbol{X}_k - \boldsymbol{X}_k(t)\right] \right.$$
$$\left. \times\, \delta\left[\boldsymbol{V}_k - \dot{\boldsymbol{X}}_k(t)\right] \right\} \delta\left[\boldsymbol{X}_j - \boldsymbol{X}_j(t)\right] \ddot{\boldsymbol{X}}_j \frac{\partial \delta\left[\boldsymbol{V}_j - \dot{\boldsymbol{V}}_j(t)\right]}{\partial \boldsymbol{V}_j} .$$

From this, it follows, with the help of (C15) and (C143), that

$$\frac{\partial p(\boldsymbol{X}, \boldsymbol{V}, t)}{\partial t} = -\sum_{j=1}^{N} \boldsymbol{V}_j \frac{\partial \delta\left[\boldsymbol{X}_j - \boldsymbol{X}_j(t)\right]}{\partial \boldsymbol{X}_j} \delta\left[\boldsymbol{V}_k - \dot{\boldsymbol{X}}_k(t)\right]$$
$$\times \left\{ \prod_{k=1,k\neq j}^{N} \delta\left[\boldsymbol{X}_k - \boldsymbol{X}_k(t)\right] \delta\left[\boldsymbol{V}_k - \dot{\boldsymbol{X}}_k(t)\right] \right\} - \sum_{j=1}^{N} \boldsymbol{F}_j \frac{\partial \delta\left[\boldsymbol{V}_j - \dot{\boldsymbol{V}}_j(t)\right]}{\partial \boldsymbol{V}_j}$$
$$\times\, \delta\left[\boldsymbol{X}_k - \boldsymbol{X}_k(t)\right] \left\{ \prod_{k=1,k\neq j}^{N} \delta\left[\boldsymbol{X}_k - \boldsymbol{X}_k(t)\right] \delta\left[\boldsymbol{V}_k - \dot{\boldsymbol{X}}_k(t)\right] \right\} .$$

It can be easily verified that

$$\frac{\partial p(\boldsymbol{X}, \boldsymbol{V}, t)}{\partial \boldsymbol{X}_j} = \frac{\partial \delta\left[\boldsymbol{X}_j - \boldsymbol{X}_j(t)\right]}{\partial \boldsymbol{X}_j} \delta\left[\boldsymbol{V}_k - \dot{\boldsymbol{X}}_k(t)\right]$$
$$\times \left\{ \prod_{k=1,k\neq j}^{N} \delta\left[\boldsymbol{X}_k - \boldsymbol{X}_k(t)\right] \delta\left[\boldsymbol{V}_k - \dot{\boldsymbol{X}}_k(t)\right] \right\} ,$$
$$\frac{\partial p(\boldsymbol{X}, \boldsymbol{V}, t)}{\partial \boldsymbol{V}_j} = \frac{\partial \delta\left[\boldsymbol{V}_j - \dot{\boldsymbol{V}}_j(t)\right]}{\partial \boldsymbol{V}_j} \delta\left[\boldsymbol{X}_k - \boldsymbol{X}_k(t)\right]$$
$$\times \left\{ \prod_{k=1,k\neq j}^{N} \delta\left[\boldsymbol{X}_k - \boldsymbol{X}_k(t)\right] \delta\left[\boldsymbol{V}_k - \dot{\boldsymbol{X}}_k(t)\right] \right\} .$$

As a result, one obtains

$$\frac{\partial p(\boldsymbol{X}, \boldsymbol{V}, t)}{\partial t} = -\sum_{i=1}^{N} \boldsymbol{V}_i \frac{\partial p(\boldsymbol{X}, \boldsymbol{V}, t)}{\partial \boldsymbol{X}_i} - \sum_{i=1}^{N} \boldsymbol{F}_i \frac{\partial p(\boldsymbol{X}, \boldsymbol{V}, t)}{\partial \boldsymbol{V}_i} . \tag{C146}$$

During the discussion of a system of material points in statistical physics, one normally introduces, as generalized coordinates, the ordinary coordinates $\boldsymbol{X}_i$ and the momenta $m_i \boldsymbol{V}_i$. If the generalized coordinates $\boldsymbol{Z}_i$ are defined as coordinates $\boldsymbol{X}_i$ and momenta per unit particle mass $\boldsymbol{V}_i$, then (C146) is transformed to

$$\frac{\partial p(Z,t)}{\partial t} = -\sum_{i=1}^{N} A_i \frac{\partial p(Z,t)}{\partial Z_i} , \tag{C147}$$

where $\boldsymbol{A}_i$ is a generalized vector consisting of $\boldsymbol{V}_i$ and $\boldsymbol{F}_i$.

C.16 Stochastic Differential Equations and the Langevin Equation

A stochastic differential equation describes a function that fluctuates quickly and randomly with respect to time and space. Therefore, its solution is a random function. Examples of stochastic differential equation are the Langevin equation, which describes the random trajectories of particle movement under the influence of a random external force, and the diffusion equation of a chemical reaction process, which lead to the fluctuation of parameters. We now discuss these equations in detail.

The Langevin Equation

The most frequently used equation is

$$\frac{d\boldsymbol{X}}{dt} = \boldsymbol{a}(\boldsymbol{X},t) + b(\boldsymbol{X},t)\xi(t) , \tag{C148}$$

where $\boldsymbol{a}(\boldsymbol{X},t)$ and $b(\boldsymbol{X},t)$ are known functions, and $\xi(t)$ is a randomly and rapidly fluctuating vector function.

This equation describes the Markovian process of the movement of a particle under the influence of a random force $\xi(t)$ whose average vanishes,

$$\langle \xi(t) \rangle = 0 , \tag{C149}$$

and which shows no correlation between its values at different times, that is, the correlation of $\xi(t)$ and $\xi(t')$ is a delta function:

$$\left\langle \xi(t)\xi(t') \right\rangle = \Xi\,\delta(t-t') . \tag{C150}$$

Both (C148) as well as condition (C150) only apply to the idealized process. In reality, the particle movement is defined according to Newton's second law via the following equation:

$$m\frac{d^2\boldsymbol{X}}{dt^2} = -h\frac{d\boldsymbol{X}}{dt} + \boldsymbol{F} + \boldsymbol{f} , \tag{C151}$$

where $-h(d\boldsymbol{X}/dt)$ is the drag force exerted on the particle by the surrounding fluid, $\boldsymbol{F}$ is the deterministic external force, for example, gravity, electrodynamic force and so on, and $\boldsymbol{f}$ is the stochastic force.

We assume that the particle is very small. Then, one can neglect its inertia in a first approximation, that is, the term on the left-hand side of (C151) is much smaller than the terms on the right-hand side. Finally, the equation assumes the form (C148). Another approximation is the condition (C150), which results in an infinitely large dispersion for $t' = t$, which is certainly impossible.

Condition (C150) is similar to the assumption of white noise in electrical engineering. Therefore, the representation of the correlation function as a delta function is the natural idealization of the process during the transition from small to large time scales in which the process, for example, the Brownian motion, is considered.

This means, however, that one has to deal very carefully with differential equations that contain delta correlated random functions since the usual numerical methods will not always be useful in this case.

The Diffusion Equation

For a deterministic description of processes in continuous media, one uses conservation equations. To close these systems, one uses phenomenological laws (Fick's law for the diffusion process, Fourier's law for the heat transport process, Navier–Stokes law in the hydromechanics of the Newtonian fluid, the kinetic equations of chemical reactions and so on). This leads to the question as to how the fluctuations of the corresponding quantities can be taken into account. Presently, we will give the answer for the example of the diffusion equation.

According to Fick's law, the diffusive current $\boldsymbol{j}(\boldsymbol{X})$ is proportional to the gradient of the concentration:

$$\boldsymbol{j}(\boldsymbol{X}, t) = -D\nabla C(\boldsymbol{X}, t) . \tag{C152}$$

On the other hand, the equation of mass conservation (equation of continuity) reads

$$\frac{\partial C}{\partial t} + \nabla \cdot \boldsymbol{j}(\boldsymbol{X}, t) = 0 . \tag{C153}$$

From (C152) and (C153), one obtains the ordinary diffusion equation

$$\frac{\partial C}{\partial t} = \nabla \cdot \left[D\nabla C(\boldsymbol{X}, t) \right] . \tag{C154}$$

By supplementing Fick's law (C152) with a fluctuation term, one obtains

$$\boldsymbol{j}(\boldsymbol{X}, t) = -D\nabla C(\boldsymbol{X}, t) + \boldsymbol{j}^{fl}(\boldsymbol{X}, t) . \tag{C155}$$

In analogy with the Langevin equation, one assumes the following properties for the fluctuation term:

$$\left\langle \boldsymbol{j}^{fl}(\boldsymbol{X}, t) \right\rangle = 0 , \tag{C156}$$

$$\left\langle j_i^{fl}(\boldsymbol{X}, t) j_k^{fl}(vecX', t') \right\rangle = K_1(\boldsymbol{X}, t) \delta_{ik} \delta(\boldsymbol{X} - \boldsymbol{X}') \delta(t - t') . \tag{C157}$$

The latter property means that different components of the fluctuation current vector $\boldsymbol{j}^{fl}(\boldsymbol{X}, t)$ at the same point are assumed to be independent as well as those at different times or in different points. This can also be described by saying that the fluctuations are local.

From (C153) and (C155), one obtains

$$\frac{\partial C}{\partial t} = \nabla \cdot \left[D \nabla C(\boldsymbol{X}, t) \right] - \nabla \cdot \boldsymbol{j}^{fl}(\boldsymbol{X}, t) . \tag{C158}$$

One notices that taking fluctuations into account results in an additional term in the form of a divergence in the diffusion equation. This term has the following statistical properties:

$$\left\langle \nabla \cdot \boldsymbol{j}^{fl}(\boldsymbol{X}, t) \right\rangle = 0 , \tag{C159}$$

$$\left\langle \nabla \cdot \boldsymbol{j}^{fl}(\boldsymbol{X}, t) \nabla' \boldsymbol{j}^{fl}(\boldsymbol{X}', t') \right\rangle = \nabla \cdot \nabla' \left[K_1(\boldsymbol{X}, t) \delta(\boldsymbol{X} - \boldsymbol{X}') \right] \delta(t - t') . \tag{C160}$$

Diffusion Equation Taking Chemical Reactions into Account

The deterministic diffusion equation with a source term due to chemical reactions has the following form:

$$\frac{\partial C}{\partial t} + \nabla \cdot \boldsymbol{j}(\boldsymbol{X}, t) = F\left[C(\boldsymbol{X}, t) \right] . \tag{C161}$$

The term on the right-hand side represents the intensity of production or consumption of the substance due to chemical reactions.

A chemical reaction involving the production or consumption of a substance is accompanied by a fluctuation of the substance concentration which can be taken into account by adding to the deterministic term $F\left[C(\boldsymbol{X}, t)\right]$, the fluctuation term $g^{fl}(\boldsymbol{X}, t)$. This term has the following properties:

$$\left\langle g^{fl}(\boldsymbol{X}, t) \right\rangle = 0 , \tag{C162}$$

$$\left\langle g^{fl}(\boldsymbol{X}, t) g^{fl}(\boldsymbol{X}', t') \right\rangle = K_2(\boldsymbol{X}, t) \delta(\boldsymbol{X} - \boldsymbol{X}') \delta(t - t') . \tag{C163}$$

The latter property expresses locality (lack of correlations between fluctuations at different points) as well as the Markovian character of chemical reactions.

From (C155), (C161)–(C163) follow the stochastic equation

$$\frac{\partial C}{\partial t} + \nabla \cdot \boldsymbol{j}(\boldsymbol{X}, t) = F\left[C(\boldsymbol{X}, t)\right] + G^{fl}(\boldsymbol{X}, t) , \tag{C164}$$

where

$$G^{fl}(\boldsymbol{X}, t) = -\nabla \cdot \boldsymbol{j}^{fl}(\boldsymbol{X}, t) + g^{fl}(\boldsymbol{X}, t) ; \quad \left\langle G^{fl}(\boldsymbol{X}, t) G^{fl}(\boldsymbol{X}, t)\right\rangle = 0 ;$$

$$\left\langle G^{fl}(\boldsymbol{X}, t) G^{fl}(\boldsymbol{X}', t')\right\rangle = \{K_2(\boldsymbol{X}, t)\delta(\boldsymbol{X} - \boldsymbol{X}')$$
$$+\nabla \cdot \nabla'\left[K_1(\boldsymbol{X}, t)\delta(\boldsymbol{X} - \boldsymbol{X}')\right]\} \delta(t - t') .$$

Equations C157 and (C161) are equations of the Langevin type. The functions $K_1(\boldsymbol{X}, t)$ and $K_2(\boldsymbol{X}, t)$ must be known beforehand. They can be obtained via a microscopic investigation of the process in a similar manner as was done for the derivation of the Chapman–Kolmogorov equation. Thus, it can be shown that the diffusion coefficients are the components of the dispersion tensor. The difference between (C157) and (C161) is the term $F\left[C(\boldsymbol{X}, t)\right]$ on the right-hand side of (C161), which is not a function but a functional.

Brownian Motion of a Particle in the Hydrodynamic Medium

The movement of a particle in a fluctuating hydrodynamic medium which is at rest at infinity under the influence of the average viscous force due to the surrounding fluid $\boldsymbol{F}$ and the fluctuating force $\boldsymbol{F}^{fl}$, a force that can, for example, be caused by thermodynamic fluctuations in the fluid, is described by the following Langevin equation:

$$m\frac{d\boldsymbol{U}}{dt} = -\boldsymbol{F} + \boldsymbol{F}^{fl} . \tag{C165}$$

We now determine the statistical characteristics of the random force $\boldsymbol{F}^{fl}(t)$. For this, we require the hydrodynamic equations. Since the particle is sufficiently small, one can use Stokes equations

$$\nabla \cdot \boldsymbol{u} = 0 , \quad \nabla \cdot \boldsymbol{T} = 0 . \tag{C166}$$

Here, $\boldsymbol{T}(\boldsymbol{r}, t) = \boldsymbol{T}^{\sigma} + \boldsymbol{T}^{fl}$ is the stress tensor with the terms $\boldsymbol{T}^{\sigma} = -p\boldsymbol{I} + 2\mu\boldsymbol{E}$ (the deterministic component and $\boldsymbol{T}^{fl}$ (the fluctuating component) where $-p\boldsymbol{I}$ is the isotropic spherical tensor (with the unit tensor $\boldsymbol{I}$) and $2\mu\boldsymbol{E} = \mu\left(\nabla\boldsymbol{u} + \nabla\boldsymbol{u}^T\right)$ is the symmetric part of viscous stress tensor (deviator) with zero trace. $\boldsymbol{E}$ is also called the strain velocity tensor.

In a statistically homogeneous medium, the fluctuating tensor $\boldsymbol{T}^{fl}$ is also symmetric with zero trace. One usually assumes that the components of the tensor $\boldsymbol{T}^{tl}$ are characterized by a Gaussian distribution with mean values $\langle T_{ik}^{fl}\rangle = 0$ and the correlation matrix

$$\left\langle T_{ik}^{fl}(\boldsymbol{r}_1, t_1) T_{lm}^{fl}(\boldsymbol{r}_2, t_2) \right\rangle$$
$$= 2kT\left(\delta_{il}\delta_{km} + \delta_{im}\delta_{kl} - \frac{2}{3}\delta_{ik}\delta_{lm}\right) \times \delta\left(\boldsymbol{r}_1 - \boldsymbol{r}_2\right)\delta\left(t_1 - t_2\right), \qquad \text{(C167)}$$

where k is the Boltzmann constant.

Then, one can rewrite the equations as

$$\frac{\partial u_i}{\partial X_i} = 0\,, \qquad \frac{\partial T_{ij}^{\sigma}}{\partial X_j} = -\frac{\partial T_{ij}^{fl}}{\partial X_j}\,. \qquad \text{(C168)}$$

We look for a solution of these equations in the form

$$u_i = \langle u_i \rangle + \tilde{u}_i\,, \quad T_{ij}^{\sigma} = \left\langle T_{ij}^{\sigma} \right\rangle + \tilde{T}_{ij}^{\sigma}\,, \qquad \text{(C169)}$$

where $\langle u_i \rangle$ and $\langle T_{ij}^{\sigma} \rangle$ are average values and $\tilde{u}_i$ and $\tilde{T}_{ij}^{\sigma}$ are small fluctuating additions.

By substituting the expressions in (C169) into (C168) and neglecting small terms of higher order, one obtains equations for the average quantities $\langle u_i \rangle$ and $\langle T_{ij}^{\sigma} \rangle$, and the fluctuating additional terms $\tilde{u}_i$ and $\tilde{T}_{ij}^{\sigma}$:

$$\frac{\partial \langle u_i \rangle}{\partial X_i} = 0\,, \qquad \frac{\partial \left\langle T_{ij}^{\sigma} \right\rangle}{\partial X_j} = 0\,, \qquad \frac{\partial \tilde{u}_i}{\partial X_i} = 0\,, \qquad \frac{\partial \tilde{T}_{ij}^{\sigma}}{\partial X_j} = 0\,. \qquad \text{(C170)}$$

The boundary conditions are the no-slip conditions $\langle u_i \rangle = U_i, \quad \tilde{u}_i = 0$ on the surface of the body and the state of rest $\langle u_i \rangle = 0$, $\tilde{u}_i = 0$ at infinity, where U_i are the components of the body velocity. The drag force on the body also consists of the deterministic force $\boldsymbol{F}$ and the fluctuating force $\boldsymbol{F}^{f l|}$, whose components are

$$\langle F_i \rangle = \int\limits_{\Sigma} \left\langle T_{ij}^{\sigma} \right\rangle n_j\, d\Sigma\,, \quad F_i^{fl} = \int\limits_{\Sigma} \tilde{T}_{ij}^{\sigma} n_j\, d\Sigma\,, \qquad \text{(C171)}$$

where n_j are the components of the external body normal and Σ is the surface of the body.

The solution of the first equation in (C171) is

$$\langle F_i \rangle = R_{ij} U_j\,, \qquad \text{(C172)}$$

where R_{ij} are the components of the resistance tensor $\boldsymbol{R}$.

To define the statistical characteristics of the random force $\boldsymbol{F}^{fl}$, we use the equation that follows from the Gaussian integral theorem, the equations (C170) and the boundary conditions:

$$\int_V \left(\langle u_i \rangle \frac{\partial \tilde{T}^{\sigma}_{ij}}{\partial X_j} - \tilde{u}_i \frac{\partial \left\langle T^{\sigma}_{ij} \right\rangle}{\partial X_j} \right) dV = \int_V \left[\frac{\partial}{\partial X_j} \left(\langle u_i \rangle \tilde{T}^{\sigma}_{ij} - \tilde{u}_i \left\langle T^{\sigma}_{ij} \right\rangle \right) \right] dV$$
$$= \int_\Sigma \left(\langle u_i \rangle \tilde{T}^{\sigma}_{ij} - \tilde{u}_i \left\langle T^{\sigma}_{ij} \right\rangle \right) n_j \, d\Sigma = - \int_\Sigma \langle u_i \rangle \frac{\partial \tilde{T}^{\sigma}_{ij}}{\partial X_j} d\Sigma$$
$$= \int_\Sigma U_i \tilde{T}^{\sigma}_{ij} n_j \, d\Sigma = U_i \tilde{F}_i \, .$$

The integration is performed over the fluid volume V. Using the above equation and the property (C167), one obtains the correlation of the the components of the fluctuating force as

$$\left\langle F_i^{fl}(t) F_j^{fl}(t') \right\rangle = \frac{1}{U_i U_j} 2kT\delta(t-t') \int_\Sigma \langle u_i \rangle \left\langle T^{\sigma}_{ij} \right\rangle n_j \, d\Sigma = 2kT R_{ij} \delta(t-t') \, , \tag{C173}$$

where k is the Boltzmann constant.

Appendix D
Basics of Complex Analysis

D.1
Complex Numbers

Complex numbers are called the numbers $x + iy$, where x and y are real numbers and i is a symbol called the *imaginary unit*. The numbers x and y are called *real* and *imaginary* parts of the complex number $x + iy$, and are specified by

$$x = \mathrm{Re}(x + iy)\,, \quad y = Im(x + iy)\,. \tag{D1}$$

Let us define, on the set of complex numbers, the notion of equality and the simplest operations with complex numbers. We say that complex numbers $x_1 + iy_1$ and $x_2 + iy_2$ are equal if and only if $x_1 = x_2$ and $y_1 = y_2$. The complex number $x_2 + iy_2$ with $x_2 = x_1$ and $y_2 = -y_1$ is called a conjugate complex number to $x_1 + iy_1$ and is designated by $\overline{x_1 + iy_1}$. Therefore,

$$\overline{x + iy} = x - iy\,. \tag{D2}$$

D.1.1
Operations with Complex Numbers

Addition

The sum $z_1 + z_2$ of complex numbers $z_1 = x_1 + iy_1$ and $z_2 = x_2 + iy_2$ is called the complex number

$$z = z_1 + z_2 = (x_1 + x_2) + i(y_1 + y_2)\,. \tag{D3}$$

From this definition follows the *commutative law* $z_1 + z_2 = z_2 + z_1$ and *associative law* $z_1 + (z_2 + z_3) = (z_1 + z_2) + z_3$.

Difference

Addition admits the inverse operation: for any two complex numbers $z_1 = x_1 + iy_1$ and $z_2 = x_2 + iy_2$, one can find such a complex number z that $z_2 + z = z_1$. This number is called the difference of numbers z_1 and z_2, and is denoted by $z_1 - z_2$. It is evident that

$$z = z_1 - z_2 = (x_1 - x_2) + i(y_1 - y_2)\,. \tag{D4}$$

Hydromechanics. Theory and Fundamentals. Emmanuil G. Sinaiski

ISBN: 978-3-527-41026-2

Multiplication

Multiplication $z_1 z_2$ of complex numbers $z_1 = x_1 + i y_1$ and $z_2 = x_2 + i y_2$ is called the complex number

$$z = z_1 z_2 = (x_1 x_2 - y_1 y_2) + i(x_1 y_2 + y_1 x_2) . \tag{D5}$$

From this definition follows the *commutative law* $z_1 z_2 = z_2 z_1$, *associative law* $z_1(z_2 z_3) = (z_1 z_2) z_3$ and distributive (relative addition) law $(z_1 + z_2) z_3 = z_1 z_3 + z_2 z_3$. At $z_1 = z_2 = i$ from multiplication definition it follows

$$i \cdot i = -1 \tag{D6}$$

It should be noted that from (D5), we get

$$z \bar{z} = x^2 + y^2 \geqslant 0 , \tag{D7}$$

that is, multiplication of a complex number with its conjugate number is always non-negative. The multiplication admits inverse operation if only the multiplier is nonzero. Let $z_2 \neq 0$. Then, one can find such a complex number z, that $z_2 z = z_1$. To do this, one must, in accordance with (D5), solve the following system of equations

$$\begin{aligned} x_2 x - y_2 y &= x_1 , \\ y_2 x + x_2 y &= y_1 , \end{aligned} \tag{D8}$$

which for $z_2 \neq 0$, can always be uniquely resolved since its determinant $x_2^2 + y_2^2 > 0$. The number z is called the quotient of two numbers z_1 and z_2 and is designated by symbol z_1/z_2. By solving the system of (D8), we obtain

$$z = \frac{z_1}{z_2} = \frac{x_1 x_2 + y_1 y_2}{x_2^2 + y_2^2} + i \frac{y_1 x_2 - x_1 y_2}{x_2^2 + y_2^2} . \tag{D9}$$

Taking the n-th power

The n-th power of the number z is defined as multiplication of $n$equal numbers $z$and is denoted by z^n:

$$z^n = \underbrace{z \ldots z}_{n} . \tag{D10}$$

The inverse operation, that is, *taking the n-th root*, is defined as follows: the number $w = \sqrt[n]{z}$ is called the complex root of n-th power extracted from the number z if $w^n = z$. At $n = 2$, it is usually written as $\sqrt{z}$, while for any $z \neq 0$, the root $\sqrt[n]{z}$ has n different values. For example, the equality (D6) can now be written as $i^2 = -1$.

D.1.2
Geometrical Interpretation of Complex Numbers

Consider the plane of Cartesian coordinates xOy, and depict a complex number $z = x + i y$ as a point with coordinates (x, y) (see Figure D.1). Thus, the real

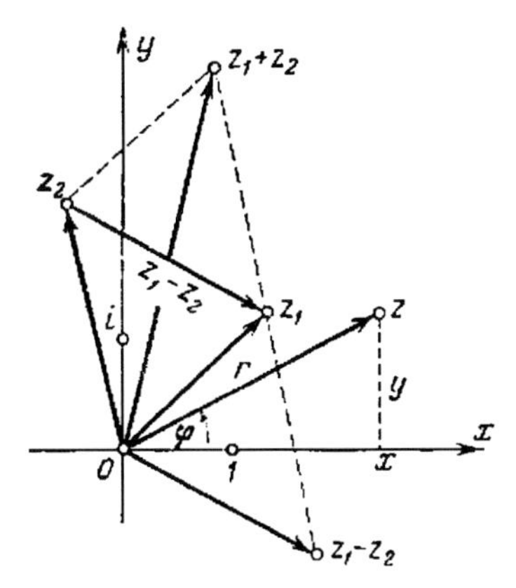

Figure D.1 Geometrical meaning of complex numbers.

part of the number will be depicted by points at the axis x (real axis), whereas purely imaginary parts will be depicted by points at the axis y (imaginary axis). In particular, the point (0, 1) at the imaginary axis serves as an image of the imaginary unit i. It is evident that the correspondence between the set of complex numbers $z = x + iy$ and the set of points (x, y) at the Cartesian plane xOy is one-to-one. To each point (x, y), corresponds a definite radius vector end and vice versa. Therefore, in what follows, we shall represent complex numbers as radius vectors on the plane.

Along with the representation of complex numbers in Cartesian coordinates, it is useful to represent them in polar coordinates r, ϕ where r is the polar radius and ϕ is the polar angle of the complex number z (see Figure D.1). Then, we have

$$z = x + iy = r\cos\phi + i\sin\phi \ . \tag{D11}$$

The polar radius r is called the module of the complex number z and is designated by $|z|$, and the angle ϕ, that is, the argument of the complex number, is designated by $\arg z$. Whereas, the module of the complex number is determined to be single-valued, namely, $|z| = \sqrt{x^2 + y^2} \geqslant 0$, the argument of the complex number is determined only up to a multiple of 2π:

$$\varphi = \arg z = \begin{cases} \arctan\frac{y}{x} + 2k\pi \ , & \text{1st and 4th quadrants} \ , \\ \arctan\frac{y}{x} + (2k+1)\pi \ , & \text{2nd and 3rd quadrants} \ ; \end{cases} \ . \tag{D12}$$

Here, $\arctan(y/x)$ is the principal value of $\arctan z$, that is, $-\pi/2 < \arctan(y/x) < \pi/2$ and k is an arbitrary integer.

For example, if we multiply two complex numbers z_1 and z_2,we obtain

$$\begin{aligned} z_1 z_2 = r_1 r_2 \big[&(\cos\varphi_1\cos\varphi_2 - \sin\varphi_1\sin\varphi_2) + i(\sin\varphi_1\cos\varphi_2 \\ &+ \sin\varphi_2\cos\varphi_1)\big] = r_1 r_2\left[\cos(\varphi_1+\varphi_2) + i\sin(\varphi_1+\varphi_2)\right] . \end{aligned} \tag{D13}$$

From (D13), it follows that during the multiplication of complex numbers, their modules are multiplied and their arguments are added.

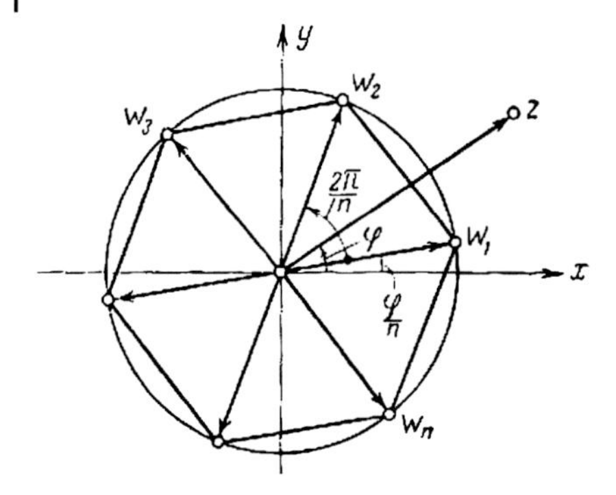

Figure D.2 n-th root of a complex number for $n = 6$.

To determine roots of $w = \sqrt[n]{z}$, one has to use the definition of the root of n-th order and the formula for multiplication, resulting in

$$|w| = \sqrt[n]{|z|}, \quad \arg w = \frac{\arg z}{n} . \tag{D14}$$

Hence, it follows that the n-th root of a complex number $z \neq 0$ has $n$different values and these numbers are located in the vortices of the regular n-polygon inscribed in the circle $|w| = \sqrt[n]{|z|}$ (see Figure D.2).

D.2 Complex Variables

D.2.1 Geometrical Notions

A *region* on the complex plane is called a set of points D having the following properties: 1) together with each point D a small enough circle with its center in this point belongs to this set (property of an *open set*); 2) any two points of D can be connected by a polygonal line consisting of points belonging to D (*connectivity* property). For example, such regions can serve as neighborhoods of points on D.

Under a *ε-neighborhood* of the point a, we understand an open circle with its center in this point, that is, a set of points z satisfying the inequality $|z - a| < \varepsilon$. The *boundary point* of the region D is called such a point which itself does not belong to D, though any neighborhood of this point contains points of D. The set of boundary points of D is called the *boundary* of this region. The region D together with its boundary is designated by $\bar{D}$ and called the *closed region*. The boundary of the region consists of a finite number of closed lines, cuts and points (see Figure D.3).

In the case of a bounded region D, the number of connected parts into which the boundary is divided is called the *order of connectivity*. In particular, when the boundary of the region D is connected, that is, it consists of one connected part, D is called the *simply connected region*.

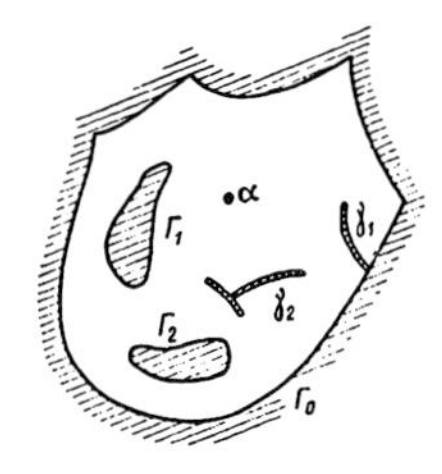

Figure D.3 Closed region: Γ_0, Γ_1, Γ_2, closed lines; γ_1, γ_2, cuts; α_1 point.

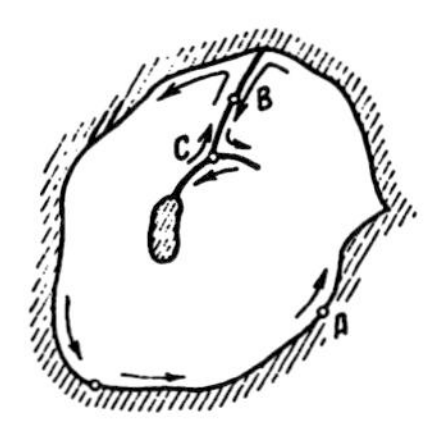

Figure D.4 Orbit of the region boundary.

Let D be a simply connected region and Γ be its boundary. Choose on Γ a point, and beginning from this point, let us go along Γ in a positive orbit. As the *positive orbit* of the region boundary Γ, the region is always on the left-hand side of the orbit. Thus, some points of Γ will be orbited only once, as for example, as in A on Figure D.4, or more than once, for example, B – two times, or C – three times. The points of the first kind are called simple points and the points of the second kind are multiple points of the contour Γ.

D.2.2
Functions of a Complex Variable

It is said that on the set M of points of the plane z, a function

$$w = f(z) \tag{D15}$$

is given, by which a correspondence law between each point z of M with a definite point or a set of points on w is specified. In the first case, the function $w = f(z)$ is called a *single-valued function*, while in the second case, it is called the *multi-valued function*. The set M is called the *domain of definition* of the function $f(z)$ and the set N of all values of w which $f(z)$ takes on M is the *domain of variability*. If we take $z = x + iy$ and $w = u + iv$, than the specification of a function of complex variable $w = f(z)$ would be equivalent to the specification of two functions of two real variables:

$$u = u(x, y)\,, \quad v = v(x, y)\,. \tag{D16}$$

If we plot values of zonto one complex plane and values of w on another one, then the function $w = f(z)$ can be considered as a mapping of the set M of the plane z into the set N of the plane w. When the function $w = f(z)$ is single-valued on the set M and thus two different points of M always correspond to different points in N, such mapping a is called one-to-one or univalent mapping. A function $z = \phi(w)$, setting up in correspondence to each point w from Na set of all points z having

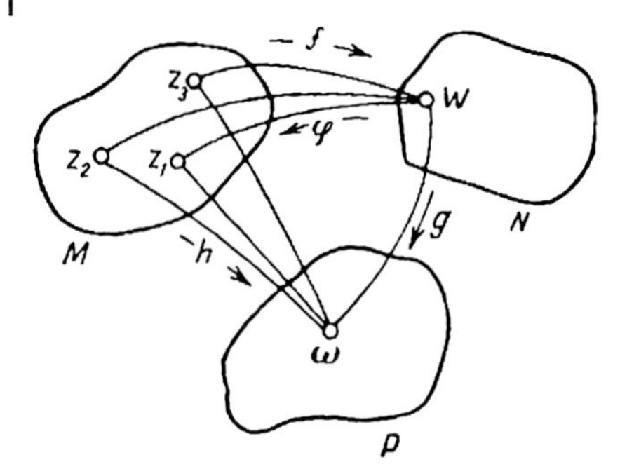

Figure D.5 Mapping with the composite function.

been mapped into the point w, is called the inverse of the function $w = f(z)$. It is clear that mapping $w = f(z)$ would be one-to-one if and only if both functions f and ϕ are one-valued functions.

Let the function $w = f(z)$ map the set M into N and $\omega = g(w)$ map N into P. Then, the function $\omega = h(w) = g\left[f(z)\right]$ mapping M into P is called the *composite function* corresponding to mapping h, which in its turn is called the superposition of mappings f and g (see Figure D.5). In particular, if the mapping $w = f(z)$ is one-to-one and the function $z = \phi(w)$ is the inverse function to f, then $\phi\left[f(z)\right] = z$.

D.2.3
Differentiation and Analyticity of Complex Functions

Let the function $w = f(z)$ be single-valued in a neighborhood of a point $z_0 = x_0 + iy_0$, except it may be at the point z_0. We shall say that there exists a limit of the function $f(z)$ at $z \to z_0$, designated as $\lim_{z \to z_0} f(z) = u_0 + iv_0 = w_0$, if the limit

$$\lim_{x \to x_0, y \to y_0} u(x, y) = u_o \quad \text{and} \quad \lim_{x \to x_0, y \to y_0} v(x, y) = v_o \tag{D17}$$

exists.

Since our definition reduces to the ordinary definition of the real function limit, the main properties of the passage to the limit are conserved for functions of complex variable. It should be noticed that in accordance with the definition, the function $f(z)$ tends to its limit *independent of the way that z approaches z_0*.

The function $f(z)$ is the *continuous function at the point z_0* if it is defined in some neighborhood of z_0 including z_0 and

$$\lim_{z \to z_0} f(z) = u_0 + iv_0 = w_0 \, . \tag{D18}$$

The function $f(z)$ is called the *continuous function in the region D* if it is continuous at each point of this region. Let the function $f(z)$ be defined in some neighborhood of the point z. We shall say that $f(z)$ is differentiable at the point z if there exists the limit

$$\lim_{h \to 0} \frac{f(z+h) - f(z)}{h} = f'(z) \tag{D19}$$

which is the derivative of the function $f(z)$ at point z. The condition of differentiability of the function $f(z)$ in terms of real functions $u(x, y)$ and $v(x, y)$ gives the *D'Alembert–Euler theorem.* Let the function $f(z) = u(x, y) + iv(x, y)$ be defined in some neighborhood of the point z and the functions $u(x, y)$ and $v(x, y)$ be differentiable at this point. Then, for the differentiability of the function $f(z)$, it is necessary and sufficient that the function obeys the following conditions (*D'Alembert–Euler conditions*):

$$\frac{\partial u}{\partial x} = \frac{\partial v}{\partial y}, \quad \frac{\partial u}{\partial y} = -\frac{\partial v}{\partial x}. \tag{D20}$$

The derivative of the function $f(z)$ that satisfies the D'Alembert–Euler conditions can be represented by the following formulas:

$$f'(z) = \frac{\partial u}{\partial x} + i\frac{\partial v}{\partial x} = \frac{\partial v}{\partial y} - i\frac{\partial u}{\partial y} = \frac{\partial u}{\partial x} - i\frac{\partial u}{\partial y} = \frac{\partial v}{\partial y} + i\frac{\partial v}{\partial x}. \tag{D21}$$

Since the ordinary properties of algebraic operations and the passage to the limit apply, the rules of function differentiation also apply

$$\begin{gathered} (f+g)' = f' + g', \quad (fg)' = f'g + fg', \quad \left(\frac{f}{g}\right)' = \frac{f'g - fg'}{g^2}, \\ \{f[g(z)]\}' = f'[g(z)]\,g'(z), \; f'(z) = \frac{1}{\varphi'(w)}. \end{gathered} \tag{D22}$$

In polar coordinates (r, ϕ), the D'Alembert–Euler conditions take on the form

$$\frac{\partial u}{\partial \varphi} = -r\frac{\partial v}{\partial r}, \quad r\frac{\partial u}{\partial r} = \frac{\partial v}{\partial \varphi}. \tag{D23}$$

A function $f(z)$ which is differentiable at each point of the region D is called an analytical (regular) function in this region. It should be noted that our definition of the analytical function suggests it to be single-valued in D because above-cited definitions of the limit and the derivative are only suited to single-valued functions.

D.3 Elementary Functions

D.3.1 Functions

Consider $w = z^n$ and $w = \sqrt[n]{z}$, where n is a positive integer.

The first of these functions

$$w = z^n \tag{D24}$$

is a single-valued function for all z. If we take polar coordinates $z = r(\cos\phi + i\sin\phi)$, $w = \rho(\cos\theta + i\sin\theta)$ in planes z and w, (D24) can be written in the form

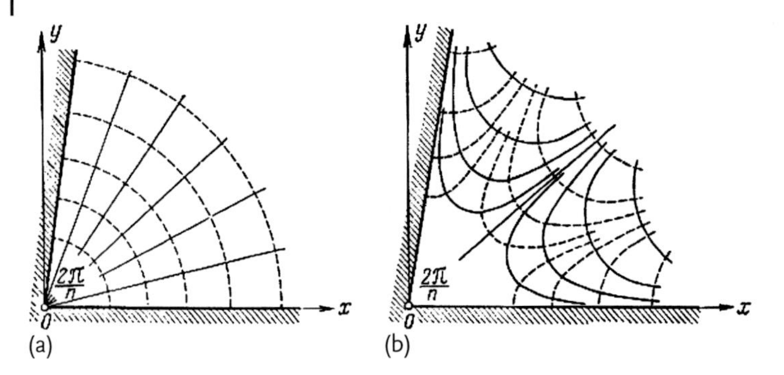

Figure D.6 (a) The inverse image of one sector of the plane (D27). (b) Curves (D29).

of two equalities

$$\rho^n = r^n\,, \quad \theta = n\varphi \tag{D25}$$

connecting real quantities.

From (D25), it follows that the mapping realized by function $w = z^n$ reduces to the rotation of each vector $z \neq 0$ by the angle $(n-1)\arg z$ and by extension of vector length by a factor of $|z|^{n-1}$. It is evident that the points z_1 and z_2 with equal modules and arguments differing from each other by an integer multiple $2\pi/n$, and only such points are mapped by (D24) to a single point. So, in order for the mapping $w = z^n$ to be univalent in a region D, it is necessary and sufficient that D does not contain two points z_1 and z_2 connected by relations

$$|z_1| = |z_2|\,; \quad \arg z_1 = \arg z_2 + \frac{2\pi k}{n}\,, \quad k \neq 0 \text{ integer}\,. \tag{D26}$$

This condition is satisfied, for example, by the sectors

$$k\frac{2\pi}{n} < \varphi < (k+1)\frac{2\pi}{n}\,, \quad (k = 0, 1, 2, \ldots)\,, \tag{D27}$$

which $w = z^n$ transforms into the plane, less the positive semi-axis. Thus, all rays with vertices at point $z = 0$ transform into rays with vertex $w = 0$ (but turned by some angle), and all arcs of circles centered at $w = 0$, but may be with another radius. Figure D.6a illustrates the inverse image of one such sector of the plane z of the polar coordinates of the plane w.

From the formula

$$w = u + iv = r^n(\cos n\varphi + i \sin n\varphi) \tag{D28}$$

equal to (D24), it follows that the two straight lines $u = u_0$ and $v = v_0$ correspond to the following curves written in polar coordinates

$$r = \sqrt[n]{\frac{u_0}{\cos n\varphi}}\,, \quad r = \sqrt[n]{\frac{v_0}{\sin n\varphi}}\,. \tag{D29}$$

These curves are shown in Figure D.6b (the first – by dotted line, the second – by solid line, at $n = 2$ they are ordinary hyperbolas).

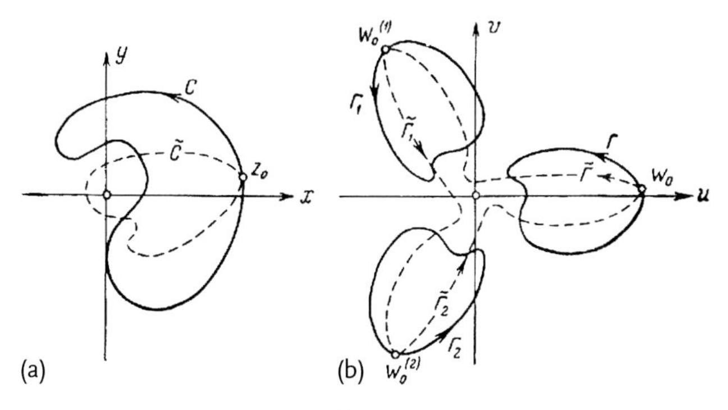

Figure D.7 The function $w = \sqrt[n]{z}$.

Finally, note that the function $w = z^n$ is analytical in the whole plane z because for any z,

$$\lim_{h\to 0}\frac{(z+h)^n - z^n}{h} = \lim_{h\to 0}\frac{nz^{n-1}h + h^2(\ldots)}{h} = nz^{n-1}\,. \tag{D30}$$

The function

$$w = \sqrt[n]{z} \tag{D31}$$

is the inverse of the function $w = z^n$ and n-valued at $z \neq 0$. As follows from Section D.1.2, the value of the root $w = \sqrt[n]{z}$ is determined by the value of the argument chosen for the point z. Denote through $\arg z_0$ one of the argument values at point $z_0 \neq 0$ and let the point describe a continuous curve C not going through the origin of coordinates. By $\arg z$, we shall designate such an argument value which changes continuously beginning from $\arg z_0$. By view of the continuity of $\arg z$ and $|z|$, the value of $w = \sqrt[n]{z}$ will also continuously change.

Now, suppose the curve C is closed and does not include the point $z = 0$. Then, during one complete orbit of C around z, the point $w = \sqrt[n]{z}$, where $\sqrt[n]{z}$ is the chosen value of the root, describes a closed curve Γ returning to its initial value $\arg z_0$. The values of the root determined by another choice of the initial value $\arg z_0$, being distinguished from the previous one by a multiple of 2π, also describe the closed curves Γ_k during a complete orbit of C, different from Γ only by a rotation $(2k\pi)/n$, $k = 1, 2, \ldots, n-1$ (see Figure D.7, solid lines).

Consider a closed curve $\tilde{C}$ without self-intersection points. It contains the point $z = 0$ and a point z_0. Then, during a complete orbit of $\tilde{C}$ in a positive direction beginning from z_0, the point corresponding to $w = \sqrt[n]{z}$ does not return back to its initial location, but occupies a new place $w_0^{(1)} = (\cos(2\pi)/n + i\sin(2\pi)/n)w_0$ where $\sqrt[n]{z_0}$ is distinct from w_0. It is explained by $\arg z$ that in the course of orbiting $\tilde{C}$, it undergoes an increment of 2π. The point $w = \sqrt[n]{z}$ comes back to its initial position only after n-fold orbit of the curve $\tilde{C}$ (see Figure D.10, solid lines).

Hence, it follows that in a region D which does not contain a closed curve, when orbiting the point $z = 0$, one can distinguish n continuous and single-valued functions, each taking one of the values of $\sqrt[n]{z_0}$. These functions are called *branches* of the multi-valued function $w = \sqrt[n]{z}$. Their values are distinguished from each other by the factor $\cos(2k\pi)/n + i\sin(2k\pi)/n$ at each fixed point. Each branch will evidently perform an univalent mapping of the region D. Thus, at each point of this region, the theorem is applicable regarding the derivative of inverse function in accordance with which there exists a definite value of the derivative

$$(\sqrt[n]{z})' = \frac{1}{(w^n)'} = \frac{1}{n}\frac{\sqrt[n]{z}}{z} \quad \text{or} \quad \left(z^{\frac{1}{n}}\right)' = \frac{1}{n}z^{\frac{1}{n}-1}. \tag{D32}$$

Thus, any of the obtained branches is an analytical function in D. The infinitely-valued function $\arg z$ in the region D of the above-considered type consists of an infinite set of continuous and single-valued branches. We shall denote each of these branches as $\arg z$, and each time indicates how it stands out.

If the region D contains at least one closed curve orbiting the point z, the branches of the function $\sqrt[n]{z}$ can not be separated from each other. Namely, if in the vicinity of some point $z \neq 0$in Dwe start along any branch, then moving along the curve orbiting the point $z \neq 0$ we come to another curve. Hence, in such a region of D it is impossible to consider $\sqrt[n]{z}$ as a set of separated (single-valued) analytical functions, as in the previous case. The point $z \neq 0$ in whose neighborhood one cannot separate n branches of the function $\sqrt[n]{z}$, that is, where the branches are connected to each other, is called *ramification point.*

D.3.2
Joukowski Function

Consider the function $w = 1/2(z + 1/z)$. This function is determined, single-valued and analytical for all $z \neq 0$. Let us find the regions of univalence of the mapping

$$w = 1/2\,(z + 1/z)\,. \tag{D33}$$

Suppose that two points z_1 and z_2 are mapped into one point w. Then we have $z_1 + 1/z_1 = z_2 + 1/z_2$ or $(z_1 - z_2)(1 - 1/(z_1 z_2)) = 0$ from which follows $z_1 = z_2$ or $z_1 z_2 = 1$. Hence, for the univalence mapping with the Joukowski function in the region Dit is necessary and sufficient that Ddoes not contain two points z_1 and z_2 connected by the relation $z_1 z_2 = 1$. This condition is fulfilled, for example, in the interior of the unit circle $|z| < 1$ or the exterior $|z| > 1$ of the unit circle. In order to study a picture of the mapping described in (D33), let us use polar coordinates $z = r(\cos\phi + i\sin\phi)$, $w = u + iv$ and separate real and imaginary parts. Then we get

$$u = \frac{1}{2}\left(r + \frac{1}{r}\right)\cos\varphi\,, \quad v = \frac{1}{2}\left(r - \frac{1}{r}\right)\sin\varphi\,, \tag{D34}$$

and we see that each circle $|z| = r_0 < 1$ transforms into the curve

$$u = \frac{1}{2}\left(r_0 + \frac{1}{r_0}\right)\cos\varphi\,, \quad v = -\frac{1}{2}\left(r_0 - \frac{1}{r_0}\right)\sin\varphi\,, \tag{D35}$$

that is, into an ellipse with semi-axes $a = 1/2(r_0 + 1/r_0)$, $b = 1/2(r_0 - 1/r_0)$ orbited in the negative direction. At $r_0 \to 1$ this ellipse contracts into the segment

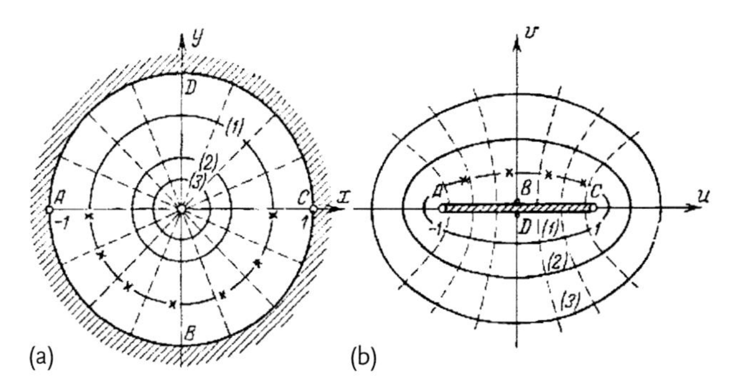

Figure D.8 Mapping of the ellipse.

$[-1, 1]$ of the axis u, at $r_0 \to 0$ it tends to the infinity. Consequently, the function (D33) maps the circle interior $|z| < 1$ into the exterior of the segment $[-1, +1]$ (see Figure D.8). All inner points of this segment are double points and the segment may be considered as if consisting of two shores: the function (D33) transforms the upper semicircle into the lower shore, and the lower semicircle into the upper shore. The radii $\arg z$ with $0 < r < 1$ pass into branches of hyperbolas

$$\frac{u^2}{\cos^2 \varphi_0} - \frac{v^2}{\sin^2 \varphi_0} = 1 \tag{D36}$$

(see Figure D.8).

The foci of these hyperbolas as well as of the ellipses are located at the ends of the segment $[-1, +1]$. It follows from the relations (D34) that the circles $|z| = r_0 > 1$ transform into ellipses with semi-axes $a = 1/2(r_0 + 1/r_0)$, $\quad b = 1/2(r_0 - 1/r_0)$. These ellipses coincide with those transformed from the circles $|z| = r_0 < 1$, with the exception that the latter are orbited in positive direction. Thus, the function (D33) also maps the exterior of the circle $|z| > 1$ into the exterior of the segment $[-1, +1]$ of the axis u. Here the upper semicircle transfers to the upper shore of the segment, and the lower semicircle into the lower shore of the segment. A more detailed description of elementary complex functions including exponential, logarithmic, trigonometric and hyperbolic functions can be found in Lavrentiev and Shabat, 1965.

D.4 Integration of Complex Variable Functions

D.4.1 Integral of Complex Variable Functions

Consider an oriented curve C and a function of complex variable $f(z)$ on it. The integral of $f(z)$ along C is called, by definition,

$$\lim_{n\to\infty} \sum_{k=0}^{\infty} f(\varsigma_k)(z_{k+1} - z_k) = \int_C f(z)dz\,, \tag{D37}$$

where $z_0 = a, z_1, z_2, \ldots, z_{n+1} = b$ are successive points dividing C into n parts, while a and b are the designated ends of C, ς_k is a point of the part $[z_k, z_{k+1}]$ of C, and the limit is taken on the assumption that $\max|z_k - z_{k+1}| \to 0$. The integral (D37) always exists if C is supposed to be a piecewise smooth curve and $f(z)$ a piecewise continuous function.

By dividing real and imaginary parts in (D37), we get

$$\int_C f(z)dz = \int_C udx - vdy + i\int_C udy + vdx\,. \tag{D38}$$

One can see that with the help of(D37), that calculating the integral of a complex variable function reduces to calculations of integrals of real variable functions. The derivative and the integral of a complex function of a real variable $w(t) = \phi(t) + i\psi(t)$ is represented by the following linear combinations

$$w'(t) = \varphi'(t) + i\psi'(t)\,, \quad \int_\alpha^\beta w(t)dt = \int_\alpha^\beta \varphi(t)dt + \int_\alpha^\beta \psi(t)dt\,. \tag{D39}$$

Let $z = z(t) = x(t) + iy(t)$ be the parametric representation of the curve C and $z(\alpha) = a$, $z(\beta) = b$. Then, the calculation of the integral of $f(z)$ along C reduces to the calculation of a complex function of a real variable:

$$\int_C f(z)dz = \int_\alpha^\beta f\left[z(t)\right] z'(t)dt\,. \tag{D40}$$

From (D38), it follows that ordinary properties of curvilinear integrals can be extended to integrals of a complex variable:

$$\int_C \left[a f(z) + bg(z)\right] dz = a\int_C f(z)dz + b\int_C g(z)dz\,,$$
$$\int_{C_1+C_2} f(z)dz = \int_{C_1} f(z)dz + \int_{C_2} f(z)dz\,, \quad \int_C f(z)dz = -\int_{C^-} f(z)dz\,,$$

where a and b are complex constants, while, by $C_1 + C_2$, we denote the curve consisting of two curves C_1 and C_2, C^- is a curve coinciding with C, but orbited in the opposite direction.

D.4.2
Some Theorems of Integral Calculus in Simply Connected Regions

Theorem D.1 Cauchy theorem

If the function $f(z)$ is analytic in a single-valued region D, then for all curves C being located in this region and having common ends, the integral $\int_C f(z)dz$ has one and the same value.

This theorem means that the value of integral is independent of the path of integration. As it is known from the theory of mathematical analysis, for the independence of the integral $\int_C P dx + Q dy$, where P and Q are functions having continuous partial derivatives , from the path of integration in a single-valued region is necessary and sufficient for P and Q to obey the condition $\partial P/\partial y = \partial Q/\partial x$. As with integrals entering in the right part of (D38), this condition reduces to

$$\frac{\partial u}{\partial y} = -\frac{\partial v}{\partial x}\,, \quad \frac{\partial v}{\partial y} = \frac{\partial u}{\partial x}\,, \tag{D41}$$

which coincide with the D'Alembert–Euler conditions. In view of this theorem, for analytic functions in single-valued regions, one can write $\int_{z_0}^{z} f(\varsigma)d\varsigma$ instead of $\int_C f(z)dz$, where z_0 and z denote the endpoints of the curve.

From theorem one:

Theorem D.2

If the function $f(z)$ is analytic in a single-valued region D, the integral

$$\int_{z_0}^{z} f(\varsigma)d\varsigma = F(z) \tag{D42}$$

considered as depending on its upper limit, is also the analytic function in D, with

$$F'(z) = \frac{d}{dt}\int_{z_0}^{z} f(\varsigma)d\varsigma = f(z)\,. \tag{D43}$$

The function whose derivative is equal to a given function $f(z)$ is called the antiderivative of this function.

Theorem D.3

Any two antiderivatives of one and the same function are distinctive from each other no more than by the constant term.

Theorem D.4

If $F(z)$ is an antiderivative of analytic function $f(z)$, then

$$\int_{z_0}^{z} f(\varsigma)d\varsigma = F(z) - F(z_0)\,. \tag{D44}$$

Theorem D.5

If the function $f(z)$ is analytic in a single-valued region D, then its integral along any closed contour C lying in the region D is equal to zero:

$$\int_C f(z)dz = 0 \,. \tag{D45}$$

Theorem D.6

If the function $f(z)$ is analytic in a single-valued region D and continuous in the closed region $\bar{D}$, then the integral of $f(z)$ taken along the boundary C_D of this region is equal to zero:

$$\int_{C_D} f(z)dz = 0 \,. \tag{D46}$$

D.4.3
Extension of Integral Calculus to Multiply Connected Regions

The Cauchy theorem and its results are generally incorrect when applied to multiply connected region. For example, the function $f(z) = 1/z$ is analytic everywhere in the ring $1/2 < |z| < 2$, but the integrals from -1 to 1 along the upper and lower half-circles of $|z| = 1$ are distinctive from each other. Along the upper part C_1 of the circle $|z| = 1$, where $z = e^{i\varphi}$, $0 < \varphi < \pi$, we have:

$$\int_{C_1} \frac{1}{z} dz = \int_{\pi}^{0} \frac{i e^{i\varphi}}{e^{i\varphi}} d\varphi = -i\pi \,,$$

while along the lower part of the semi-circle C_2, where $z = e^{i\varphi}$, $-\pi < \varphi < 0$, we get

$$\int_{C_2} \frac{1}{z} dz = \int_{-\pi}^{0} \frac{i e^{i\varphi}}{e^{i\varphi}} d\varphi = i\pi \,.$$

However, when in a multiply connected region, the curves C_1 and C_2 with common ends are located so that they restrict a simply connected region belonging to D, the integrals along these curves are evidently equal to each other. From this, it follows that *the value of an integral of an analytical function in a multiply connected region remains unchanged if the contour of integration is continuously deformed so that its ends remain immovable inside the region D.*

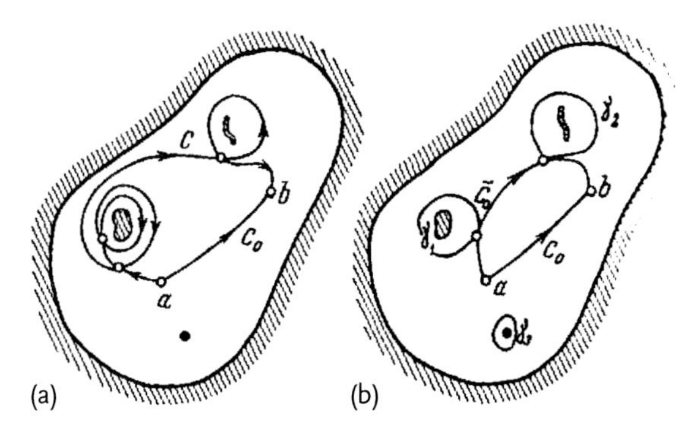

Figure D.9 Integration in a multiply connected region.

Assume that in a multiply connected region D, we are given two points a and b, and a simple, that is, without self-interaction, curve C_0 which connects these points. Let C be another curve connecting these points (Figure D.9a). In accordance with the previously mentioned remark, it is possible without changing the magnitude of the integral to deform the curve C to another curve $\tilde{C}$ one located in D and consisting of: 1) the curve $\tilde{C}_0$ which together with C_0 restrict the simply connected region belonging to D; 2) the set of simple closed curves γ_k $(k = 1, 2, \dots, m)$, any of which contains one connected part of the boundary D (Figure D.9b. Now, the curves γ_k can be orbited several times and in different directions (in Figure D.4b, the curve γ_1 is orbited thrice clockwise and γ_2, once counter-clockwise). For convenience, let us denote by γ_k $(k = 1, 2, \dots, m)$ the curves which are orbited counter-clockwise, and also introduce the curves γ_k $(k = m + 1, \dots, n)$ which surround the connected parts of the region boundary, and which are not inside of $\tilde{C}$ as, for example, γ_3 in Figure D.9b.

Let us introduce the notation

$$\Gamma_k = \int\limits_{\gamma_k} f(z) dz \quad (k = 1, 2, \dots, n) . \tag{D47}$$

During a continuous deformation of γ_k, when these curves are remaining inside D, the integrals (D47) are not changed and the quantities Γ_k are determined only by the function $f(z)$ and the region D. Let N_k be integers showing how much and in what direction the γ_k are orbited as a part of the curve $\tilde{C}$. These numbers can be positive, negative or zero, for example, in the Figure D.9 they are $N_1 = -3$, $N_2 = 1$, and $N_3 = 0$. As a result, we have

$$\int\limits_a^b f_C(z) dz = \int\limits_a^b f_{\tilde{C}}(z) dz = \int\limits_a^b f_{C_0}(z) dz + N_1\Gamma_1 + N_2\Gamma_2 + \dots + N_n\Gamma_n . \tag{D48}$$

Here, $\int_a^b f_C(z) dz$ denotes the integral from a up to b along C, and the quantities Γ_k are called *integral periods* or *cyclic constants* of the function $f(z)$ in multiply connected region D.

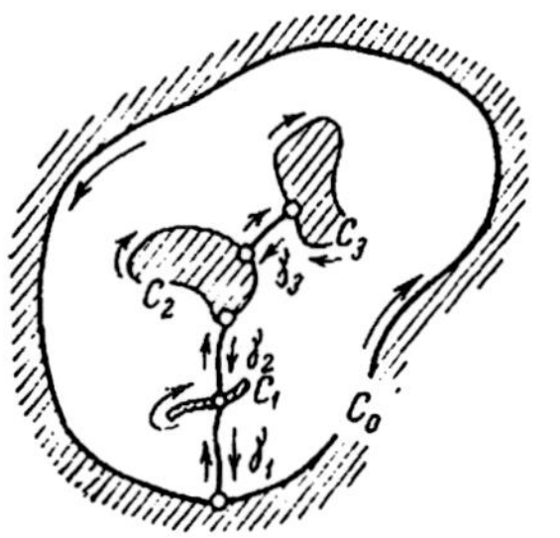

Figure D.10 The Cauchy theorem in a multiply connected region.

It should be noted that the Cauchy theorem one of the previous section can be imparted so that it holds to be true even for multiply connected regions. Let the function $f(z)$ be analytic in a multiply connected region D bounded by curves $C_0, C_1, \ldots, C_n$ (see Figure D.10) and continuous in $\bar{D}$. Make cuts $\gamma_1, \ldots, \gamma_n$ which transform D into the simply connected region D^*, and denote through C^*, the boundary of the obtained region, that is, the curve consisting of the parts of curves C_k and γ_k. In this case, the latter are orbited twice in the opposite directions (marked in Figure D.10 by arrows). Then, if the function $f(z)$ is analytic in the simply connected region D^* and continuous in $\bar{D}$, then

$$\int_{C^*} f(z)dz = \int_{C_0} f(z)dz + \sum_{k=1}^{n} \int_{C_k} f(z)dz = 0 . \tag{D49}$$

Hence, for regions of any connectivity, the Cauchy theorem is true in the following form: *If the function $f(z)$ is analytic in the region D and continuous in $\bar{D}$, then its integral along the boundary of this region orientated in such a manner so that the region D always remains at one and the same side of the integratin path is equal to zero.*

D.4.4 Cauchy Formula

Let the function $f(z)$ be analytic in the n-connected region D and continuous in $\bar{D}$. Then, for any inner point z of this region, the so called Cauchy formula applies:

$$f(z) = \frac{1}{2\pi i} \int_{C} \frac{f(\varsigma)d\varsigma}{\varsigma - z} , \tag{D50}$$

where C is the boundary of D being orbited so that the region D always remains on the left. Since we enter only values of $f(z)$ at the boundary C of the region D in the right part of (D47), the values of the function $f(z)$ at the interior points are entirely determined by its values at the boundary. Thus, the Cauchy formula allows one to calculate values of the function $f(z)$ at any point of the region D by given values of this function at the boundary. In the particular case when the boundary C is a circle $|\varsigma - z| = R$ with radius R, then taking $\varsigma - z = \mathrm{R}e^{i\varphi}$, we obtain from the Cauchy formula

$$f(z) = \frac{1}{2\pi} \int_0^{2\pi} f(z + \mathrm{Re}^{i\varphi}) d\varphi \, .$$

This formula expresses the so called theorem of the mean:

Theorem D.7

If the function $f(z)$ is continuous in a closed circle and analytic inside this circle, then its value at the center of the circle is equal to the arithmetical mean of this function at the circle.

Cauchy theorem for higher derivatives:

Theorem D.8

If the function $f(z)$ is analytic in D and continuous in $\bar{D}$, it has at each point in D derivatives of all orders, and the n-th derivative is represented by the formula

$$f^{(n)}(z) = \frac{n!}{2\pi i} \int_C \frac{f(\varsigma) d\varsigma}{(\varsigma - z)^{n+1}} \, , \tag{D51}$$

where C is the boundary of the region D.

This theorem can be also formulated as follows: *If the function $f(z)$ is continuous on the boundary C of the region D, then the function*

$$f(z) = \frac{1}{2\pi i} \int_C \frac{f(\varsigma) d\varsigma}{(\varsigma - z)^{n+1}} \tag{D52}$$

represented by the Cauchy formula is analytic in this region.

We should also mention the following two theorems:

Theorem D.9 Cauchy and Liouville

If the function $f(z)$ is analytic in the whole plane and is bounded, then it is constant.

Theorem D.10 Inverse theorem of Cauchy and Liouville

If the function $f(z)$ is continuous in a simply connected region D and the integral $f(z) = \int_C f(z) dz$ along any closed contour lying in D is equal to zero, then $f(z)$ is analytic in this region.

D.5
Representation of a Function as a Series

D.5.1
Taylor Series

We begin with the generalization of the Taylor series well known from the theory of mathematical analysis to functions of a complex variable: *any analytical function can be represented in the vicinity of a point a as Taylor series*

$$f(z) = f(a) + \frac{f'(a)}{1!}(z-a) + \cdots + \frac{f^{(n)}(a)}{n!}(z-a)^n + R_n\,, \tag{D53}$$

where the residual term is

$$R_n = \frac{(z-a)^{n+1}}{2\pi i}\int\limits_C \frac{f(\varsigma)d\varsigma}{(\varsigma - z)(\varsigma - a)^{n+1}}\,.$$

D.5.2
Laurent Series

The Taylor series are convenient to represent functions being analytic in circular regions. However, it is important to be able to represent functions in the form of a power series in regions of a different shape. For example, when considering a function being analytic in the vicinity of a point a everywhere except the point a, one should consider annular regions of the form $0 < |z-a| < R$. It turns out that for a function being analytic in annular regions $r < |z-a| < R$, $r \geqslant 0$, $R \leqslant \infty$, one can get expansions in terms of positive and negative powers of $(z-a)$:

$$f(z) = \sum_{n=-\infty}^{\infty} c_n (z-a)^n\,. \tag{D54}$$

Let the function $f(z)$ be analytic in a ring K: $r < |z-a| < R$, where $r \geqslant 0$, $R \leqslant \infty$. Choose arbitrary numbers r', R' and a number k, $0 < k < 1$, and consider the ring $r'/k < |z-a| < kR'$. In an inner point of this ring, the function f(z) can be represented in accordance with the Cauchy formula (D48) as follows:

$$f(z) = \frac{1}{2\pi i}\int\limits_C \frac{f(\varsigma)d\varsigma}{\varsigma - z} - \frac{1}{2\pi i}\int\limits_c \frac{f(\varsigma)d\varsigma}{\varsigma - z}\,, \tag{D55}$$

where both circles $C : |z-a| = R'$ and $c : |z-a| = r'$ are being orbited in a counter clockwise direction. We have for the first integral

$$\left|\frac{z-a}{\varsigma - a}\right| < \frac{kR'}{R'} = k < 1\,,$$

and consequently, the fraction in it can be expanded as

$$\frac{1}{\varsigma - z} = \frac{1}{\varsigma - a}\frac{1}{1-\frac{z-a}{\varsigma - a}} = \frac{1}{\varsigma - a} + \frac{z-a}{(\varsigma - a)^2} + \cdots + \frac{(z-a)^n}{(\varsigma - a)^{n+1}} + \ldots$$

By multiplying both parts of this relation by $1/(2\pi i)\, f(\varsigma)$ and integrating the result term by term by ς, we get the expansion of the first term of (D55) in the power series

$$f_1(z) = \frac{1}{2\pi i} \int\limits_C \frac{f(\varsigma) d\varsigma}{\varsigma - z} = \sum_{n=0}^{\infty} c_n (z-a)^n \,, \tag{D56}$$

where

$$c_n = \frac{1}{2\pi i} \int\limits_C \frac{f(\varsigma) d\varsigma}{(\varsigma - a)^{n+1}} \quad (n = 0, 1, 2, \ldots) \,. \tag{D57}$$

For the second integral, we have

$$\left| \frac{\varsigma - a}{z - a} \right| < \frac{k r'}{r'} = k < 1 \,,$$

and

$$\frac{1}{\varsigma - z} = -\frac{1}{z-a} \frac{1}{1 - \frac{\varsigma - a}{z - a}} = -\frac{1}{z-a} - \frac{\varsigma - a}{(z-a)^2} - \cdots - \frac{(z-a)^{n-1}}{(\varsigma - a)^n} - \cdots$$

Thus, we get the expansion of the second integral in a negative power series

$$f_2(z) = -\frac{1}{2\pi i} \int\limits_c \frac{f(\varsigma) d\varsigma}{\varsigma - z} = \sum_{n=0}^{\infty} c_{-n} (z-a)^{-n} \,, \tag{D58}$$

where

$$c_{-n} = \frac{1}{2\pi i} \int\limits_c f(\varsigma)(\varsigma - a)^{n-1} d\varsigma \quad (n = 1, 2, 3 \ldots) \,. \tag{D59}$$

By combining both representations of coefficients c_n and c_{-n}, we can write

$$f(z) = f_1(z) + f_2(z) = \sum_{n=-\infty}^{\infty} c_n (z-a)^n \,, \tag{D60}$$

where

$$c_n = \frac{1}{2\pi i} \int\limits_\gamma \frac{f(\varsigma) d\varsigma}{(\varsigma - a)^{n+1}} \quad (n = 0, \pm 1, \pm 2, \ldots) \,, \tag{D61}$$

and γ is a circle $|z - a| = \rho$ with $r' < \rho < R$.

The expansion (D60) of the function $f(z)$ over positive and negative powers of $(z - a)$ with coefficients (D61) is called the Laurent series of the function $f(z)$ with a center at point a. The series (D56) is called a true series and (D58) is the principal part of this series.

Thus, we have

Theorem D.11 Laurent theorem

In a ring K ($r < |z-a| < R$), in which the function $f(z)$ is analytic, one can represent $f(z)$ by its Laurent series (D60) with coefficients (D61).

D.6 Singular Points

This method of Laurent expansion allows one to understand the behavior of analytic functions in the neighborhood of points in which the analyticity of the function breaks down. Such points are called *isolated singular points.*

The point a is called an isolated singular point of the function $f(z)$ if there exists a vicinity $r < |z-a| < R$ of this point, with the exception of itself, in which $f(z)$ is analytic.

Note that we deal with points in whose neighborhood the function is single-valued. There are three types of isolated singular points depending on the behavior of the function $f(z)$ in the neighborhood of these points.

1. The point a is called a *removable singular point* if there exists a finite limit $\lim_{z\to a} f(z)$.
2. The point is called *pole* if $f(z)$ tends to infinity as z is approached a, that is, if $\lim_{z\to a} |f(z)| = \infty$.
3. The point a is called a *singular exceptional point* if $\lim_{z\to a} f(z)$ does not exist.

Let us present the main properties of the behavior of functions at singular points. If a is an isolated singular point of the function $f(z)$, then in accordance with the Laurent theorem of the previous section, this function can be expanded in the Laurent series in the ring of its analyticity $r < |z-a| < R$:

$$f(z) = \cdots + \frac{c_{-n}}{(z-a)^n}(z-a) + \cdots + \frac{c_{-1}}{z-a} + c_0 + c_1(z-a) + \cdots + c_n(z-a)^n + \cdots \tag{D62}$$

This expansion has a different form depending on the form of the singular point. Let us present three relevant theorems.

Theorem D.12

In order for a singular point of $f(z)$ to be removable, it is necessary and sufficient that the Laurent expansion of $f(z)$ at the vicinity of this point does not contain the principal part, that is, the expansion should be presented as

$$f(z) = c_0 + c_1(z-a) + \cdots + c_n(z-a)^n + \cdots \tag{D63}$$

Theorem D.13

In order to be a pole of $f(z)$, it is necessary and sufficient that the principal part of the Laurent expansion of $f(z)$ in the vicinity of this point would only contain a finite number of terms, that is,

$$f(z) = \ldots + \frac{c_{-n}}{(z-a)^n}(z-a) + \ldots + \frac{c_{-1}}{z-a} + \sum_{k=0}^{\infty} c_k (z-a)^k \,. \tag{D64}$$

Thus, the number of the highest negative power of the expansion would coincide with the degree of the pole.

Theorem D.14

The point a is a singular exceptional point if and only if the principal part of the Laurent expansion of $f(z)$ in the vicinity of a contains an infinite number of expansion terms.

In accordance with the nature of singular points, there are two simple classes of single-valued analytical functions:

1. *integer or holomorphic* functions, that is, functions which have no singular points. Such functions are represented as $\sum_{n=0}^{\infty} c_n z^n$. Examples of integer functions are polynomials, exponential function, $\sin z$, $\cos z$ and so on.
2. *Fractional or meromorphic* functions, that is, functions which only have poles as singular points. From this, the meromorphic function can have only a finite number of poles in any bounded region. Examples of meromorphic functions are integer functions, fractional-rational function, trigonometric functions and others.

D.6.1 Theorem about Residues

The residue of the function $f(z)$, denoted as $\operatorname{res} f(a)$, at an isolated singular point a is called the number

$$\frac{1}{2\pi i} \int_{\gamma} f(z) dz \,, \tag{D65}$$

where γ is a small circle $|z-a| = \rho$. In accordance with (D61), for $n = -1$, we have

$$\operatorname{res} f(a) = \frac{1}{2\pi i} \int_{\gamma} f(z) dz = c_{-1} \,, \tag{D66}$$

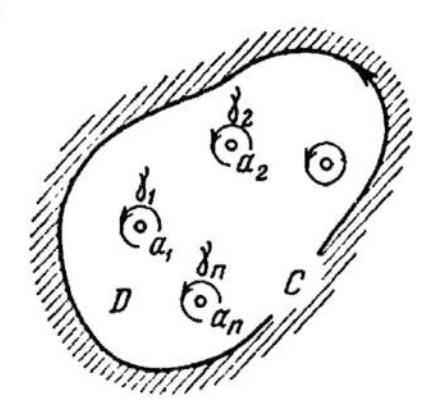

Figure D.11 The theorem of residues.

that is, *the residue of the function at the singular point a is equal to the coefficient by the term with power −1 in the Laurent expansion of $f(z)$ in the vicinity of a.*

Thus, the residue of the function at a removable singular point is always equal to zero. The finding of the residue in a pole of the order n is facilitated by the formula

$$\operatorname{res} f(a) = \frac{1}{(n-1)!} \lim_{z \to a} \frac{d^{n-1}}{dz^{n-1}} \left[(z-a)^n f(z)\right]. \tag{D67}$$

In particular, for the pole of order one, (D67) takes an especially simple form

$$\operatorname{res} f(a) = \lim_{z \to a} \left[(z-a) f(z)\right]. \tag{D68}$$

If the function $f(z)$ in the vicinity of a has the form of a quotient of two analytic functions

$$f(z) = \frac{\varphi(z)}{\psi(z)}$$

with $\varphi(a) \neq 0$ *and* $\psi(z)$ having at a the zero of the first order, that is, $\psi(a) = 0$, and $\psi'(a) \neq 0$, (D68) can be changed to

$$\operatorname{res} f(a) = \lim_{z \to a} \frac{\varphi(z)}{\psi(z)}(z-a) = \lim_{z \to a} \frac{\varphi(z)}{\frac{\psi(z)-\psi(a)}{z-a}} = \frac{\varphi(z)}{\psi'(a)}. \tag{D69}$$

The application of the theory of residues is based chiefly on the following theorem.

Theorem D.15

Let the function $f(z)$ be continuous on the boundary C of the region D and analytic in this region except for a finite numbers of singular points $a_1, a_2, \ldots, a_n$. Then,

$$\frac{1}{2\pi i} \int_C f(z) dz = 2\pi i \sum_{k=1}^{n} \operatorname{res} f(a_k). \tag{D70}$$

This theorem follows from the application of the Cauchy theorem to the region D^* and a set of circles γ_k surrounding singular points (see Figure D.11)

Let the point a be a pole of the function $f(z)$ of order n. Then, the function $g(z) = 1/f(z)$ has at this point a zero of the same order. Introduce the notion of the *logarithmic derivative*

$$\left[\ln f(z)\right] = \frac{f'(z)}{f(z)}.$$

Under *logarithmic residue* of the analytic function, we shall understand the residue of the logarithmic derivative of this function. Since

$$\left[\ln f(z)\right]' = -\left[\ln g(z)\right]' ,$$

then if the function $g(z) = 1/f(z)$ has at point a, a zero of order n, the logarithmic derivative $\left[\ln f(z)\right]'$ has at this point a pole of the order one with residue equal to $-n$.

Theorem D.16

At zeros and poles of the function $f(z)$, its logarithmic derivative $f'(z)/f(z)$ has poles of the first order, at a zero of the function the logarithmic residue is equal to the order of the zero and at the pole minus the pole order.

Theorem D.16 above and the theorem of residues allow one to use logarithmic residues to calculate the number of zeros and poles of analytic function in given regions. Let the function $f(z)$ be analytic inside the region D everywhere except for a finite number of poles $b_1, b_2, \ldots, b_m$ of multiplicity $p_1, p_2, \ldots, p_m$, continuous at the boundary C of this region, not equal to 0 on C and additionally, $f'(z)$ must be continuous on C. Denote the zeros of $f(z)$ in D by $a_1, a_2, \ldots, a_n$ and their multiplicity by $n_1, n_2, \ldots, n_l$. By applying theorem two and the theorem of residues to the logarithmic derivative, we obtain

$$\frac{1}{2\pi i}\int\limits_C \frac{f'(z)}{f(z)} dz = (n_1 + n_2 + \cdots + n_l) - (p_1 + p_2 + \cdots + p_m) = N - P , \tag{D71}$$

where N and P are the total numbers of zeros and poles. Thus, each zero and pole is calculated as many times as its order. Let us find out the geometric meaning of the left-hand side of (D71). Since $f'(z)/f(z)dz = d\left[\ln f(z)\right]$ and $\ln z = \ln|z| + i \arg z$, we have

$$\frac{1}{2\pi i}\int\limits_C d\left[\ln f(z)\right] = \frac{1}{2\pi i}\int\limits_C d\left[\ln|f(z)|\right] + \frac{1}{2\pi i}\int\limits_C d\left[\arg f(z)\right], \tag{D72}$$

where ln and arg denote any branches of theses functions. As far as on orbit of the closed contour C, the function $\ln|f(z)|$ comes back to its initial value, and the first integral in the right side of (D72) vanishes. On the other hand, if the point $w = 0$ lies inside the contour described by the point $w = f(z)$, and when z is orbiting C, the final value of $\arg f(z)$ can be distinguished from the initial one (see Figure D.12). As a result, the second term of (D72) will differ from zero and will be equal to the change of the argument of the function $f(z)$ when orbiting C divided by 2π, having the meaning of the *number of revolutions* around the beginning $w = 0$ of the vector $f(z)$ during a round trip of C or what is the same of the curve Γ

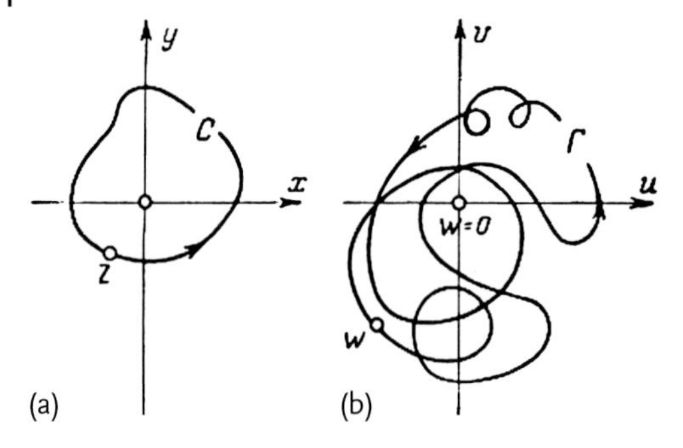

Figure D.12 The argument principle.

corresponding to C under the mapping $w = f(z)$ (in Figure D.12, this number is equal to one)

$$\frac{1}{2\pi}\int_C d\left[\arg f(z)\right] = \frac{1}{2\pi}\Delta_C \arg f(z).$$

Equations D71 and (D72) give the so called argument principle:

Theorem D.17

Let the function $f(z)$ be analytic everywhere inside the region D except for a finite number of poles, continuous together with $f'(z)$ at the boundary C of this region and nonzero on C. Then, the difference between the total number of zeros and poles of this function inside C is equal to the number of revolutions of the vector w by passing round the curve Γ corresponding to C at mapping $w = f(z)$, or similarly, to the sum of logarithmic residues of $f(z)$ in the region D:

$$N - P = \frac{1}{2\pi}\Delta_C \arg f(z) = \frac{1}{2\pi i}\int_C \frac{f'(z)}{f(z)} dz\,. \tag{D73}$$

D.6.2
Infinitely Remote Point

Up to this point, we only considered finite points of the complex variable plane. However, sometimes, it is useful to introduce the infinitely remote point. This point can be clearly illustrated by *stereographic projection* of the plane z on a sphere touching the plane at its south pole (see Figure D.13). Such a projection sets up a correspondence between a point z of the complex plane and a point Z of the sphere which is obtained by intersection of the sphere with a straight line connecting z with the north pole of the sphere. The stereographical projection sets up a point-to-point correspondence between the complex plane and the sphere except for the north pole. The points Z are taken to be spherical images of complex numbers z and the sphere itself is called the *numerical* sphere.

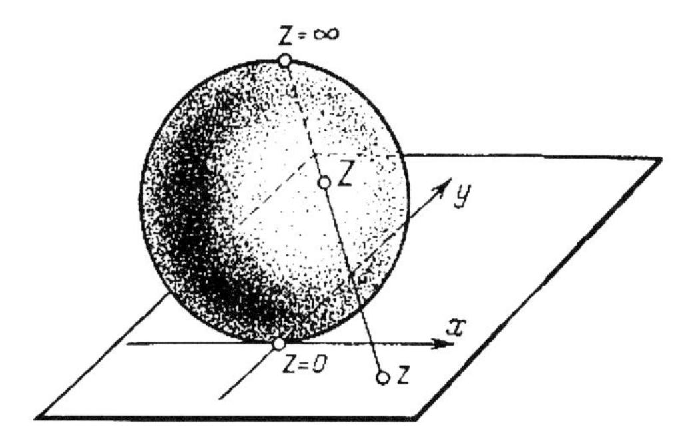

Figure D.13 The numerical sphere.

In order to extend the correspondence to the whole sphere, one has to introduce the *conditional infinitely remote point* (complex number $z = \infty$) and take it to be equivalent to the north pole. The number $z = \infty$ does not take part in arithmetic operations as ordinary complex numbers. However, it is suited to such operations as passage to the limit. The plane of complex variables together with the infinitely remote point is called the *complete complex plane*, while the plane without this point is called the *open plane*. Under a *neighborhood of an infinitely remote point*, one understands a circle on the sphere with a center at its north pole or, in other words, a set of points z obeying the inequality $|z| > R$, including the infinitely remote point.

Let the function $f(z)$ be analytic in a vicinity of the infinitely remote point except itself $z = \infty$. To such a function, one can extend, without change, the definition of a singular point. It is said that $z = \infty$ is a singular removable point, pole or essentially singular point of the function $f(z)$depending on whether $\lim_{z\to\infty} f(z)$ is finite, infinite or does not exist. However, the criterion of a singular point connected with the Laurent expansion would be changed according to the following reasoning. Let us take $z = 1/\varsigma$ and $f(z) = f(1/\varsigma) = \varphi(\varsigma)$. Then, $\varphi(\varsigma)$ would be analytic in a vicinity of the point $\varsigma = 0$. The latter point will be a singular point for $\varphi(\varsigma)$ of the same kind as for $z = \infty$ because $\lim_{z\to\infty} f(z) = \lim_{\varsigma\to 0} \varphi(\varsigma)$. The Laurent expansion of $f(z)$ in the vicinity of $\varsigma = 0$ can be obtained by a simple change of variable $z = 1/\varsigma$ in the Laurent expansion of $\varphi(\varsigma)$ in the vicinity of $\varsigma = 0$. However, at such a replacement, the true part is changed to the principal part and vice versa.

Theorem D.18

In the case of a removable singularity at the infinitely remote point, the Laurent expansion of the function $f(z)$ in the vicinity of this point does not contain any positive power of z; in the case of a pole, it contains a finite number of positive powers of z; in the case of a essentially singularity, it contains an infinite number of positive powers of z.

Now, the Cauchy–Liouville theorem (Section D.4.4) can be formulated as follows.

Theorem D.19

If the function $f(z)$ is analytic in the complete complex plane z, then it is constant.

Let the function $f(z)$ be analytic in a vicinity of the infinitely remote point $z = \infty$, except it may be at the point itself. Under residue of this function at an infinitely remote point, it is understood that

$$\operatorname{res} f(\infty) = \frac{1}{2\pi i} \int_{\gamma^-} f(z) dz \,, \tag{D74}$$

where γ^- is a large enough circle $|z| = \rho$ orbited clockwise so that the vicinity of the point $z = \infty$ is kept to the left. From this definition, it follows that the *residue of the function f(z) at infinity is equal to the coefficient by z^{-1} in the Laurent expansion of this function in the vicinity of $z = \infty$ taken with opposite sign.*

From Theorem D.19 and the last statement, follows;

Theorem D.20

If the function $f(z)$ has in the complete complex plane a finite number of singular points, the sum of all residues of this function, including the residue at infinity, is equal to zero, that is,

$$\frac{1}{2\pi i} \int_{\gamma} f(z) dz + \frac{1}{2\pi i} \int_{\gamma^-} f(z) dz$$
$$= \operatorname{res} f(a_1) + \cdots + \operatorname{res} f(a_n) + \operatorname{res} f(\infty) = 0 \,. \tag{D75}$$

D.7 Conformal Transformations

Generated and developed on the basis of physical problems, the method of conformal transformation (mapping) has found a lot of applications in different regions of physics, for example, in hydro- and aeromechanics.

D.7.1 Notion of Conformal Transformation

Assume that one is given a continuous and one-to-one transformation of the region D to the region D^*:

$$W = f(z) = u(x, y) + iv(x, y) \,. \tag{D76}$$

Additionally, suppose that the functions $u(x, y)$ and $v(x, y)$ are differentiable in this region. Let us fix a point z_0 in D and approximate the increments of functions u and v in the vicinity of this point by their differentials. By definition, the increments can be represented as

$$u - u_0 = \frac{\partial u}{\partial x}(x - x_0) + \frac{\partial u}{\partial y}(y - y_0) + \eta_1 \Delta r ,$$
$$v - v_0 = \frac{\partial v}{\partial x}(x - x_0) + \frac{\partial v}{\partial y}(y - y_0) + \eta_2 \Delta r ,$$

where partial derivatives are taken at z_0, $\Delta r = \sqrt{(x - x_0)^2 + (y - y_0)^2}$, and η_1, η_2 tend to zero at $\Delta r \to 0$. The replacement of increments by differentials is equal to changing the transformation $w = f(z)$ by

$$u - u_0 = \frac{\partial u}{\partial x}(x - x_0) + \frac{\partial u}{\partial y}(y - y_0) ,$$
$$v - v_0 = \frac{\partial v}{\partial x}(x - x_0) + \frac{\partial v}{\partial y}(y - y_0) , \tag{D77}$$

which is called the *principal linear part* of the transformation (D76). The transformation (D77) can be rewritten as

$$u = ax + by + l , \quad v = cx + dy + m \tag{D78}$$

where

$$a = \frac{\partial u}{\partial x} ; \quad b = \frac{\partial u}{\partial y} ; \quad c = \frac{\partial v}{\partial x} ; \quad d = \frac{\partial v}{\partial y} ;$$
$$l = u_0 - \frac{\partial u}{\partial x} x_0 - \frac{\partial u}{\partial y} y_0 ; \quad m = v_0 - \frac{\partial v}{\partial x} x_0 - \frac{\partial v}{\partial y} y_0$$

does not depend on x and y. It represents the so called linear transformation of the plane (x, y). The transformation (D78) for $\Delta = ad - bc \neq 0$ realizes a *one-to-one transformation of the whole plane z to the whole plane w*. It is known that such a transformation transforms squares on the plane z into parallelograms on the plane w, a circle $(x - x_0)^2 + (y - y_0)^2 = r^2$ with a center at z_0 into an ellipse with a center at w_0.

A circle in the z-plane is transformed into a circle in the w-plane for

$$bd + ac = 0 \quad \text{and} \quad a^2 + b^2 = c^2 + d^2 . \tag{D79}$$

The first relation of (D79) gives $a/d = b/c = \lambda$ and the second one gives $\lambda^2 = 1$ or $\lambda = \pm 1$.

The case $\lambda = 1$ gives

$$a = -d, b = -c . \tag{D80}$$

These relations rewritten in complex form yield the function of a complex variable

$$w = Az + B \tag{D81}$$

where

$$A = \sqrt{\Delta} e^{i\alpha}, \; B = l + im \,. \tag{D82}$$

Linear transformations of the type (D81) obeying conditions (D79) are called *orthogonal transformations.* This transformation defined by (D82) conserves orbit direction (or orientation) of closed contours, whereas the transformation

$$w = \sqrt{-\Delta} e^{i\alpha} \bar{z} + B \tag{D83}$$

with $\Delta = -a^2 - b^2 < 0$ changes the orientation to the opposite.

Hence, the one-to-one transformation

$$w = f(z) = u(x, y) + iv(x, y) \tag{D84}$$

of D into D^* is called a conformal transformation if in the vicinity of any point of D, the principal linear part of this transformation is orthogonal and conserves the orientation. From this definition follows the two main properties of the conformal transformation:

1. *a conformal transformation transforms infinitesimal circles into circles accurate to a small quantity of higher order;*
2. *a conformal transformation conserves angles between curves at points of their intersection.*

Using (D78) and (D80) we can write the conformity conditions of the transformation (D84) as

$$\frac{\partial u}{\partial x} = \frac{\partial v}{\partial y} \,, \quad \frac{\partial u}{\partial y} = -\frac{\partial v}{\partial x} \,, \tag{D85}$$

with

$$\Delta = \left(\frac{\partial u}{\partial x}\right)^2 + \left(\frac{\partial v}{\partial y}\right)^2 = \left|f'(z)\right|^2 \neq 0 \,. \tag{D86}$$

The conditions (D85) coincide with the D'Alembert–Euler conditions (see (D20)) of analyticity of the function $f(z)$ in the region D. The inequality (D86) shows that $f'(z)$ should be nonzero.

Further, we have

$$\frac{\partial u}{\partial x} = \frac{\partial v}{\partial y} = \sqrt{\Delta} \cos \alpha \,, \quad \frac{\partial v}{\partial x} = -\frac{\partial u}{\partial y} = \sqrt{\Delta} \sin \alpha \,,$$

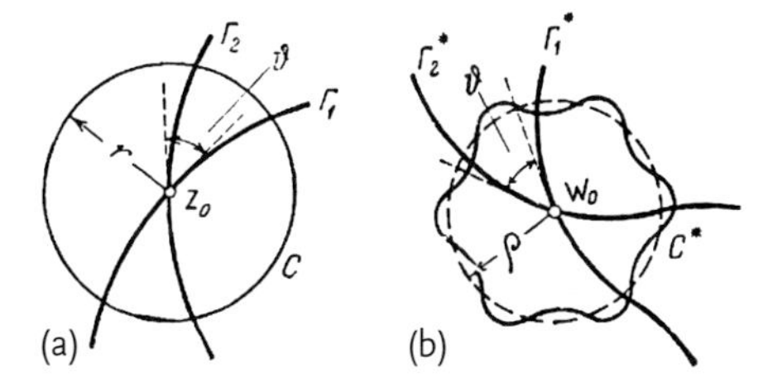

Figure D.14 Geometric interpretation of the complex variable derivative.

from which with

$$\left|f'(z)\right| = \sqrt{\Delta}\,; \quad \arg\left|f'(z)\right| = \alpha \tag{D87}$$

follows the geometric interpretation of the derivative of a complex variable: (see Figure D.14)

modulus and argument of the derivative $f'(z)$ determine the extension coefficient and angle of rotation of the principle linear part of the transformation $w = f(z)$ at point z or rather, the extension coefficient and angle of rotation of the transformation $w = f(z)$ itself at point z.

So, in order for the function $w = f(z)$ to realize a conformal transformation in the region D, it is necessary and sufficient that this function would be univalent, analytic and $f'(z) \neq 0$ everywhere in D.

D.7.2 Main Problem

The main problem in the theory of conformal mapping (transformation) consists of the following: *For given regions D and D^*, it is required to obtain a function realizing conformal mapping of one of these regions on the other.*

As far as there isn't any simple general algorithm to solve this problem, the solution of this problem in a concrete case is performed by the following steps:

- general conditions of existence and uniqueness for conformal transformation are revealed;
- different particular classes of regions are determined for which maps could be performed with the help of the combination of elementary functions;
- with the help of general properties of analytical functions, different properties of conformal transformations depending on the forms of regions to be mapped are studied
- if necessary, approximate methods of conformal mapping are developed.

The basis for all of these methods lies the following theorem:

Theorem D.21 Riemann theorem

For two simply connected regions D and D^*, with boundaries consisting of more than one point, and for any choice of points z_0 in D, w_0 in D^* and real number α_0, there is one and only one conformal transformation $w = f(z)$ of region D into region D^* such that

$$f(z_0) = w_0, \arg f'(z_0) = \alpha_0 \ .$$

D.7.3
Correspondence of Boundaries

Now, consider the correspondence of boundaries for the conformal transformation of regions. For the sake of convenience, let us introduce at the boundary C of the region D the real parameter s, that is, the arc length measured from a fixed point of the curve C so that along C for any function $\varsigma(z)$ we have $\varsigma = \varsigma(s)$. If a function $f(z)$ is continuous in a closed region D, we can take at the boundary C of this region $f(\varsigma) = f\left[\varsigma(s)\right] = \varphi(s)$ and name it the *boundary function*. Now, we give *the theorem regarding the correspondence of boundaries*:

Theorem D.22

Let the function $w = f(z)$ realize the conformal mapping of regions D and D^*. Then,

1. if the boundary of D^* has no infinite branches, then $f(z)$ is continuous at the boundary of D and the boundary function $f(\varsigma) = f\left[\varsigma(s)\right] = \varphi(s)$ performs continuous and one-to-one correspondence of boundaries of regions D and D^*;
2. if boundaries of regions D and D^* do not contain infinite branches, and are continuous at each point, and hence a bounded curvature, the boundary function $\varphi(s)$ is continuously deformed.

D.7.4
Linear Fractional Function

Transformations of the form

$$w = \frac{az + b}{cz + d} \ , \tag{D88}$$

where a, b, c and $d$are complex constants with $ad - bc \neq 0$ are called linear fractional transformations. Since

$$\frac{dw}{dz} = \frac{ad - bc}{(cz + d)^2}$$

exists where ever $z \neq -d/c$, the function (D88) is analytic everywhere on the whole plane z except at a the point $z = -d/c$ at which it has pole of the first order. Equation D88 can be uniquely solved for z, yielding

$$z = \frac{-dw + b}{cw - a} \tag{D89}$$

and is determined on the complete plane w. It is equal to ∞ at $w = a/c$ and to $-d/c$ at $w = \infty$. Therefore, the linear fractional function performs a univalent transformation of the complete plane z into the complete plane w. It is easy to see that the linear fractional function (D88) is the only function having such a property. Therefore, the following theorem holds:

Theorem D.23

If a function is univalent and analytic in the complete plane z, except at the point $C = -d/c$, then it is a linear fractional function.

Now, we will discuss the geometric properties of the linear fractional function (D88). For $c = 0$, it reduces to the integer function of the form $f(z) = az + b$. Let $k = |a|$, $\alpha = \arg a$, that is, $a = k(\cos\alpha + i\sin\alpha)$ and represent the function $f(z)$ as a complex function made up by functions $z_1 = (\cos\alpha + i\sin\alpha)z$; $z_2 = kz$; $W = z_1 + b$. The first and the second transformations are reduced respectively to the rotation of the plane z on the angle α and the similarity transformation of the plane z_1 on the similarity factor k. The third transformation means displacement on the constant vector b. In order to study geometrical properties of the function (D88) at $c \neq 0$, we represent it in the form

$$w = A + \frac{B}{z - C}\,, \tag{D90}$$

where A, B and C are constants. One must consider this transformation as a composite consisting of three transformations:

$$1)\ z_1 = z - C\,;\quad 2)\ z_2 = \frac{1}{z_1}\,;\ 3)\quad W = A + Bz_2\,. \tag{D91}$$

The transformation 1) reduces to displacement, and 2) to displacement and rotation together with extension. As for the displacement 3), we have to first consider the transformation $w = 1/z$ which in polar coordinates $z \neq re^{i\varphi}$, $w = \rho e^{i\vartheta}$ may be rewritten as follows: $\rho = 1/r$, $\vartheta = -\varphi$. The last transformation consists of two more descriptive transformations, namely,

$$\text{a)}\ \rho_1 = \frac{1}{r}\,,\quad \vartheta_1 = \varphi\,;\quad \text{b)}\ \rho = \rho_1\,, \vartheta = -\vartheta_1\,.$$

The transformation b) is the *symmetry transformation* with respect to the real axis, whereas the transformation b) – inversion, is the symmetry transformation relative unit circle. We shall call points z and z^* symmetric with respect to the circle $C_0 : |z - z_0| = R_0$ if they lie on one straight line going

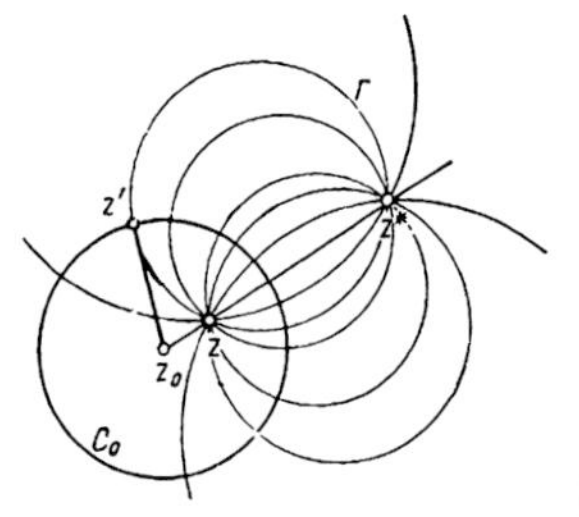

Figure D.15 The inversion transformation.

through z_0 and $|z - z_0| \cdot |z^* - z_0| = R_0^2$. The transformation transforming each point $z$of the complex plane into point z^* being symmetric relative circle C_0 is called the *symmetry relative to this circle or inversion* (see Figure D.15). The main property of this transformation is as follows: *Points z and z^* are then and only then symmetric relative circle C_0 when they are vertices of a bundle of circles Γ orthogonal to the circle C_0 (circular property)*. From this property, it follows that when the circle C_0 degenerates into a straight line, the symmetry relative to the circle turns into an ordinary symmetry. The inversion relative on an arbitrary circle C_0 is a conformal transformation of the second kind, that is, the transformation changing orientation. Other important properties of the inversion are: *the inversion transforms any circle C of the complete plane again into a circle; inversion transforms any pair of points z_1 and z_2 which are symmetric relative to a circle C into a pair of points z_1^* and z_2^* which are symmetric relative to a circle C^*, which is the image of the circle C (property of conservation of symmetric points)*.

We point out some additional properties of the linear fractional functions:

1. *The linear fractional function conserves angles in the complete plane.*
2. *The linear fractional function $w = (az + b)/(cz + d)$ with $ad - bc \neq 0$ realizes a univalent conformal transformation of the complete z-plane into a complete w-plane. This transformation:*

   a. *transforms any circle of the complete z-plane into a circle of the complete w-plane;*
   b. *transforms any pair of points symmetric relative to the circle C into a pair of points symmetric relative to the image of the circle C.*

D.7.5
Particular Cases

Now, we must consider several important examples of the linear fractional functions.

1. *Transformation of the upper half-plane into the unit circle.*
   Let a point a of the upper half-plane be transformed into the center of the circle $w = 0$. Then, in accordance with the property of the conservation of symmetry

of the complex conjugated point $\bar{a}$ which is symmetric to the point a, the relative real axis should transfer to the point $w = \infty$ which is symmetric to the point $w = 0$ relative to the unit circle (see Figure D.16). Therefore, the desired transformation should be in the form

$$w = k\frac{z - a}{z - \bar{a}} , \tag{D92}$$

where k is a constant. Let us determine k from the condition that the circle has unit radius. Then, the point $z = 0$ has to transfer to the point of unit circle and $\left|k\frac{a}{\bar{a}}\right| = |k| = 1$. Hence, $k = e^{i\alpha}$ and

$$w = e^{i\alpha}\frac{z - a}{z - \bar{a}} , \tag{D93}$$

where α is any real number.

2. *Transformation of a band into the unit circle* (see Figure D.17)
   Let the plane z be given on the band D: $-\pi/4 < \Re z < \pi/4$, which should be conformally transformed into the circle $|w| < 1$ under conditions $f(\pm\pi/4) = \pm 1$, $f(i\infty) = i$, where $i\infty$ denotes the upper infinitely remote point of the band. At first, let us turn the band through the right angle and widen it by two via $z_1 = 2iz$. Then, use the exponential function $z_2 = e^{z_1}$ to transform the obtained band $-\pi/2 < \Im z_1 < \pi/2$ into the right half-plane $\Re z_2 > 0$. We must now transform this half-plane into the unit circle so that the points $z_2 = i, -i, 0$ corresponding to points $z = \pi/4,\ \pi/4,\ i\infty$ are transformed into the points $w = 1, -1, i$. As a result, we get $(w-1)/(w-i)(1+i) = (z_2 - i)/(z_2)$, or $w = 1/i(z_2 - i)/(z_2 + 1)$. By inserting expressions here for z_2 and z_1, we obtain

$$w = \frac{1}{i}\frac{e^{2iz} - i}{e^{2iz} + 1} = \tan z . \tag{D94}$$

3. *Joukowski transformation*
   We begin with the transformation of the exterior of an arc on the exterior of a circle. Suppose that ends of the arc AB on the plane z lie at points $z = \pm a$ and the circle in plane w goes through the same points. Also, assume that the middle of the arc lies at the point $z = ih$ so that the tangent line to the arc at point $z = a$ makes an angle $\alpha = \arctan h/a$ with negative axis x and the tangent line to the circle at point $w = a$ makes an angle $\beta = \pi/2 - \alpha/2$ with the positive axis u (see Figure D.18).

With the help of the linear fractional transformation $z_1 = (z - a)/(z + a)$, we transform the exterior of the arc AB on the exterior of the straight line. Since $[dz_1/dz]_{z=a} > 0$, the angle of inclination of this straight line to the negative axis is also equal to α. Let us further look for a transformation of the exterior of a given circle in the plane w on the exterior of the straight line obtained before. In order to do this, we again use the linear fractional function $w_1 = (w - a)/(w + a)$ which transforms the circle into the plane and its circle into a straight line. For

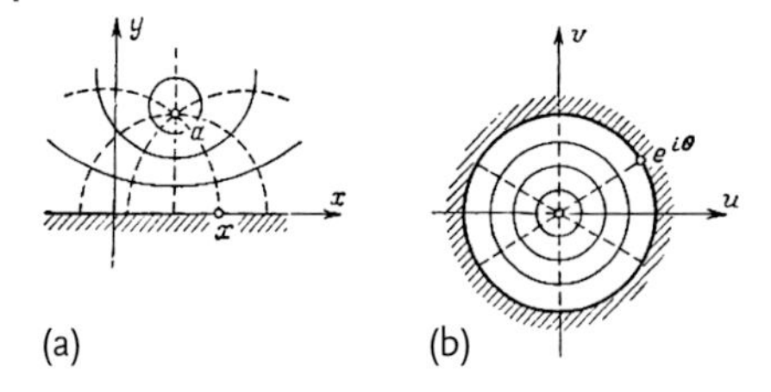

Figure D.16 Transformation of the upper half-plane into a unit circle.

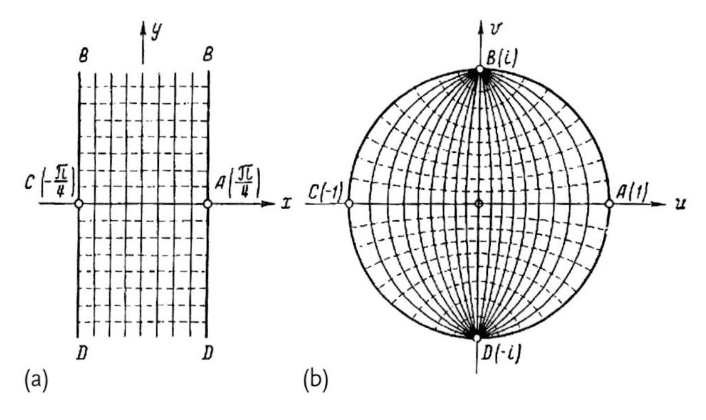

Figure D.17 Transformation of a band into a unit circle.

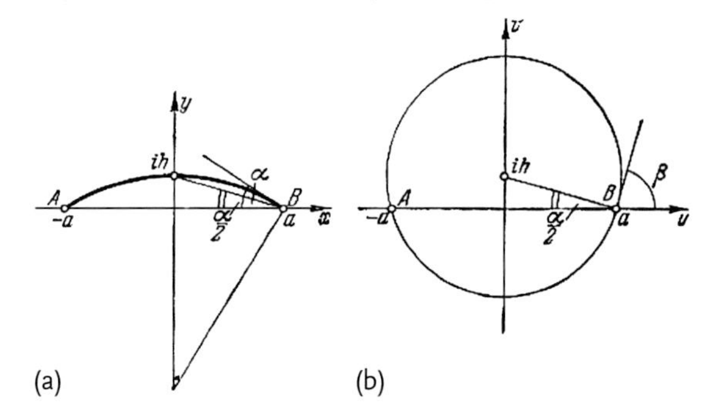

Figure D.18 Transformation of an arc exterior into the exterior of a straight line.

$[dw_1/w]_{w=a} > 0$, the angle of inclination of this straight line to the positive axis is equal to β. Thus, the transformation $z_1 = w_1^2 = ((w-a)/(w+a))^2$ transforms our circle into the exterior of the straight line forming the angle $2\beta = \pi - \alpha$ with the positive axis. Hence, this straight line coincides with the one obtained by the initial transformation $z_1 = (z-a)/(z+a)$. By excluding z_1, we get

$$\left(\frac{w-a}{w+a}\right)^2 = \frac{z-a}{z+a} \,. \tag{D95}$$

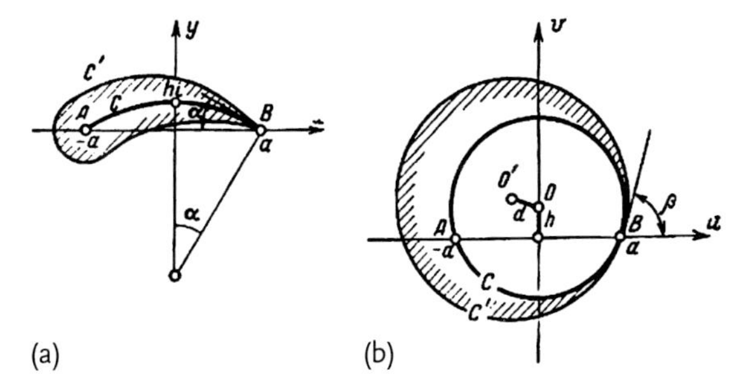

Figure D.19 Transformation of the Joukowski profile.

From (D95), we obtain

$$z = \frac{1}{2}\left(w + \frac{a^2}{w}\right) \quad \text{or} \quad w = z + \sqrt{z^2 - a^2}\,. \tag{D96}$$

This function transforms any circle C' touching the circle C at point $w = a$, into a closed curve embracing the arc AB and having at point $B(z = a)$ the cuspidal point. This curve looks like an airfoil section. That is why the function (D94) is called the *Joukowski transformation*. The form of the Joukowski *transformation* depends on three parameters: a characterizing the width of the airfoil; h – the curvature of the airfoil and d – the distance between centers of circles C and C' characterizing the thickness of the airfoil (see Figure D.19).

D.8
Application of the Theory of Complex Variables to Boundary-Value Problems

As previously mentioned, the theory of complex variables and especially its geometrical part, the theory of conformal transformation has found widespread application to problems of hydro- and aeromechanics. We begin the account of some problems with a brief summary of the theory of harmonic functions.

D.8.1
Harmonic Functions

A *harmonic function* in the region D is called a real function $u(x, y)$ of two real variables which has, in this region, continuous partial derivatives of the second order and satisfies the *Laplace differential equation*

$$\frac{\partial^2 u}{\partial x^2} + \frac{\partial^2 u}{\partial y^2} = 0\,. \tag{D97}$$

This equation is a linear differential equation. Therefore, any linear combination of harmonic functions $\sum_{k=1}^{n} \alpha_k u_k(x, y)$ satisfies (D97).

Potentials of the most important vector fields in physics are harmonic functions so that each harmonic function can be physically considered as a potential of some vector field. That is why harmonic functions are often called *potentials* and the theory of harmonic functions are *potential theory*.

The connection between analytic and harmonic functions follows from the following theorems:

Theorem D.24

Real and imaginary parts of any single-valued and analytically function $f(z) = u(x,y) + iv(x,y)$ in region D are harmonic functions in this region. This theorem follows directly from D'Alembert–Euler conditions.

Two harmonic functions in region D connected by the D'Alembert–Euler conditions are called *conjugate functions*.

Theorem D.25

For any function $u(x,y)$ being harmonic in a simply connected region, a harmonic function $v(x,y)$ conjugated to $u(x,y)$ can be found. The function $v(x,y)$ is given up to a constant C by the formula

$$v(x,y) = \int_{z_0}^{z} -\frac{\partial u}{\partial y} dx + \frac{\partial u}{\partial x} dy + C . \tag{D98}$$

It should be noted that in the multiply connected region D, (D98) has another form (see (D46)):

$$v(x,y) = \int_{z_0}^{z} -\frac{\partial u}{\partial y} dx + \frac{\partial u}{\partial x} dy + N_1 \Gamma_1 + \ldots + N_n \Gamma_n + C , \tag{D99}$$

where the N_k are arbitrary integers and the quantities Γ_k are integrals along closed circles γ_k, each containing one connected part of the boundary of D:

$$\Gamma_k = \int_{\gamma_k} -\frac{\partial u}{\partial y} dx + \frac{\partial u}{\partial x} dy . \tag{D100}$$

Boundary theorem of uniqueness: On the boundary C of the region D, give a function $u(\varsigma)$ piecewise continuous at finite points $\varsigma_1, \varsigma_2, \ldots, \varsigma_n$ of discontinuity of the first kind. Then, in D, exists no more than one harmonic and bounded function $u(z)$, which at points $\varsigma \neq \varsigma_k$ of the boundary, takes given values $u(\varsigma)$.

D.8.2 Dirichlet Problem

The *first Dirichlet problem* is as follows: *it is required to find harmonic function $u(z)$ which at the boundary of the region D, takes given continuous values $u(\varsigma)$.*

The condition of the function continuity is too strained. Sometimes, one happens to consider the generalized Dirichlet problem: *At the boundary of the region D, it is a given function $u(\varsigma)$ being continuous everywhere except for a finite number of points $\varsigma_1, \varsigma_2, \ldots, \varsigma_n$, at which this function has discontinuities of the first kind. It is required to find the harmonic and bounded function $u(z)$ in the region D, taking values $u(\varsigma)$ at all points of this function continuity.*

The solution of the generalized Dirichet problem can be reduced to the solution of an ordinary problem with the help of a special method which, for the sake of simplicity, will be illustrated for the example of a simply connected region. Let us denote by $u^-(\varsigma_k)$ and $u^+(\varsigma_k)$ limiting values of the boundary function $u(\varsigma)$ at $\varsigma \to \varsigma_k$ along C in positive and negative directions respectively, and through $h_k = u^+(\varsigma_k) - u^-(\varsigma_k)-$, the jump of $u(\varsigma)$ at point ς_k. For generality, assume that ς_k is an angular point of the contour C, and by $\varphi^-(\varsigma_k)$ and $\varphi^+(\varsigma_k)$, denote angles between x-axis and tangents to C at point ς_k (see Figure D.20). In addition, define $\alpha_k = \varphi^+(\varsigma_k) - \varphi^-(\varsigma_k)$ (if ς_k is not an angle point, then $\alpha_k = -\pi$). Take the function $u_k = h_k/\alpha_k \arg(z - \varsigma_k)$, where *arg* is the argument with an appropriately chosen branch.

It is evident, that this function is harmonic in D and continuous in $\bar{D}$ everywhere except at the point $\varsigma = \varsigma_k$. If $\varsigma \to \varsigma_k$ along the way, to which the tangent at point ς_k makes an angle $\theta\,(\varphi^- < \theta < \varphi^+)$ with the axis x, this function tends to the limit $h_k/\alpha_k\theta$. In going along the curve C in a positive direction through the point ς_k, the function $u_k(\varsigma)$ gets a jump $h_k/\alpha_k\varphi^+ - h_k/\alpha_k\varphi^- = h_k$.

Let $u(z)$ be a solution of the general Dirichlet problem at given boundary values of $u(\varsigma)$. Consider the function

$$U(z) = u(z) - \sum_{k=1}^{n} \frac{h_k}{\alpha_k} \arg(z - \varsigma_k)\,. \qquad \text{(D101)}$$

It is harmonic in D and continuous in $\bar{D}$ because $u(z)$ and all functions $u(z) = \sum_{k=1}^{n} \frac{h_k}{\alpha_k} \arg(z - \varsigma_k)$ are harmonic in D. The limiting values of $U(z)$ at $z \to \varsigma \neq \varsigma_k$ are equal to $U(z) = u(z) - \sum_{k=1}^{n} u_k(\varsigma)$ and the function $U(\varsigma)$ remains continuous while going through each point ς_k.

Hence, the solution of the generalized Dirichlet problem $u(z)$ can be represented as a sum of the function $U(z)$ solving the Dirichlet problem with continuous boundary values $U(\varsigma) = u(\varsigma) - \sum_{k=1}^{n} u_k(\varsigma)$, and a function $\sum_{k=1}^{n} u_k(\varsigma)$:

$$u(z) = U(z) + \sum_{k=1}^{n} \frac{h_k}{\alpha_k} \arg(z - \varsigma_k)\,. \qquad \text{(D102)}$$

For some applications, it is important to consider, along with the Dirichlet problem, the second boundary-value problem, that is, the Neumann problem: *It is re-*

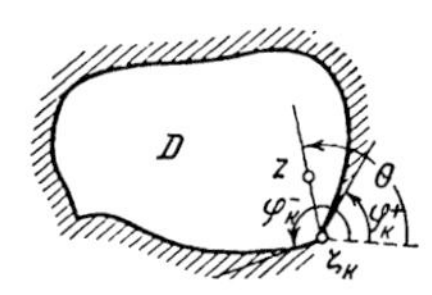

Figure D.20 The general Dirichlet problem.

quired to obtain in the region D a harmonic function $u(z)$ by the given values of its normal derivative at the boundary C

$$\frac{\partial u}{\partial n} = \frac{\partial u}{\partial x} \cos \alpha + \frac{\partial u}{\partial y} \sin \alpha = g(\varsigma) \tag{D103}$$

and the value $u(z)$ at a certain point z_0 of the region $\bar{D}$. It is assumed that in (D103), it is considered the external normal and the angle α is an angle between normal and x-axis. The function $g(\varsigma)$ can have on C a finite number of the first kind of discontinuities and the function $u(z)$ itself and its derivatives of the first order are supposed to be bounded. For solvability of the Neumann problem, it is necessary that the condition

$$\int_C g(\varsigma) ds = 0 \tag{D104}$$

is satisfied.

Under the additional assumption that partial derivatives are continuous in $\bar{D}$, the solution of the Neumann problem reduces to the solution of the Dirichlet problem for the conjugate harmonic function.

D.9 Physical Representations and Formulation of Problems

D.9.1 Plane Field and Complex Potential

Let us consider the *stationary plane-parallel vector fields*, meaning, the fields whose vectors do not depend on time and are parallel to a plane S_0, and at all points of any straight line the vectors of the field are perpendicular to S_0 and equal in magnitude and in direction. It is evident that the pattern of the field is the same on all planes parallel to S_0 and the field is completely described by the *plane field* of vectors lying in the plane S_0.

Let us introduce in the plane S_0, the Cartesian coordinate system (x, y). Then, each field vector $\boldsymbol{A}$ with components (A_x, A_y) will be characterized by complex number $\boldsymbol{A} = A_x + iA_y$ where A_x and A_y are known function of x and y: $A_x = A_x(x, y)$, $A_y = A_y(x, y)$ or what is the same of the complex variable $z = x + iy$. Hence, stationary vector fields are described by complex numbers and functions of complex variables. For the most important fields, in practice, it is possible to obtain an analytical function of the so called field complex potential owing to which problems connected with such fields are more amenable to study and to calculate. We first recall some main concepts of the vector analysis for plane fields.

The flux of a vector field $\boldsymbol{A}$ through a curve C is called the integral

$$N = \int_C (\boldsymbol{A}, \boldsymbol{n}^0) ds\,, \tag{D105}$$

where $(\boldsymbol{A}, \boldsymbol{n}^0)$ is the scalar product of the vector $\boldsymbol{A}$ and unit normal vector $\boldsymbol{n}^0$ to the curve C. If we denote the differentials along C by dx and dy, that is, we take $\boldsymbol{s}^0 ds = dx + idy$, then $\boldsymbol{n}^0 ds = -i\boldsymbol{s}^0 ds = dy - idx$, $ds = dy - idx$, $(\boldsymbol{A}, \boldsymbol{n}^0)ds = A_x dy - A_y dx$ and (D105) takes on the form

$$N = \int_C (A_x dy - A_y dx) . \tag{D106}$$

The *surface density of the flux*, that is, the limit of the ratio of a flux through a closed curve C to the area S bounded by this curve at the condition that S tends to the point z is called *divergence* of the field at point z:

$$\text{div}\boldsymbol{A} = \nabla \cdot \boldsymbol{A} = \lim_{C \to z} \frac{1}{S} \int_C (\boldsymbol{A}, \boldsymbol{n}^0) ds . \tag{D107}$$

It is known that

$$\text{div}\boldsymbol{A} = \frac{\partial A_x}{\partial x} + \frac{\partial A_y}{\partial y} . \tag{D108}$$

The point at which $\nabla \cdot \boldsymbol{A} > 0$ is called the *source*, whereas at $\nabla \cdot \boldsymbol{A} < 0$, it is called the *sink*. If $\nabla \cdot \boldsymbol{A} = 0$, the field is called the *solenoidal field*. In such a field, the flux through any closed line c, the interior d of which belongs to the field is equal to zero. The latter follows from the *Gauss theorem*

$$\int_C (\boldsymbol{A}, \boldsymbol{n}^0) ds = \iint_d \nabla \cdot \boldsymbol{A} dS . \tag{D109}$$

By the same theorem, the flux through any cross section of the so called *tube of flow*, that is, the region bounded by two *streamlines*, both of which are tangent at any point to the field vector (see Figure D.21). The condition $\nabla \cdot \boldsymbol{A} = 0$ shows that the expression $-A_y dx + A_x dy$ is a differential of a function $v(x, y)$ called a stream function. Since $dv = -A_y dx + A_x dy$, we have

$$\frac{\partial v}{\partial x} = -A_y , \quad \frac{\partial v}{\partial y} = A_x$$

and the function $v(x, y)$ can be restored by its total differential dv as follows

$$v(x, y) = \int_{z_0}^{z} -A_y dx + A_x dy + \text{const} . \tag{D110}$$

In view of the condition $\nabla \cdot \boldsymbol{A} = 0$, the integral (D110) in the simply connected region D is independent of the path of integration and thus determines the single-valued function, whereas in multiply connected region, it has cyclic constants and determines a multiply connected function. In a solenoidal field, the flux through

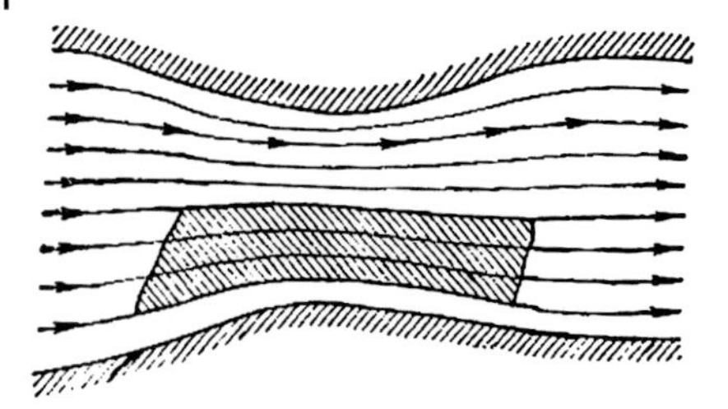

Figure D.21 Streamlines and flow tube.

the line C is in accordance with formulae (D106) and (D110), and is equal to the function increment at the ends of C:

$$N = \int_{z_1}^{z_2} -A_y dx + A_x dy = \int_{z_1}^{z_2} dv = v(z_2) - v(z_1) . \tag{D111}$$

Thus, if the region D is multiply connected, one should take the branch of $v(z)$ continuous on C.

The circulation of the field along the closed contour is expressed by the following integral

$$\boldsymbol{\Gamma} = \int_C (\mathbf{A}, \mathbf{s}^0) ds = \int_C A_x dx - A_y dy . \tag{D112}$$

The surface density of circulation is called curl and is equal to

$$\text{curl}\mathbf{A} = \nabla \times \mathbf{A} = \lim_{C \to z} \frac{1}{S} \int_C (\mathbf{A}, \mathbf{s}^0) ds . \tag{D113}$$

It is known that

$$\nabla \times \boldsymbol{A} = \frac{\partial A_y}{\partial x} - \frac{\partial A_x}{\partial y} . \tag{D114}$$

A point at which $\nabla \times \boldsymbol{A} \neq 0$ is called the *vortex point* or shortly, the *vortex*. If $\nabla \times \boldsymbol{A} = 0$ at each point of the region D, the field in this region is called irrotational or potential.

In this case, the circulation along any closed line c, the interior d of which belongs to the field, is equal to zero. This conclusion follows from the Riemann–Green formula

$$\int_C (\mathbf{A}, \mathbf{s}^0) ds = \iint_d \nabla \times \mathbf{A} dS . \tag{D115}$$

The condition (D114) shows that the expression $A_x dx + A_y dy$ is the differential of a function $u(x, y)$ called the *potential function* or shortly, the *potential* of the field.

This name follows from the relations

$$A_x dx + A_y dy = du\,, \quad A_x = \frac{\partial u}{\partial x} \quad \text{and} \quad A_y = \frac{\partial u}{\partial x} \quad \text{or} \quad \mathbf{A} = \nabla u\,. \tag{D116}$$

The potential function is restored from its differential with the help of the integral

$$u(x, y) = \int_{z_0}^{z} A_x dx + A_y dy + \text{const}\,. \tag{D117}$$

In view of the condition $\nabla \times \mathbf{A} = 0$, this integral in the simply connected region D is independent of the path of integration, whereas it has cyclic constants in a multiply connected region. If the field in the region D is simultaneously solenoidal and irrotational, then from $\nabla \cdot \mathbf{A} = 0$ and $\mathbf{A} = \nabla u$, we get

$$\frac{\partial u}{\partial x} = \frac{\partial v}{\partial y}\,, \quad \frac{\partial u}{\partial y} = -\frac{\partial v}{\partial x}\,, \tag{D118}$$

which coincides with the D'Alembert–Euler equations. Hence, we have the following *theorem*: in a plane field without sources and vortices, the stream function and the potential are conjugate harmonic functions. From this theorem, such field streamlines and equipotential lines form orthogonal families follow.

The function of a complex variable

$$f(z) = u(x, y) + i v(x, y) \tag{D119}$$

is called the *complex potential* of the field. The complex potential in multiply connected region of the field, for example, when the field has sources and (or) vortices which should be excluded from consideration, can be a multivalued function. With the help of the complex potential, all main quantities characterizing the field can be expressed.

For example, using the formula for the derivative of an analytic function and (D116) and (D118), we obtain

$$\mathbf{A} = \frac{\partial u}{\partial x} + i\frac{\partial u}{\partial y} = \frac{\partial u}{\partial x} - i\frac{\partial v}{\partial x} = \overline{f'(z)}\,. \tag{D120}$$

Since $f'(z)dz = (A_x - iA_y)(dx + idy)$, we can rewrite (D106) and (D112) as follows

$$N = \Im \int_C f'(z)dz\,, \quad \Gamma = \text{Re} \int_C f'(z)dz\,. \tag{D121}$$

Combining both formulae, we get

$$\Gamma + iN = \int_C f'(z)dz\,. \tag{D122}$$

D.9.2
Examples of Plane Fields

1. *Point Source* (see Figure D.22)
   Consider a single point source located at the origin of coordinates and assume that vortices are absent. For reasons of symmetry, it is clear that the field vector should have the form

$$A = \varphi(r)\boldsymbol{r}^0 , \tag{D123}$$

where $r = |z|$ is the distance of the point from the origin of coordinates and $\boldsymbol{r}^0 = \frac{z}{|z|}$ is the unit vector directed from the origin of coordinates to the point z. The flux of the vector through any circle $|z| = r$ with a center at the origin of coordinates that is equal to

$$N = \int_{|z|=r} (A, \boldsymbol{r}^0) ds = \varphi(r) 2\pi r .$$

Then, $\varphi(r) = N/(2\pi r)$. The quantity N is called the *source intensity*. By substituting this expression in (D123), we obtain

$$A = \frac{N}{2\pi r}\boldsymbol{r}^0 = \frac{N}{2\pi}\frac{z}{|z|^2} = \frac{N}{2\pi}\frac{1}{\bar{z}} . \tag{D124}$$

Using (D120), we get the derivative of the complex potential

$$f'(z) = \frac{N}{2\pi}\frac{1}{z}$$

and then the complex potential itself,

$$f(z) = \frac{N}{2\pi}\log z + c . \tag{D125}$$

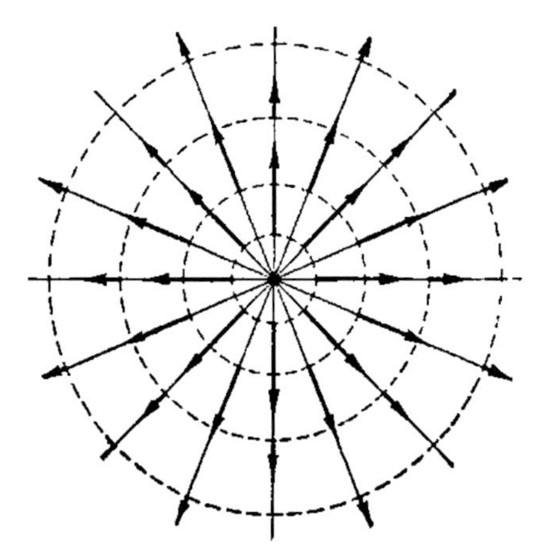

Figure D.22 Point source: streamlines are shown as solid lines; equipotential lines are shown as dotted lines.

By separating real and imaginary parts of the complex potential, we receive a potential function and a stream function respectively

$$u = \frac{N}{2\pi} \log |z| + c_1 , \quad v = \frac{N}{2\pi} \arg z + c_2 . \tag{D126}$$

2. *Point vortex* (see Figure D.23)
Using the same reasoning, we obtain the vector field from the point vortex located at the origin of coordinates

$$\boldsymbol{A} = \frac{\Gamma}{2\pi} \frac{1}{\bar{z}} , \tag{D127}$$

where Γ is the vortex intensity, that is, the circulation of vector $\boldsymbol{A}$ along any closed contour surrounding the vortex. The complex potential differs from the previous one by a factor i and the potential function and stream function change places:

$$f(z) = i\left(\frac{\Gamma}{2\pi} \log z + c\right), \quad u = -\frac{\Gamma}{2\pi} \arg z + c_1 , \quad v = \frac{\Gamma}{2\pi} \log |z| + c_2 . \tag{D128}$$

3. *Vortex-source*
Suppose that at the origin of coordinates is a concentrated source of intensity N and a vortex of intensity Γ. The field vector and the complex potential are obtained by addition of expressions (D124) and (D127), and of (D125) and (D128), respectively:

$$\boldsymbol{A} = \frac{N + i\Gamma}{2\pi} \frac{1}{z} , \quad f(z) = \frac{N + i\Gamma}{2\pi} \log z + c . \tag{D129}$$

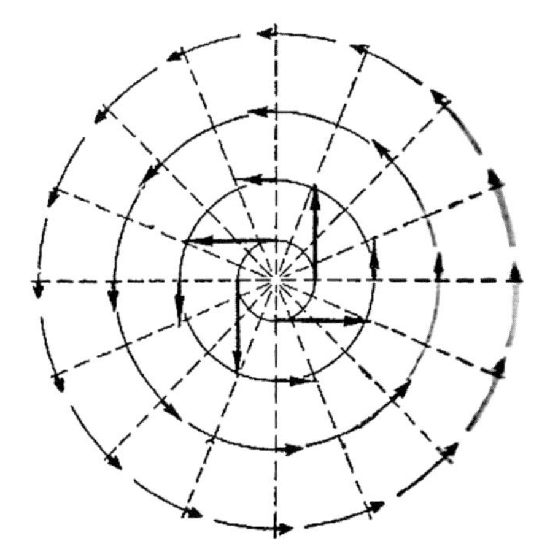

Figure D.23 Point vortex: streamlines are shown as solid lines; equipotential lines are shown as dotted lines.

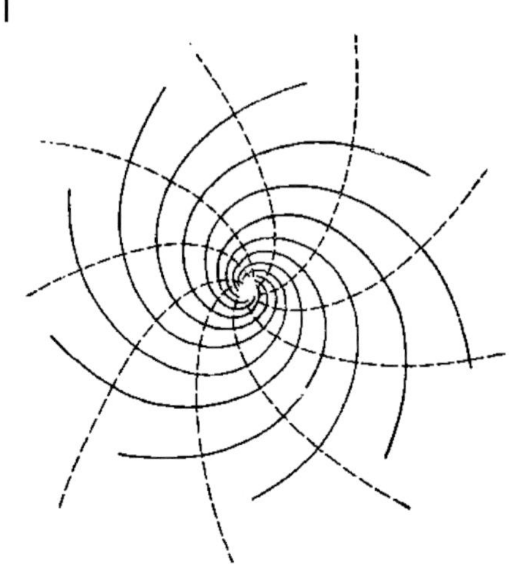

Figure D.24 Vortex source: streamlines are shown as solid lines; equipotential lines are shown as dotted lines.

Streamlines and equipotential lines in polar coordinates $z = re^{i\varphi}$ are represented by equations

$$\Gamma \ln r + N\varphi = c_1 , \quad N \ln r - \Gamma\varphi = c_2 .$$

These equations give a family of logarithmic spirals (see Figure D.24).

4. *Dipole*
   Consider a system consisting of a source and a sink of intensities $\pm N$ located respectively at points $z_1 = -h$ and $z_2 = 0$ (see Figure D.25). The complex potential of this system can be obtained by addition of a source potential and a sink potential

$$f_h(z) = \frac{N}{2\pi} \log(z + h) - \frac{N}{2\pi} \log z . \tag{D130}$$

   Consider, the limiting case when $h \to 0$ and $N \to \infty$, but the product $Nh \to p$, simultaneously remains finite, yielding the moment p (see Figure D.26). The complex potential is obtained by passage to the limit in (D130) at $h \to 0$:

$$f(z) = \lim_{h \to 0} \frac{Nh}{2\pi} \frac{(\log(z + h) - \log z}{h} = \frac{p}{2\pi} \frac{d}{dz} \log z = \frac{p}{2\pi z} . \tag{D131}$$

5. *Simple layer*
   Assume that sources are located on a line C with linear density $\rho(\varsigma)$. Denoting $r = |\varsigma - z|\,\rho(\varsigma)$, we get from (D128) the potential $\rho(\varsigma)/(2\pi)(\ln r)ds$ of elementary source $\rho(\varsigma)ds$ located at point ς. Integration gives the potential of a simple layer

$$u(z) = \frac{1}{2\pi} \int\limits_C \rho(\varsigma) \log r ds . \tag{D132}$$

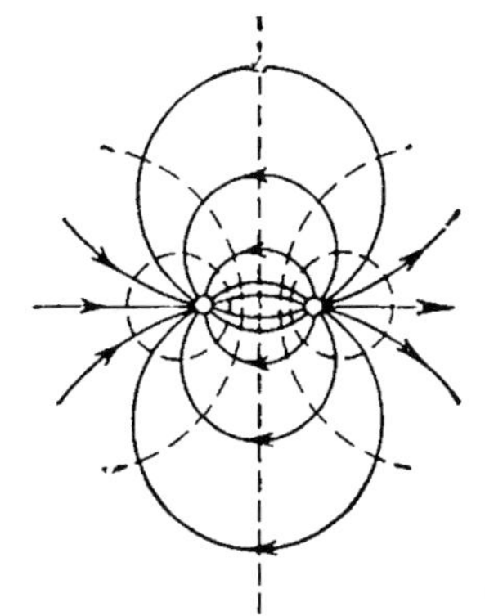

Figure D.25 Dipole with separated source and sink.

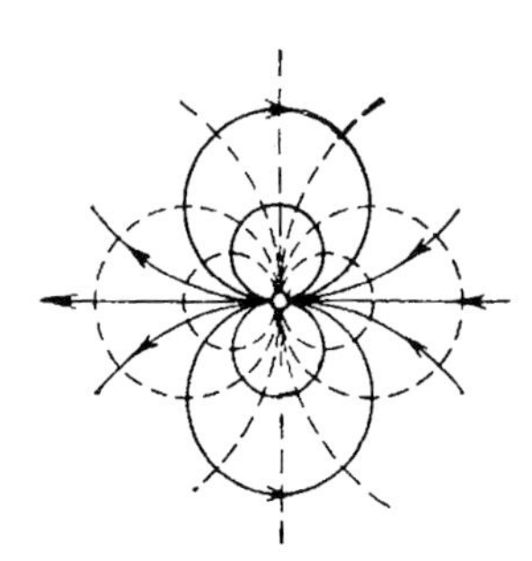

Figure D.26 Point dipole.

6. *Double layer*
Suppose that accompanied by the line C, carrying sources with density $\rho(\varsigma)$, there is one more line C' obtained from C if small segments of equal length h are plotted along all normals in some direction. Let the density of source distribution at C' be such that at its length element ds', a source of intensity $\rho' ds' = -\rho ds$ is located (the right and left part are taken respectively at points ς and ς', (see Figure D.27). The limiting layer which results at $h \to 0$ and $\rho(\varsigma) \to \infty$, so that $h\rho(\varsigma)$ remains finite and $h\rho(\varsigma) \to \mu(\varsigma)$ is called the *double layer* with density of moments μ. The potential of the double layer is found as follows. At fixed $h \neq 0$, we get from (D130)

$$u(z) = \frac{1}{2\pi} \int\limits_C \rho(\varsigma) \ln r ds + \frac{1}{2\pi} \int\limits_{C'} \rho' \ln r' ds' ,$$

where $r' = |r - \varsigma'|$. For small h, neglecting terms of higher order than h, we get $r' = r - h\partial r/\partial n$, where $\partial/\partial n$ is the derivative in the direction of the normal to C and pointing against C'. From this, follows

$$\ln r' = \ln r + \ln\left(1 - \frac{h}{r}\frac{\partial r}{\partial n}\right) = \ln r - h\frac{\partial}{\partial n} \ln r .$$

Taking into account $\rho' ds' = -\rho ds$, we obtain

$$u_h(z) = \frac{1}{2\pi} \int\limits_C h\rho(\varsigma) \frac{\partial}{\partial n} (\ln r) ds .$$

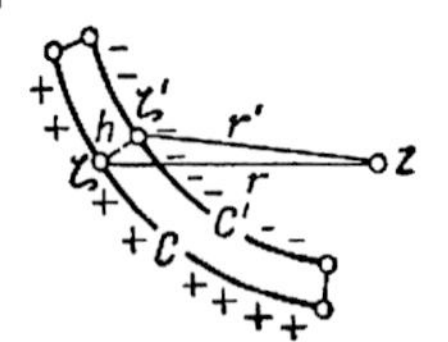

Figure D.27 Double layer.

Going now to the limit at $h \to 0$, we finally get the potential of double layer

$$u(z) = \frac{1}{2\pi} \int_C \mu(\varsigma) \frac{\partial}{\partial n} (\ln r) ds . \quad \text{(D133)}$$

Let us show that any harmonic function can be interpreted as a potential of some plane field. For the sake of simplicity, we restrict the consideration to the case of a simply connected region bounded by a closed curve C. Let the function $u(z)$ be given and harmonic in D. Take the conjugate function $v(z)$ and apply to the function $f(z) = u(z) + i v(z)$ the Cauchy integral formula:

$$u(z) = \frac{1}{2\pi i} \int_C \frac{f(\varsigma)}{\varsigma - z} d\varsigma .$$

Then, define $\varsigma - z = r e^{i\varphi}$. Then, by differentiating it with respect to ς at constant z, we get

$$\frac{d\varsigma}{\varsigma - z} = d \ln(\varsigma - z) = d \ln r + i d\varphi .$$

By substituting this expression into the Cauchy formula and separating the real part, we obtain

$$u(z) = \frac{1}{2\pi} \int_C u(\varsigma) d\varphi + \frac{1}{2\pi} \int_C v(\varsigma) d \ln r . \quad \text{(D134)}$$

On the line C, we have $d\varphi = \partial\varphi/\partial s ds$. In view of the D'Alembert–Euler conditions applied to the function $\ln(\varsigma - z)$ which is analytic on C, we can write $\partial\varphi/\partial s = -\partial \ln r/\partial n$, where by $\partial/\partial n$ we denote differentiation with respect to the inner normal to C. Therefore, the first integral in (D134) can be written as

$$u_1(z) = -\frac{1}{2\pi} \int_C u(\varsigma) \frac{\partial}{\partial n} \ln r ds .$$

It represents a double layer potential with density of moments $-u(\varsigma)$.

We now assume that $v(\varsigma)$ has a continuous derivative. Then, by integrating the second term of (D134) by parts, we obtain:

$$u_2(z) = -\frac{1}{2\pi} \int_C \frac{\partial v}{\partial s} \ln r ds$$

because the off-integral term vanishes, owing to closure of the contour C. Hence, the second term is a simple layer potential with density $-\partial v/\partial s$. Therefore, we obtain an important property of any harmonic function. *Any harmonic function $u(z)$ in a simply connected region D can be represented in the form of a sum of potentials of simple and double layers distributed along the boundary of D.*

7. *Velocity field of fluid flow*
   Let the vector $\boldsymbol{V}$ represent the velocity of fluid particles in the stationary plane flow of an incompressible fluid. The flux of the velocity vector

$$N = \int_C (\boldsymbol{V}, \boldsymbol{n}^{(0)})\, ds \qquad \text{(D135)}$$

means the amount of fluid flowing in a unit time through the curve C. The circulation is the integral of tangential components V_s of the velocity vector along a closed curve C. Sources and vortex points are interpreted as points in the vicinity of which the flux and the circulation, that is, $\nabla \cdot \boldsymbol{V}$ and $\nabla \times \boldsymbol{V}$, along a closed curve surrounding the point are nonzero. If in some region D, vortices and sources are absent, one can get in this region an analytic function $f(z) = u(z) + iv(z)$, complex field potential, such that $\boldsymbol{V} = \overline{f'(z)}$.
Now, consider the hydrodynamic interpretation of the simplest singular points of analytical function $f(z)$. From example three, we can see that the *logarithmic ramification point* a, in whose vicinity $f(z)$ has form $f(z) = c\log(z-a) + g(z)$, where $g(z)$ is regular function, can be interpreted as a vortex-source point of the intensity $N + i\Gamma = 2\pi c$. By generalizing example four a little, we see that the pole a of the first order with residue c_{-1} in which vicinity $f(z) = c_{-1}/(z-a) + g(z)$ is interpreted as a dipole obtained by merging of two vortex-source points of intensities $\pm(N + i\Gamma)$ located at points z_1 and z_2. Thus, $c_{-1} = 1/2\pi \lim\limits_{h\to 0}(N + i\Gamma)h$. In similar fashion, *poles of higher order* can be interpreted. Consider, for example, two dipoles with moments $\mp 2\pi c_{-2}/h$ located respectively at points $z_1 = a - h$ and $z_2 = a$. The limiting formation of this system at $h \to 0$ is called the *quadrupole* with moment $2\pi c_{-2}$. The complex potential of the quadrupole is equal to

$$\lim_{h\to 0} -\frac{c_{-2}}{h}\left(\frac{1}{z-a+h} - \frac{1}{z-a}\right) = -c_{-2}\frac{d}{dz}\frac{1}{z-a} = \frac{c_{-2}}{(z-a)^2}\,.$$

Hence, the pole of the second order in whose vicinity

$$u(z) = \frac{c_{-2}}{(z-a)^2} + \frac{c_{-1}}{z-a} + g(z)$$

is interpreted as a combination of a dipole and a quadrupole with moments depending on coefficients c_{-1} and c_{-2}.
Generally, the pole of n-th order of the function $f(z)$ is interpreted as a set of *multipoles* of orders $2, 4, \ldots, 2n$ with moments depending on coefficients of the principal part of the expansion of $f(z)$ in the neighborhood of this pole. Thus,

a multipole of n-th order is defined as the limiting formation obtained at $h \to 0$ from a set of multipoles of orders $2k-2$ with moments $\mp p_{2k-2}/h$ located at points $z_1 = a - h$ and $z_2 = a$.

Multiple points of a complex potential at which it is equal to zero are *ramification points* of streamlines and equipotential lines. These points are called *critical points of the flow*. The velocity at these points is equal to zero.

In conclusion, we derive the Chaplygin formula for the vector of the lifting force acting on a cylindrical body in plane-parallel flow. Consider the motion of an airfoil with constant velocity $-\mathbf{V}_\infty$. For the velocity not approaching the velocity of sound, the air can be taken as ideal incompressible gas and the vortex formation around the airfoil can be neglected. In addition, we shall consider the airfoil as infinitely long cylinder with generatrices perpendicular to the velocity vector. Then, the velocity field can be taken as plane-parallel and we can restrict ourself by studying the plane field in any cross section perpendicular to the cylinder generatrices. Finally, for the sake of convenience, we assume that the airfoil is at rest and the air flows by the airfoil with constant velocity $\mathbf{V}_\infty$ at infinity (see Figure D.28).

The pressure in a stationary vortex-free flow is determined by the Bernoulli formula $p = A - \rho V^2/2$, where A is a constant, ρ – air density and $V = |\boldsymbol{V}|$ – the velocity of the flow. Using this formula, we can calculate the total force acting on the contour C of the wing, that is, the (*lifting force*). Since the pressure at Cis directed inside along the normal, the force acting on the element $d\varsigma$ of the airfoil contour C is vectorially equal to $p i d\varsigma = A i d\varsigma - \rho V^2/2 i d\varsigma$, and the total force acting on C is

$$P = X + iY = \int_C p i d\varsigma = -\frac{\rho i}{2} \int_C V^2 d\varsigma \qquad \text{(D136)}$$

(the integral over closed contour vanishes). As far as the air flow around the contour C is concerned , the velocity of flow at points of C is directed along the tangent:

$$V = \overline{f'(\varsigma)} = V e^{i\varphi} ,$$

where $\varphi = \arg d\varsigma$. From here, it follows that the velocity is equal to $V = \overline{f'(\varsigma) e^{-i\varphi}}$ and (D134) takes the form

$$P = -\frac{\rho i}{2} \int_C \left|\overline{f'(\varsigma)}\right|^2 e^{-2i\varphi} d\varsigma = -\frac{\rho i}{2} \int_C \left|\overline{f'(\varsigma)}\right|^2 \overline{d\varsigma} ,$$

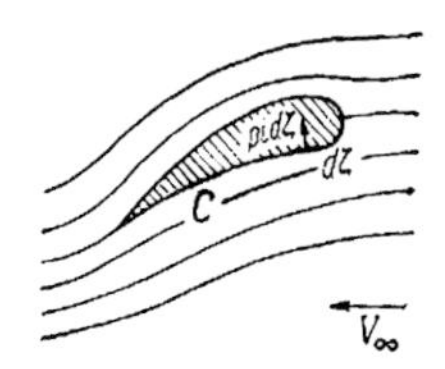

Figure D.28 Flow around an airfoil.

because $e^{-2i\varphi} d\varsigma = e^{-i\varphi} |d\varsigma| = \overline{d\varsigma}$. Passing to conjugate quantities, we obtain the vector complex conjugate to the vector of lifting force:

$$\bar{P} = X - iY = \frac{\rho i}{2} \int_C |f'(z)|^2 \, d\varsigma \, . \tag{D137}$$

This formula is the Chaplygin formula.

References to Appendix

1 Aris, R. (1962) *Vectors, Tensors, and Basic Equations of Fluid Mechanics.* Prentice-Hall, Inc., Englewood Cliffs, N.J.

2 Batchelor, G.K. (1953) *The Theory of Homogeneous Turbulence.* Cambridge, University Press, Cambridge.

3 Batchelor, G.K. (1970) *An Introduction to Fluid Dynamics.* Cambridge Univ. Press, Cambridge.

4 Cercignani, C. (1969) *Mathematical Methods in Kinetic Theory.* Macmillan, London.

5 Gardiner, C.W. (1985) *Handbook of Statistic Methods,* 2nd ed., Springer-Verlag, Berlin, Heidelberg, New York.

6 Hinze, J.O. (1959) *Turbulence.* McGraw-Hill, New York.

7 Klyatskin, V.I. (1975) *Statistical Description of Dynamic Systems with Fluctuating Parameters.* Nauka, Moscow, 1975.

8 Kochin, N.E. (1938) *Vektorrechnung und Tensorrechnungsbeginn.* GONTI, Moscow.

9 Kolmogorov, A.N. (1941) Local Structure of Turbulence in Incompressible Fluids at very High Reynolds Numbers. *Dokl. Akad. Nauk SSSR* **30**(4), 299–303.

10 Landau, L.D. and Lifshitz, E.M. (1964) *Statistical Physics.* Nauka, Moscow.

11 Landau, L.D. and Lifshitz, E.M. (1988) *Theoretical Physics: Volume VI: Hydrodynamics.* Moscow.

12 Lavrentiev, M.A. and Shabat, B.V. (1965) *Methods of complex function variable.* Nauka, Moscow.

13 Loitszyanskiy, I.G. (1970) *Mechanics of Liquid and Gas.* Nauka, Moscow.

14 Monin, A.C. and Yaglom, A.M. *Statistical Fluid Mechanics: Mechanics of Turbulence,* Vol. 1 (1971), Vol. 2 (1975), MIT Press, Cambridge MA.

15 Pobedrya, B.E. (1986) *Lectures on Tensor Analysis.* Moscow State University, Moscow.

16 Pope, S.B. (2000) *Turbulent Flows.* Cambridge Univ. Press, Cambridge.

17 Prandtl, L. (1956) Führer durch die Strömungslehre, Verlag Vieweg & Sohn, Braunschweig.

18 Rashevskiy, P.K. (1956) *Course of Differential Geometry.* GITTL, Moscow.

19 Sedov, L.I. (1973) *Mechanics of Continuous Media* (in 2 Vol.), Nauka, Moscow.

20 Serrin, J. (1959) *Mathematical Principles of Classical Fluid Mechanics, Handbuch der Physik,* Band VIII/1 Strömungsmechanik, Berlin Göttingen Heidelberg.

Index

Hydromechanics. Theory and Fundamentals. Emmanuil G. Sinaiski

ISBN: 978-3-527-41026-2

Y